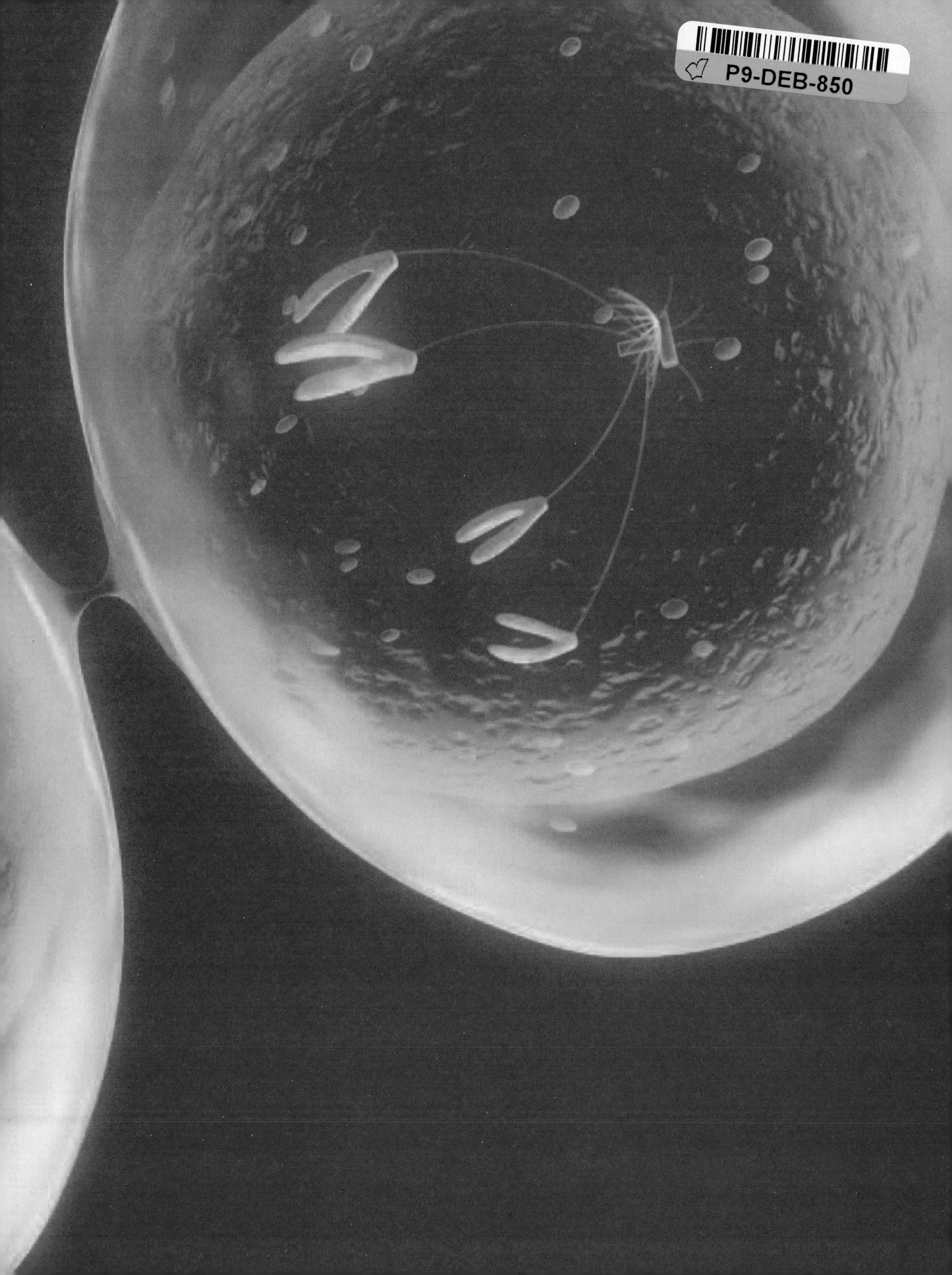

# THE HUMAN BODY

First published by Parragon in 2012

Parragon
Queen Street House
4 Queen Street
Bath BA1 1HE, UK
www.parragon.com



Produced by Guy Croton and Neil Adams



ISBN 978-1-4454-8472-3
Printed in China

# THE HUMAN BODY

## A FAMILY REFERENCE GUIDE

Parragon

Bath • New York • Singapore • Hong Kong • Cologne • Delhi
Melbourne • Amsterdam • Johannesburg • Shenzhen

# CONTENTS

## 1 THE MIRACLE OF LIFE

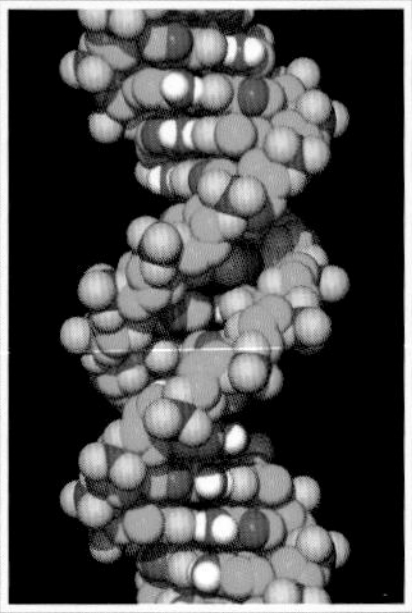
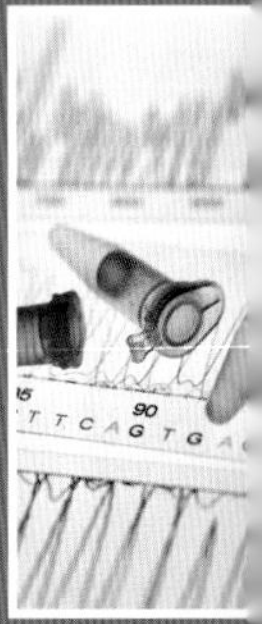

## 2 A PERFECT MACHINE

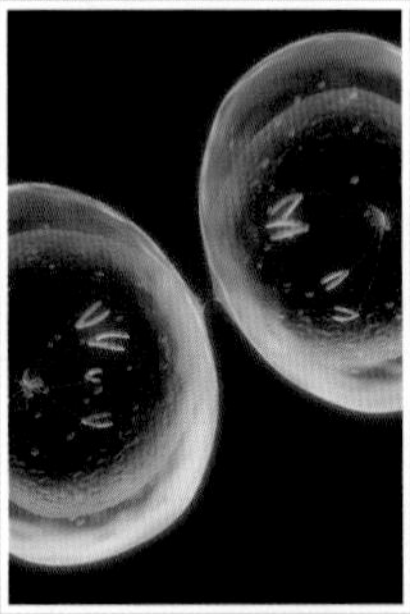
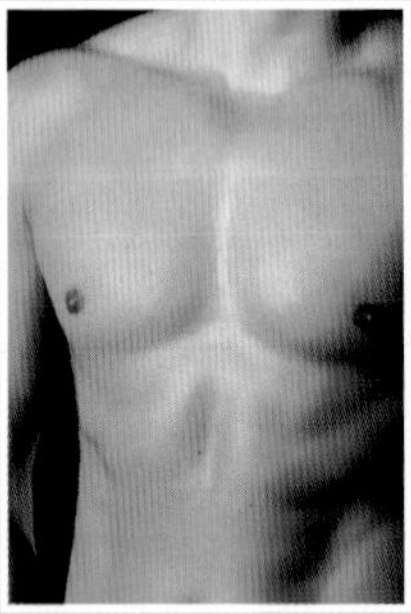

## 3 FIGHTING DISEASE AND DISORDERS

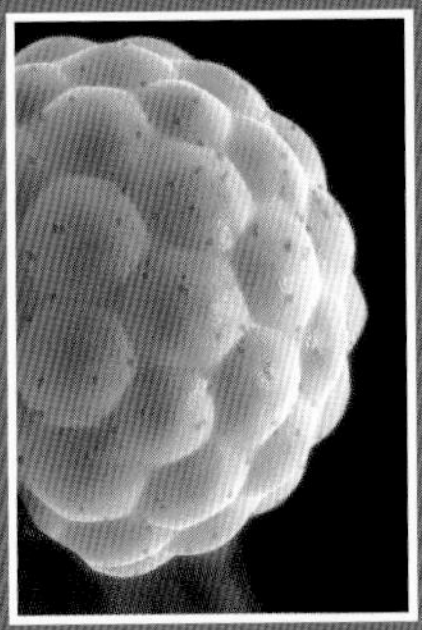
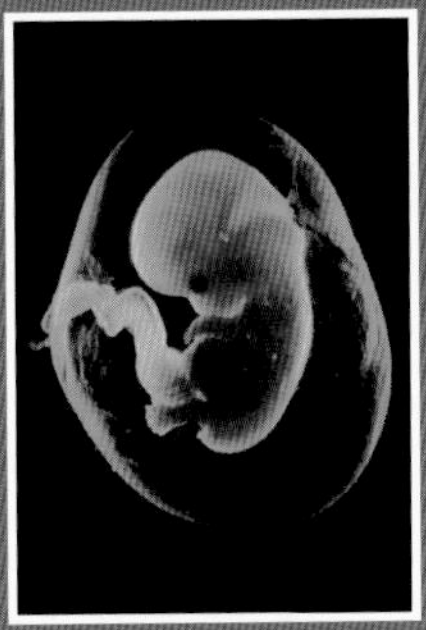

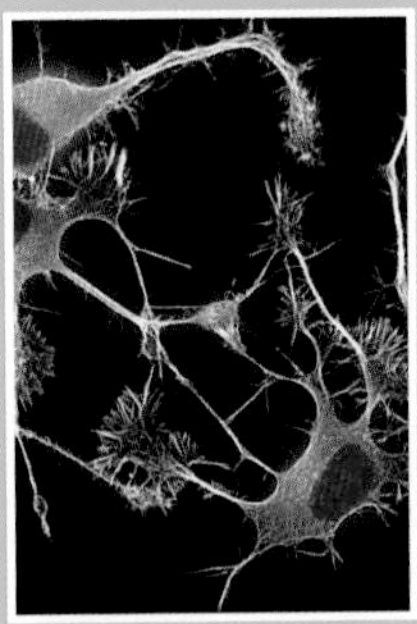



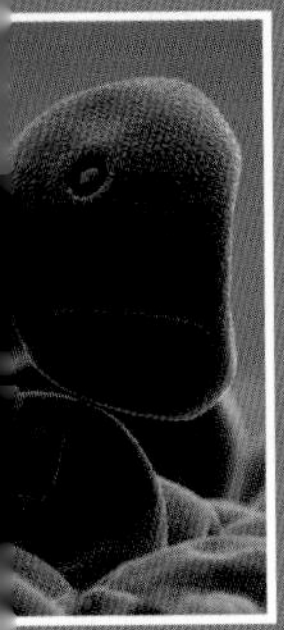

# INTRODUCTION

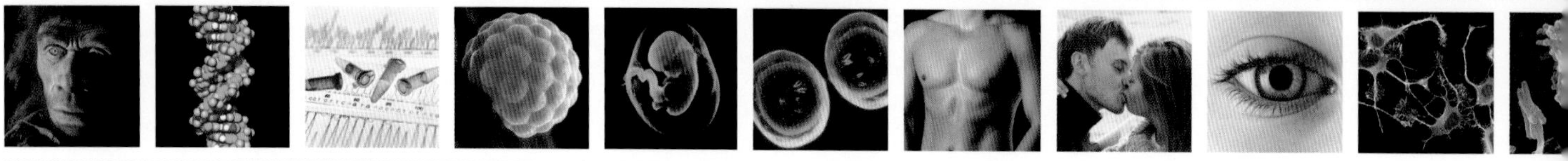

**The human body is arguably the most complex organism that exists on this planet. Made up of more than 100,000 billion cells, each with its own unique identity, the human anatomy is a fabulously synchronized and integrated living machine that can perform a remarkable range of activities and functions.**

A human being consists of a single structure, but this one entity is made up of billions of smaller structures of four major kinds. These are cells, tissues, organs, and systems.

Cells have long been recognized as the simplest units of living matter that can maintain life and reproduce themselves. The term "cell" was first applied by Robert Hooke, a 17th-century English scientist, who compared the internal structure of a piece of cork to the cells inhabited by monks in a monastery. The human body, which is made up of numerous cells, begins as a single, newly fertilized cell. The characteristics common to all living cells include the ability to reproduce, breathe, move, react to external stimuli, and create or utilize energy in order to perform their tasks. As human beings have evolved, many cells of the body have become increasingly more specialized, performing ever more intricate and incredible functions. One example is in

**THE CELL**
It is the smallest unit of the human body—and of all living organisms—able to function autonomously. It is so small that it can be seen only with a microscope. In a body such as that of a human being millions of cells are organized into tissues and organs.

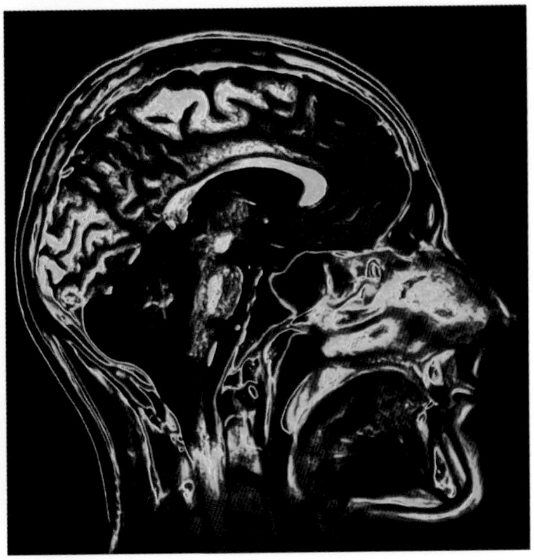

**THE BRAIN**
The brain is the body's control center. Underneath its folds more than 100 billion neurons organize and examine incoming information and act as a guide for the organism.

the retina of the eye, in which there are two kinds of cone cell, some of which react to red light and some to blue or green.

When similar kinds of cell are grouped together in the body, they form tissues, such as the epithelial cells that make up the protective coverings of body surfaces and the linings of the lungs and the intestines. Tissues are somewhat more complex units than cells. This is because, by their very nature, they are organizations of a great many similar cells with varying amounts and types of nonliving, intercellular substance between them.

Organs are yet more complex units than tissues. An organ is an organization of several different kinds of tissues arranged together in such a way that they can perform a special function. For example, the stomach is an organization of muscle, connective, epithelial, and nervous tissues. Muscle and connective tissues form its wall, epithelial and connective tissues form its lining, and nervous tissue extends throughout both its wall and its lining.

Finally, ten distinct systems comprise the most complex of the component units of the human body. A system is an organization of varying numbers and kinds of organs so arranged that

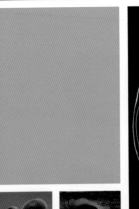

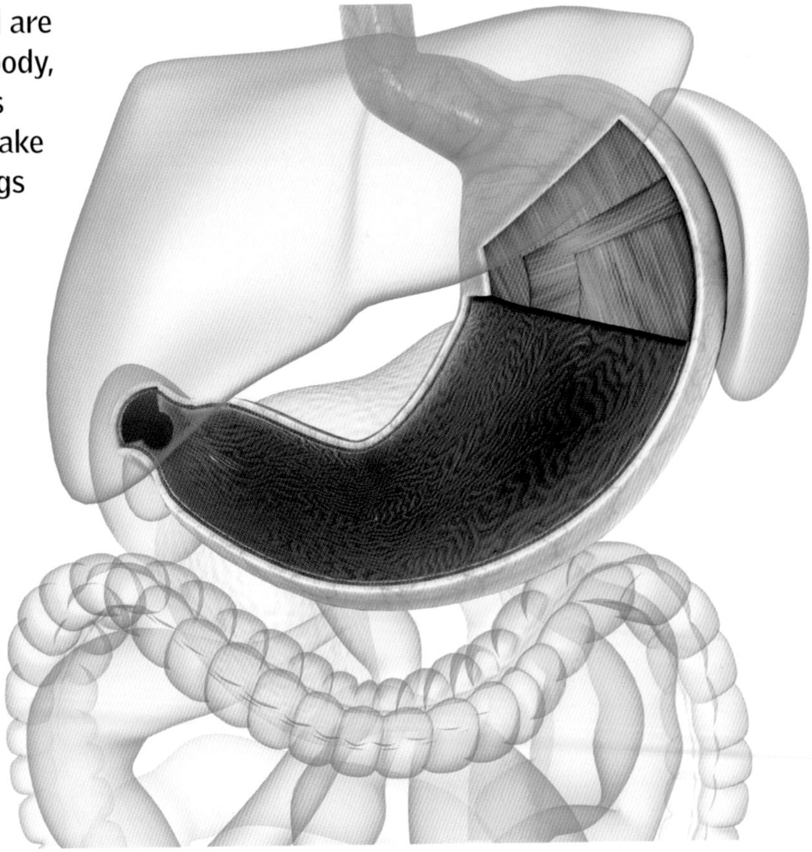

**THE STOMACH**
The stomach is the best known of the internal body organs, but it is also the most misunderstood. Its work consists of starting the digestion process, storing semi-digested food, and releasing the food slowly and continuously into the digestive system.

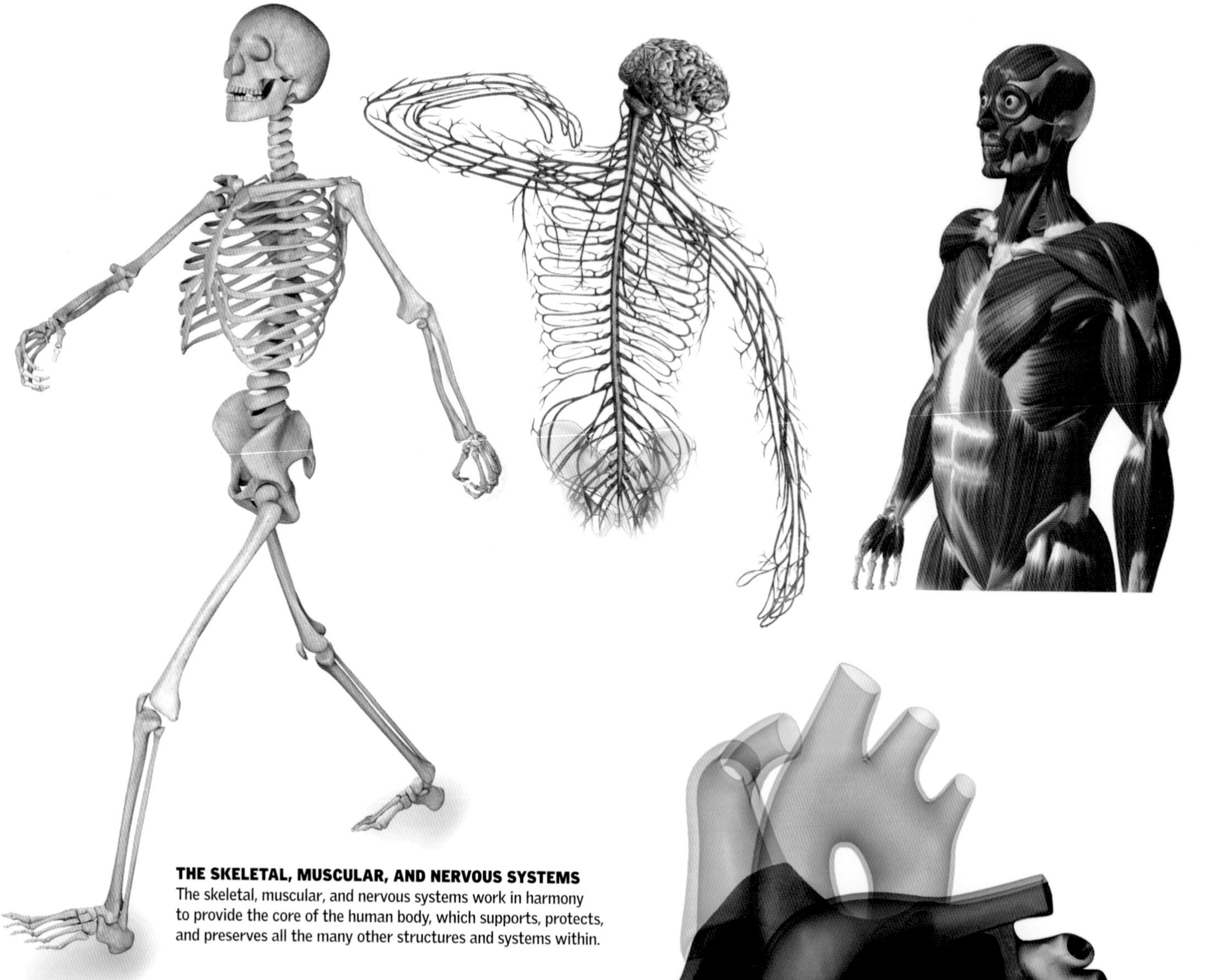

**THE SKELETAL, MUSCULAR, AND NERVOUS SYSTEMS**
The skeletal, muscular, and nervous systems work in harmony to provide the core of the human body, which supports, protects, and preserves all the many other structures and systems within.

together they can perform complex functions for the body. These systems interact with one another as well as the innumerable other structures within the body to ensure that the human anatomy functions as one incredible, totally synchronized and integrated entity. The ten key systems that make up the human body are the skeletal system; the muscular system; the nervous system; the endocrine system; the cardiovascular system; the lymphatic system; the respiratory system; the digestive system; the urinary system; and the reproductive system.

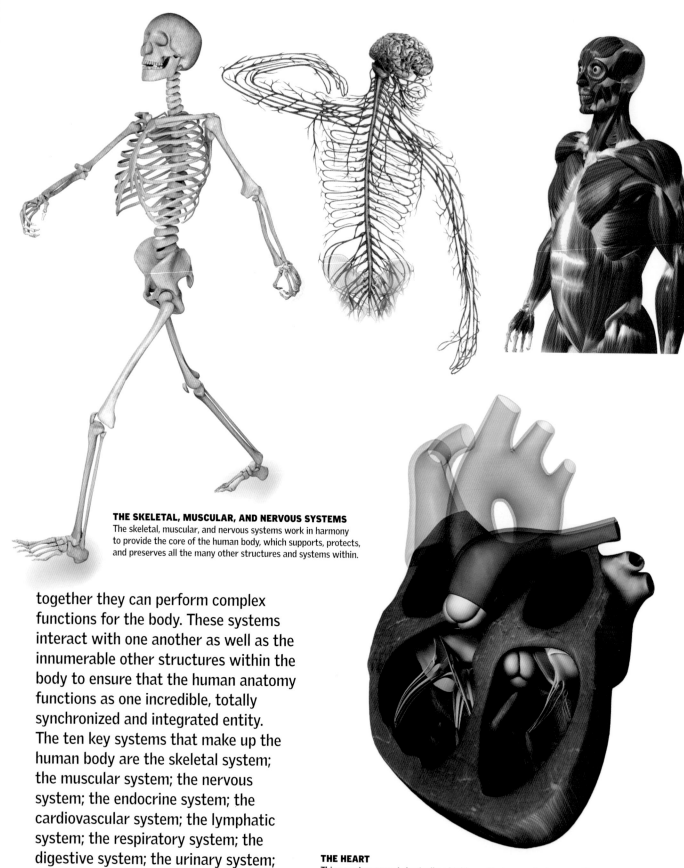

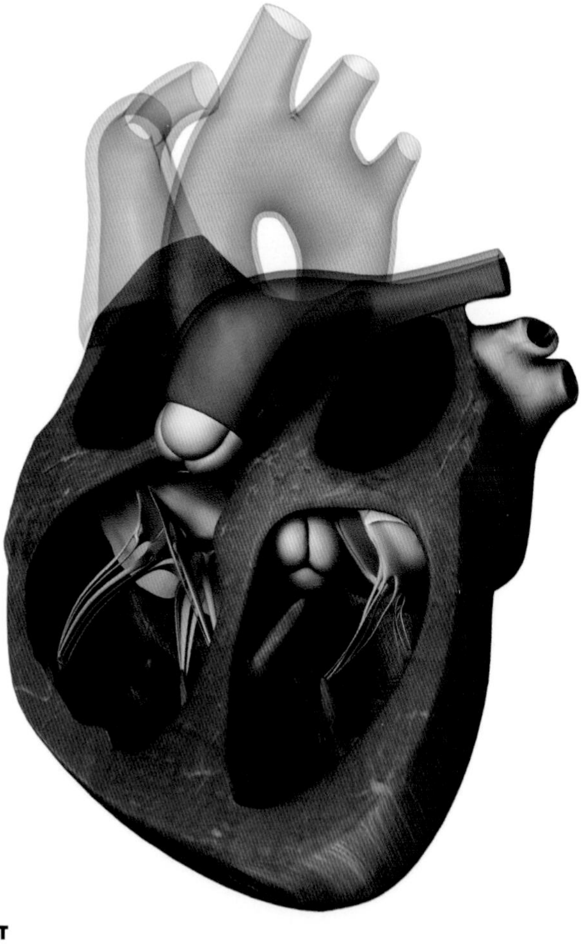

**THE HEART**
This amazing organ is basically a highly resilient pump that powers the human body every second of every day—usually for a number of decades.

## How This Book Works

**ICE AGE HUMAN**
*Homo neanderthalensis* was able to use fire and diverse tools. They used skins to cover themselves from cold and to build shelter, and stones and wood were key materials in the weapons that they used for hunting.

The remarkable book that you hold in your hands is a fully illustrated and entirely comprehensive guide to the structure, function, and disorders of the human body. Arranged in three discrete sections, *The Human Body* defines and explains every aspect of the human anatomy and how it works. Part I, "The Miracle of Life," is a detailed tour through human evolution over the millennia. Opening with an account of how our ancestors developed and survived in the ancient world, the content progresses to explore the mechanisms of heredity that define every human being who is alive today. A full introduction to the science of DNA and genetics is followed by a fascinating exploration of how a human being comes into existence. The chapter entitled "From Zygote to Embryo" explains the miracle of birth in

**THE NEOLITHIC CITY OF ÇATAL HÜYÜK**
Some 10,000 years ago, climatic change and a gradual increase in temperatures brought a modification to the life of humans. Instead of roaming from place to place to hunt, people began to create societies based on sedentary life, agriculture, and the domestication of animals. Some villages grew so much that they became true cities, such as Çatal Hüyük in southern Turkey.

crystal clear and accessible language that will thrill and captivate parents and young readers alike. This is followed by an extensive chapter explaining exactly how a fetus develops over time into a fully grown human adult.

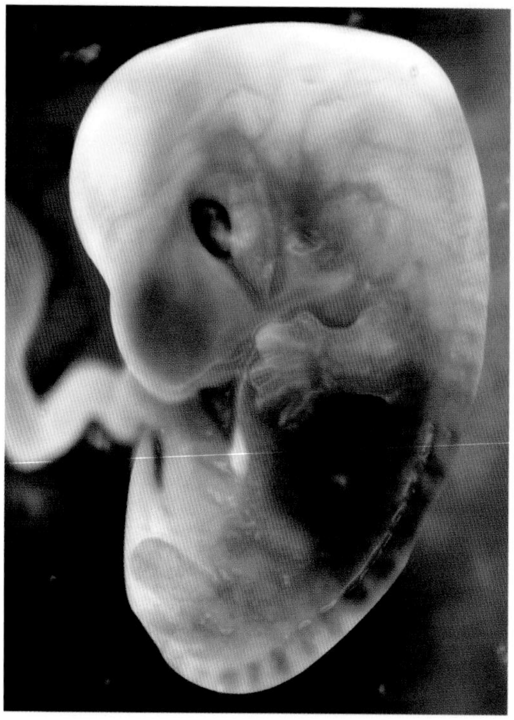

**HUMAN EMBRYO**
36 days after fertilization of the egg, it is still impossible to see a human shape, but the brain, eyes, ears, other vital organs and limbs have started to develop, and the heart has begun to beat.

Part 2, "A Perfect Machine," celebrates the extraordinary amalgam of cells, tissues, structures, and systems that is the human anatomy. Employing only the most straightforward and easy to follow terminology—the first chapter is entitled "What Are We Made Of?"—this section explains the formation and function of the many bones and muscles of the body, followed by a fascinating introduction to the extraordinary systems that allow your body to perform its everyday tasks, both involuntarily and as you intended.

The final section of the book, Part 3, "Fighting Disease and Disorders," examines what happens when the human body is beset by illness or infirmity. The bacteria and other deleterious life-forms that cause disease are covered in detail in a chapter entitled "Microlife." This is followed by a comprehensive analysis of the most common diseases that afflict human beings, ranging from cancer to allergies, AIDS, circulatory conditions, bone degeneration, respiratory infections, and digestive problems. Finishing on a remarkably positive note, *The Human Body* celebrates the tremendous advances

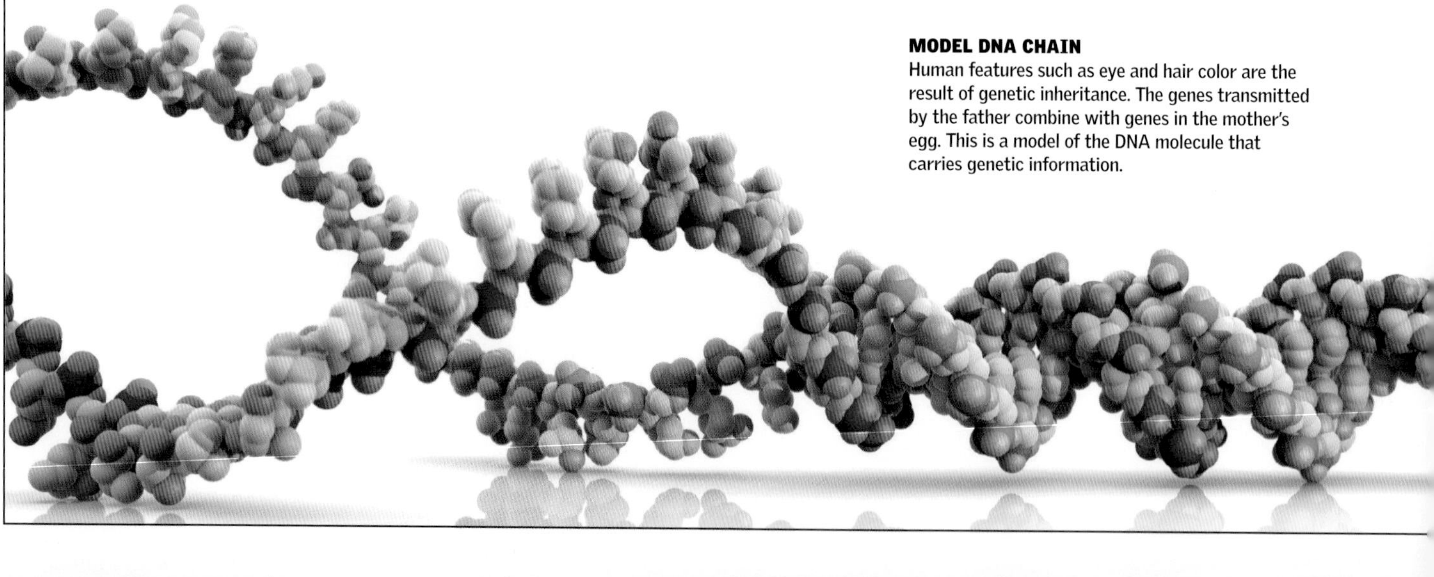

**MODEL DNA CHAIN**
Human features such as eye and hair color are the result of genetic inheritance. The genes transmitted by the father combine with genes in the mother's egg. This is a model of the DNA molecule that carries genetic information.

that we have seen in anatomical understanding and medicine during the last few decades. In an intriguing chapter entitled "Advanced Technology," all the latest developments in the study of anatomy are explored in detail, from early diagnosis to artificial organs, laser surgery, nanomedicine, and the latest methods employed in transplantation. The book closes with an analysis of the feasibility of eternal human life in the future, which is based on the latest scientific and medical research and prognoses.

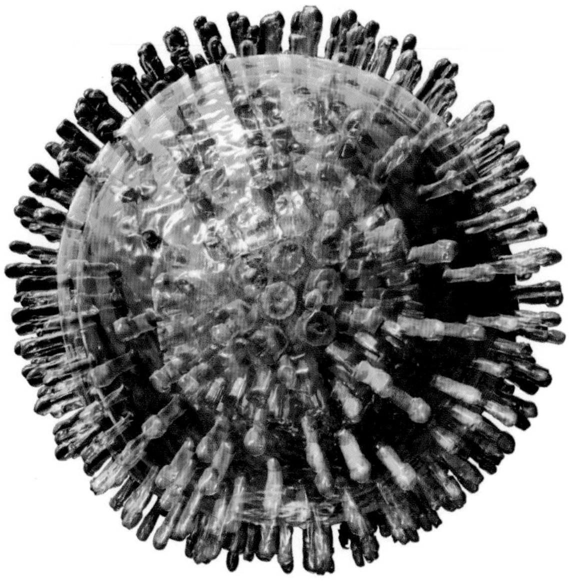

**HERPESVIRUS**
The herpesvirus is widely used in gene therapy.

## Clarity and Comprehension

Whereas many family medical encyclopaedias and guides to the human body are encumbered with difficult terminology and hard-to-grasp concepts, this book is written in refreshingly clear and easy-to-follow language. However, the truly great advantage of the book that you hold in your hands is the extraordinary illustrative material that informs every sentence and paragraph you will read. From detailed cutaway diagrams to superb three-dimensional artworks of the principal organs and structures of the body, this book offers a visual feast that will explain everything you ever wanted to know about how your body works. In addition to the sumptuous and highly colorful artwork that you will find throughout the book, there are many absorbing and instructive photographs to enjoy, which illustrate all aspects of human anatomy and living. Turn the page and begin enjoying your extraordinary journey through the human body!

**AFTER CHILDBIRTH**
Once a baby is born, many changes take place in the child and in the mother. After the umbilical cord is cut, the baby begins to breathe on its own, and its circulatory system is autonomous. For the mother—in pain, with breasts full of milk, and a crying baby—the situation can be stressful. At the same time, a deep and intense bond with the child usually develops.

# 1 THE MIRACLE OF LIFE

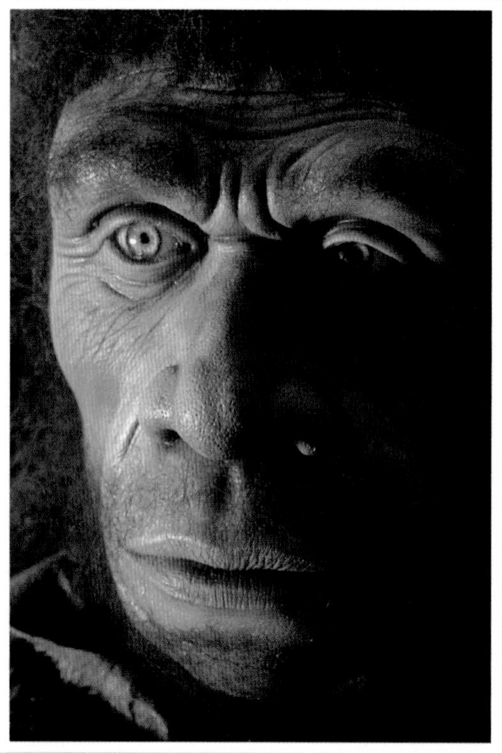

16

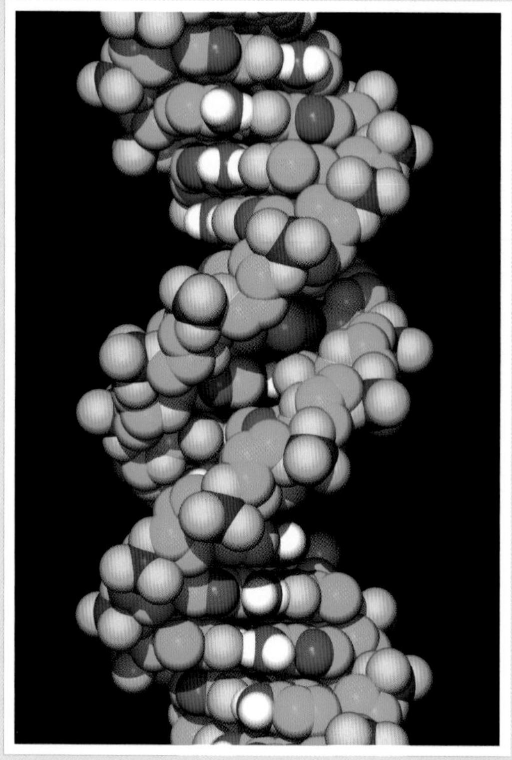

32

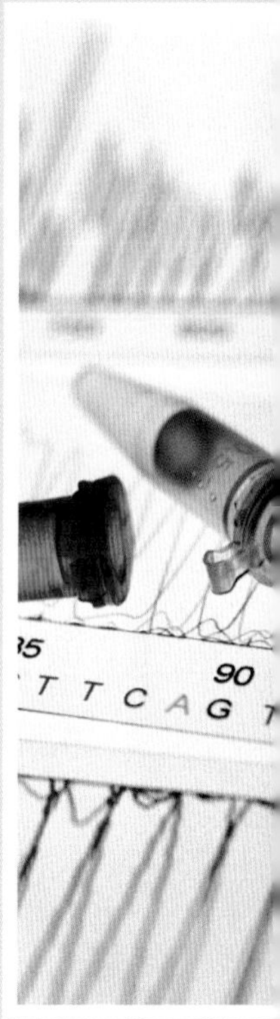

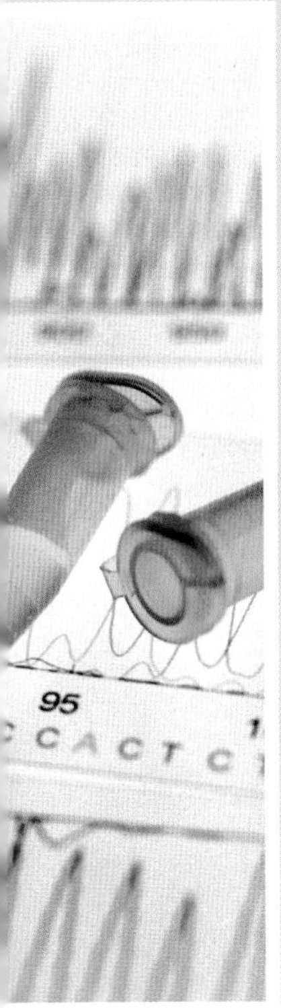


52

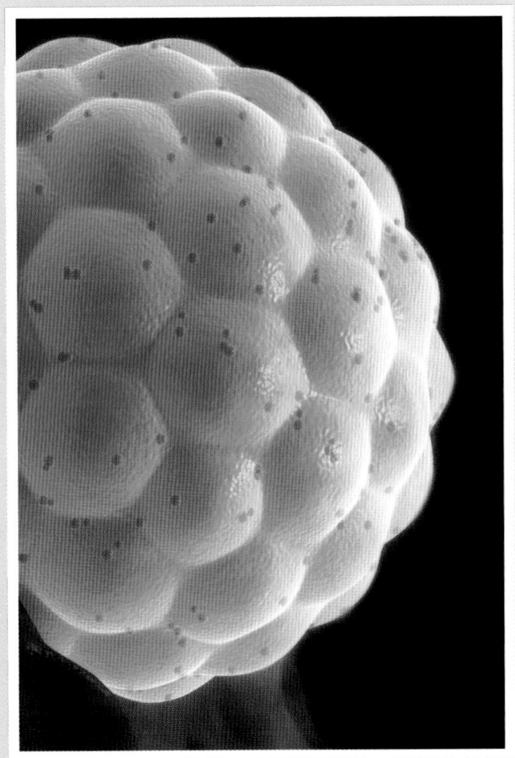

66

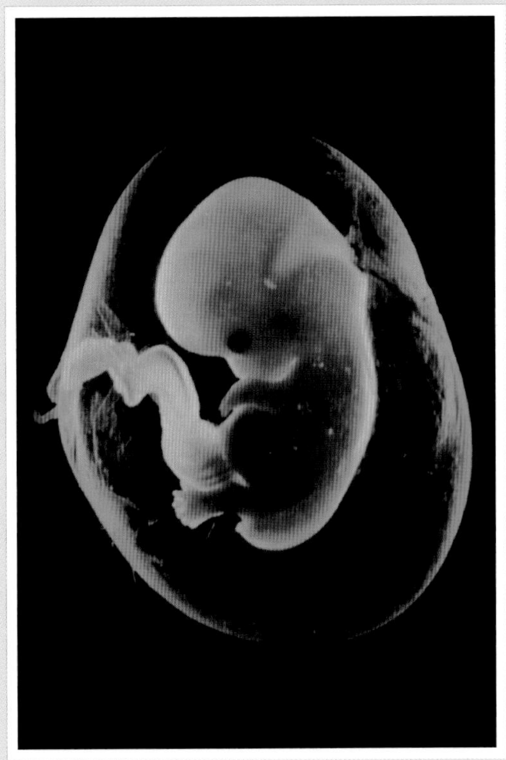

82

# THE MIRACLE OF LIFE

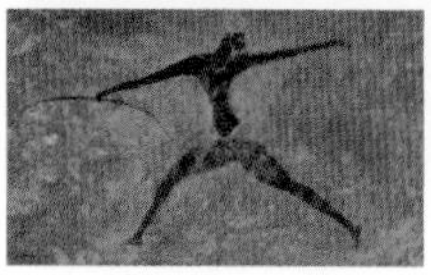
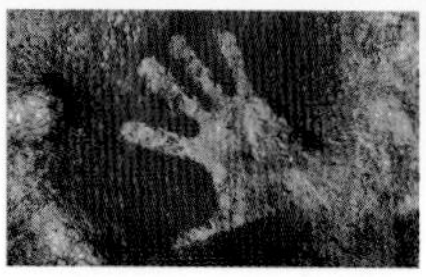
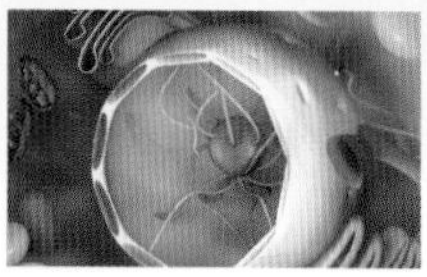

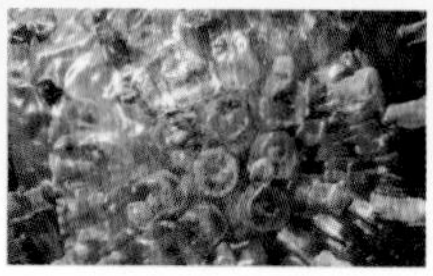

**When did humans appear? What is it that makes us different from the rest of the animals? In what way did language develop? Why is it so important to have deciphered the sequence of the human genome?**

This book offers answers to these and many other questions about the mysteries and marvels of human evolution. Scientists maintain that modern humans originated in Africa because that is where they have found the oldest bones. In addition, genetics has just arrived at the same conclusion, since the DNA studies have confirmed that all humans are related to the African hunter-gatherers who lived some 150 million years ago. Studying the fossils, the experts also found that human skulls from two million years ago already show the development of two specific protuberances that in the present-day brain control speech, the capability that perhaps was as important for early humans as the ability to sharpen a rock or throw a spear. Today thanks to science it is possible to affirm that the brain has changed drastically in the evolutionary course of the species, reaching a greater complexity in humans. This has facilitated, among other things, the capacity to store information and the flexibility in behavior that makes a human an incredibly complex individual.

The purpose of this book is to tell you and show you in marvelous images many of the answers that people have found throughout history, through their successes, failures, and new questions. These new questions have served to shape the world in which we live, a world whose scientific, technological, artistic, and industrial development surprises and at times frightens us.

History is full of leaps. For thousands of years nothing may happen, until all of a sudden some new turn or discovery gives an impulse to humankind. For example, with the domestication of animals and the cultivation of plants, a profound societal revolution occurred.

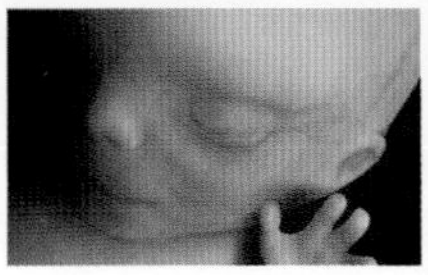

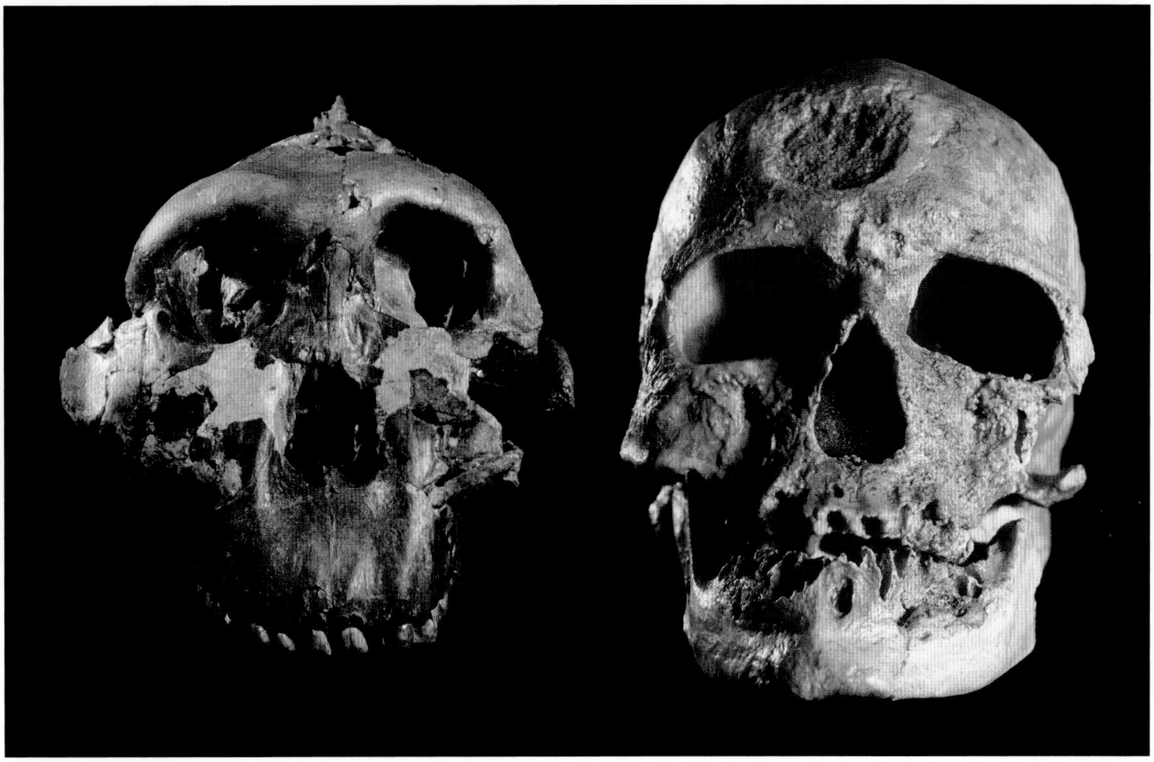

**FACES OF THE PAST**
The skull of *Australopithecus* (left) shows a reduced cerebral portion and a strong jaw. To the right, Cro-Magnon, a representative of modern humans, exhibits a more evolved skull with greater cerebral capacity.

This period of prehistory, called the Neolithic, which dates to 10 million years ago, opened the way for the development of civilization. With the possibility of obtaining food without moving from place to place, the first villages were established and produced great demographic growth.

The book that you have in your hands explains all this in an accessible way. Here you will also find information about the latest discoveries related to the structure of DNA, the molecule of heredity, that opens new areas of investigation. It contributes to the study of clinical and forensic medicine and posits new questions about the origin of life and where we are headed as humans.

The possibility of untangling the sequence of the human genome is not only important in trying to explain why we are here and to explore our evolutionary past, but it also offers the possibility of altering our future.

A guide for parents as well as young people, this section of this remarkable book also recounts in exhaustive detail the almost magical way in which human babies evolve from the very moment that fertilization occurs through to childbirth. Thanks to new technological advances, incredible photos show, day by day, how the embryo evolves, when the heart begins to beat, and even when the brain, eyes, legs, arms, mouth, and teeth of the fetus are formed.

# Human Evolution

*Homo sapiens*, the name that scientifically designates our species, is the result of a long evolutionary process that began in Africa during the Pliocene Epoch. Very few fossils have been found, and there are no clear clues about what caused the amazing development of the culture. Some believe that a change in the brain or vocal apparatus permitted the emergence of a complex language. Other theories hypothesize that a change in the architecture of the human mind allowed *Homo sapiens* to use imagination. What is certain is that hunting and gathering was a way of life for 10,000 years until people formed settlements after the Ice Age and cities began to emerge.



**NEANDERTHAL** (opposite)
Our close cousin was strong, an able hunter, and an excellent artisan. Nobody can explain why the Neanderthals disappeared.

# Primates Develop

Perhaps motivated by climatic change, some five million years ago the species of primates that inhabited the African rainforest subdivided, making room for the appearance of the hominins, our first bipedal ancestors. From that time onward, the scientific community has tried to reconstruct complex phylogenetic trees to give an account of the rise of our species. DNA studies on fossil remains allow us to determine their age and their links with different species. Each new finding can put into question old theories about the origin of humans.

**FUNCTION OF SPEECH**

In humans, speech has a semantic character. Upon speaking, a human always addresses other people with the object of influencing them, changing their thoughts, enriching them mentally, or directing their conduct toward something specific. Some scientists believe that a change in the brain or vocal apparatus allowed the development of complex language, which facilitated creativity and the acquisition of knowledge.

## Primates That Talk

The rise of symbolic language, which is a unique ability of humans, is a mystery. But the evolution of the speech apparatus in humans has been decisive. The human larynx is located much lower than in the rest of the mammals. This characteristic makes it possible to emit a much greater variety of sounds.

**THE PHYLOGENETIC TREE**

This cladogram (map of emergence of new species from previous ones) shows the relationship of the Homo genus to the other species of primates.

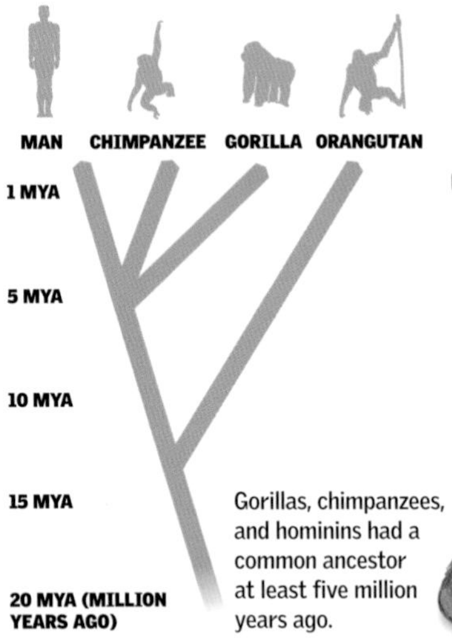


## *Australopithecus*

**PRECURSOR**

This ape was the first true hominin but is extinct today.

## *Homo habilis*

**THE GREAT LEAP**

Its brain was much greater, and there were substantial anatomical changes.

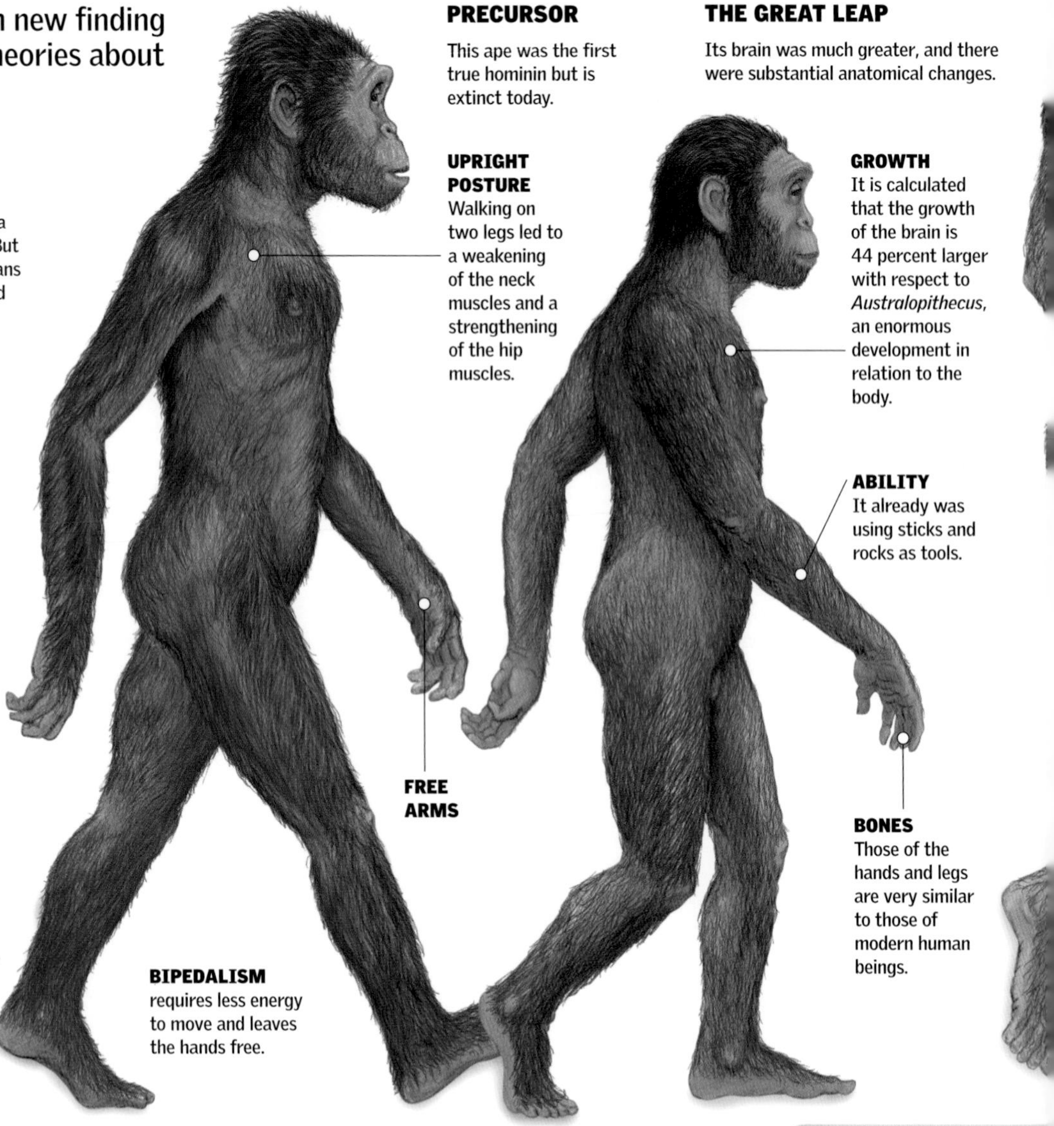

UPRIGHT POSTURE
Walking on two legs led to a weakening of the neck muscles and a strengthening of the hip muscles.
FREE ARMS
BIPEDALISM
requires less energy to move and leaves the hands free.
GROWTH
It is calculated that the growth of the brain is 44 percent larger with respect to *Australopithecus*, an enormous development in relation to the body.
ABILITY
It already was using sticks and rocks as tools.
BONES
Those of the hands and legs are very similar to those of modern human beings.

**NOT-SO-DISTANT RELATIVES**

There are various uncertainties and disagreements among paleontologists about how the evolutionary tree for hominins branches out. This version is based on one created by paleoanthropologist Ian Tattersall.

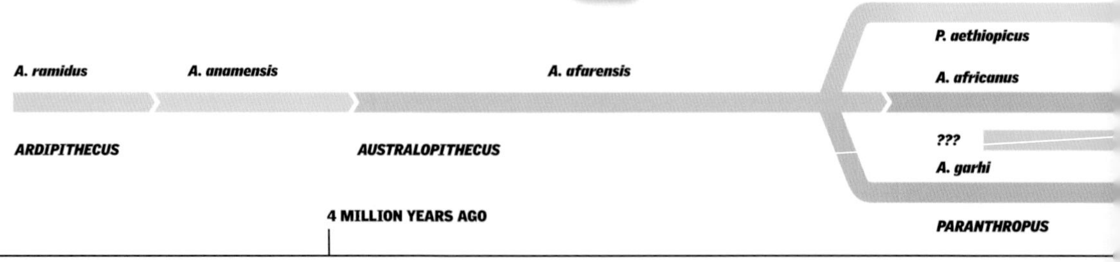

**TOOLS FOR SPEAKING**
The larynx of humans is located much lower than in chimpanzees and thus allows humans to emit a greater variety of sounds.

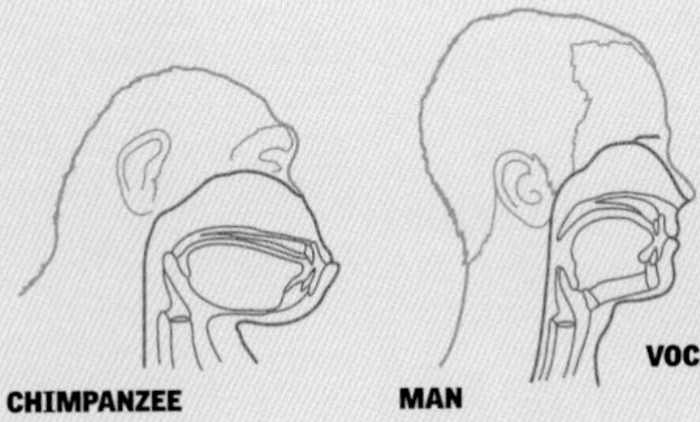


**AND FOR THINKING**
The evolution of the brain has been essential for the development of language and other human capacities. Greater cranial capacity and nutrition have had physiological influences.

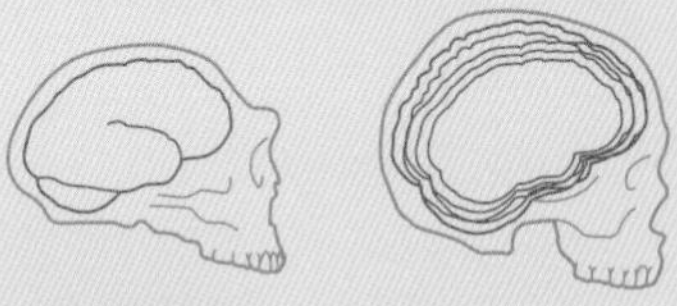


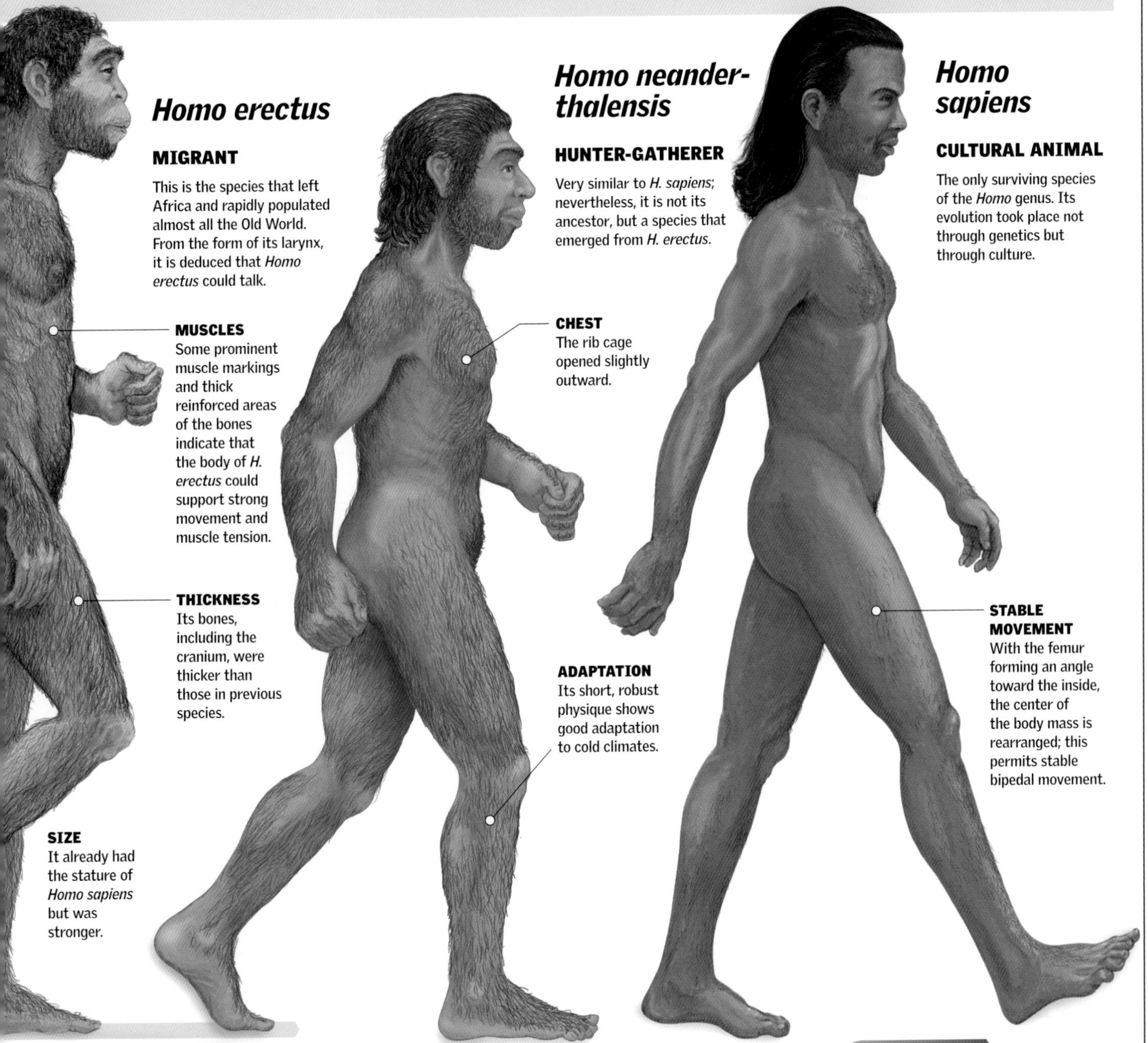

## *Homo erectus*

**MIGRANT**

This is the species that left Africa and rapidly populated almost all the Old World. From the form of its larynx, it is deduced that *Homo erectus* could talk.

**MUSCLES**
Some prominent muscle markings and thick reinforced areas of the bones indicate that the body of *H. erectus* could support strong movement and muscle tension.

**THICKNESS**
Its bones, including the cranium, were thicker than those in previous species.

**SIZE**
It already had the stature of *Homo sapiens* but was stronger.

## *Homo neanderthalensis*

**HUNTER-GATHERER**

Very similar to *H. sapiens*; nevertheless, it is not its ancestor, but a species that emerged from *H. erectus*.

**CHEST**
The rib cage opened slightly outward.

**ADAPTATION**
Its short, robust physique shows good adaptation to cold climates.

## *Homo sapiens*

**CULTURAL ANIMAL**

The only surviving species of the *Homo* genus. Its evolution took place not through genetics but through culture.

**STABLE MOVEMENT**
With the femur forming an angle toward the inside, the center of the body mass is rearranged; this permits stable bipedal movement.

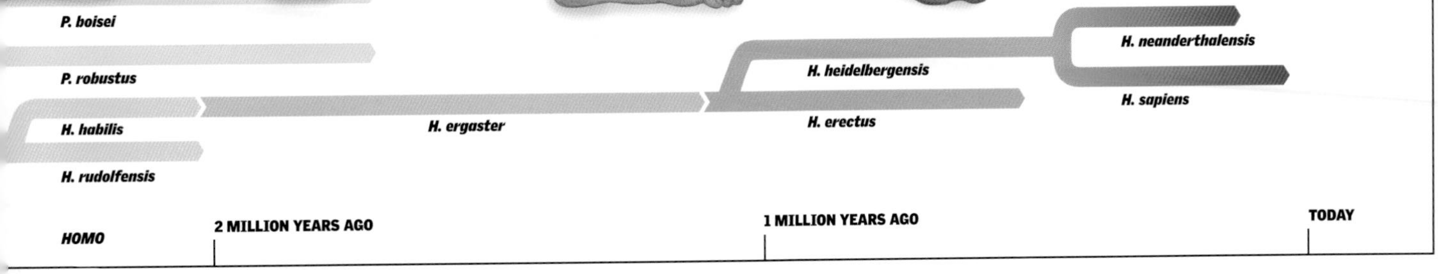

# First Humans

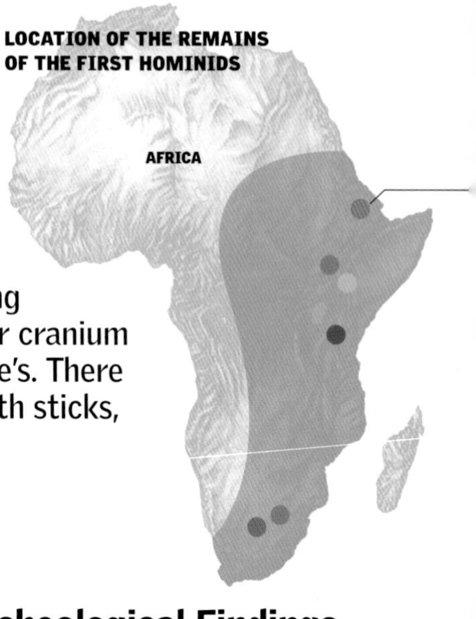


The *Australopithecus* were the first humanlike creatures who could walk in an upright posture with their hands free, as indicated by the fossils found in Tanzania and Ethiopia. It is believed that climatic changes, nutritional adaptations, and energy storage for movement contributed to bipedalism. In any case, their short legs and long arms are seen as indications that they were only occasional walkers. Their cranium was very different from ours, and their brain was the size of a chimpanzee's. There is no proof that they used stone tools. Perhaps they made simple tools with sticks, but they lacked the intelligence to make more sophisticated utensils.

## Adaptation to the Environment

The climatic changes that occurred during the Miocene probably transformed the tropical rainforest into savannah. Various species of hominins left their habitats in the trees and went down to the grasslands in search of food. It is conjectured that the first hominins began to stand up to see over the grasslands.

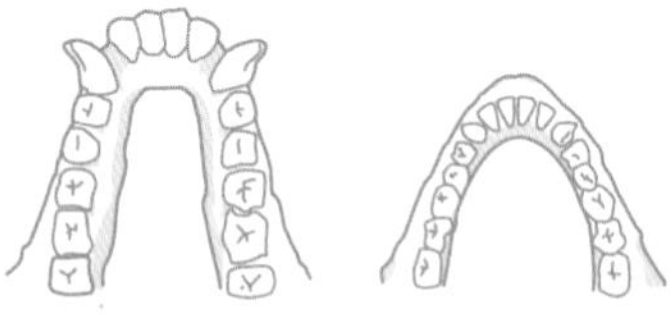


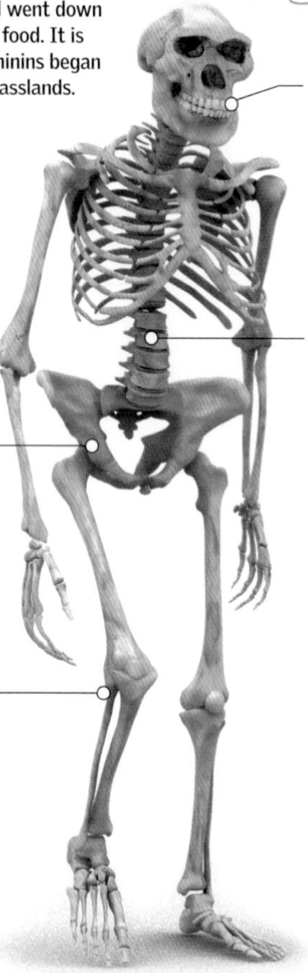

**SPECIAL TEETH**
They had large incisors like spatulas in front, and the teeth became arranged in the form of an arch.

**BIPEDALISM**
By walking on two feet, they were able to free their upper limbs while they moved.

**DORSAL SPINE**
Had many curves to maintain balance. Given that monkeys do not have lumbars, the weight of the body falls forward.

**ADAPTED PELVIS**
Morphological changes in the pelvis, sacrum, and femur made these bones similar to those in modern humans.

**TOE**
Whereas in chimpanzees the big toe is used to grasp, the position of the big toe and the foot arch in hominins supported movement in a bipedal posture.

**KNEE**
Unlike chimpanzees, the rim of the femur had an elliptical shape like that in the human knee.

***AUSTRALOPITHECUS AFARENSIS***

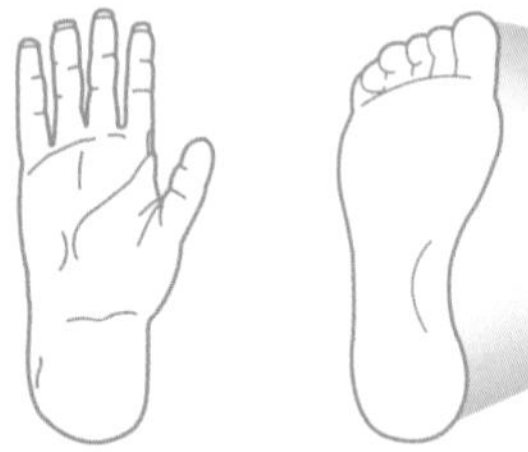


## Archeological Findings

The fossil skull of a child was found in 1924 in the Taung mine (South Africa). The remains included the face with a jaw and tooth fragments as well as skull bones. The brain cavity had been replaced with fossilized minerals. Later, in 1975, footprints of hominins were found in Laetoli (Tanzania). It is believed that more than three million years ago, after a rain that followed a volcanic eruption, various specimens left their tracks in the moist volcanic ash.

**SKULL OF TAUNG**
This had a round head and strong jaw. Its cranial cavity could house a brain (adult) of 26 cubic inches (440 cu cm).

2.5
million
years ago

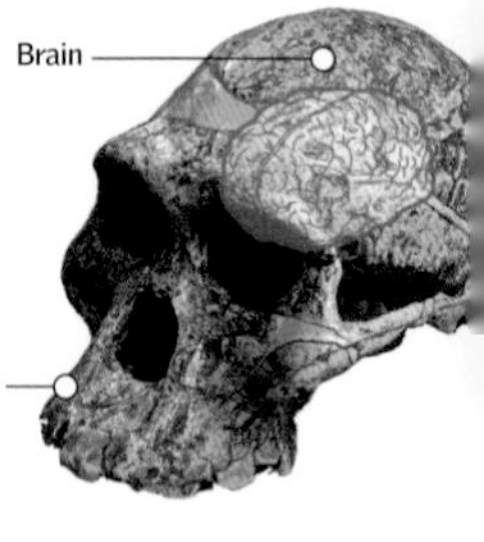

**LAETOLI**
In 1975 in Laetoli (Tanzania), tracks of hominins that archeologists found in fossilized volcanic ash provided evidence of hominins walking on two legs (bipedalism).

3.6
million
years ago

***AUSTRALOPITHECUS AFARENSIS***

| AUSTRALOPITHECUS ANAMENSIS | AUSTRALOPITHECUS AFRICANUS | PARANTHROPUS AETHIOPICUS |
|---|---|---|
| 4.2 to 3.9 million years ago. Primitive hominin with wide molars. | 3 to 2.5 million years ago. Globular skull with greater cerebral capacity. | Approximately 2.5 million years ago. Robust skull and solid face. |

***AUSTRALOPITHECUS ANAMENSIS*** | ***PARANTHROPUS AETHIOPICUS*** | ***AUSTRALOPITHECUS AFRICANUS*** | ***PARANTHROPUS ROBUSTUS*** | ***PARANTHROPUS BOISEI***

## *Australopithecus afarensis*

Considered the oldest hominin, it inhabited eastern Africa between three and four million years ago. A key aspect in human evolution was the bipedalism achieved by *A. afarensis*. The skeleton of "Lucy," found in 1974, was notable for its age and completeness.

3 million years ago

**COMPARATIVE SIZE**

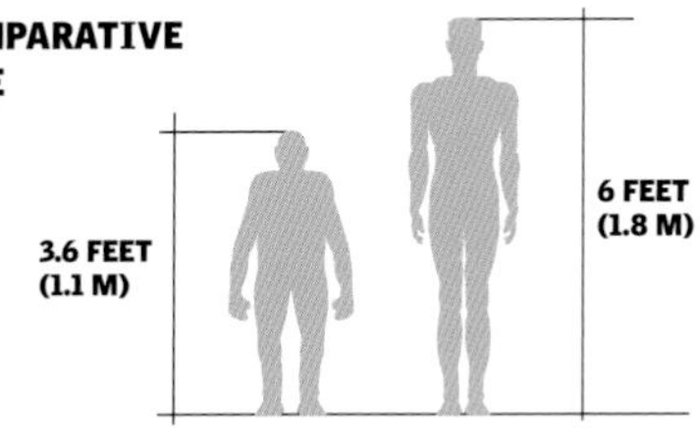


**THE SKELETON OF LUCY**

This hominid found in Ethiopia had the size of a chimpanzee, but its pelvis allowed it to maintain an upright position.

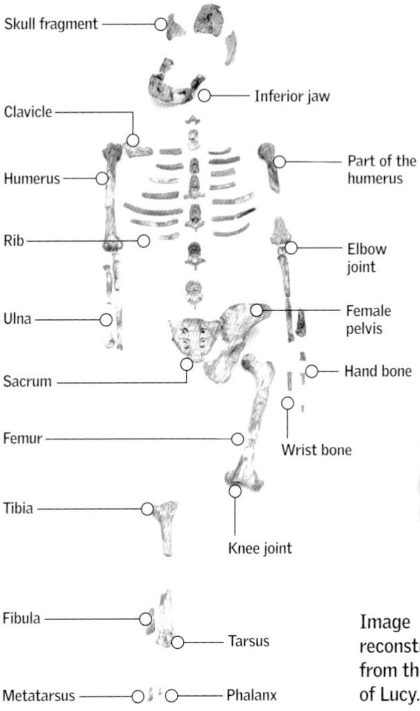

Image reconstructed from the bones of Lucy.

**PARANTHROPUS BOISEI**

2.2 to 1.3 million years ago. Skull adapted for consumption of tough vegetables.

**PARANTHROPUS ROBUSTUS**

1.8 to 1.5 million years ago. Very robust, bony appearance.

# Use of Tools

The emergence of *Homo habilis*, which had a more humanlike appearance than *Australopithecus*, in eastern Africa, showed important anatomical modifications that allowed advancement, especially in the creation of various stone tools, such as flaked pebbles for cutting and scraping and even hand axes. The bipedal posture for locomotion was established, and the first signs of language appeared. Stone technology became possible thanks to the notable increase in brain size in *Homo habilis*. In turn, the anatomic development of *Homo erectus* facilitated its migration toward areas far from its African origins, and it appears to have populated Europe and Asia, where it traveled as far as the Pacific Ocean. *Homo erectus* was capable of discovering fire, a vital element that improved human nutrition and provided protection from the cold.

## *Homo habilis*

The appearance of *Homo habilis* in eastern Africa between 2 and 1.5 million years ago marked a significant advancement in the evolution of the human genus. The increased brain size and other anatomical changes together with the development of stone technology were substantive developments in this species, whose name means "handy man." Although it fed on carrion, it was still not capable of hunting on its own.

**THE BRAIN**
The cranial cavity of *Homo habilis* was larger than that of *Australopithecus*, reaching a cerebral development of between 40 and 50 cubic inches (650-800 cu cm). It is believed that this characteristic was key in developing the capacity of making tools, considering that it had half the brain size of modern humans.

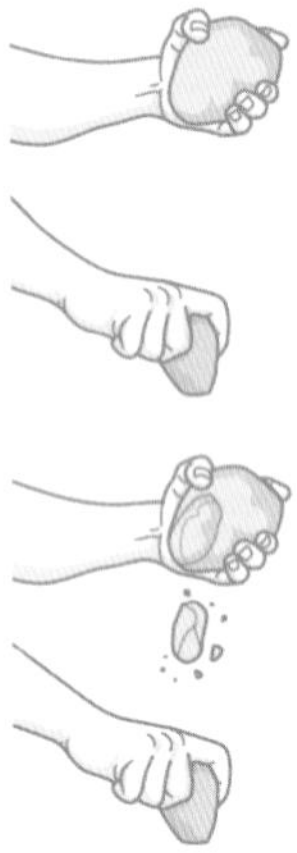

**1**

**CARVING**
The first step was to select rocks and scrape them until sharp.

**2**

**REMOVING**
A "stone hammer" was used to sharpen the edges of the tools.

**THIS CARVED ROCK IS THE OLDEST KNOWN TOOL.**

**2 MILLION YEARS AGO**
Appearance of *Homo habilis* in eastern Africa.

**1.7 MILLION YEARS AGO**
*Homo erectus* is the first hominin to leave its habitat.

**1.5 MILLION YEARS AGO**
*Homo habilis* disappears because of unknown causes.

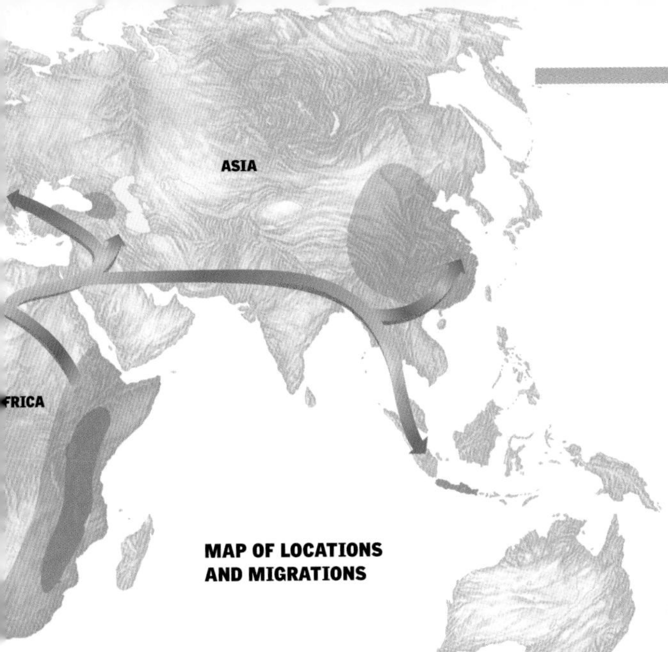


MAP OF LOCATIONS AND MIGRATIONS

HOMO HABILIS

HOMO ERECTUS

## *Homo erectus*

The "erect man" is native to East Africa, and its age is estimated at 1.8 million years. It was the first hominin to leave Africa. In a short time it populated a great part of Europe. In Asia it reached China to the east and the island of Java to the southeast. Much of what is known about this species was learned from a finding called Turkana Boy near Lake Turkana, Kenya, in 1984. This species was tall and had long limbs. The brain of this specimen was larger than that of *Homo habilis*, and it could have made the fundamental discovery of making fire.

**COMPARATIVE SIZES**

***HOMO HABILIS*** **5 FEET (1.3 M)**

***HOMO ERECTUS*** **5.3 FEET (1.6 M)**

***HOMO SAPIENS*** **6 FEET (1.8 M)**

**ARCHEOLOGICAL FINDINGS**

The first being known as *Homo habilis* was found in 1964 in the Olduvai Gorge, located in the Serengeti Plain (Tanzania). The later discovery of the Turkana Boy (Kenya) revealed many of the physical particularities of *Homo erectus*.

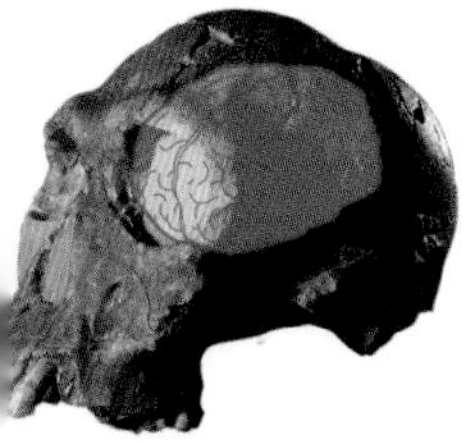

**SKULL OF *HOMO HABILIS* FOUND IN OLDUVAI (TANZANIA)**

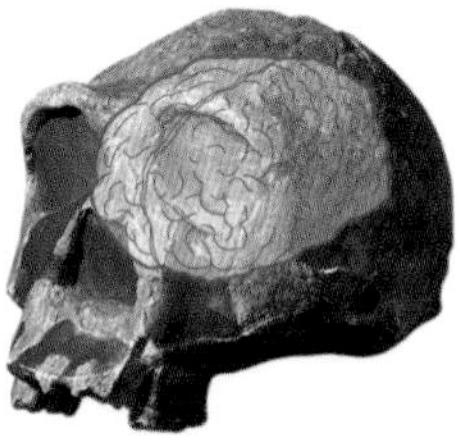

**SKULL OF *HOMO ERECTUS* FOUND IN KOOBI FORA (KENYA)**

**HAND AX IN THE SHAPE OF A DROP**

**FIRE**

One of the major discoveries in the evolution of humans. It was used not only for protection from the cold but also to treat wood and cook food. The first evidence of the use of fire is some 1,500,000 years ago.

***HOMO ERECTUS***

**ABOUT 1.5 MILLION YEARS AGO**

First use of fire by *Homo erectus*, in southern Africa.

# Able Hunters

Descendants of *Homo heidelbergensis*, the Neanderthals were the first inhabitants of Europe, western Asia, and northern Africa. Diverse genetic studies have tried to determine whether it is a subspecies of *Homo sapiens* or a separate species. According to fossil evidence, Neanderthals were the first humans to adapt to the extreme climate of the glacial era, to carry out funerals, and to care for sick individuals. With a brain capacity as large or larger than that of present-day humans, Neanderthals were able to develop tools in the style of the Mousterian culture. The cause of their extinction is still under debate.

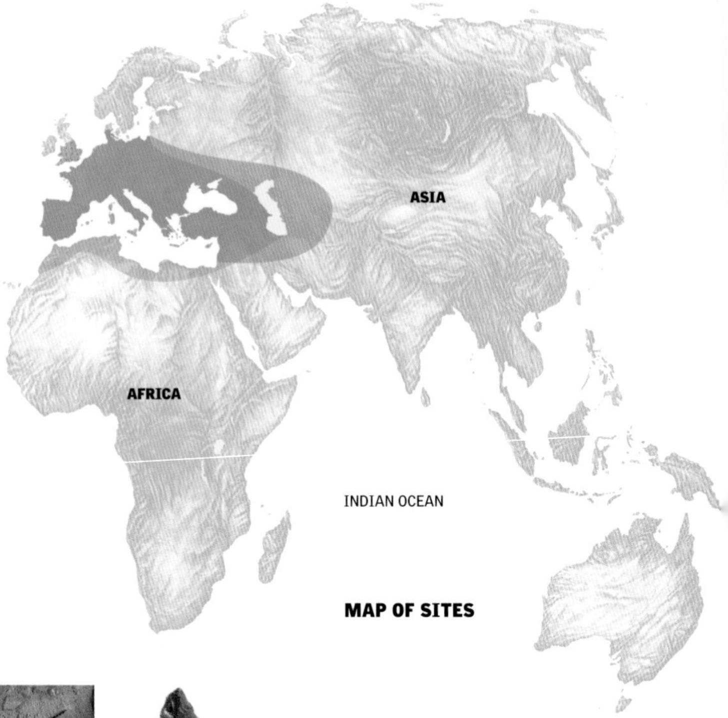


**MAP OF SITES**

## *Homo neanderthalensis*

The Middle Paleolithic (400,000 to 30,000 years ago) is dominated by the development of *Homo neanderthalensis*. In the context of the Mousterian culture, researchers have found traces of the first use of caves and other shelters for refuge from the cold. Hunters by nature, *H. neanderthalensis* created tools and diverse utensils, such as wooden hunting weapons with sharpened stone points.

60,000 years ago

**THE AGE OF SOME NEANDERTHAL DISCOVERIES.**

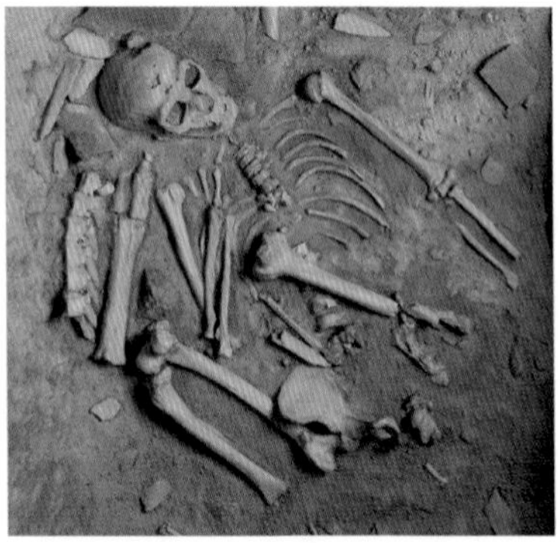

**Graves**

Much is known about the Neanderthals because they buried their dead.

**Man—hunter**

Males were dedicated to the search for food, while the women looked after children. It is believed that Neanderthals hunted large prey over short distances. They used wooden spears with stone points and probably jumped on the prey.

100,000 years ago

**TOOLS FOUND**

**Rocks for cutting and scraping**

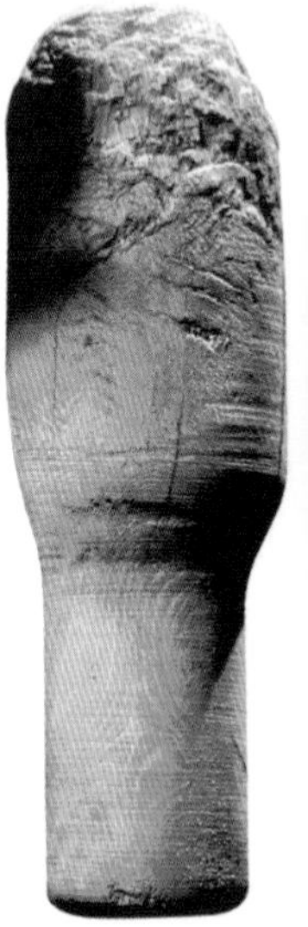

**Tools for tanning hides**

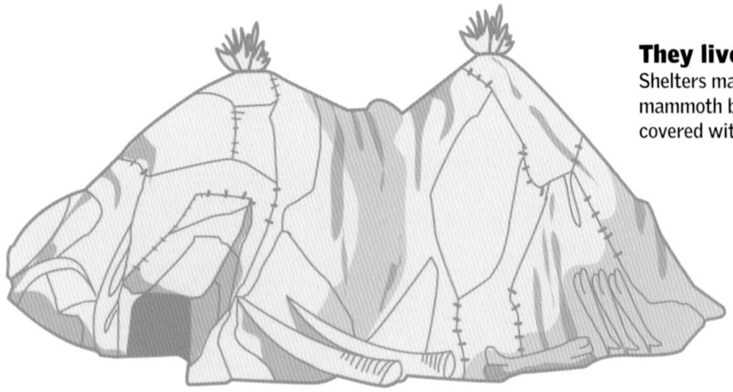

**They lived in**

Shelters made of mammoth bones and covered with skins.

**600,000 YEARS AGO**

*Homo heidelbergensis* is in Europe, part of Asia, and Africa.

**400,000 YEARS AGO**

Wooden spears found in Germany and the United Kingdom date back to this time.

**160,000 YEARS AGO**

*Homo neanderthalensis* lives in the Ice Age in Europe and western Asia.

HOMO NEANDERTHALENSIS

HOMO HEIDELBERGENSIS

## Humans of the Ice Age

Characterized as the caveman of the Ice Age, *Homo neanderthalensis* was able to use fire and diverse tools that allowed it to work wood, skins, and stones, among other materials. They used the skins to cover themselves from cold and to build shelter, and the stones and the wood were key materials in the weapons used for hunting. The bone structure of their fossils reveals a skull with prominent ciliary arcs, sunken eyes, a wide nose, and large upper teeth, probably used to grasp skins and other objects during the process of rudimentary manufacture.

**PHYSICAL CONTEXT**

The bones in the hand made it possible to grasp objects much more strongly than modern man can.

**COMPARATIVE SIZE**

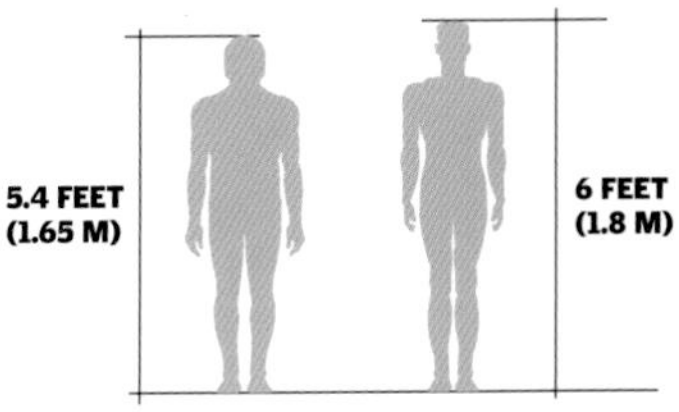


**GREATER CRANIAL CAPACITY**

In comparison to modern humans, Neanderthals had a larger brain capacity.

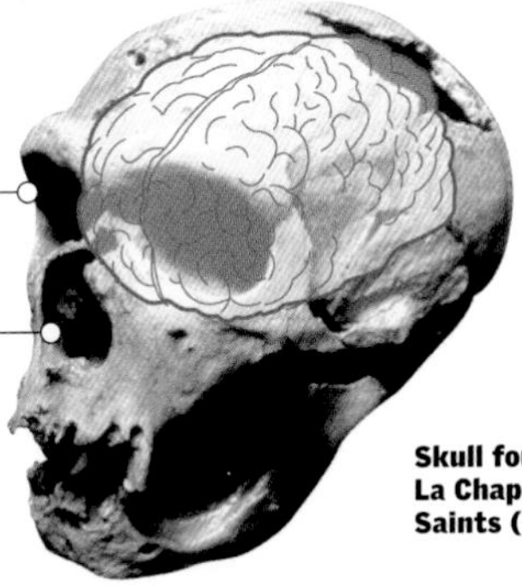


**Skull found in La Chapelle-aux-Saints (France)**

98 cubic inches
(1,600 cu cm) cranial capacity

**150,000 YEARS AGO**

First *Homo sapiens* found in Africa.

**25,000 YEARS AGO**

*Homo neanderthalensis* becomes extinct from unknown causes.

# Direct Ancestors

The origin of the human species is still in debate, even though scientists have been able to establish that *H. sapiens* is not directly related to the Neanderthals. The most accepted scientific studies for dating Neanderthal fossils places the oldest specimens some 195,000 years ago in Africa. New genetic studies based on mitochondrial DNA have corroborated that date and have also contributed to determining the possible migration routes that permitted the slow expansion of *H. sapiens* to other continents. Meanwhile, the new discoveries raise unanswered questions about what happened in the course of the 150,000 years that preceded the great cultural revolution that characterizes *H. sapiens* and that occurred some 40,000 years ago with the appearance of Cro-Magnon in Europe.

## *Homo sapiens sapiens*

It is believed that Cro-Magnon arrived in Europe some 40,000 years ago. Evidence of prehistoric art, symbolism, and ritual ceremonies distinguish this advanced culture from other species of hominins that preceded it. It was well adapted to its environment, lived in caves, and developed techniques of hunting in groups. It captured large animals with traps and small ones with rocks.

**TOOLS**
*Homo sapiens* invented multiple tools for various uses that were usually made from stone, bone, horns, and wood.

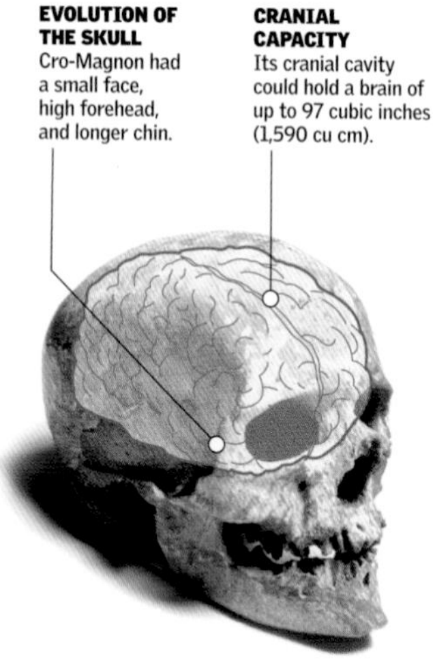
EVOLUTION OF THE SKULL
Cro-Magnon had a small face, high forehead, and longer chin.
CRANIAL CAPACITY
Its cranial cavity could hold a brain of up to 97 cubic inches (1,590 cu cm).

**150,000 YEARS AGO**
The "Mitochondrial Eve" is the common ancestor of all people.

**120,000 YEARS AGO**
*Homo sapiens* begins to extend through Africa.

## Theories of Expansion

There is no agreement among scientists about how the expansion of *Homo sapiens* to the entire world took place. It is believed that the "Mitochondrial Eve," the most recent common ancestor, lived in Africa, because the people of that continent have greater genetic diversity than those of the other continents. From there, in various migratory waves, *Homo sapiens* would have reached Asia, Australia, and Europe. However, some scientists think that there were no such migrations but that modern humans evolved more or less simultaneously in various regions of the ancient world.

KEY — GENERAL ROUTE — 40,000 YEARS AGO: DATE OF MIGRATION

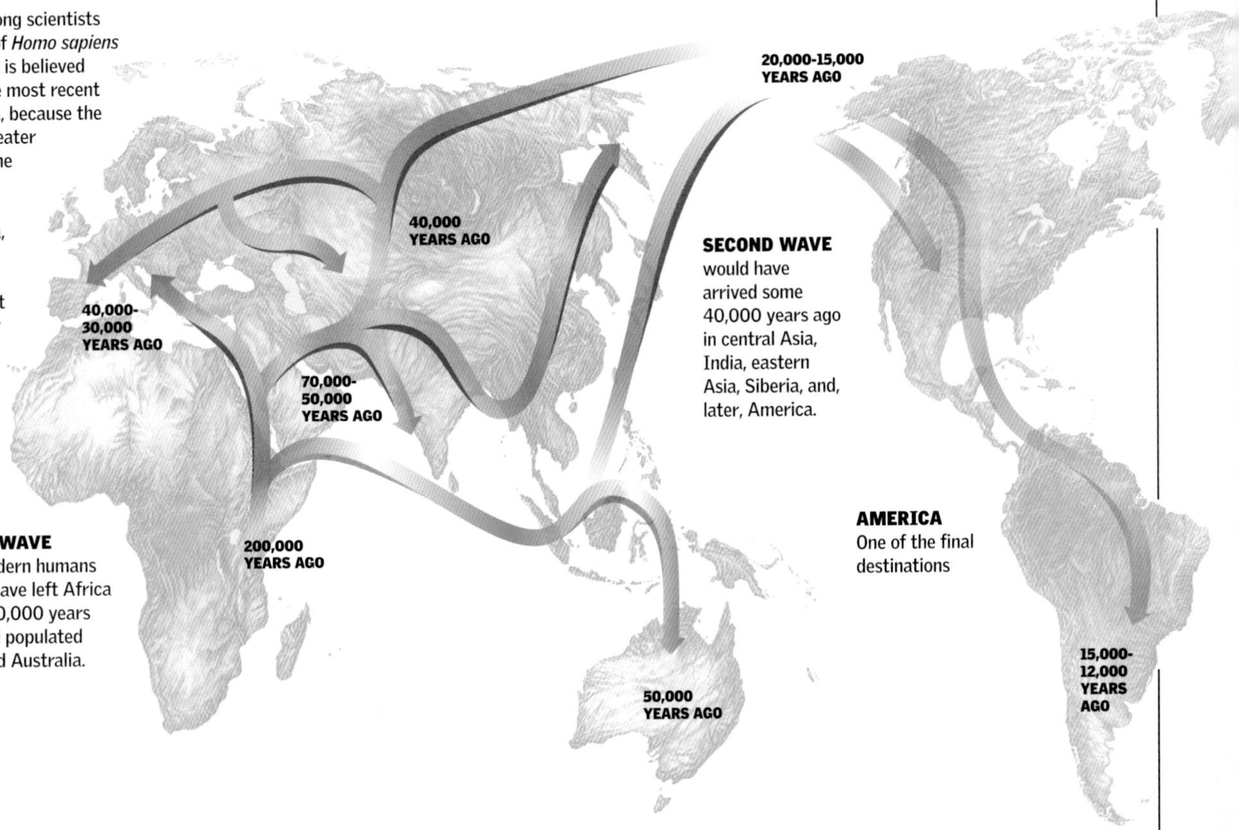


**SECOND WAVE**
would have arrived some 40,000 years ago in central Asia, India, eastern Asia, Siberia, and, later, America.

**FIRST WAVE**
The modern humans would have left Africa some 60,000 years ago and populated Asia and Australia.

**AMERICA**
One of the final destinations

**AFRICAN CRADLE**
The majority of paleoanthropologists and geneticists agree that humans of today emerged in Africa. It is there they have found the oldest bones.

### Out of Africa

According to this theory, modern man is an evolution of the archaic *Homo sapiens* that emerged in Africa. From there it would have extended to the rest of the world, overrunning the Neanderthals and primitive *Homo sapiens*. The anatomical differences between the races would have occurred in the last 40,000 years.

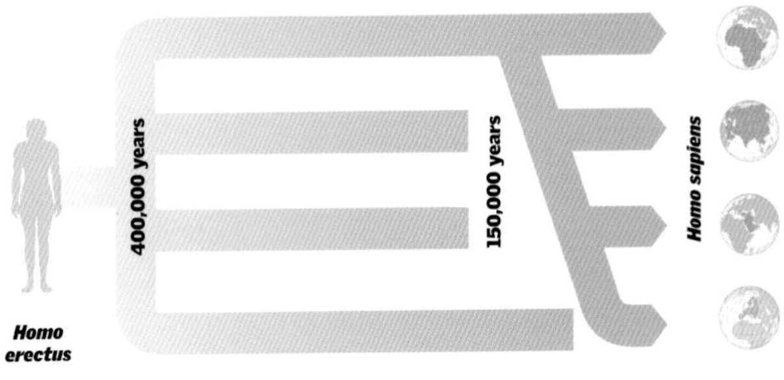


### Multiregional Evolution

The theory of regional continuity, or multiregional evolution, states that the modern human developed simultaneously in diverse regions of the world, like the evolution of local archaic *Homo sapiens*. The last common ancestor would be a primitive *Homo erectus* that lived in Africa some 1.8 million years ago.

| 90,000 YEARS AGO | 60,000 YEARS AGO | 40,000 YEARS AGO |
|---|---|---|
| "Nuclear Adam" was the common ancestor of all the men of the world. | Traces of *Homo sapiens* in China. | Cro-Magnon (type of *Homo sapiens*) appears in Europe. |

# Culture, the Great Leap

Although questions remain about how culture originated, it is almost impossible to determine which things of the human world are natural and which are not. Scientists of many disciplines are trying to answer these questions from the evidence of prehistoric life found by paleontologists. The subspecies of mammals to which man belongs, *Homo sapiens sapiens*, appeared in Africa some 150,000 years ago, disseminated through the entire Old World some 30,000 years ago (date that the oldest signs of art were found), and colonized America 11,000 years ago; but the first traces of agriculture, industry, population centers, and control over nature date from barely the last 10,000 years. Some believe that the definitive leap toward culture was achieved through the acquisition of a creative language capable of expressing ideas and sentiments more advanced than the simple communication of *Homo erectus*.

## The First Artists

Cave paintings, like those of the caves of Altamira (Spain) and Lascaux (France), leave no doubt that those who made them truly possessed the attributes of human beings. Architecture had not arrived, but paintings had, engraved and sculptured in stone or bone. There exist various theories about the function of cave painting that consider the aesthetic, the magical, the social, and the religious—not much different from the questions about art today.

### CAVE-PAINTING TECHNIQUES

**GEOMETRIC DESIGNS**
Dotted and lineal geometric designs, along with mythical chimeras, have been found among European cave paintings similar to the rock art of Aboriginal Australians.

**BLOWING**
One technique consisted of blowing pigment through a rod or hollow bone.

**COLORS**
The pigments used were of natural origin, such as vegetable charcoal, red ocher, and brown ocher.

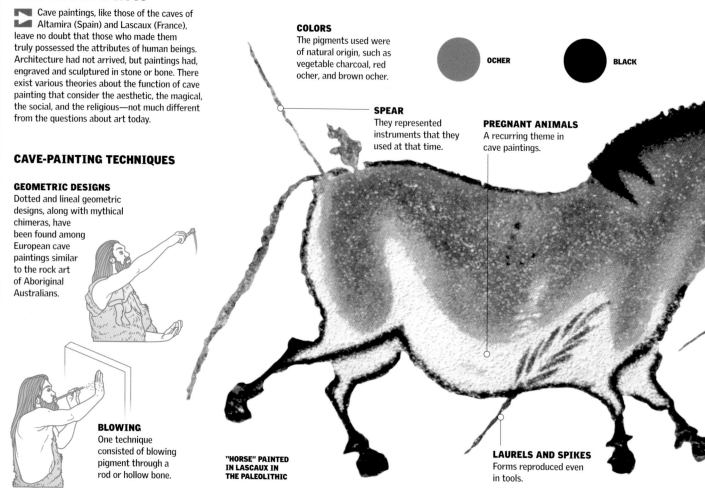

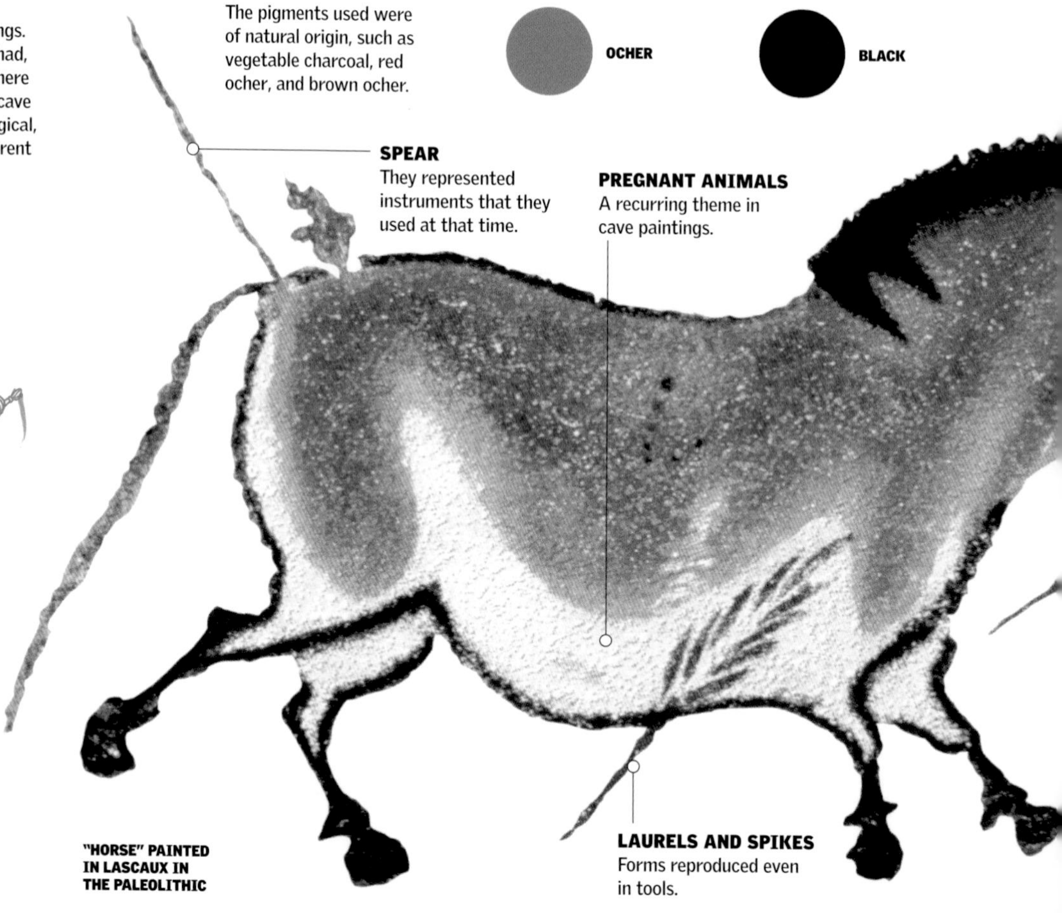


**SPEAR**
They represented instruments that they used at that time.

**PREGNANT ANIMALS**
A recurring theme in cave paintings.

**LAURELS AND SPIKES**
Forms reproduced even in tools.

**"HORSE" PAINTED IN LASCAUX IN THE PALEOLITHIC**

**WÜRM GLACIATION**
**35,000 YEARS AGO**
The Upper Paleolithic begins.

**AURIGNACIAN**
**30,000 YEARS AGO**
Tools of mammoth tusk, flake tools.

**PERIGORDIAN**
**27,000 YEARS AGO**
Well-cut tools, including a multiangle graver.

**ART ON THE WALLS**

Cave painting is a phenomenon that was found mainly in the current regions of France and Spain. In France, there are more than 130 caves; the most famous are located in the Aquitaine region (Lascaux, Pech-Merle, Laugerie, La Madeleine) and in the Pyrenees (Niaux, Le Tucs d'Audubert, Bedeilhac). Spain has some 60 caves in the Cantabria region to the north, among them the cave of Altamira, and 180 caves farther south. Examples from other regions include caves at Addaura, Italy, and Kapova, Russia. Portable art, on the other hand, was abundant in all Europe.

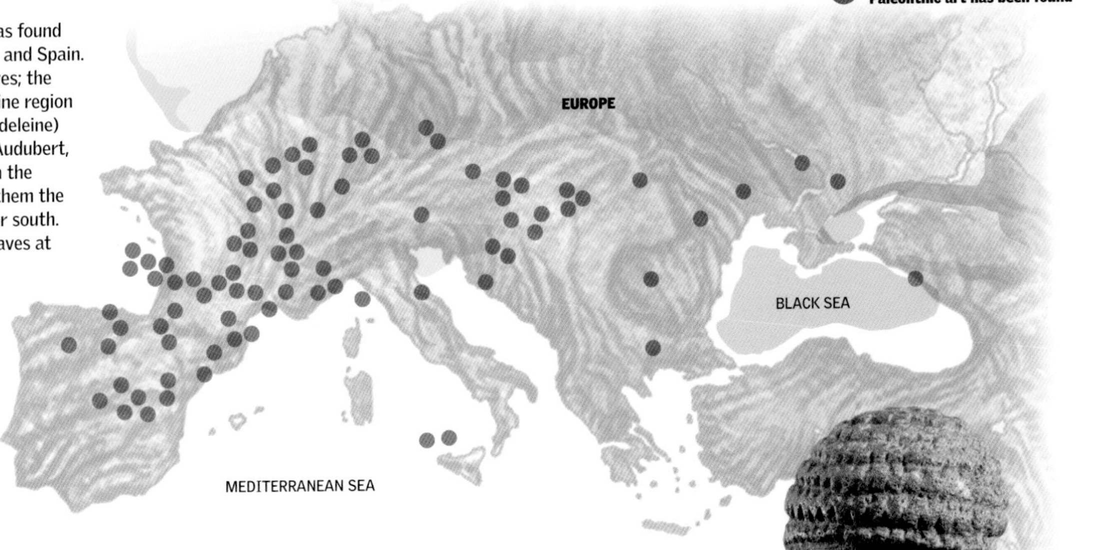


14,000 years old

**THE PAINTINGS OF ALTAMIRA.**

## Builders of Objects

*Homo sapiens sapiens* distinguished itself from its ancestors, who were already making rudimentary tools, through the growing use of such new materials as bone and above all for the specialization of new tools. Mortars, knives, boring tools, and axes had forms and functions continually more sophisticated. There also appeared, in addition to utensils and tools, objects with ornamental and representative functions that attested to humans' increasing capacity for symbolism. These manifestations, through which the art could leave the caves, are known as portable art. It produced objects that were utilitarian, luxurious, or ceremonial, like the Paleolithic "Venus" figurines.

**MICROCEPHALY**
The head is small in relation to the rest of the animal's body.

**OTHER THEMES AND MOTIVES**

**HUNTING SCENES IN THE CAVE OF TASSILI-N-AJJER, ALGERIA**

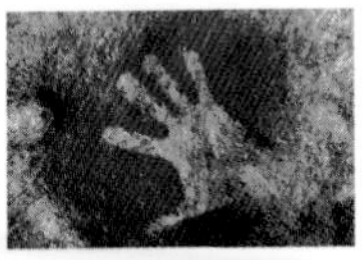

**HANDS IMPRINTED AS A NEGATIVE APPEAR IN MULTIPLE PLACES.**

**SYMBOLISM**
The "Venus of Willendorf" measures 4 inches (11 cm) in height and was found in Austria.

24,000 years old

**IS THIS LITTLE STATUE.**

**PALEOLITHIC TOOLS**

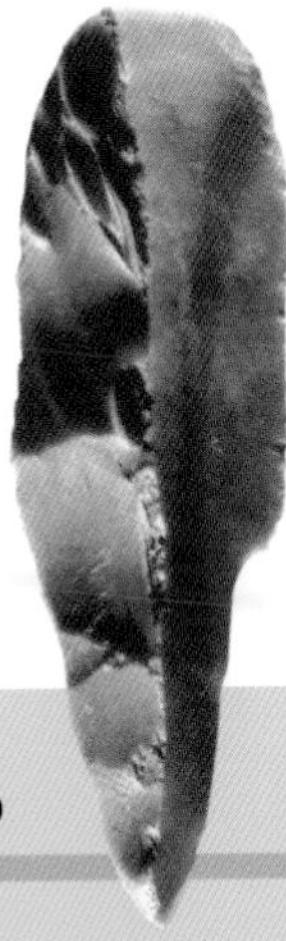

**TWO-SIDED KNIFE**
Its invention presaged the most important cultural revolution of the Upper Paleolithic.

**HARPOON**
This complex instrument of bone dates from some 11,000 years ago (Magdalenian Period, France).

**POLISHED AX**
Found in Wetzlar, Germany, it shows the polishing technique of 20,000 years ago.

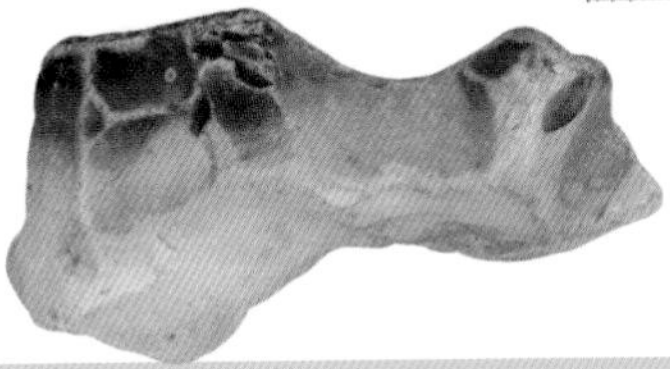

**SOLUTREAN**
**20,000-50,000 YEARS AGO**

Use of oxide to paint, pointed instruments.

**MAGDALENIAN**
**15,000 YEARS AGO**

The greatest flourishing of cave art in southern Europe.

**END OF PALEOLITHIC**
**9,000 BC**

End of the glaciations, with an improvement of the global climate.

# Urban Revolution

Some 10,000 years ago, there was an interglacial period on Earth that caused a gradual increase in temperatures and an overall climatic change that brought a modification to the life of humans. Instead of roaming from place to place to hunt, people began to create societies based on sedentary life, agriculture, and the domestication of animals. Some villages grew so much that they became true cities, such as Çatal Hüyük in southern Turkey. In the ruins of this city, considered one of the milestones of modern archeology, were found a good number of ceramics and statues of the so-called mother goddess—a woman giving birth—that belonged to a fertility ritual. In addition, there are signs that the inhabitants practiced funeral rights and built dolmens for collective graves.

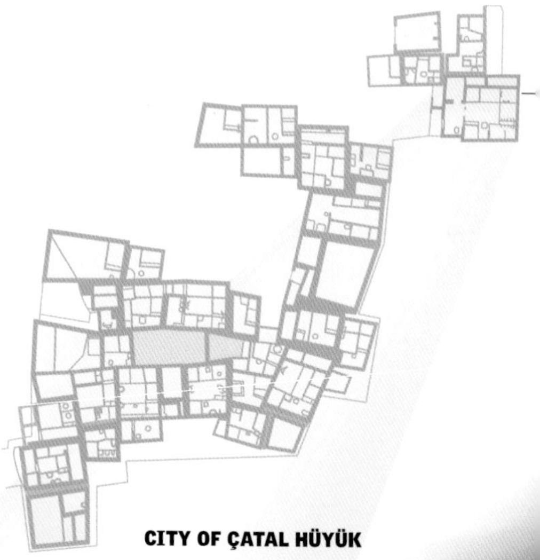

**CITY OF ÇATAL HÜYÜK**

## The Neolithic City of Çatal Hüyük

Çatal Hüyük is located in southern Anatolia (Turkey). Houses were built side by side, sharing a common wall. There were no exterior windows or openings, and the buildings had flat-terraced roofs. People entered through the roof, and there were usually one or two stories. The walls and terraces were made of plaster and then painted red. In some main residences, there were paintings on the walls and roof. The houses were made of mud bricks and had a sanctuary dedicated to the mother goddess. During the excavation, many religious articles were uncovered: the majority were ceramic figures in relief depicting the mother goddess and heads of bulls and leopards.

**270 square feet (25 sq m)**
**WAS THE AVERAGE SIZE OF A HOUSE.**

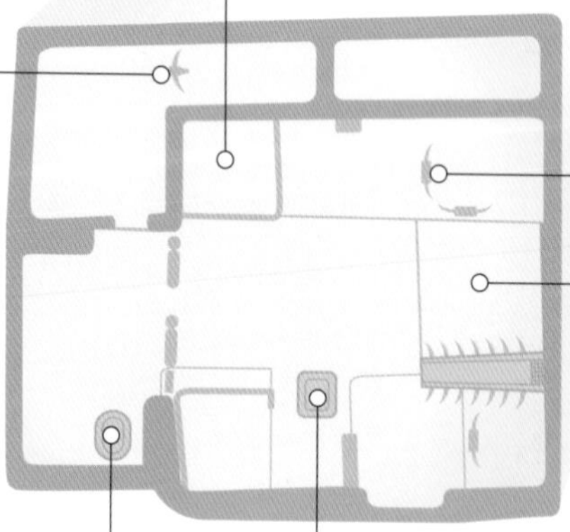


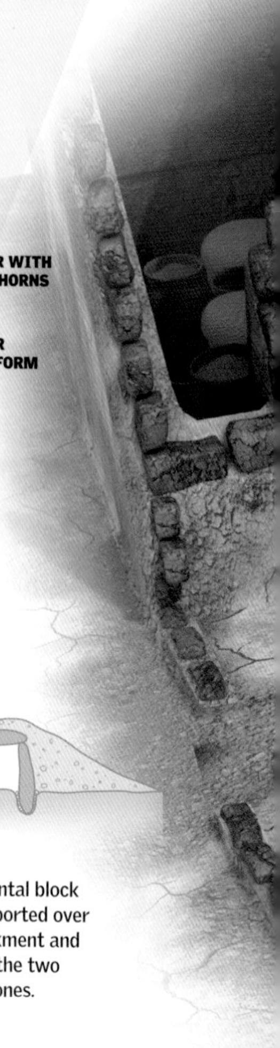

### OTHER TYPES OF CONSTRUCTION

The process of carrying out a megalithic construction began in a quarry, where large blocks of stone were extracted.

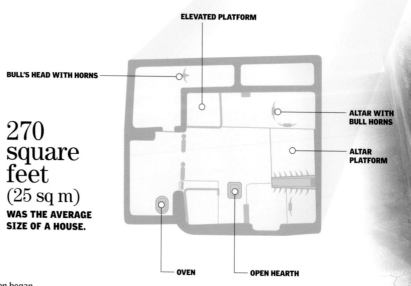

**1 Transport**
The stones were transported on rollers to the place chosen for the erection of the monument.

**2 Erection**
The blocks were dropped into a hole and placed in a vertical position.

**3 Earthworks**
Embankments were made for the construction of a dolmen.

**4 Trilith**
The horizontal block was transported over an embankment and placed on the two upright stones.

| 8000 BC | 7000 BC | 6000 BC |
|---|---|---|
| First indications of agricultural activities. | Expansion of agriculture. Complex funerary rites. | Stable settlements in the Persian Gulf. |

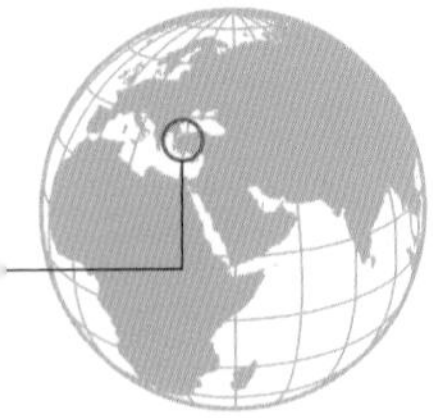

**LOCATION OF ÇATAL HÜYÜK**

| | |
|---|---|
| **Country** | Turkey |
| **Year** | 7000 BC |
| **Type of City** | Farming-livestock |

**CROPS**
In the fields near Çatal Hüyük, the inhabitants grew wheat, sorghum, peas, and lentils. They gathered apples, pistachios, and almonds.

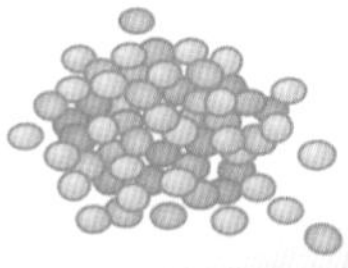

**LENTILS**

**APPLES**

**WHEAT**

# 6000 years BC

**ÇATAL HÜYÜK WAS ONE OF THE FIRST CITIES.**

**CULTS**
There is a direct relationship between the emergence of agriculture and the cult of the feminine because of the importance of fertility. Statuettes of pregnant women were found in homes in shrines decorated with molded bull heads and other figures.

| 3500 BC | AD 320 |
|---|---|
| Invention of writing in Mesopotamia. | First vehicles with wheels in Asia. |

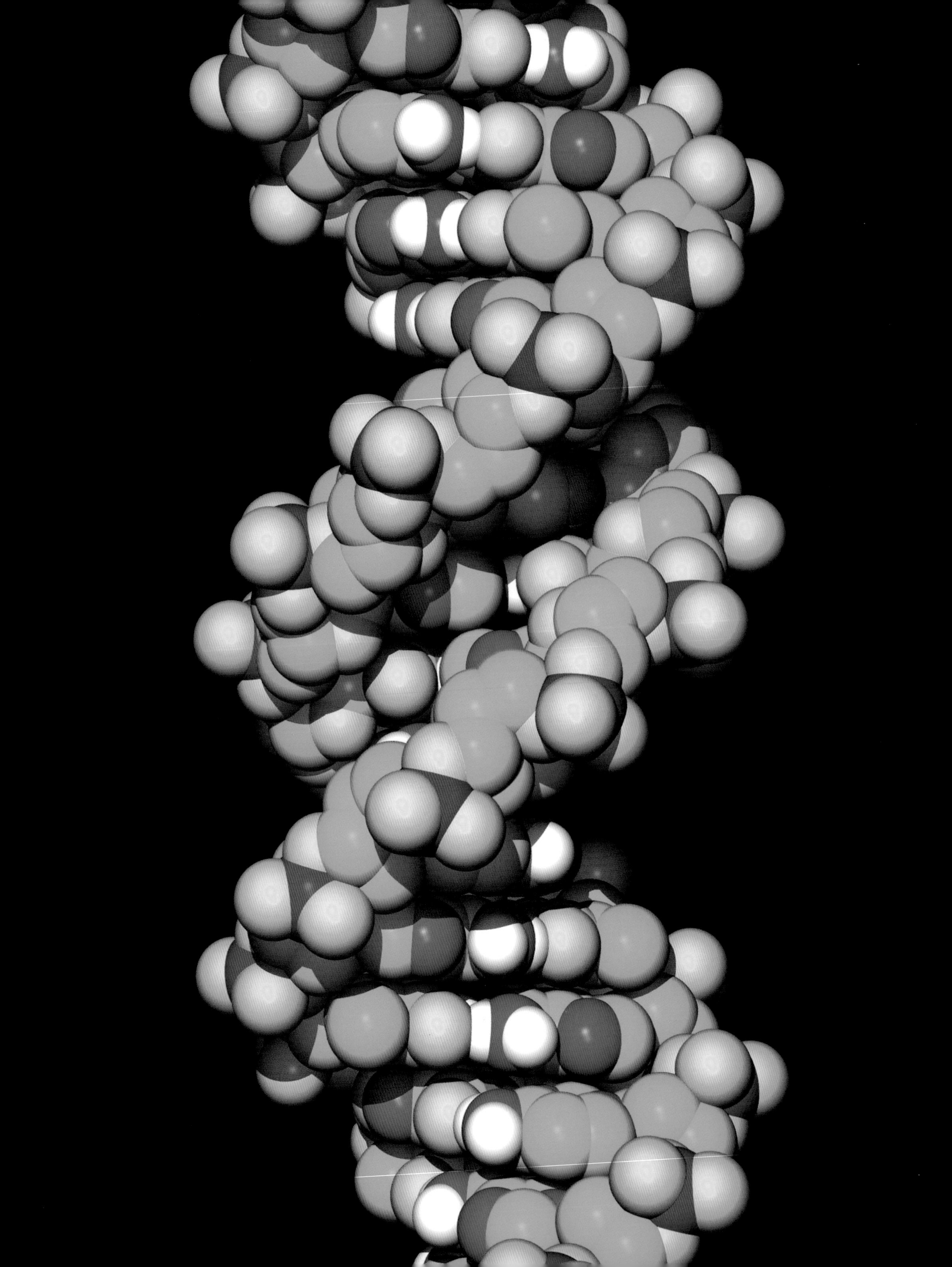

# Mechanisms of Heredity

The cells of the body are constantly dividing to replace damaged cells. Before a cell divides to create new cells, a process known as mitosis, or to form ovules or spermatozoa, a process called meiosis, the DNA included in each cell needs to copy, or replicate, itself. This process is possible because the DNA strands can open and separate. Each of the two strands of the original DNA serves as a model for a new strand. In this chapter, we will also tell you how human beings vary in height, weight, skin color, eyes, and other physical characteristics despite belonging to the same species. The secret is in the genes, and we will show it to you in a simple way.



**DNA** (opposite)
Complex macromolecule that contains a chemical code for all the information necessary for life.

# Chemical Processes

Although it is assumed today that all life-forms are connected to the presence of oxygen, life began on Earth more than three billion years ago in the form of microorganisms. They determined, and still determine today, the biological processes on Earth. Science seeks to explain the origin of life as a series of chemical reactions that occurred by chance over millions of years and that gave rise to the various organisms of today. Another possibility is that life on Earth originated in the form of microbes that reached the Earth from space, lodged, for instance, within a meteorite that fell to the Earth's surface.

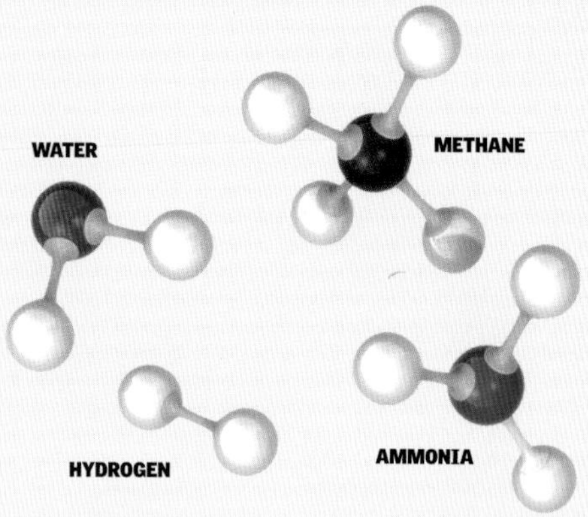

IN THE PROCESS, THE NEW SUBSTANCES COULD HAVE MADE COPIES OF THEMSELVES.

## The First Reaction

Some four billion years ago, the atmosphere contained very little free oxygen and carbon dioxide. However, it was rich in simple chemical substances, such as water, hydrogen, ammonia, and methane. Ultraviolet radiation and discharges of lightning could have unleashed chemical reactions that formed complex organic compounds (carbohydrates, amino acids, nucleotides), creating the building blocks of life. In 1953, Americans Harold Urey and Stanley Miller tested this theory in the laboratory.

## Original Cells

The origin of life on Earth can be inferred from molecular evolution. The first living organisms (prokaryotes) began to develop in groups, giving rise to a process of cooperation called symbiosis. In this way, more complex life-forms called eukaryotes emerged. Eukaryotes have a nucleus that contains genetic information (DNA). In large measure, the development of bacteria was a chemical evolution that resulted in new methods to obtain energy from the Sun and extract oxygen from water (photosynthesis).

### Prokaryotes

These were the first life-forms, with no nucleus or enveloping membranes. These single-celled organisms had their genetic code dispersed between the cell walls. Today two groups of prokaryotes survive: bacteria and archaeobacteria.

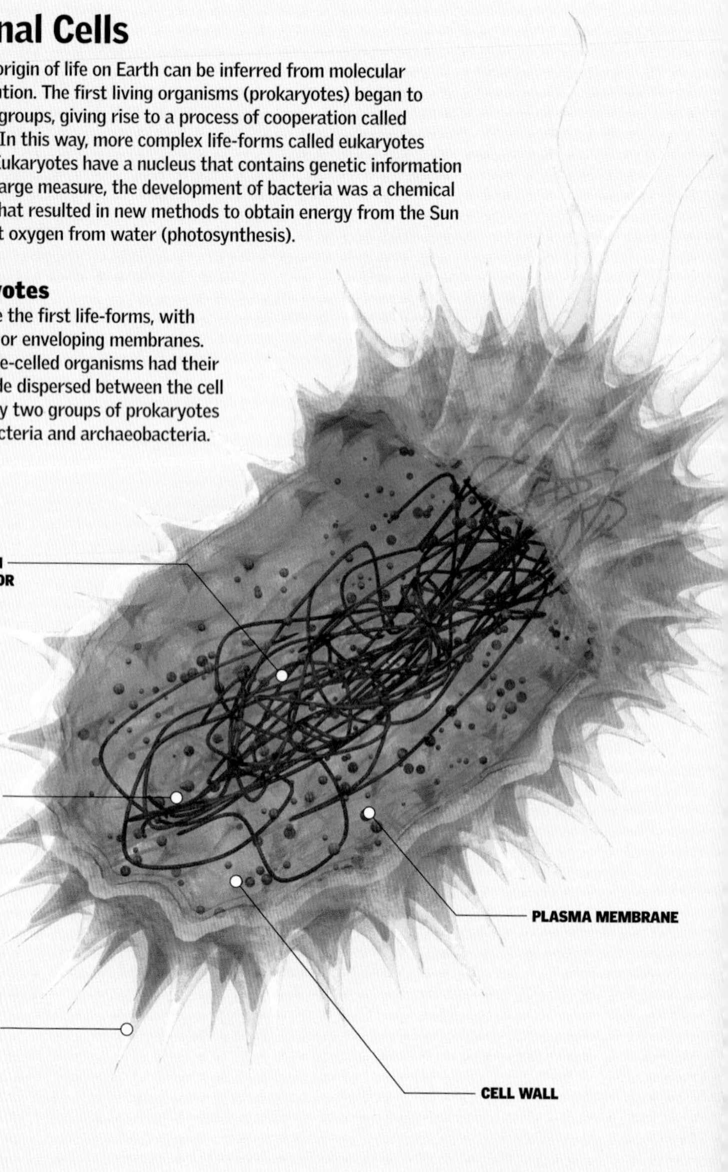


**ARCHEAN**
**4.6 BILLION YEARS AGO**

The Earth's atmosphere sets it aside from the other planets.

**4.2 BILLION YEARS AGO**

Volcanic eruptions and igneous rock dominate the Earth's landscape.

**4 BILLION YEARS AGO**

The Earth's surface cools and accumulates liquid water.

## Eukaryotes

Have a central nucleus that contains nucleic acid (DNA). The content of the nucleus is called nucleoplasm. The substance outside the nucleus is called cytoplasm, and it contains various organelles with different functions. Many are involved in generating energy for the organism's development.

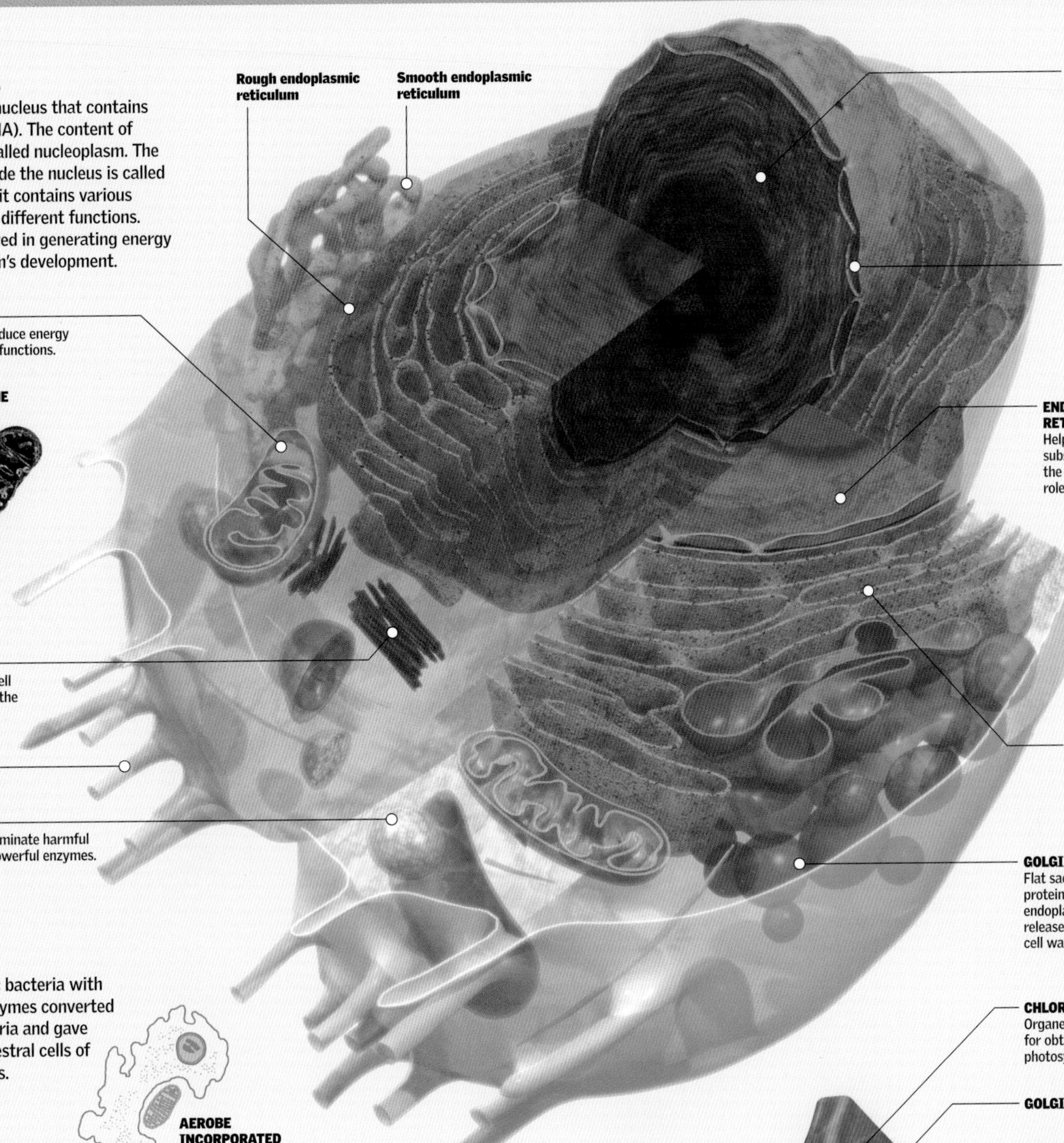

Rough endoplasmic reticulum
Smooth endoplasmic reticulum
NUCLEUS
Contains a large amount of genetic information in strands of DNA that give the cell instructions to grow, function, and reproduce.
NUCLEAR PORES
MITOCHONDRIA
Organelles that produce energy for various cellular functions.
ENDOPLASMIC RETICULUM
Helps transport substances through the cell and plays a role in fat metabolism.
CENTRIOLE
Key structure for cell division, located in the center of the cell.
MICROTUBULES
RIBOSOMES
produce the proteins that make up the cell.
LYSOSOMES
break down and eliminate harmful substances with powerful enzymes.
GOLGI BODIES
Flat sacs that receive proteins from the wrinkled endoplasmic reticulum and release them through the cell wall.

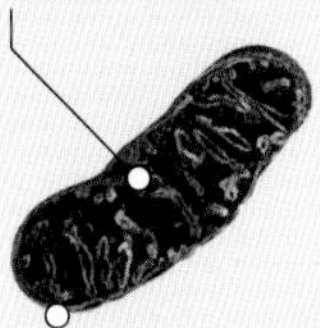

INNER MEMBRANE
OUTER MEMBRANE

**A**

### ANIMALS

Certain aerobic bacteria with respiratory enzymes converted into mitochondria and gave rise to the ancestral cells of modern animals.

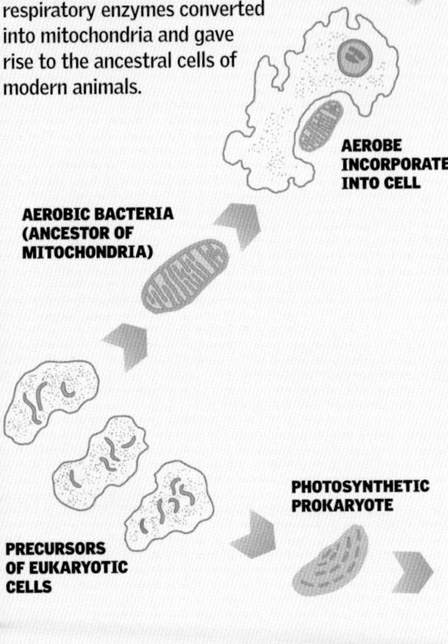


**B**

### PLANTS

Certain photosynthetic bacteria invaded eukaryotic cells and became chloroplasts, originating the ancestral plant cell.

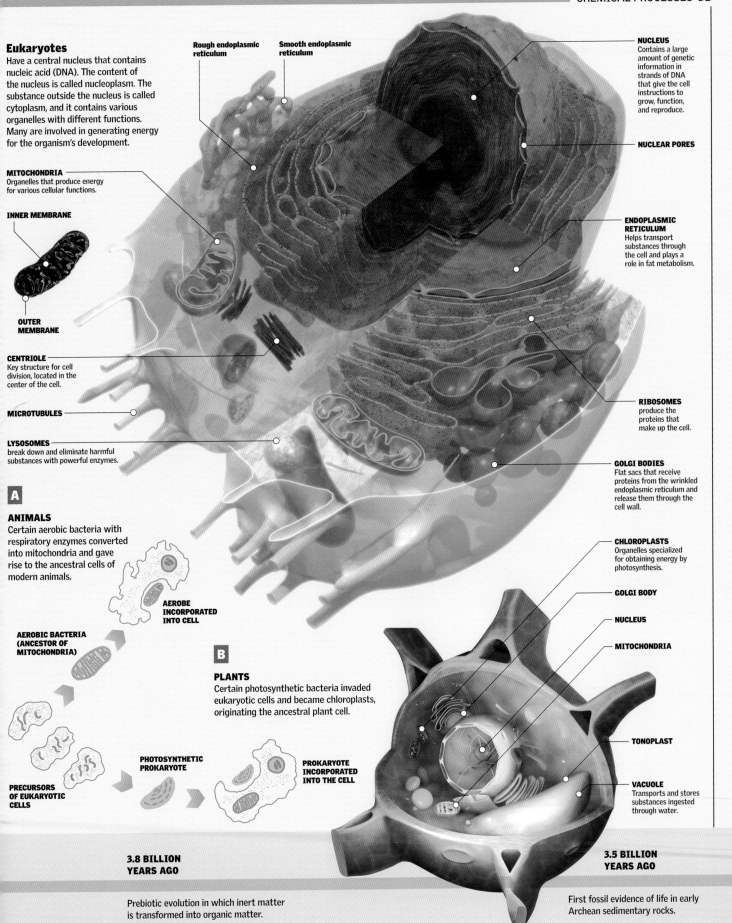

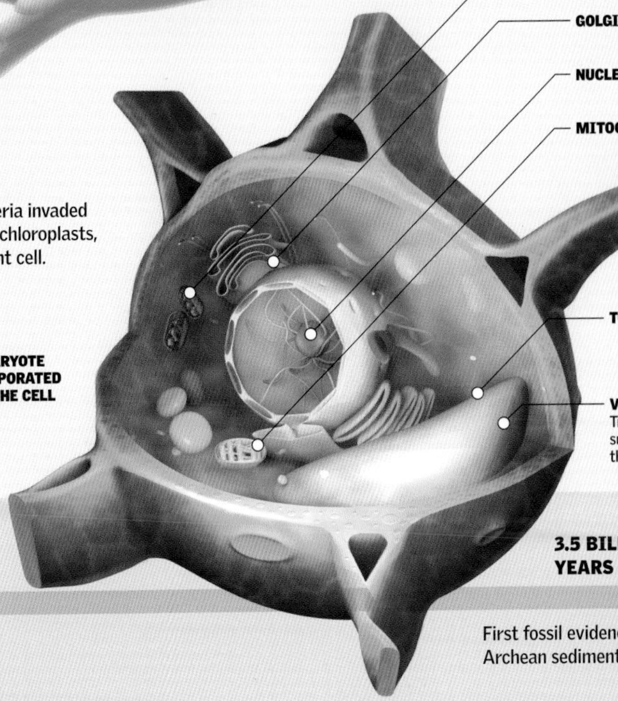


**3.8 BILLION YEARS AGO**

Prebiotic evolution in which inert matter is transformed into organic matter.

**3.5 BILLION YEARS AGO**

First fossil evidence of life in early Archean sedimentary rocks.

# The Tree of Life

Here is a visual representation to explain how all living beings are related. Unlike genealogical trees, in which information supplied by families is used, phylogenetic trees use information from fossils as well as that generated through the structural and molecular studies of organisms. The construction of phylogenetic trees takes into account the theory of evolution, which indicates that organisms are descendants of a common ancestor.

## Eukaryota

This group consists of species that have a true nucleus in their cellular structure. It includes unicellular and multicellular organisms, which are formed by specialized cells that do not survive independently.

## Animals

Multicellular and heterotrophic. Two of their principal characteristics are their mobility and their internal organ systems. Animals reproduce sexually, and their metabolism is aerobic.

**CNIDARIANS**
include species such as the jellyfish and corals.

**BILATERAL**
Symmetrical bilateral organisms.

**MOLLUSKS**
Include the octopus, snails, and oysters.

**VERTEBRATES**
Have a vertebral column, a skull that protects the brain, and a skeleton.

**TETRAPODS**
Animals with four limbs.

**CARTILAGINOUS FISH**
include the rays and sharks.

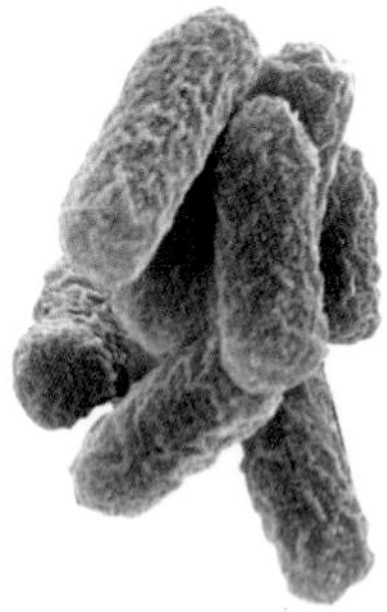

**AMPHIBIANS**
When young they are water dwellers; later they live on land.

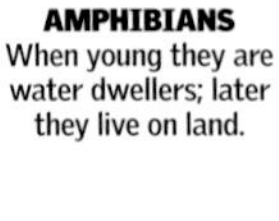

## Archaea

These organisms are unicellular and microscopic. The majority are anaerobic and live in extreme environments. About one half of them give off methane in their metabolic process. There are more than 200 known species.

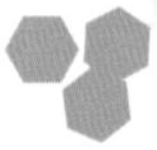

**EURYARCHAEOTA**
*Halobacteria salinarum*

**KORARCHAEOTA**
The most primitive of the archaea.

**CRENARCHAEOTA**
Live in environments with high temperatures.

## Plants

Multicellular autotrophic organisms; they have cells with a nucleus and thick cellular walls that are grouped in specialized tissues. They carry out photosynthesis by means of chloroplasts.

**NOT VASCULAR**
No internal vessel system.

**VASCULAR**
Internal vessel system.

**SEEDLESS**
They are small plants with simple tissues.

**WITH SEED**
Some have exposed seed and some have flower and fruit.

**ANGIOSPERM**
With flower and fruit. More than 250,000 species form this group.

**GYMNOSPERM**
With naked seeds; cycadophytes were examples.

## Relationships

The scientific evidence supports the theory that life on Earth has evolved and that all species share common ancestors. However, there are no conclusive facts about the origin of life. It is known that the first life-forms must have been prokaryotes, or unicellular beings, whose genetic information is found anywhere inside their cell walls. From this point of view, the archaea are prokaryotes, as are bacteria. For this reason, they were once considered to be in the same kingdom of living things, but certain characteristics of genetic transmission places them closer to the eukaryotes.

### Amniotes

The evolution of this feature allowed the tetrapods to conquer land and to adapt to its distinct environments. In amniote species the embryo is protected in a sealed structure called the amniotic egg. Among mammals, only monotremes continue to be oviparous; however, in the placental subclass, to which humans belong, the placenta is a modified egg. Its membranes have transformed, but the embryo is still surrounded by an amnion filled with amniotic fluid.

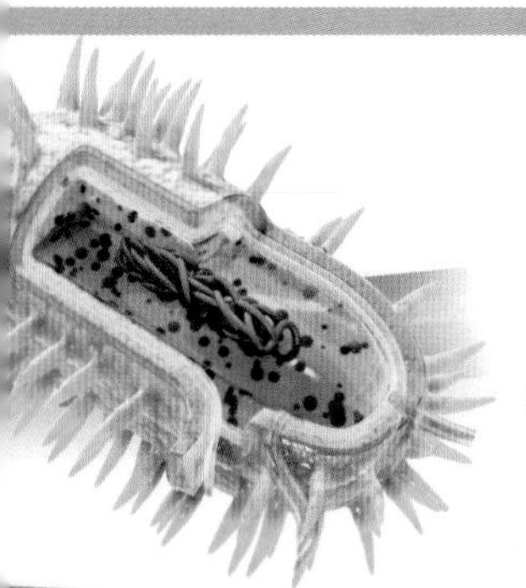

# Bacteria

Unicellular organisms that live on surfaces in colonies. Generally they have one cellular wall composed of peptidoglycans, and many bacteria have cilia. It is believed that they existed as long as three billion years ago.

**COCCALS**
The pneumococcals are an example.

**BACILLUS**
*Escherichia coli* has this form.

**SPIRILLUM**
In the form of a helicoid or spiral.

**VIBRIO**
Found in saltwater.

# Protista

A paraphyletic group, it includes the species that cannot be classified in any other group. There are, therefore, many differences among protista species, such as algae and the amoeba.

**10,000,000**

**SPECIES OF ANIMALS ARE CALCULATED TO INHABIT THE EARTH IN THEIR DISTINCT ENVIRONMENTS.**

# Fungi

Cellular heterotrophic organisms with cell walls thickened with chitin. They carry out digestion externally and secrete enzymes to reabsorb the resulting molecules.

**BASIDIOMYCETES**
Include the typical capped mushrooms.

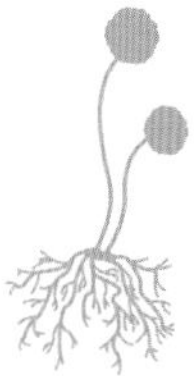

**ZYGOMYCETES**
Reproduce through zygospores.

**ABOUT 5,000**

**SPECIES OF MAMMALS ARE INCLUDED IN THREE GROUPS.**

**ASCOMYCETES**
Most species are grouped here.

**CHYTRIDIOMYCETES**
Can have mobile cells.

**DEUTEROMYCETES**
Asexual reproduction.

**ARTHROPODS**
Have an external skeleton (exoskeleton). Their limbs are jointed appendages.

**INSECTS**
The greatest evolutionary success.

**MYRIAPODS**
Millipedes and centipedes.

**ARACHNIDS**
Spiders, scorpions, and acarids.

**BONY FISH**
Have spines and a jaw.

**CRUSTACEANS**
Crabs and ocean lobsters.

## Cladistics

This classification technique is based on the evolutionary relationship of species coming from similar derived characteristics and supposes a common ancestor for all living species. The results are used to form a diagram in which these characteristics are shown as branching points that have evolved; at the same time, the diagram places the species into clades, or groups. Although the diagram is based on evolution, its expression is in present-day characteristics and the possible order in which they developed. Cladistics is an important analytical system, and it is the basis for present-day biological study. It arises from a complex variety of facts: DNA sequences, morphology, and biochemical knowledge. The cladogram, commonly called the tree of life, was introduced in the 1950s by the German entomologist Willi Hennig.

**AMNIOTES**
Species that are born from an embryo inside an amniotic egg.

**MAMMALS**
The offspring are fed with mother's milk.

**BIRDS AND REPTILES**
Oviparous species. Reptiles are ectothermic (cold-blooded).

**PLACENTAL**
The offspring are born completely developed.

**MARSUPIALS**
The embryo finishes its development outside of the mother.

**MONOTREMES**
The only oviparous mammals. They are the most primitive.

**TURTLES**
The oldest reptiles.

**CROCODILES**
Scaly and with long bodies.

**SNAKES**
Also includes lizards.

## Humans

Humans belong to the class Mammalia and specifically share the subclass of the placentals, or eutherians, which means that the embryo develops completely inside the mother and gets its nutrients from the placenta. After birth, it depends on the mother, who provides the maternal milk in the first phase of development. Humans form part of the order Primates, one of the 29 orders in which mammals are divided. Within this order, characteristics are shared with monkeys and apes. The closest relatives to human beings are the great apes.

# Self-Copying

All living organisms utilize cellular division as a mechanism for reproduction or growth. The cellular cycle has a phase called the S phase in which the duplication of the hereditary material, or DNA, occurs. In this phase, two identical sister chromatids are united into one chromosome. Once this phase of duplication is finalized, the original and the duplicate will form the structures necessary for mitosis and, in addition, give a signal for the whole process of cellular division to start.

## The Cellular Nucleus

The nucleus is the control center of the cell. Generally it is the most noticeable structure of the cell. Within it are found the chromosomes, which are formed by DNA. In human beings, each cellular nucleus is composed of 23 pairs of chromosomes. The nucleus is surrounded by a porous membrane made up of two layers.

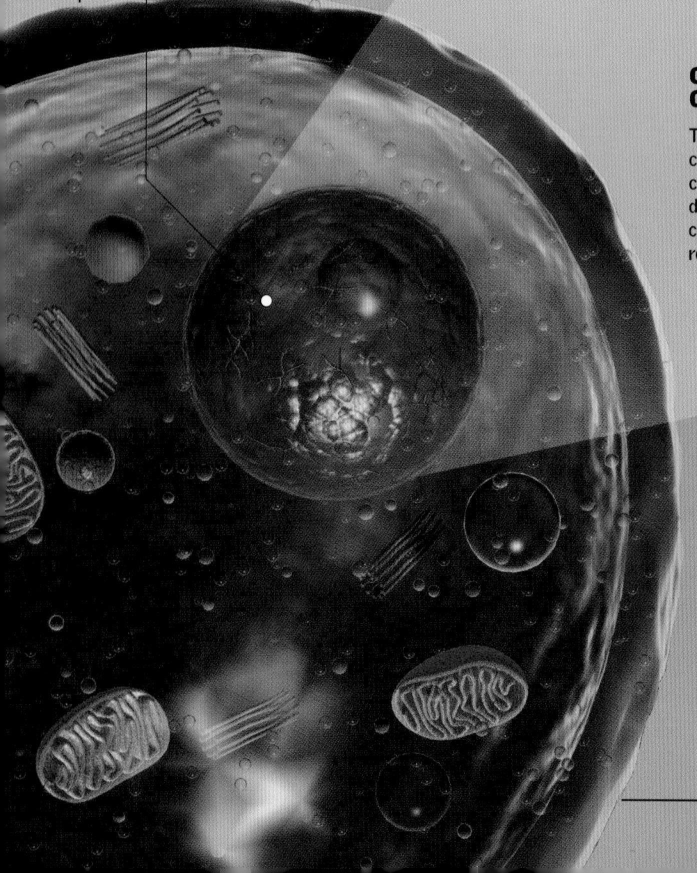

### GROWTH AND CELLULAR DIVISION

The cellular cycle includes cell growth, in which the cell increases in mass and duplicates its organelles, and cell division, in which DNA is replicated and the nuclei divide.

**1**

**PHASE G1**
The cell doubles in size. The number of organelles, enzymes, and other molecules increases.

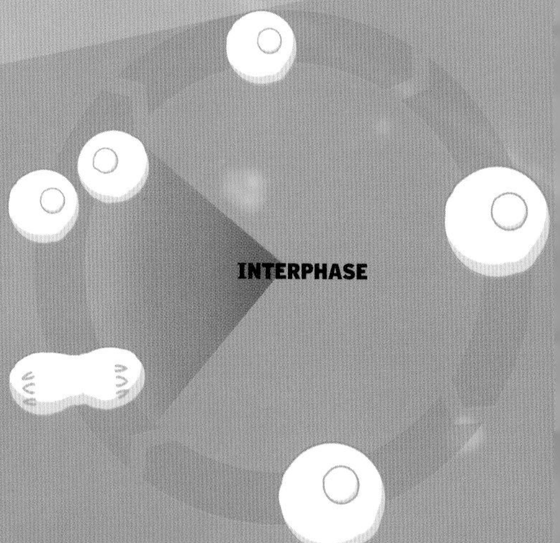


**5**

**CYTOKINESIS**
The cytoplasm of the mother cell divides and gives rise to two daughter cells identical to the mother.

**4**

**MITOSIS**
The two sets of chromosomes are distributed, one set for each nucleus of the two daughter cells.

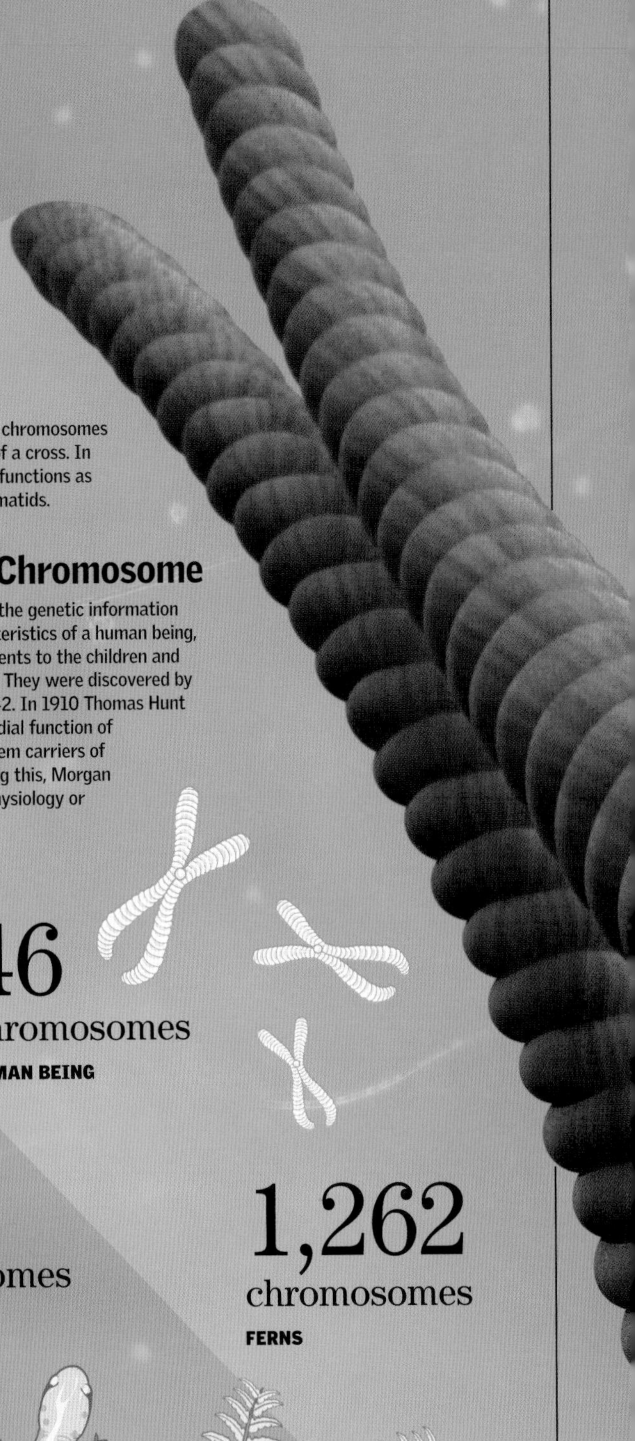

# 6.5 feet
(2 m)

**LENGTH OF DNA IN HUMAN CELL CHROMOSOMES**

### HOW THEY LOOK

Once they have duplicated, the chromosomes form a structure in the shape of a cross. In this structure, the centromere functions as the point of union for the chromatids.

## History of the Chromosome

The chromosomes carry the genetic information that controls the characteristics of a human being, which are passed from the parents to the children and from generation to generation. They were discovered by Karl Wilhelm von Nägeli in 1842. In 1910 Thomas Hunt Morgan discovered the primordial function of the chromosomes: he called them carriers of genes. Thanks to demonstrating this, Morgan received the Nobel Prize for Physiology or Medicine in 1933.

**46** chromosomes

**HUMAN BEING**

### AMOUNT OF CHROMOSOMES ACCORDING TO SPECIES

The number of chromosomes of a species varies independently of its size and complexity. A fern has thousands of chromosomes and a fly only a few pairs.

**24** chromosomes

**SALAMANDER**

**1,262** chromosomes

**FERNS**

**8** chromosomes

**FRUIT FLY**

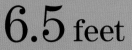

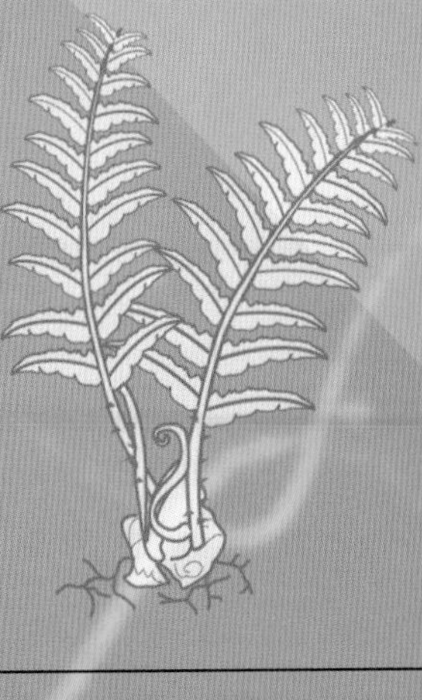

**2**

**S PHASE**
The DNA and associated proteins are copied, resulting in two copies of the genetic information.

**3**

**PHASE G2**
The chromosomes begin to condense. The cell prepares for division.

# The Chromosome

The chromosome is a structural unit that consists of a molecule of DNA associated with proteins. Eukaryote chromosomes condense during mitosis and meiosis and form structures visible through a microscope. They are made of DNA (deoxyribonucleic acid), RNA (ribonucleic acid), and proteins. The majority of the proteins are histones, small, positively charged molecules. Chromosomes carry the genes, the functional structures responsible for the characteristics of each individual.

## Karyotype

The ordering and systematic classification of the chromosomes by pairs, size, and position of the centromere. The chromosomes that are seen in a karyotype are found in the metaphase of mitosis. Each one of them consists of two sister chromatids united by their centromeres.

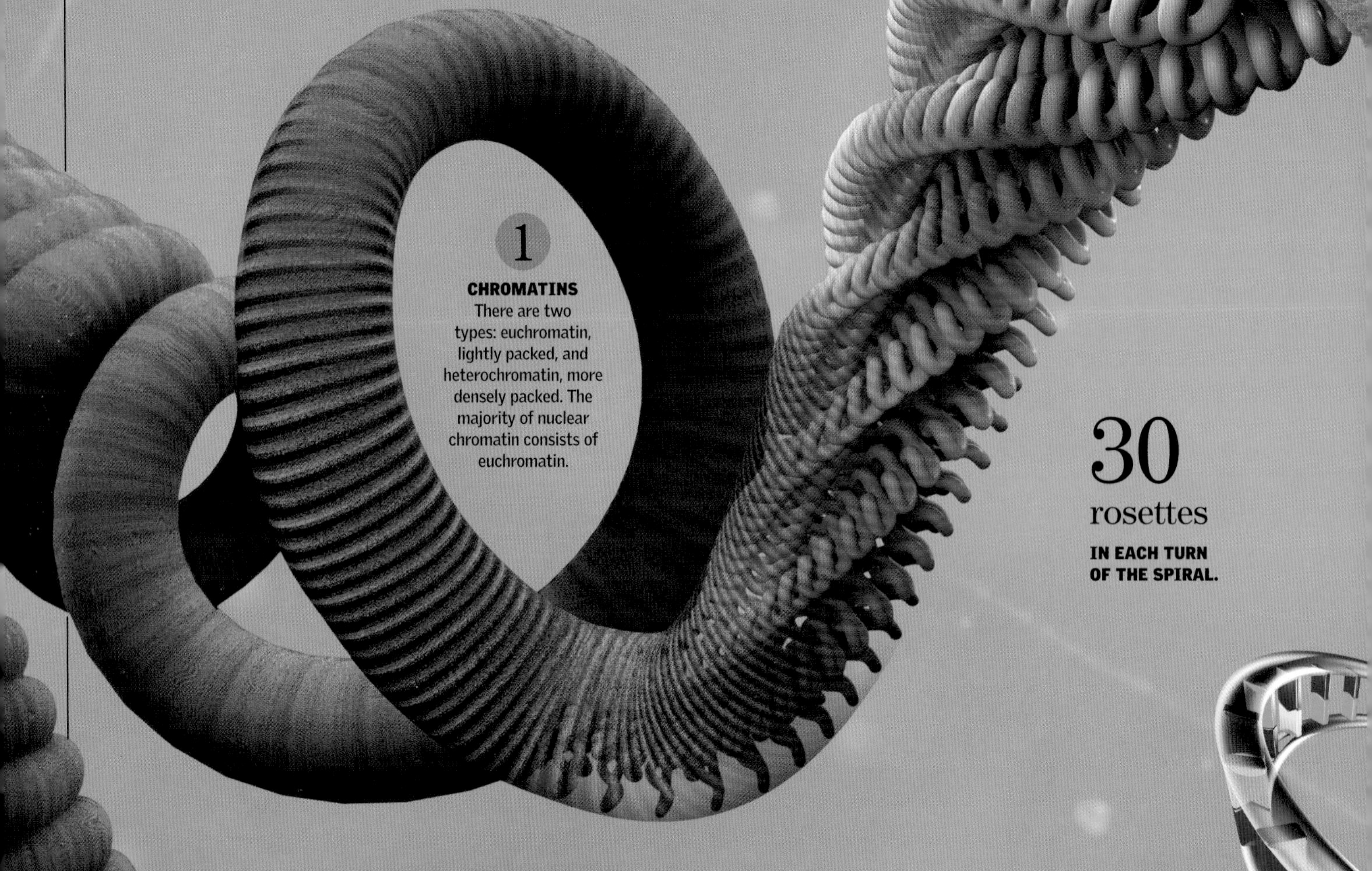

1
CHROMATINS
There are two types: euchromatin, lightly packed, and heterochromatin, more densely packed. The majority of nuclear chromatin consists of euchromatin.
30
rosettes
IN EACH TURN OF THE SPIRAL.

## Carrier of Genes

In the DNA, certain segments of the molecule are called genes. These segments have the genetic information that will determine the characteristics of an individual or will permit the synthesis of a certain protein. The information necessary for generating the entire organism is found in each cell, but only the part of the information necessary for reproducing this specific type of cell is activated. The reading and transmission of the information for use outside the nucleus is performed by messenger RNA.

**PROKARYOTE CELL**
Prokaryote cells do not have a cellular nucleus, so the DNA is found in the cytoplasm. The size of the DNA differs according to species. Prokaryotes are almost all unicellular organisms belonging to the domains of the archaea and bacteria.

## 2

**THE FRAMEWORK**
Each one of the rosettes consists of loops stabilized by the "scaffolding" of other proteins. These loops help to condense the chromatin.

# 6
loops
**IN EACH ROSETTE.**

## 3

**SOLENOID**
A group of six nucleosomes that form each turn inside the loops.

0.0000012 inch
(0.00003 mm)
**DIAMETER OF EACH SOLENOID.**

**SPACER DNA**
The nucleosomes are united by chains of base pairs of DNA 0.0000004 inch (0.00001 mm) long.

# 6
nucleosomes
**IN EACH TURN.**

**PEARL NECKLACE**
If the DNA chain is stretched and observed under a microscope, it resembles beads on a string. Nevertheless, DNA chains are generally found pressed very tightly around the nucleus.

# 60
base pairs
**THE AMOUNT OF DNA BETWEEN NUCLEOSOMES.**

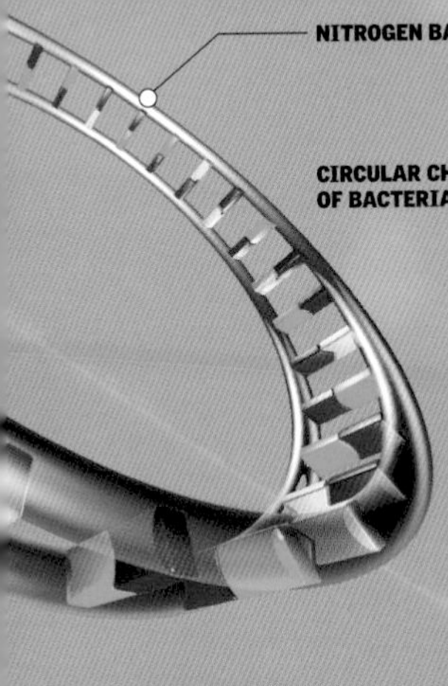


## 4

**NUCLEOSOME**
A group of eight histone molecules with two DNA spirals twisted around them. The "tails" of the histones seem to interact with the molecules that regulate genetic activity.

# The Replication of Life

In deoxyribonucleic acid—DNA—all the genetic information of a complete organism is found. It has complete control of heredity. A DNA molecule consists of two strands of relatively simple compounds called nucleotides. Each nucleotide consists of a phosphate, sugar, and one of four kinds of nitrogenous bases. The nucleotides on each strand are paired in specific combinations and connected to each other by hydrogen bonds. The two strands coil around each other in the form of a spiral, or double helix.

**REPLICATION**

The genetic information is encoded in the sequence of the bases of the DNA nucleotides aligned along the DNA molecule. The specificity of the pairing of these bases is the key to the replication of DNA. There are only two possible combinations—thymine with adenine and guanine with cytosine—to form the complementary links of the strands that make up the DNA chain.

## Complementary

Various specialized proteins called enzymes act as biological catalysts, accelerating the reactions of replication: helicase, which is in charge of opening the double helix of DNA; polymerase, which is in charge of synthesizing the new strands of DNA in one direction; and ligase, which seals and joins the fragments of DNA that were synthesized.

50

nucleotides

**PER SECOND IS THE SPEED OF DNA REPLICATION IN HUMANS.**

## Biological Revolution

Deciphering the molecular structure of DNA was the major triumph of biomolecular studies in biology. Based on work by Rosalind Franklin on the diffraction of X-rays by DNA, James Watson and Francis Crick demonstrated the double-helix composition of DNA in 1953 and for their work won the 1962 Nobel Prize for Physiology or Medicine.

1

**WEAK BRIDGES**
Helicase separates the double helix, thus initiating the replication of both chains. The chains serve as a model to make a new double helix.

2

**FREED ENERGY**
The energy to form new links is obtained from the phosphate groups. The free nitrogenous bases are found in the form of triphosphates. The separation of the phosphates provides the energy to interlace the nucleotides in the new chain that is being built.

3

**NEW CONNECTION**
The new chains of DNA couple in short segments, and the ligase joins them to form the daughter molecules.

4

**PERFECT REPLICATION**
The result is two new molecules, each with one strand from the original DNA and one new complementary strand. This is called semiconservative replication. The genetic information of the new strand is identical to that of the original DNA molecule.

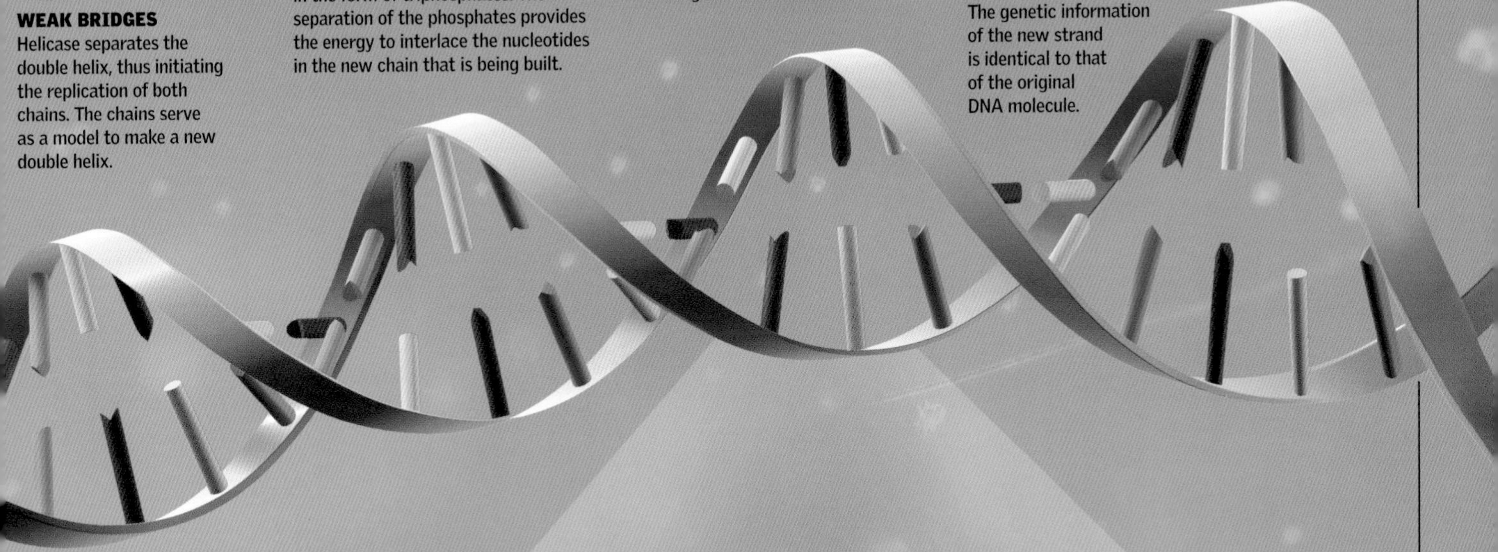

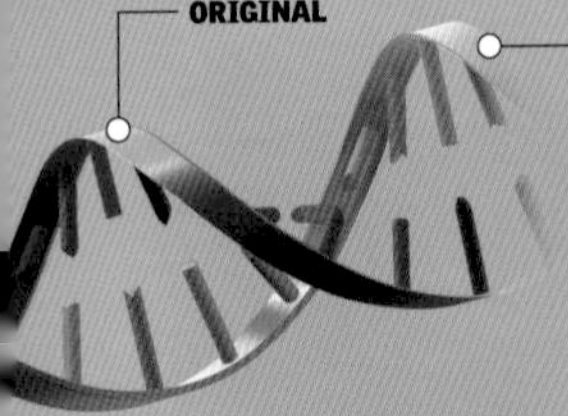


**BASIC MECHANISM**
The new bases join to make a DNA chain that is a daughter of the previous model.

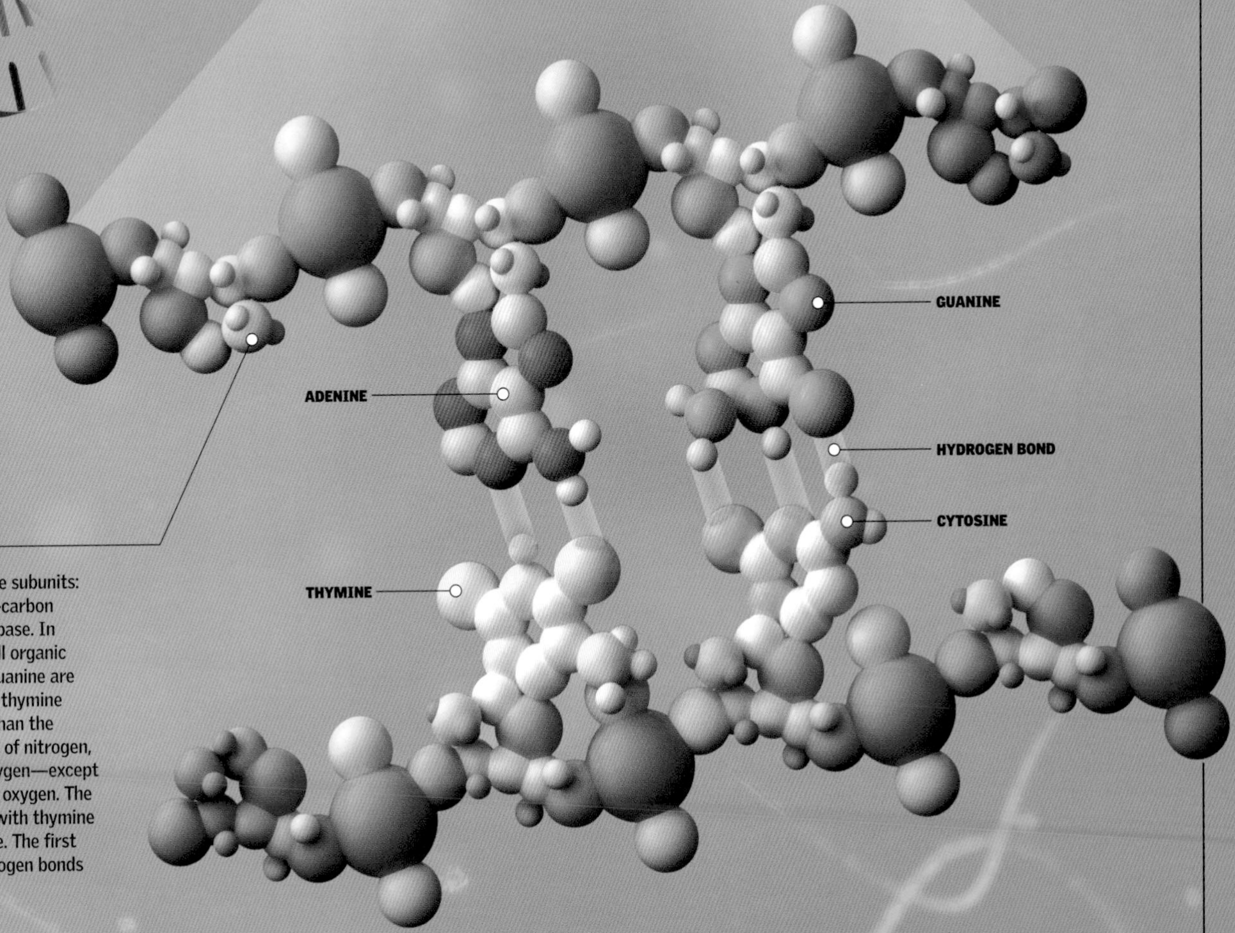


## Nucleotides

The nucleotides have three subunits: a phosphate group, a five-carbon sugar, and a nitrogenous base. In DNA these bases are small organic molecules. Adenine and guanine are purines, and cytosine and thymine are pyrimidines, smaller than the purines. All are composed of nitrogen, hydrogen, carbon, and oxygen—except for adenine, which has no oxygen. The adenine is always paired with thymine and guanine with cytosine. The first pair is joined by two hydrogen bonds and the second by three.

### DNA TRANSCRIPTION

The process of copying one simple chain of DNA is called transcription. For it to happen, the double strands separate through the action of an enzyme, permitting the enzyme RNA polymerase to connect to one of the strands. Then, using the DNA strand as a model, the enzyme begins synthesizing messenger RNA from the free nitrogenous bases that are found inside the nucleus.

1

**SEPARATION OF DNA**
When the DNA is to be transcribed, its double chain separates, leaving a sequence of DNA bases free to be newly matched.

2

**TRANSCRIPTION**
One of the chains, called the transcriptor, is replicated by the addition of free bases in the nucleus through the action of an enzyme called RNA polymerase. The result is a simple chain of mRNA (messenger RNA).

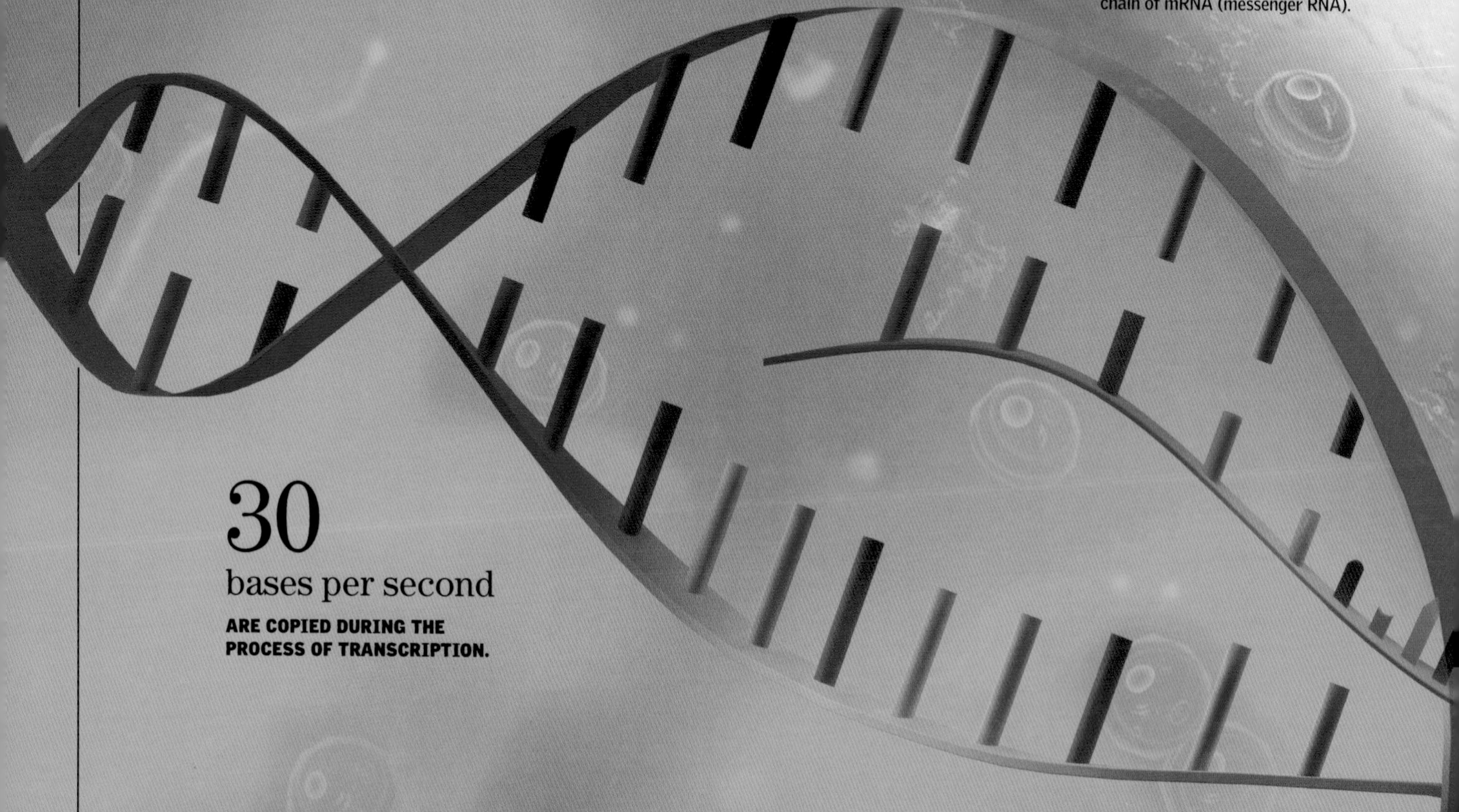

30
bases per second
**ARE COPIED DURING THE PROCESS OF TRANSCRIPTION.**

# Transcription of the Genetic Code

This complex process of translation allows the information stored in nuclear DNA to arrive at the organelles of the cell to conduct the synthesis of polypeptides. RNA (ribonucleic acid) is key to this process. The mRNA (messenger RNA) is in charge of carrying information transcribed from the nucleus as a simple chain of bases to the ribosome. The ribosome, together with transfer RNA (tRNA), translates the mRNA and assembles surrounding amino acids following the genetic instructions.

**COMPRESSION OF RNA**
In the formation of mRNA, useless parts are eliminated to reduce its size.

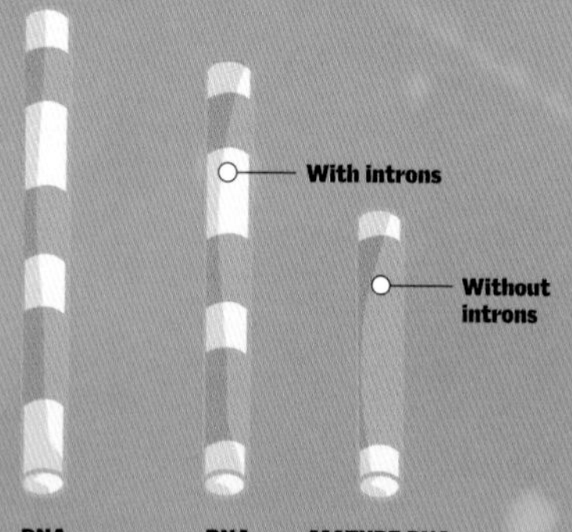

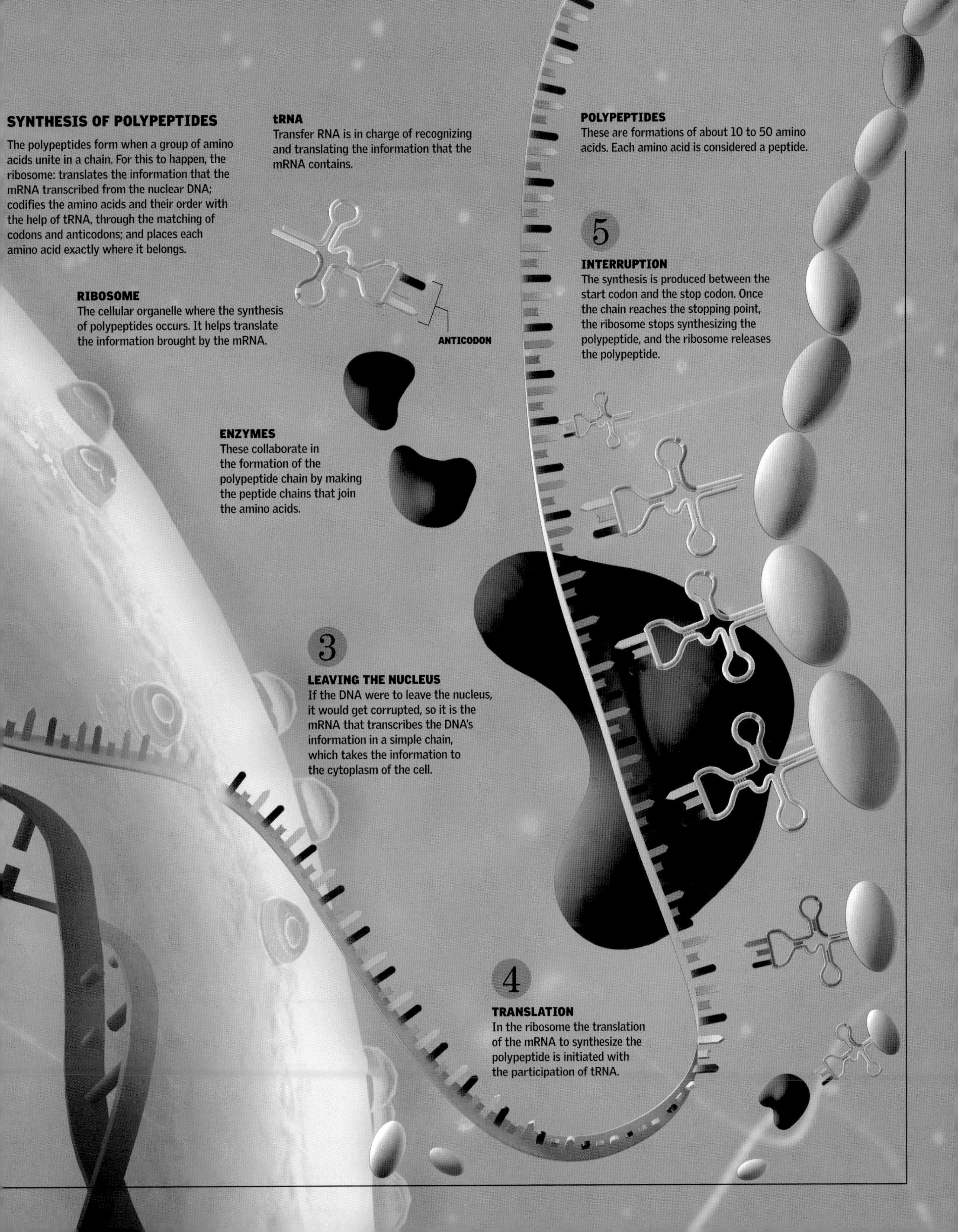


## SYNTHESIS OF POLYPEPTIDES

The polypeptides form when a group of amino acids unite in a chain. For this to happen, the ribosome: translates the information that the mRNA transcribed from the nuclear DNA; codifies the amino acids and their order with the help of tRNA, through the matching of codons and anticodons; and places each amino acid exactly where it belongs.

**tRNA**
Transfer RNA is in charge of recognizing and translating the information that the mRNA contains.

**POLYPEPTIDES**
These are formations of about 10 to 50 amino acids. Each amino acid is considered a peptide.

**RIBOSOME**
The cellular organelle where the synthesis of polypeptides occurs. It helps translate the information brought by the mRNA.

**ENZYMES**
These collaborate in the formation of the polypeptide chain by making the peptide chains that join the amino acids.

### 5 INTERRUPTION

The synthesis is produced between the start codon and the stop codon. Once the chain reaches the stopping point, the ribosome stops synthesizing the polypeptide, and the ribosome releases the polypeptide.

### 3 LEAVING THE NUCLEUS

If the DNA were to leave the nucleus, it would get corrupted, so it is the mRNA that transcribes the DNA's information in a simple chain, which takes the information to the cytoplasm of the cell.

### 4 TRANSLATION

In the ribosome the translation of the mRNA to synthesize the polypeptide is initiated with the participation of tRNA.

# The Path of the Gene

Sexual differences in the heredity of traits constitute a model known as sex-linked inheritance. The father of genetics was Gregor Mendel. He established the principle of independent segregation, which is possible only when the genes are situated on different chromosomes; if the genes are found on the same chromosome, they are linked, tending to be inherited together. Later Thomas Morgan contributed more evidence of sex-linked inheritance. Today many traits are identified in this model, such as hemophilia and color blindness.

## A MEIOSIS I

This first division has four phases, of which prophase 1 is the most characteristic of meiosis, since it encompasses its fundamental processes—pairing and crossing over, which allow the number of chromosomes by the end of this process to be reduced by half.

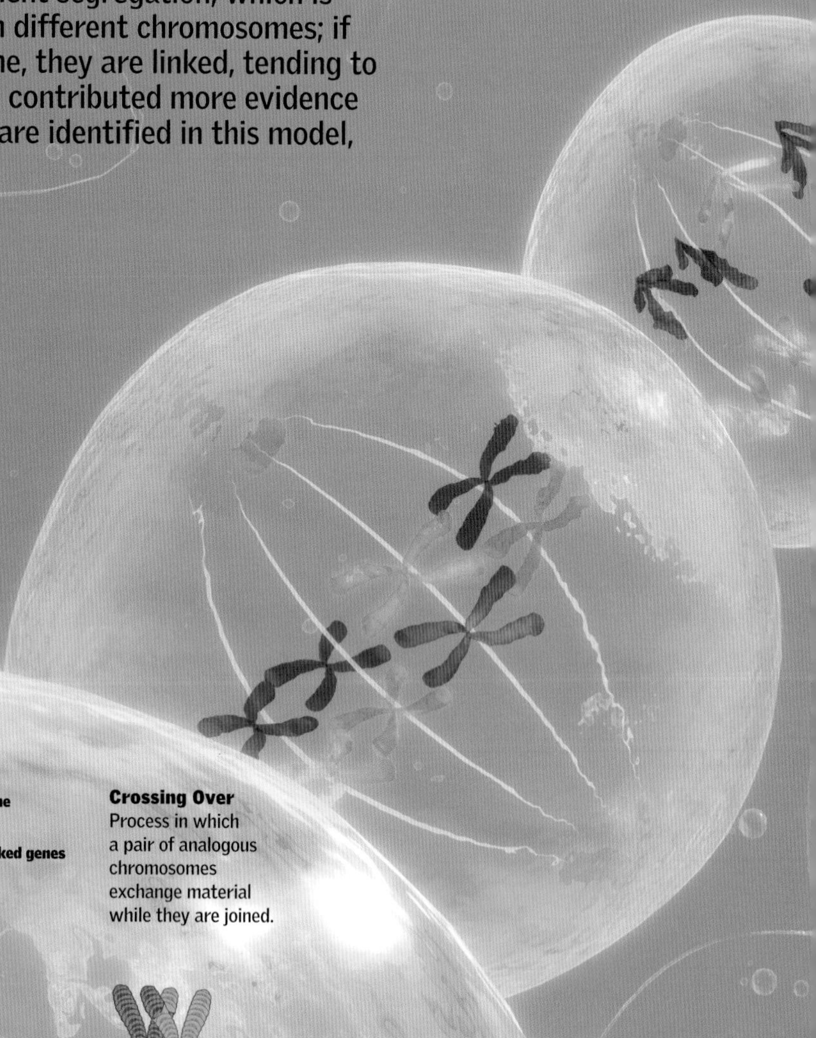

**2**

**METAPHASE I**
The nuclear membrane disappears. The chiasmata, composed of two chromosomes, align, and the centromeres move away.

**3**

**ANAPHASE I**
The chiasmata separate. The chromosomes separate from their homologues to incorporate themselves into the nucleus of the daughter cell.

**1**

**PROPHASE I**
The homologous chromosomes pair up, forming chiasmata, which are unique to meiosis.

CHROMOSOME FROM THE MOTHER

CHROMOSOME FROM THE FATHER

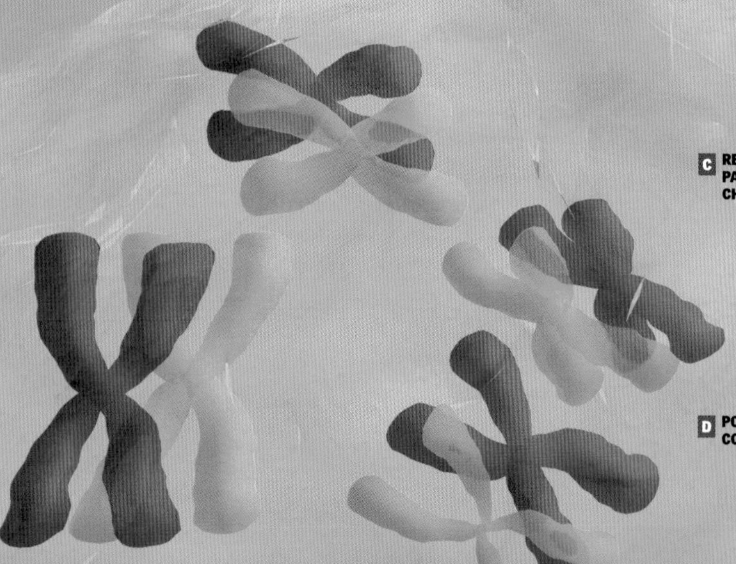

**Linkage**
The genes, arranged in linear form and on the same chromosome, are inherited as isolated units.

**Crossing Over**
Process in which a pair of analogous chromosomes exchange material while they are joined.

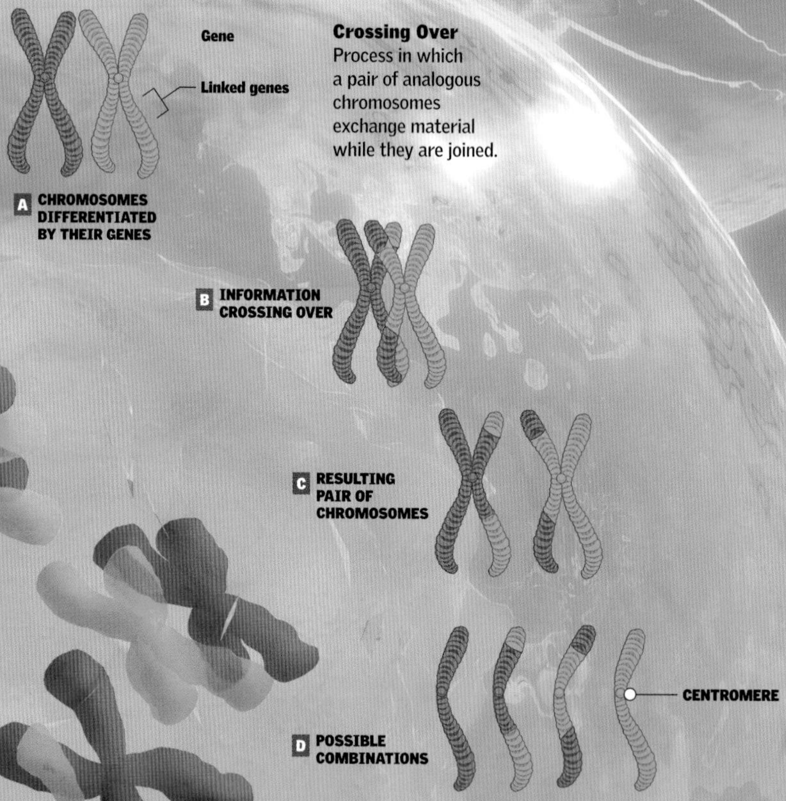

**4**

**TELOPHASE I**
The nuclear membranes reform, and the number of chromosomes enclosed in each has been reduced by half.

**5**

**PROPHASE II**
The division of the new daughter cells begins: the chromatids condense; the nuclear membranes disintegrate; and the spindles form.

## B MEIOSIS II

In the second division, the two chromatids that form each chromosome from meiosis I are separated. As a result of this double division, four daughter cells are produced that contain half the characteristic chromosomal number—i.e., 23 chromosomes each (haploid cells). Each chromosome will be composed of a chromatid.

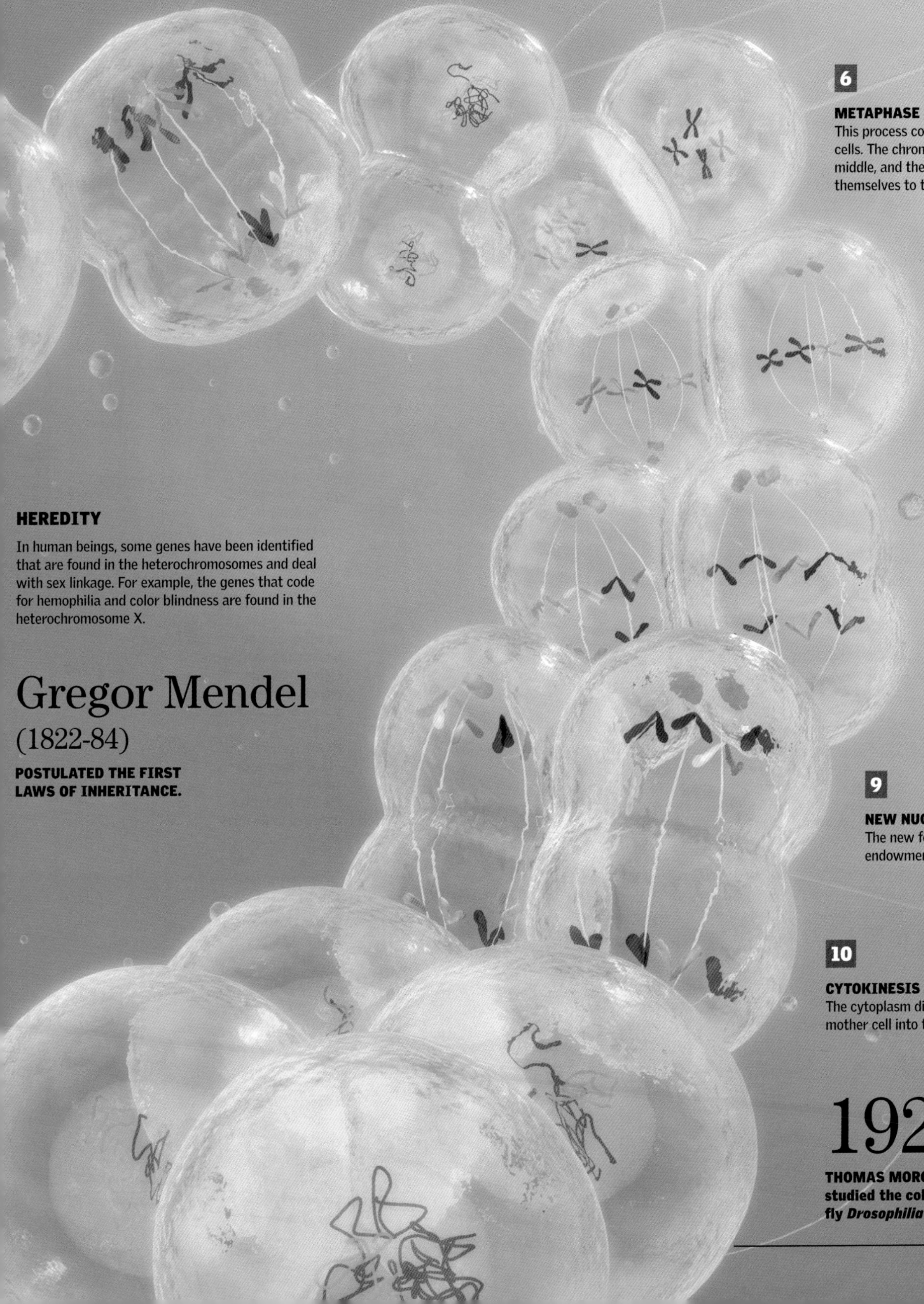

**6**

**METAPHASE II**
This process continues in the daughter cells. The chromosomes align at their middle, and the chromatids affix themselves to the fibers of the spindle.

**7**

**ANAPHASE II**
The centromeres divide again, and the sister chromatids divide, going to opposite poles.

**8**

**NUCLEUS OF TELOPHASE**
The spindle disappears and forms a membrane around each nucleus.

**9**

**NEW NUCLEI**
The new formations have a haploid endowment of chromosomes.

**10**

**CYTOKINESIS**
The cytoplasm divides, separating the mother cell into two daughter cells.

**HEREDITY**

In human beings, some genes have been identified that are found in the heterochromosomes and deal with sex linkage. For example, the genes that code for hemophilia and color blindness are found in the heterochromosome X.

# Gregor Mendel

(1822-84)

**POSTULATED THE FIRST LAWS OF INHERITANCE.**

# 1920

**THOMAS MORGAN studied the color of eyes in the fly *Drosophilia melangaster*.**

# Problems of Heredity

Toward the end of the 19th century, the form in which the physical traits of parents were transmitted to their offspring was uncertain. This uncertainty extended to the breeding of plants and animals, which posed a problem for agriculture and livestock producers. In their fields they sowed plants and raised animals without knowing what the quality of their products would be. The work of Gregor Mendel and his contributions to molecular genetics eventually led to a solution to these problems and to an understanding of how the mechanisms of heredity work.

## The legacy of Mendel

The principles proposed by Mendel are the basis of classical, or Mendelian, genetics, which reached its peak at the beginning of the 20th century. This science studies how the variants, or alleles, for a morphological trait are transmitted from one generation to the next. Later, after confirmation that the components of the nucleus are those in charge of controlling heredity, molecular genetics developed. This science studies heredity on a molecular level and analyzes how the structure of DNA and its functional units, or genes, are responsible for heredity. Molecular genetics links classical genetics and molecular biology. Its use allows us to know the relationship that exists between visible traits and the molecular hereditary information.

**DOMINANT AND RECESSIVE**

The traits of a gene in an individual are expressed according to a pair of variants, or alleles. In general, the dominant alleles are expressed even though there may be another allele for the same gene. A recessive allele is expressed only if it is the only allele present in the pair.

**DOMINANT** With two dominant alleles, the individual is homozygous dominant for this trait.

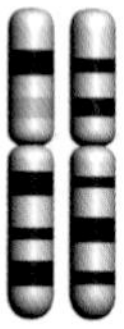

**HETEROZYGOUS** When there is an allele of each type, the individual is heterozygous for this trait.

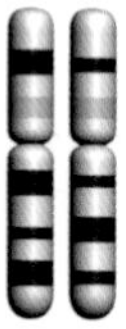

**HOMOZYGOUS** With two recessive alleles, the individual is homozygous recessive for this trait.

**HOMOZYGOUS OR HETEROZYGOUS** Brown color of the eyes is present in individuals with at least one dominant allele.

**HOMOZYGOUS RECESSIVE** Blue color of the eyes is present in individuals with two recessive alleles.

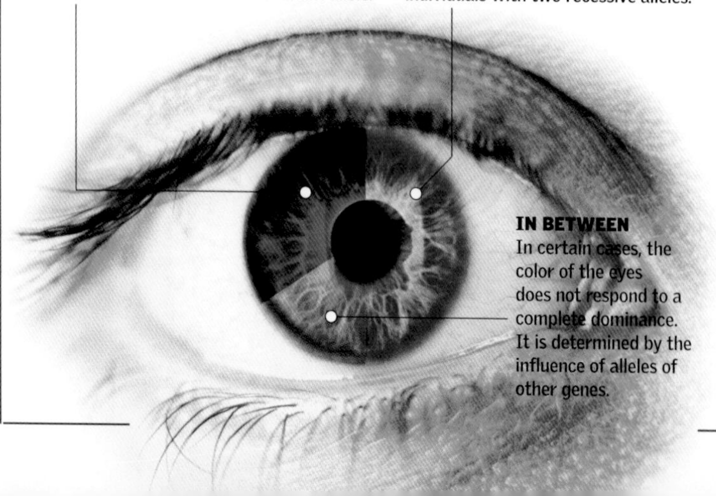

**IN BETWEEN** In certain cases, the color of the eyes does not respond to a complete dominance. It is determined by the influence of alleles of other genes.

**FROM THE GARDEN**

During the 19th century, the gardens of the Abbey of Saint Thomas were the laboratory that Mendel used for his experiments on heredity. During the 20th century, classical genetics and molecular genetics amplified our knowledge about the mechanism of heredity.

**1869**
The Austrian Augustinian monk Gregor Mendel proposes the laws that explain the mechanisms of heredity. His proposal is ignored by scientists.

**1869**
Johann Friedrich Miescher, a Swiss doctor, suggests that deoxyribonucleic acid, or DNA, is responsible for the transmission of hereditary traits.

**1889**
Wilhelm von Waldeyer gives the name "chromosomes" to the structures that form cellular DNA.

**1900**
The German Carl Erich Correns, the Austrian Erich Tschermak, and the Dutchman Hugo de Vries discover, independently, the works of Mendel.

**1926**
T.H. Morgan demonstrates that the genes are found united in different groups of linkages in the chromosomes.

**1953**
James Watson and Francis Crick propose a doublehelix polymer model for the structure of DNA.

**1973**
Investigators produce the first genetically modified bacteria.

**1977**
North American scientists for the first time introduce genetic material from human cells into bacteria.

**1982**
The United States commercializes recombinant insulin produced by means of genetic engineering.

**1990**
An international public consortium initiates the project to decipher the human genome.

**1997**
Dolly the sheep is the first cloned mammal.

**2000**
The Human Genome Project and the company Celera publish the deciphered human genome.

## The man who calculated

Gregor Johann Mendel was born in Heinzendorf, Austria, in 1822 and died in the city of Brünn, Austria-Hungary (now Brno, Czech Republic) in 1884. He was a monk of the Augustinian order who at the University of Vienna pursued, over three years, different studies in mathematics, physics, and natural sciences. This ample academic training and his great intellectual capacity permitted him to develop a series of experiments in which he used pea plants (*Pisum sativum*). He analyzed various traits, among them the appearance of flowers, fruits, stems, and leaves. In his methodology, he included an innovation: he submitted his results to mathematical calculations. His conclusions were key to understanding the mechanism of heredity.

**PEAS** The pea plants of the *Pisum sativum* species were key for the conclusions obtained by Mendel about heredity.

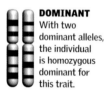

**BOTANY** This display is a botanical teaching tool. An altruistic naturalist, Mendel dedicated himself to conserving in herbariums the specimens of different species of plants.

## Uniformity

Mendel's first law, or principle, about heredity proposes that by crossing two homozygous parents (P), dominant and recessive for the same trait, its descendant, or filial 1 (F1), will be uniform. That is, all those F1 individuals will be identical for the homozygous dominant trait. In this example using the trait seed color, yellow is dominant and green is recessive. Thus, the F1 generation will be yellow.

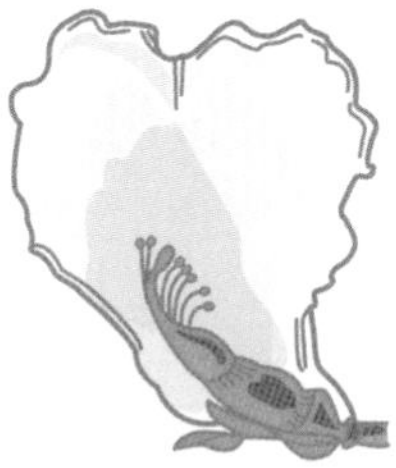

### PURE INDIVIDUALS

**1** Mendel used pure individuals, plants that he knew were homozygous dominant and recessive for a specific trait. For his experiments, Mendel carefully covered or directly cut the stamens of the flowers to prevent them from self-fertilizing.

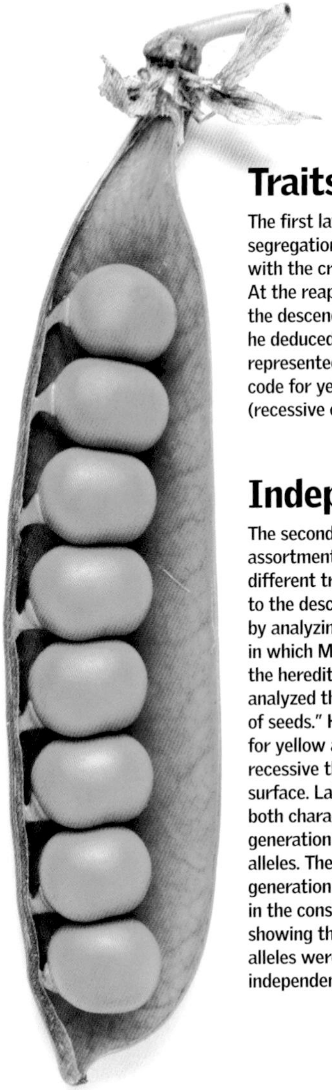

## Traits and Alleles

The first law, known as the law of segregation, comes from the results obtained with the crosses made with F1 individuals. At the reappearance of the color green in the descendants, or filial 2 (F2) generation, he deduced that the trait seed color is represented through variants, or alleles, that code for yellow (dominant color) and green (recessive color).

## Independence

The second law, called the law of independent assortment, proposes that the alleles of different traits are transmitted independently to the descendant. This can be demonstrated by analyzing the results of the experiments in which Mendel examined simultaneously the heredity of two traits. For example, he analyzed the traits "color and surface texture of seeds." He took as dominant alleles those for yellow and a smooth surface and as recessive the alleles for green and a wrinkled surface. Later he crossed pure plants with both characteristics and obtained the F1 generation that exhibited only dominant alleles. The self-fertilization of the F1 generation produced F2 individuals in the constant proportion 9:3:3:1, showing that combinations of alleles were transmitted in an independent manner.

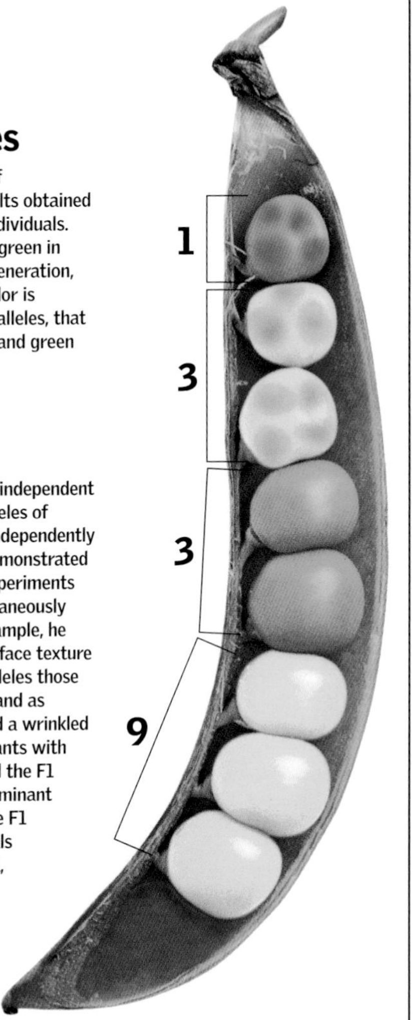


Yellow

CROSSING

Green

$F_1$

OBTAINING THE FIRST FILIAL GENERATION

Yellow

SELF-FERTILIZATION

Yellow

$F_2$

OBTAINING THE SECOND FILIAL GENERATION

### INSEMINATION

**2** Once self-fertilization was impeded, Mendel inseminated the pollen of a homozygous dominant on an ovary of a homozygous recessive and vice versa. In addition to color, he analyzed other traits, such as length of stem, appearance of seeds, and color of flowers.

TALL STEM SHORT STEM

## Yellow: 3
## Green: 1

The cross, or self-fertilization, of individuals of the F1 generation produces F2 individuals with yellow and green seeds in constant 3:1 ratio. In addition, it is deduced that the F1 generation is made up of heterozygous individuals.

**GREEN** The green seeds appear in lower proportion than the yellow.

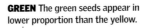

### FRUITFUL WORK

**3** When the plants produced legumes, the seeds exhibited determined colors. Upon carrying out his experiments on hundreds of individuals, he obtained much information. The monk recorded the data in tables and submitted them to probability analysis. In this way Mendel synthesized his results into the conclusions that we know today as the Mendelian laws, or principles, of inheritance.

# An Answer to the Resemblance

The baby has the mother's eyes but the father's hair color; the nose is like the grandfather's, and the mouth is like the grandmother's. These and other possible combinations are caused by genetic inheritance. The genes transmitted by the father combine with genes in the mother's egg, forming a single cell that will turn into a new human being. Through cell division during the baby's growth inside the uterus, the genes will expand, and the dominant ones will impose themselves over the recessive ones. In the case of twins, the physical resemblance results because they share the same genes.

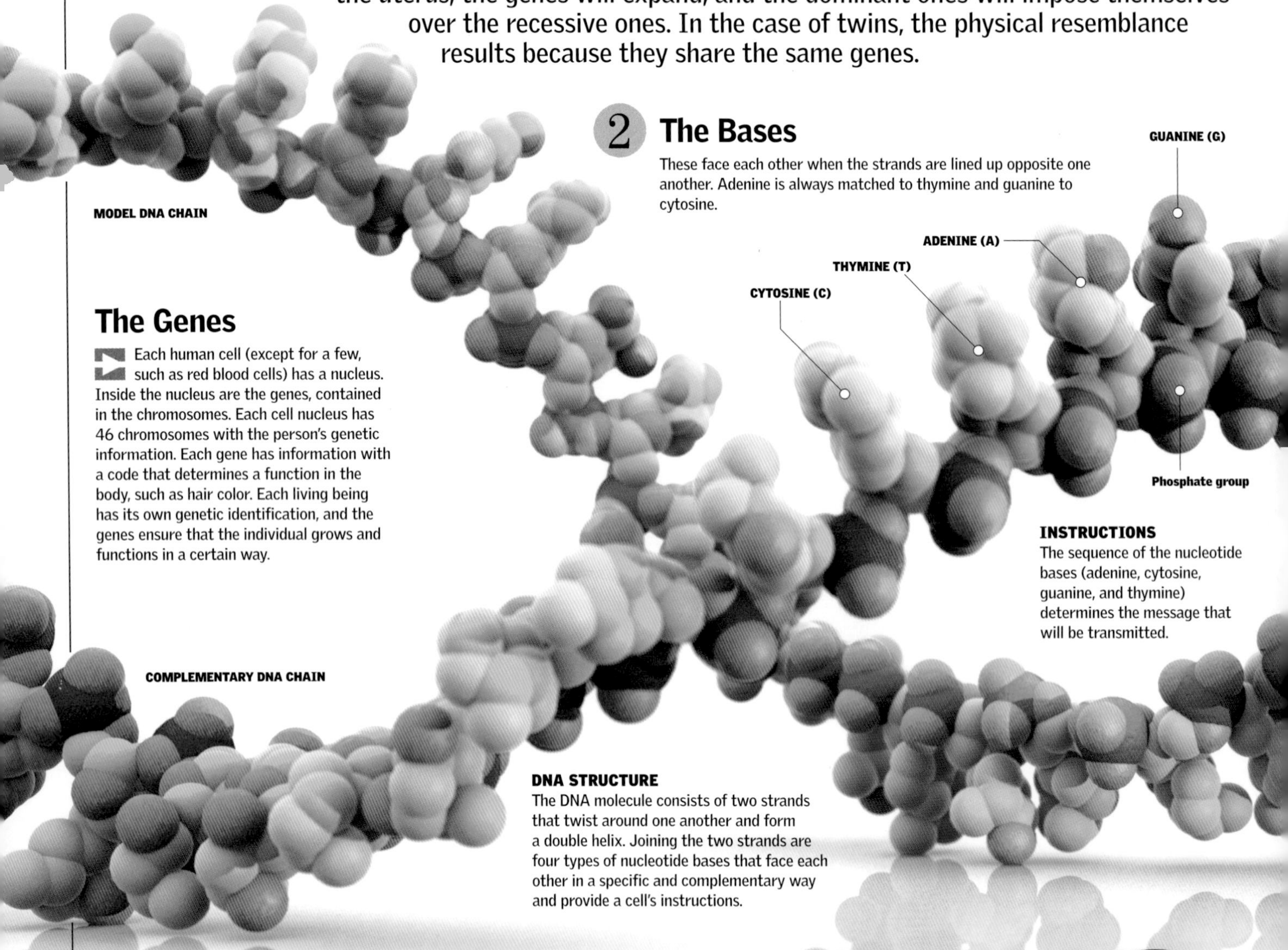


## The Genes

Each human cell (except for a few, such as red blood cells) has a nucleus. Inside the nucleus are the genes, contained in the chromosomes. Each cell nucleus has 46 chromosomes with the person's genetic information. Each gene has information with a code that determines a function in the body, such as hair color. Each living being has its own genetic identification, and the genes ensure that the individual grows and functions in a certain way.

## 2 The Bases

These face each other when the strands are lined up opposite one another. Adenine is always matched to thymine and guanine to cytosine.

**INSTRUCTIONS**
The sequence of the nucleotide bases (adenine, cytosine, guanine, and thymine) determines the message that will be transmitted.

**DNA STRUCTURE**
The DNA molecule consists of two strands that twist around one another and form a double helix. Joining the two strands are four types of nucleotide bases that face each other in a specific and complementary way and provide a cell's instructions.

## 1 DNA Strands

Every strand is made of a sequence of nucleotides. Each nucleotide is composed of a phosphate group, a sugar, and a nitrogenated base.

**25,000**

**GENES ARE CONTAINED IN THE NUCLEUS OF EACH CELL IN THE HUMAN BODY.**

## Identical and Fraternal Twins

It is estimated that one in 70 childbirths produces either identical (monozygotic) or fraternal (dizygotic) twins. Identical twins have the same genes and therefore are alike and of the same sex. They come from one fertilized egg. In some cases, twins share the placenta. Fraternal twins, on the other hand, are the same in age but not in genetic material. They come from two eggs that are released at the same time and are fertilized by different sperm.

## Chromosomes

Like long, thin threads, rolled into an X-shape, these contain DNA. The genetic information is stored inside them. Their characteristic shape helps in the transmission of genes to the next generation. Each cell contains a total of 46 chromosomes arranged in 23 pairs. To form gametes, the cell divides twice, resulting in cells with 23 chromosomes instead of 46. When the sex cells join, the cell they create is a zygote, which has the 46 chromosomes necessary to form a human being.

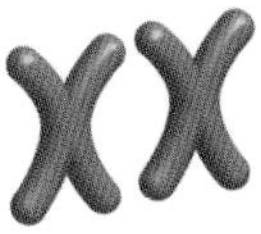

**WOMAN**
The normal karyotype of women is 46XX.

**MAN**
The normal karyotype of men is 46XY.

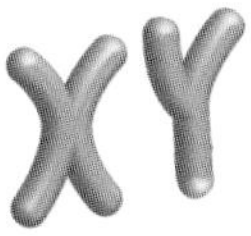

### 4 The Chromosomes

The zygote has a cell with 46 chromosomes. As the zygote grows inside the mother's uterus, the genes go about building the baby's organs. They will determine the gender as well as the structure of the body.

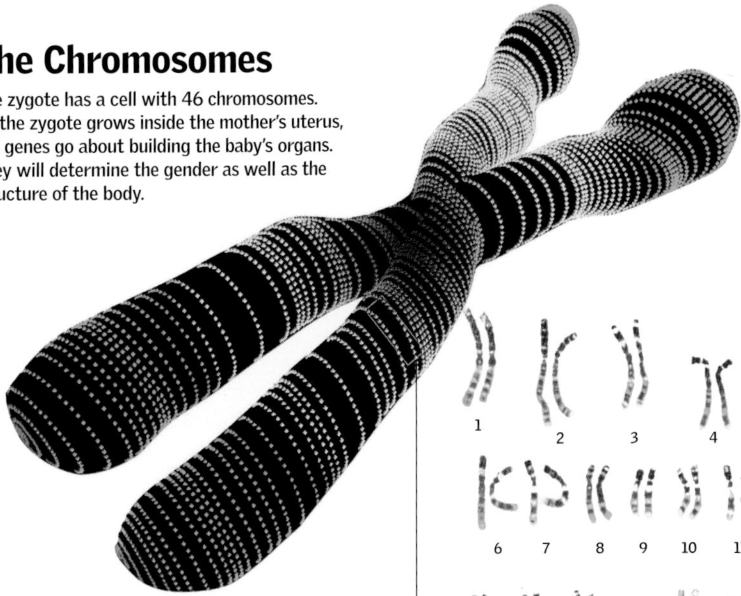

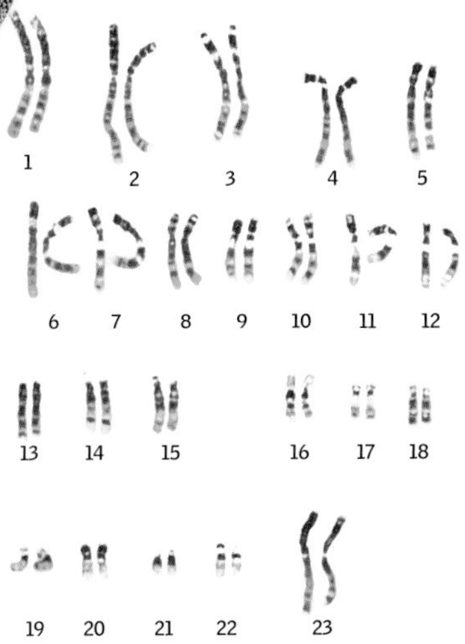


**23 PAIRS OF CHROMOSOMES**
These are classified according to their size. The largest pair is called chromosome 1, the next one chromosome 2, and so on until the last one, which is either XX or XY. In this way, the genes in each chromosome can be located and studied.

### 3 Double helix

The most common structure of DNA, a double helix, is formed from the union of two chains.

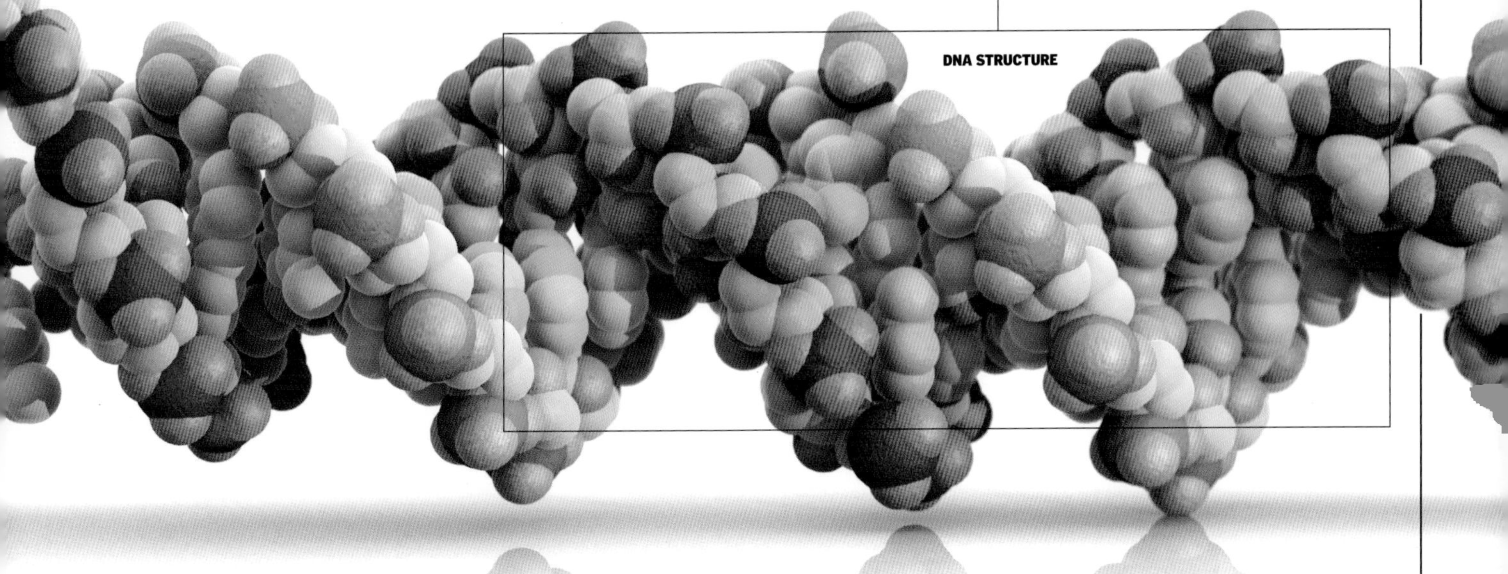


## Resemblances

If one observes different vertebrate embryos, the similarities between them are notable. These resemblances reveal that they are all descended from a common ancestor. The development of the body parts is marked by very similar genes. Morphologically all embryos possess a segmented tail, a heart with two cavities, and branchial (gill) clefts. The greatest difference appears in fish, which retain the branchial clefts. In other groups (amphibians, birds, mammals), one of the clefts transforms into the ear canal and the other into the eustachian tube. In spite of the changes in outer appearance, the observable patterns of internal organization tend to be preserved.

**GENETIC DEVELOPMENT**

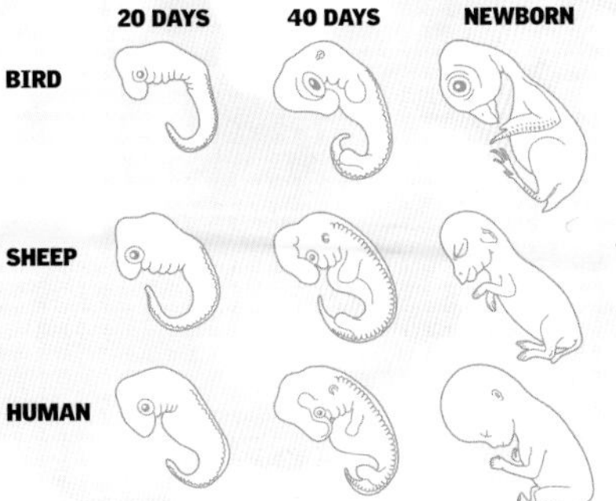


## Made-to-Order Babies

Genetics is also used to find out which genes a baby will have. If the mother and the father have a defective gene, they could opt for preimplantation genetic diagnosis to make sure that the baby will be born healthy. This controversial method can determine if the embryo will be a boy or a girl, and it also prevents hereditary health risks. In preimplantation, the mother takes a drug to produce eggs, which are then fertilized with a sperm from the father. Later a DNA test is done on the embryos' cells, and then two or three healthy embryos are selected and implanted in the mother's uterus.

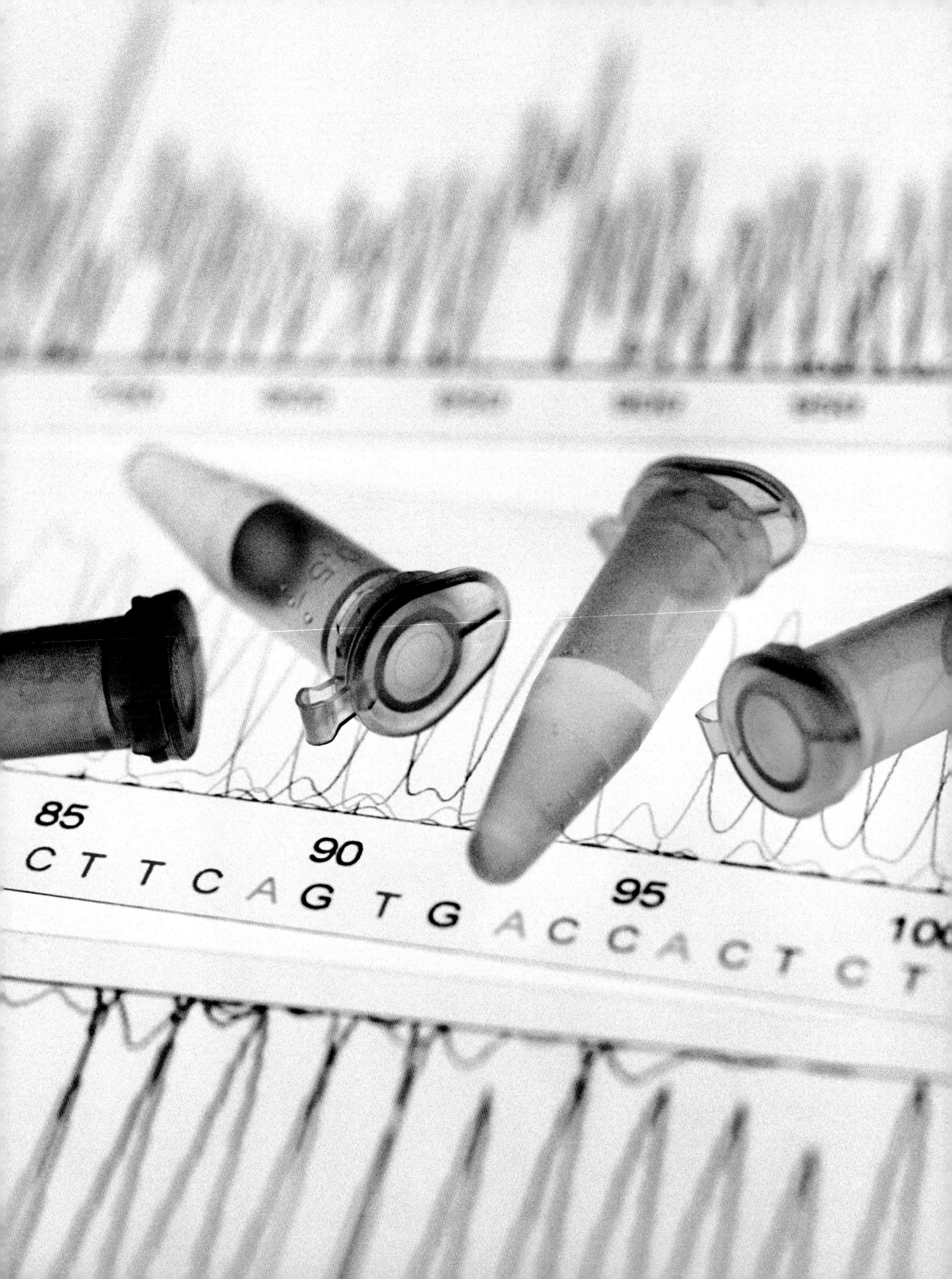
85
90
C T T C A G T G
95
A C C A C T C T

# The Age of Genetics

DNA analysis has become a common practice in diagnosing and predicting genetically inherited diseases. It is also highly useful in forensic procedures. The DNA sequence, like fingerprints, is unique to each individual. In these pages you will learn about achievements in the field of genetically modified foods and animals, the latest advances in genetic medicine, and future applications of stem cells. According to specialists, these cells could be used to regenerate damaged tissues or organs. Another technique that will surely provide a definitive cure for serious diseases will involve exchanging defective genes for healthy ones.



**DNA ANALYSIS** (opposite)
Genetic identification is a nearly infallible proof of identity used in cases of disappearance, rape, murder, and paternity suits.

# Genome in Sight!

HUMAN
**30,000**
**genes**

**O**ne of the most far-reaching and extraordinary scientific achievements is the deciphering of the human genome. This is the complete set of hereditary information contained in the DNA of human chromosomes. In less than 20 years, with a combination of original genetic techniques and the power of computers, scientists glimpsed the location of all the genes, including those that determine eye color, hair type, blood type, and even a person's sex.

## Genetic Dictionary

The 46 human chromosomes, together with mitochondrial DNA, contain all of a human being's genetic information. Knowing the location and function of each gene or group of genes is useful for several reasons. It enables us to know if an illness stems from a defect in a gene or group of genes and even to correct the illness through gene therapy. We can also better understand any potential interaction among genes that are near each other in a chromosome and the effects of that interaction. Studying the human genome can even reveal the origin of our species among the primates.

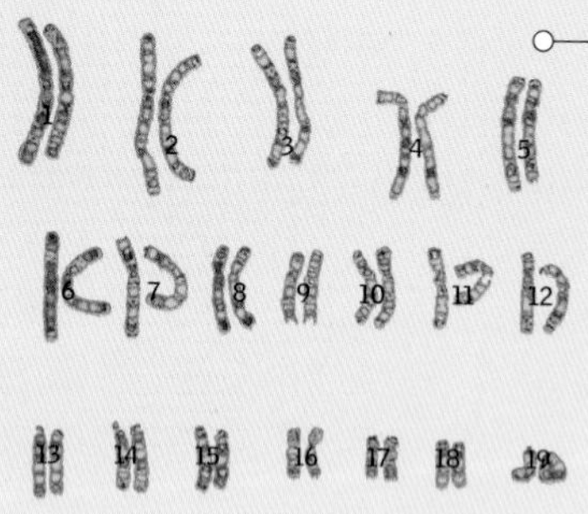


**AUTOSOMES** are the 22 pairs of human chromosomes, excluding the sex chromosomes.

**SEX CHROMOSOMES**

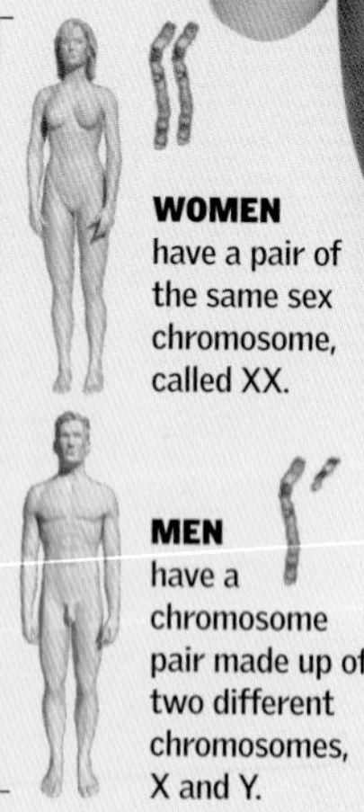

**WOMEN** have a pair of the same sex chromosome, called XX.

**MEN** have a chromosome pair made up of two different chromosomes, X and Y.

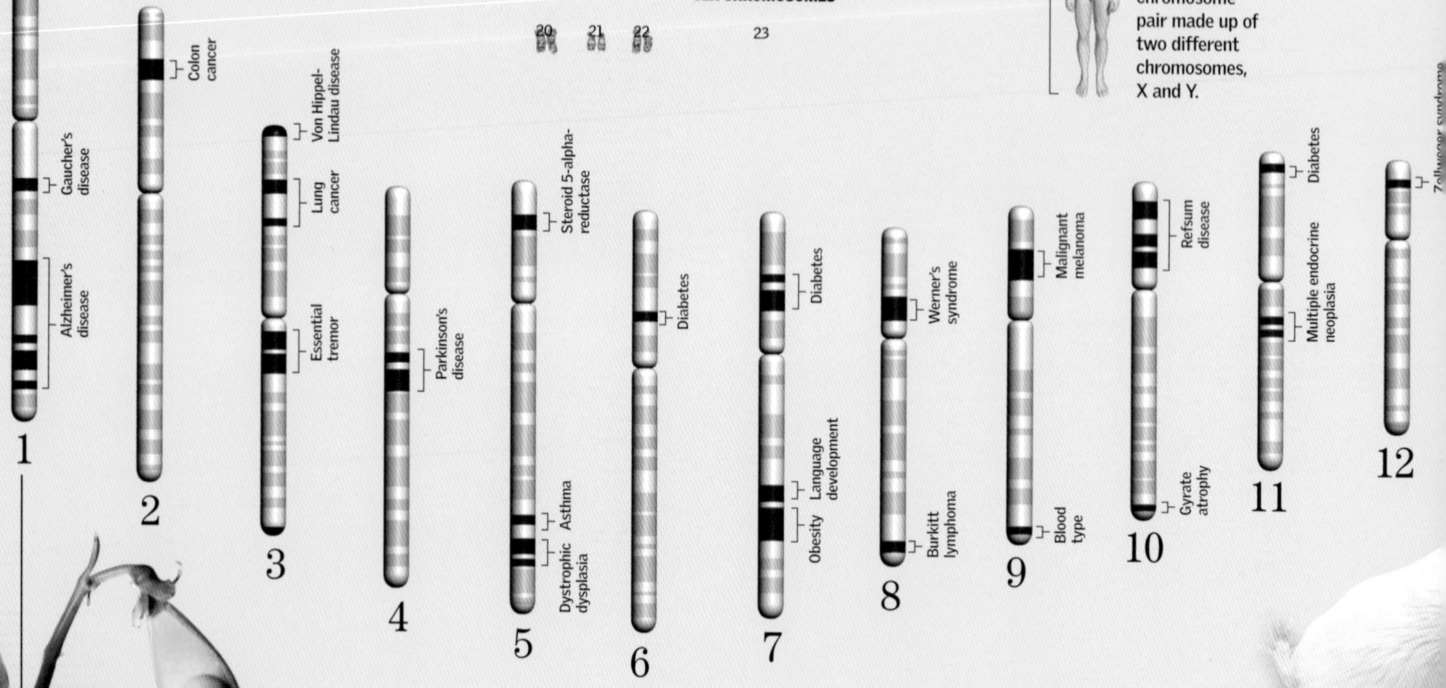

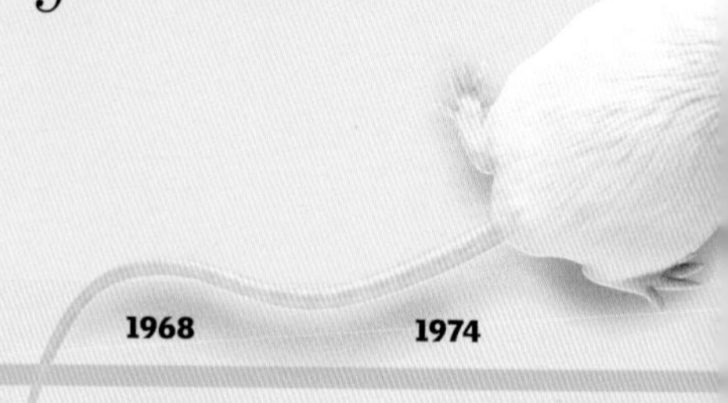

**1900**
**Gregor Mendel** is rediscovered by Tschermak, De Vries, and Correns.

**1911**
*Drosophila melanogaster*, **the fruit fly**, is the subject of experiments by T.H. Morgan based on chromosomal theory.

**1953**
James Watson and Francis Crick propose a structural model of DNA.

**1955**
Discovery that the human species has **46 chromosomes**.

**1968**
The first description of a **restriction enzyme**.

**1974**
John Gurdon first used somatic nuclei to create **clones** of an amphibian larva.

**PLANT**
25,000 genes

**EARTHWORM**
19,000 genes

**FLY**
13,000 genes

## Chromosome

This contains tightly coiled and folded DNA. It consists of sister chromatids that contain the same genes.

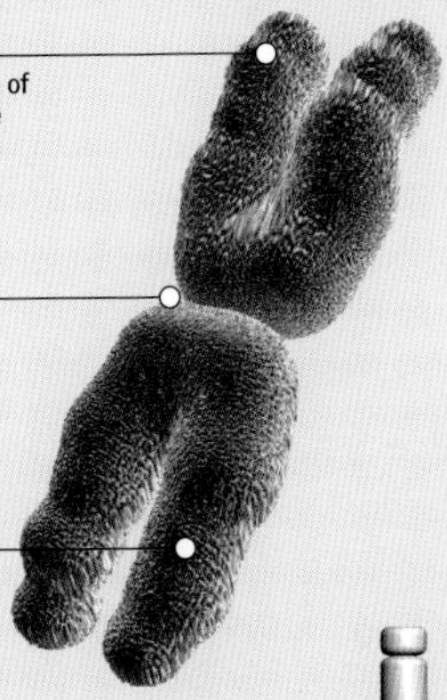


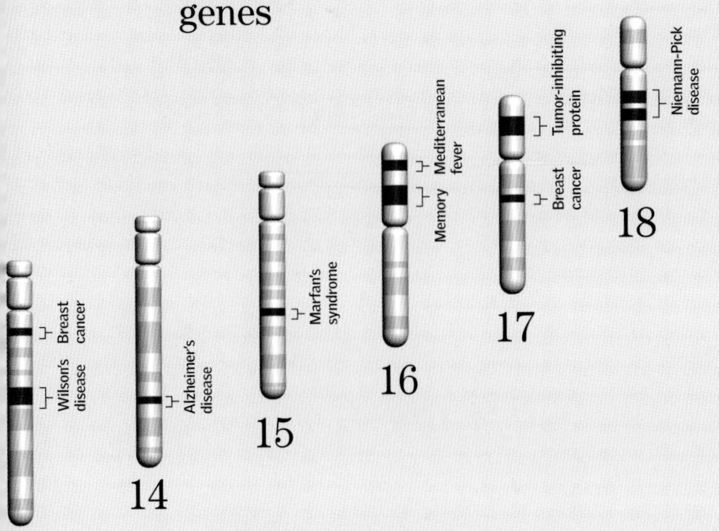


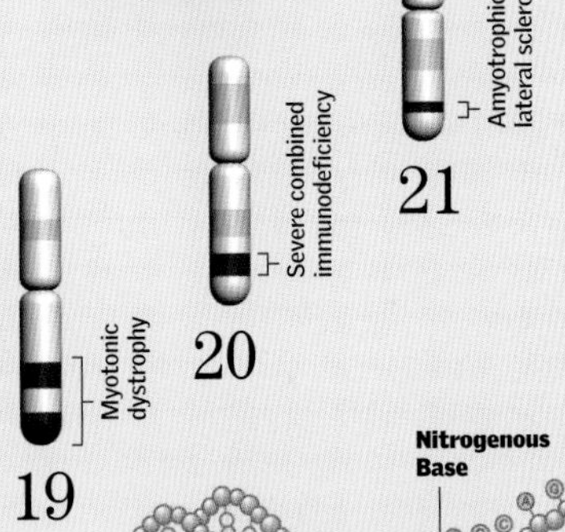


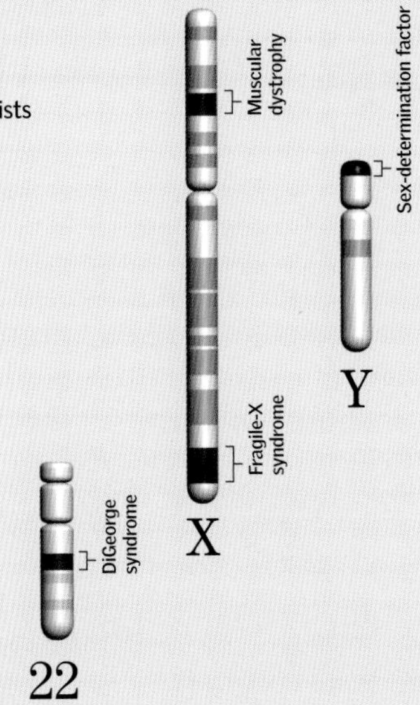


**1 MULTIPLICATION**
Each segment of DNA in which the sequence of bases is unknown is subjected to the polymerase chain reaction (PCR), which makes it possible to make thousands of copies of the same segment of DNA.

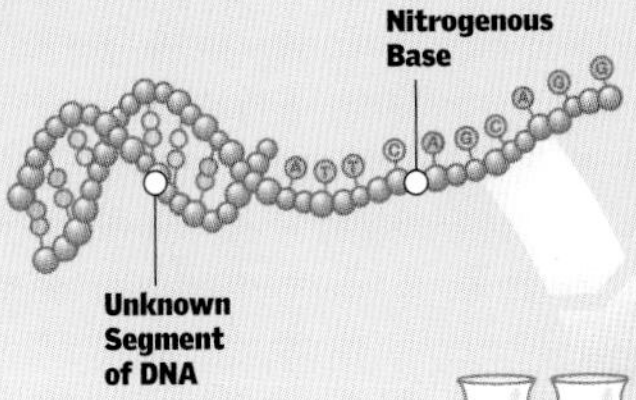


**2 IN VITRO**
Solutions with high concentration of a ddNTP, for example ddGTP, will produce copies of DNA of different length from standard nucleotides. It works because the DNA-copying process is interrupted if a ddNTP is inserted instead of a standard nucleotide.

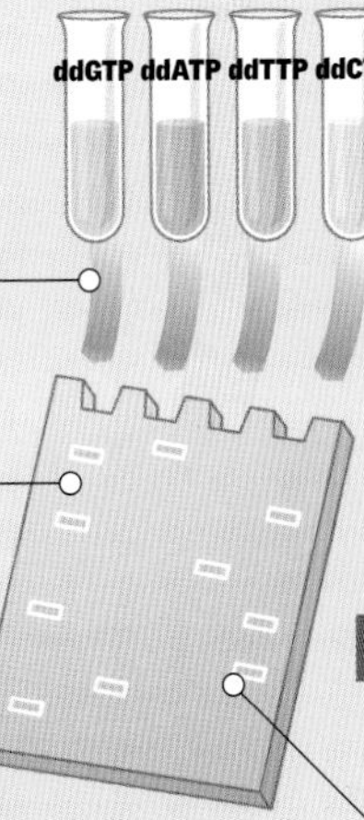


**3 ELECTROPHORESIS**
On a gel, the copies of DNA travel different distances according to their length. This movement is called electrophoresis.

The lighter-weight copies travel a greater distance in the gel.

## Sanger Method

Frederick Sanger, an English biochemist, devised an extraordinary method for deciphering the human genome by identifying the location of each nitrogenous base in the DNA. He divided human DNA into portions of different sizes and used the PCR technique to make thousands of copies. He then made in vitro copies of each DNA fragment using the cellular mechanism of DNA replication. He added his own twist to this process by using fluorescent dideoxynucleotides (ddNTP). These molecules compete with standard nucleotides during the DNA replication process.

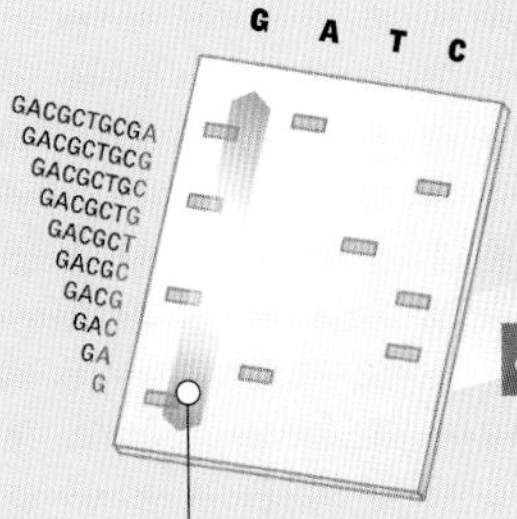


Fragments of DNA seen by fluorescence

**4 PUZZLE**
By placing the gel in front of UV light, the researcher can observe how the bases fit and form the exact sequence of bases of the unknown DNA segment.

**75**
Sanger develops echnique for ciphering the sequence bases in DNA.

**1981**
The first transgenic **rats** and **insects** are obtained.

**1983**
Kary Mullis creates the **polymerase chain reaction technique**.

**1993**
A plan is proposed to finish **sequencing** DNA in the human genome project.

**1994**
The first **transgenic tomato** is made.

**1998**
The genome sequencing of the *Caenorhabditis elegans* nematode is completed.

**2003**
The magazines *Science* and *Nature* publish **the complete sequence of the human genome**.

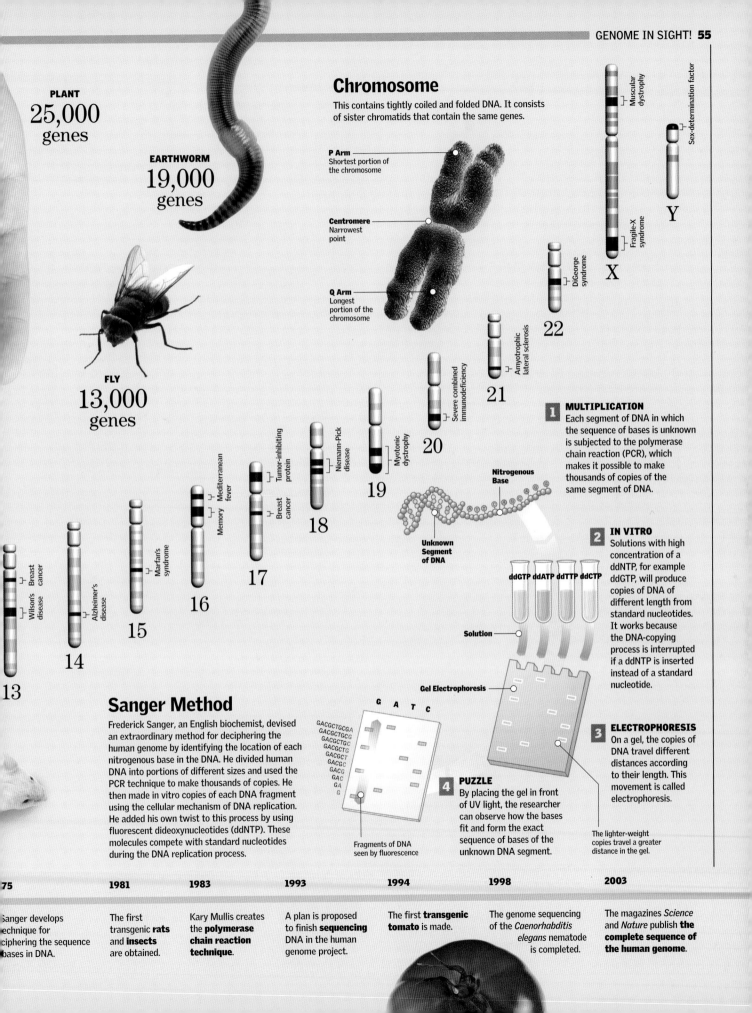

# Stem Cells

The reasoning is simple: if an organism with more than 200 different types of cells is formed from a group of embryonic cells without specialization, then manipulating the division of these original cells (called stem cells) should make it possible to generate all the human tissues and even produce autotransplants with minimal risk. Although such work is in progress, the results are far from being a medical reality. Scientists all over the world are studying its application.

**EMBRYONIC CELLS**
This photograph shows the eye of a needle with an embryo formed only by stem cells before cellular differentiation begins.

## Cellular Division

All the cells of higher organisms divide and multiply through mitosis, with the exception of the reproductive gametes. Mitosis is the process through which a cell divides to form two identical cells. For this to happen, the first cell copies its genetic material inside the nucleus, and later it slowly partitions until it fully divides, producing two cells with the same genetic material. An adult cell divides on average 20 times before dying; a stem cell does it indefinitely.

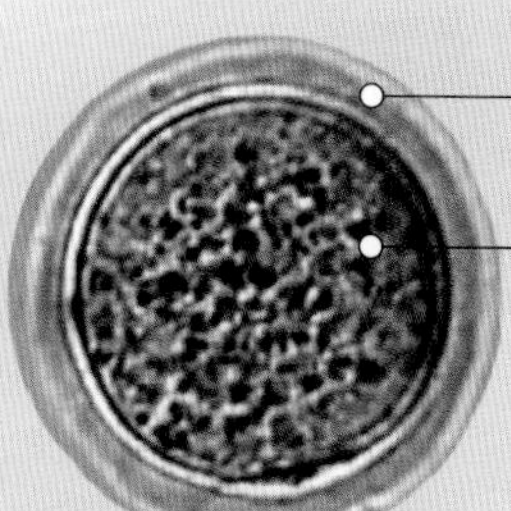

**CYTOPLASM**

**NUCLEUS**
This contains the DNA. First it duplicates the DNA, and then it divides.

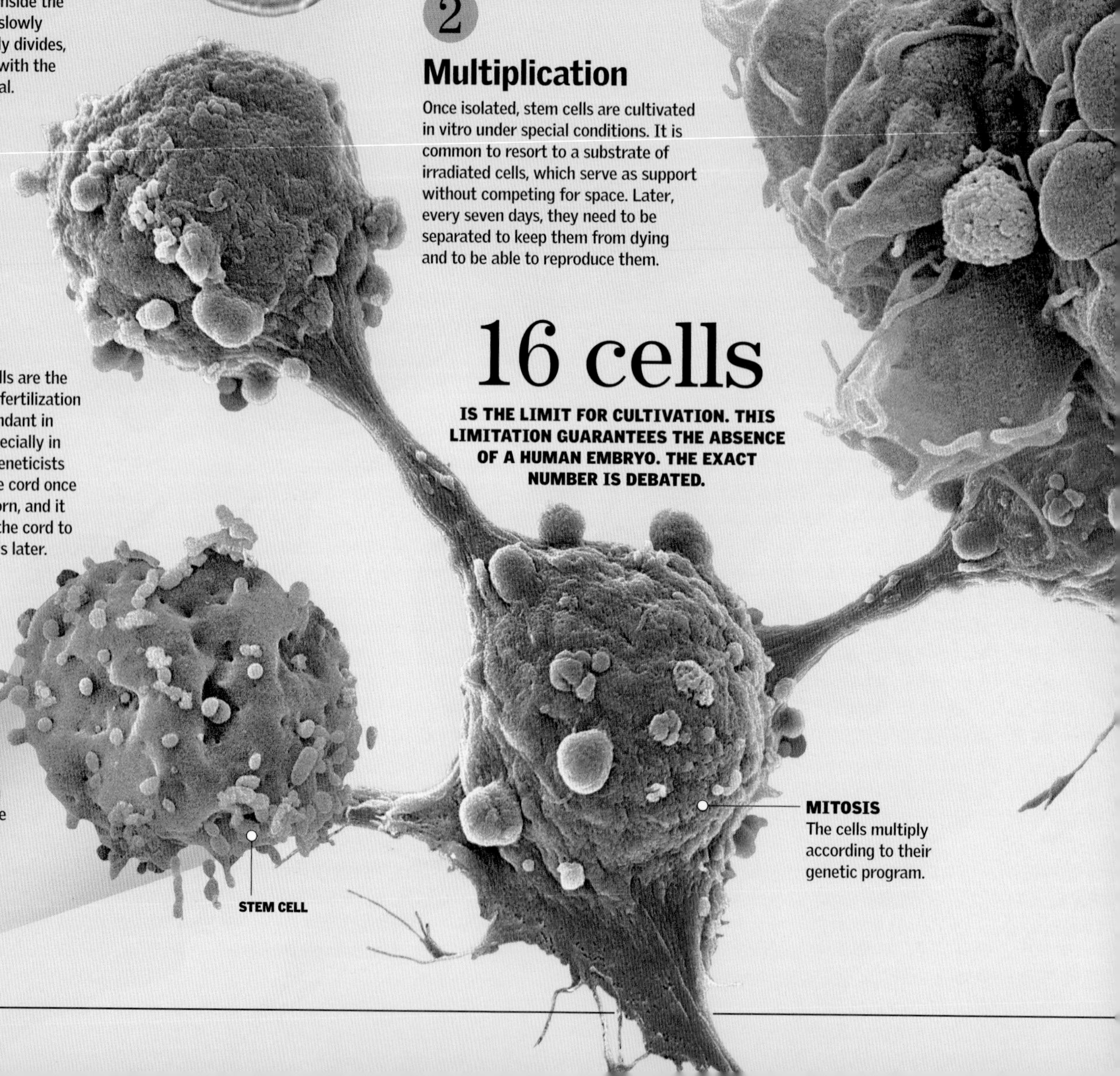

**STEM CELLS**

## 2 Multiplication

Once isolated, stem cells are cultivated in vitro under special conditions. It is common to resort to a substrate of irradiated cells, which serve as support without competing for space. Later, every seven days, they need to be separated to keep them from dying and to be able to reproduce them.

## 1 Obtaining

Because the stem cells are the first that form after fertilization occurs, they are abundant in the placenta and especially in the umbilical cord. Geneticists obtain them from the cord once the baby has been born, and it is possible to freeze the cord to harvest the stem cells later.

# 16 cells

**IS THE LIMIT FOR CULTIVATION. THIS LIMITATION GUARANTEES THE ABSENCE OF A HUMAN EMBRYO. THE EXACT NUMBER IS DEBATED.**

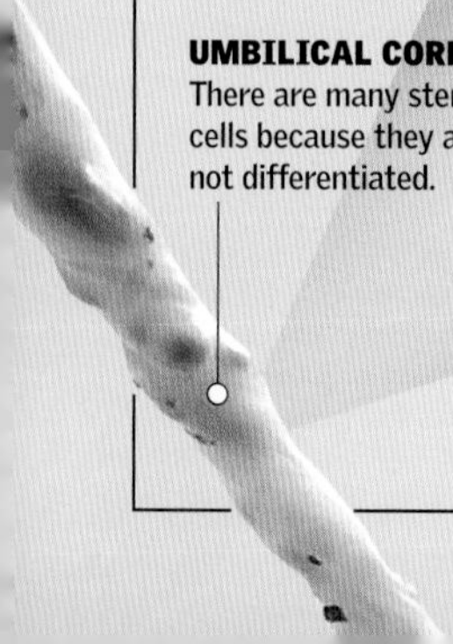

**UMBILICAL CORD**
There are many stem cells because they are not differentiated.

**STEM CELL**

**MITOSIS**
The cells multiply according to their genetic program.

**ACTIVATORS**
Chemical and hormonal activators guide the specialization.

## 3 Differentiation

The stem cells are pluripotent, which is to say that they have the power to create any of the more than 200 different cells of the body. This process happens as the embryo grows. If the optimal conditions could be created in vitro, it would be possible to form in a laboratory all the cells of the body using the genetic program of the cells. In practice, this technique is possible only with a few types of cells, in particular blood cells.

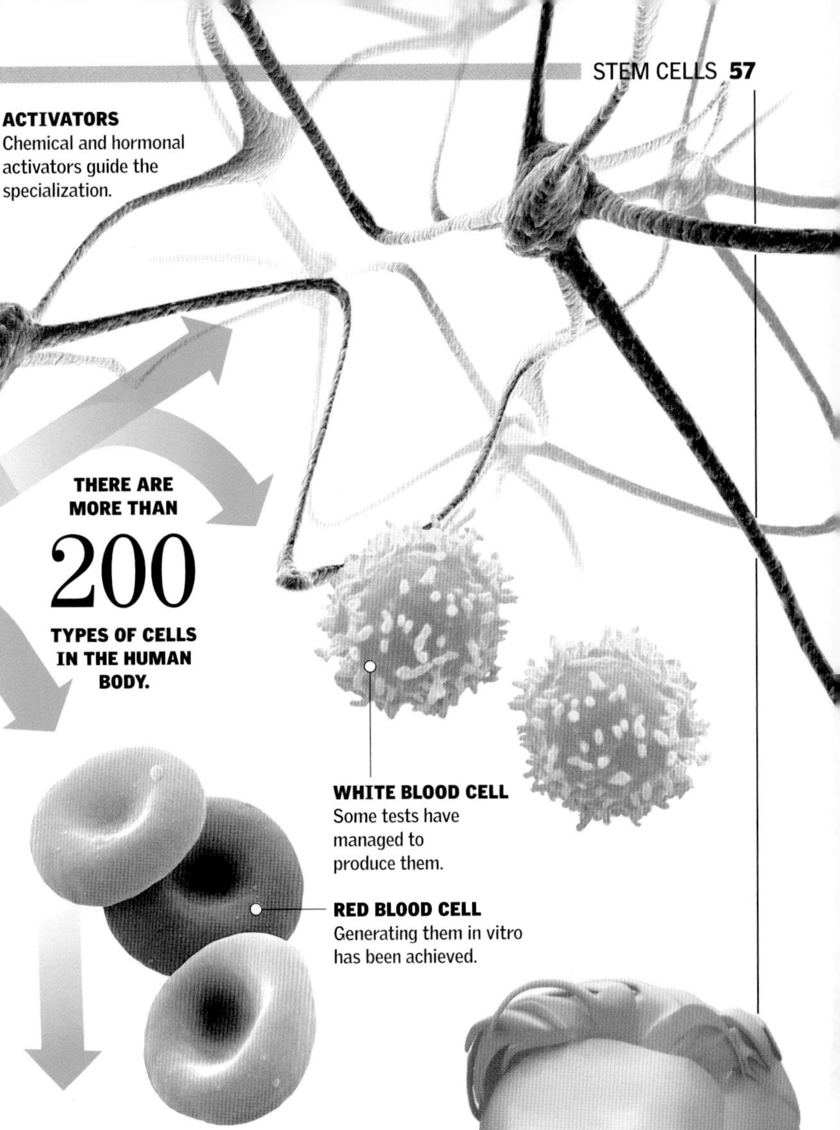

**NEURONS**
have yet to be grown in the laboratory.

**THERE ARE MORE THAN**
# 200
**TYPES OF CELLS IN THE HUMAN BODY.**

**WHITE BLOOD CELL**
Some tests have managed to produce them.

**RED BLOOD CELL**
Generating them in vitro has been achieved.

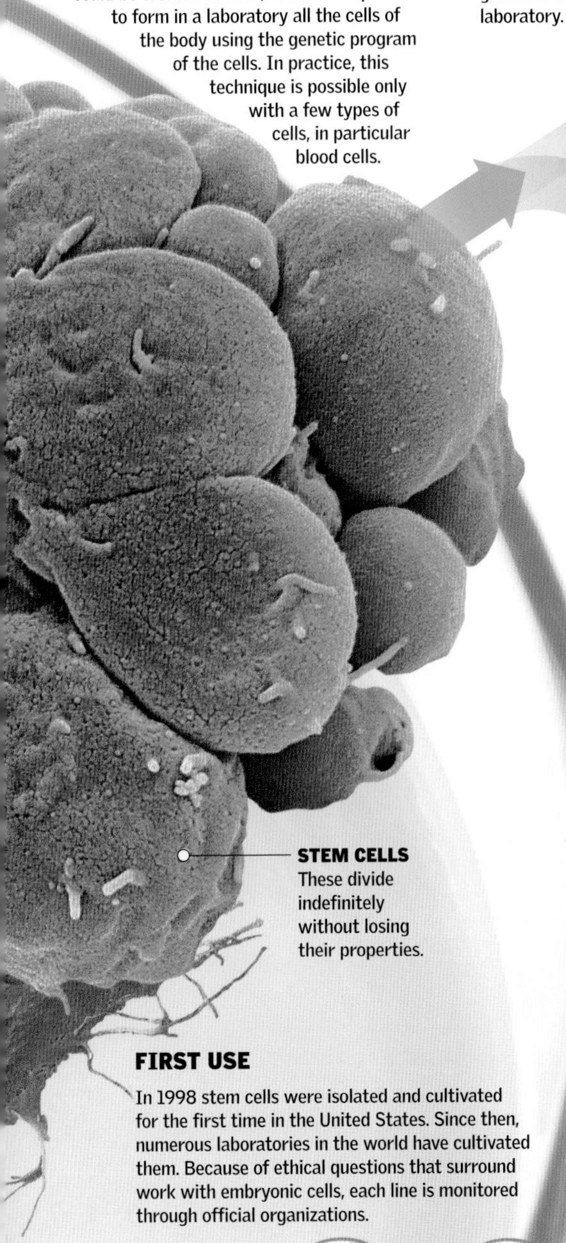

**STEM CELLS**
These divide indefinitely without losing their properties.

### FIRST USE

In 1998 stem cells were isolated and cultivated for the first time in the United States. Since then, numerous laboratories in the world have cultivated them. Because of ethical questions that surround work with embryonic cells, each line is monitored through official organizations.

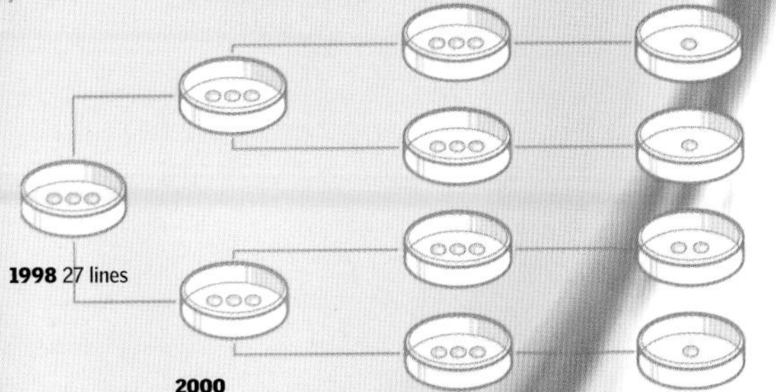


## 4 Implantation

Doctors and geneticists hope to be able to provide new pluripotent cells to damaged tissue and provoke its regeneration. To date, they have been able to introduce umbilical-cord hematopoietic stem cells into patients with dysfunctional formation of red blood cells. This is equivalent to a bone marrow transplant without surgical intervention.

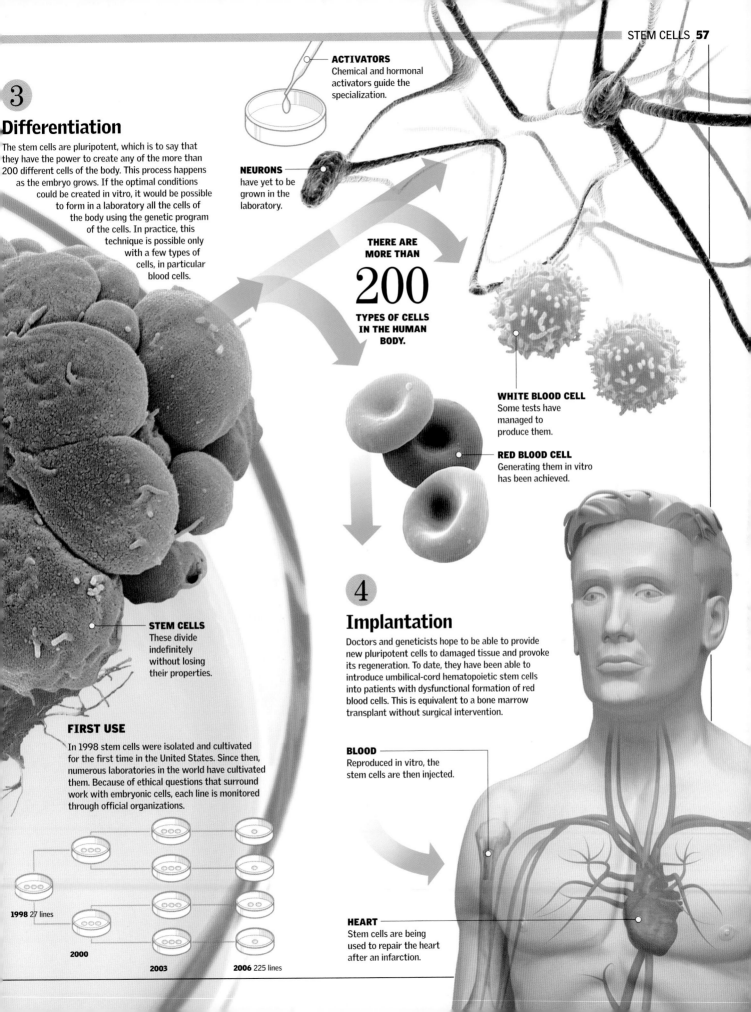

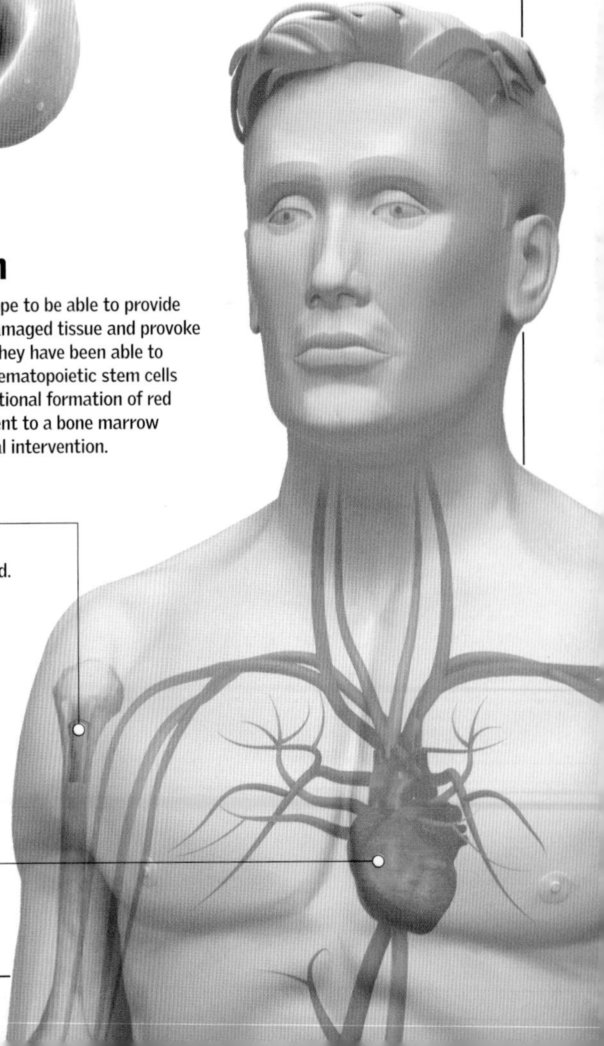

**BLOOD**
Reproduced in vitro, the stem cells are then injected.

**HEART**
Stem cells are being used to repair the heart after an infarction.

# Gene Therapy

One of the latest breakthroughs in medicine, gene therapy is used to introduce genetic material to correct deficiencies of one or more defective genes that are the cause of an illness. Several different techniques have been developed for use with human patients, almost all of which are at the research stage. The problem with illnesses with a genetic origin is that therapy must modify the cells of the affected organ. To reach all these cells, or a significant number of them, demands elaborate protocols or, as is the case for viruses, the use of nature's biological weapons to cause other illnesses.

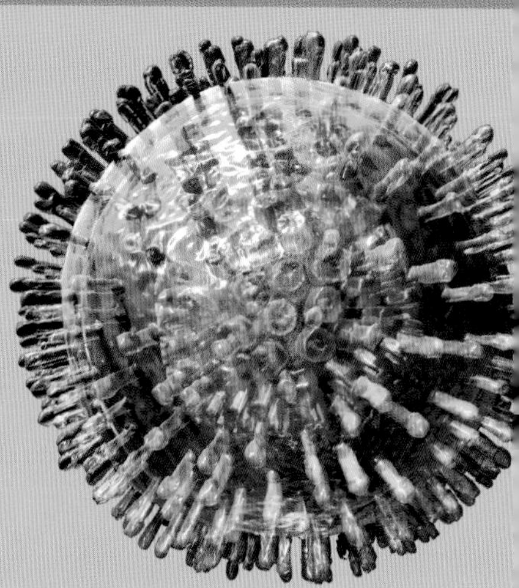

**HERPESVIRUS**
The herpesvirus is an icosahedral virus and holds a DNA sequence that needs to be modified so that it will not cause an illness. It is widely used in gene therapy.

## Treatable Illnesses

Illnesses with a genetic origin are difficult to treat, since the organism has poorly coded genes and the fault is therefore present in all its cells. Cystic fibrosis and Duchenne muscular dystrophy are examples of monogenetic illnesses that can potentially be treated with these therapies. Gene therapy has also been attempted on cancer and HIV infection, among other pathologies. A definitive cure may be found for many genetic illnesses, but the techniques for gene therapy are still in the development stage.

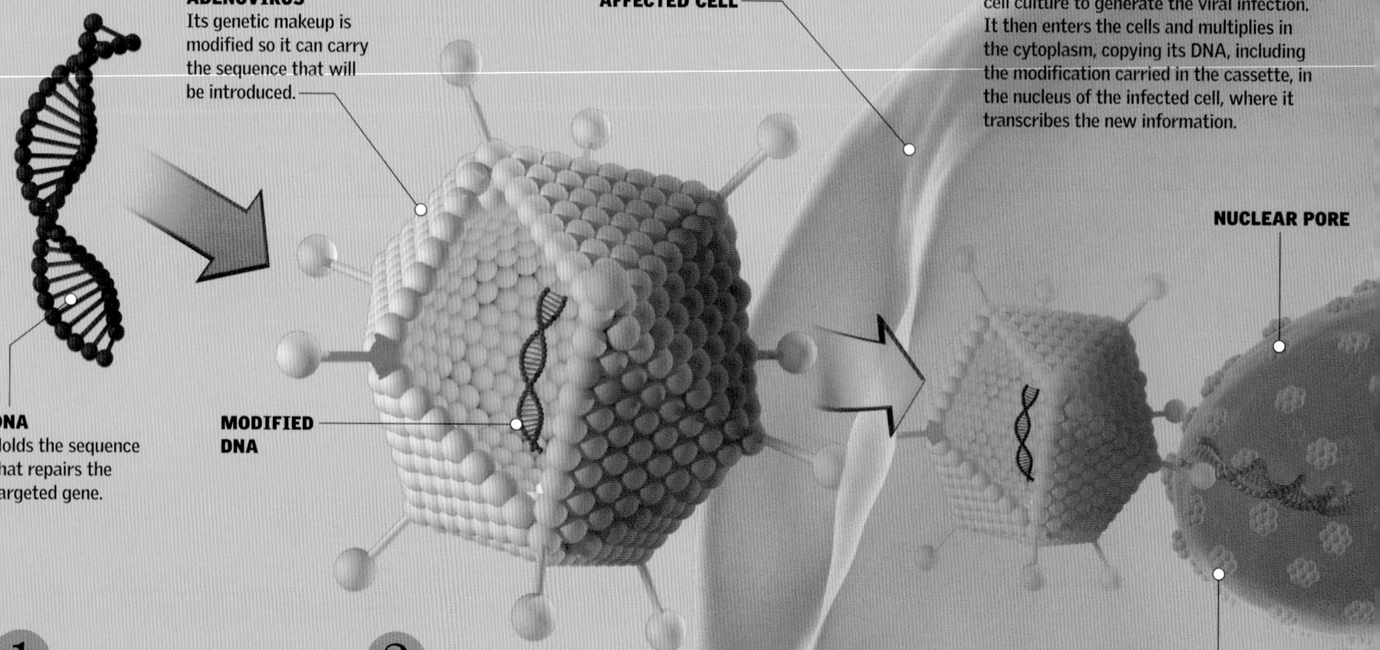


### 1 Identification

The DNA sequence that corresponds to the gene that causes the deficiency requiring treatment is identified. Then the correct sequence is isolated and multiplied to guarantee a quantity that can modify the organism. Because a monogenetic illness generally affects the function of one organ, the cell volume that is targeted for modification is large. Then a technique is chosen to transfect the cells.

### 2 Vehicle

An adenovirus is an icosahedral virus that contains double-stranded DNA and lacks an outer envelope. It is primarily the cause of a number of mild respiratory illnesses. If the virus can be modified to be nonpathogenic, it has the potential for use in transporting a modified sequence of DNA in a region called a cassette. Even though its capacity is limited, its effectiveness rate is very high.

### 3 Replacement

The modified adenovirus is inoculated in a cell culture to generate the viral infection. It then enters the cells and multiplies in the cytoplasm, copying its DNA, including the modification carried in the cassette, in the nucleus of the infected cell, where it transcribes the new information.

**DNA TRANSCRIPTION**

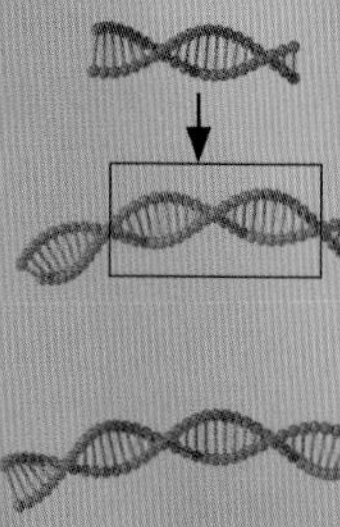

## 4 Synthesis

The infected culture cells, which have the new genetic information, can now synthesize the compound that caused the dysfunction. Generally these are proteins that cannot be synthesized because the gene for their elaboration is disassociated or damaged. The process begins once the cells divide and transcribe the gene in question. The protein that was not synthesized before is now transcribed and produced.

# Relationship

**IT IS CRITICAL THAT THE HYPOTHETICAL NUMBER OF CELLS TO BE MODIFIED AND THE NUMBER OF VIRUSES NEEDED FOR THE THERAPY TO WORK ARE IN THE CORRECT RELATIONSHIP.**

# Kilobase

**THE UNIT IN WHICH DNA AND RNA ARE MEASURED; THE CAPACITY OF A VIRUS'S CASSETTE, WHICH ON AVERAGE IS APPROXIMATELY FIVE KILOBASES.**

**NONVIRAL GENE THERAPIES**

Many are based on physical means such as electrical techniques. They have the advantage of producing material in vitro, which allows for a large transfer capacity not limited by the number of bases that can be transfected by a virus. The problem is that these methods are not efficient for reaching target cells in the organism. The most important therapies of this type are microinjection, calcium phosphate precipitation, and electroporation (the use of an electric field to increase the permeability of the cell membrane).

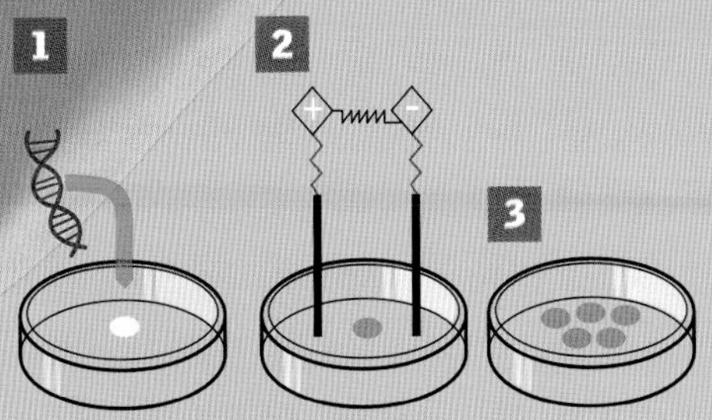

# Genetic Solution

Genetic engineering applies technologies for manipulating and transferring DNA between separate organisms. It enables the improvement of animal and plant species, the correction of defective genes, and the production of many useful compounds. For example, some microorganisms are genetically modified to manufacture human proteins, which are vital for those who do not produce them efficiently.

## Genetic Engineering

Genetic recombination consists of integrating DNA from different organisms. For example, a plasmid is used to insert a known portion of human DNA into the DNA of bacteria. The bacteria then incorporate new genetic information into their chromosomes. When their own DNA is transcribed, the new DNA is transcribed as well. Thus, the bacteria formulate both their own proteins and foreign proteins, such as human insulin.

### 3 Insertion

A culture of nonpathogenic receptor bacteria is placed in a solution that contains the recombined plasmid. The solution is then subjected to chemical and electrical stimuli to incorporate the plasmid that contains the insulin gene.

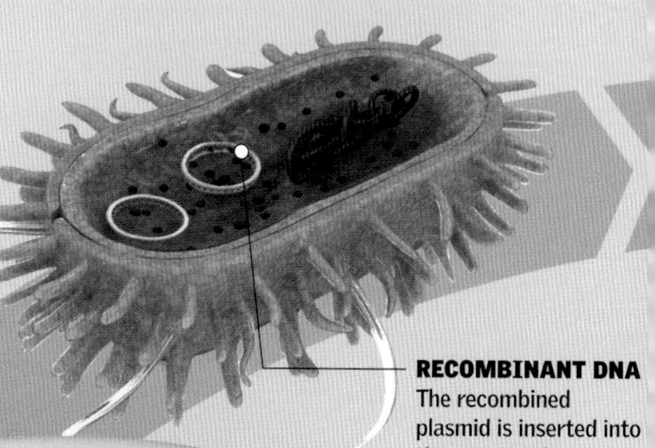

RECOMBINANT DNA
The recombined plasmid is inserted into the receptor bacteria.

### 2 Union

The human and bacterial DNA join at their free ends and form a recombined plasmid. This plasmid contains the human insulin gene.

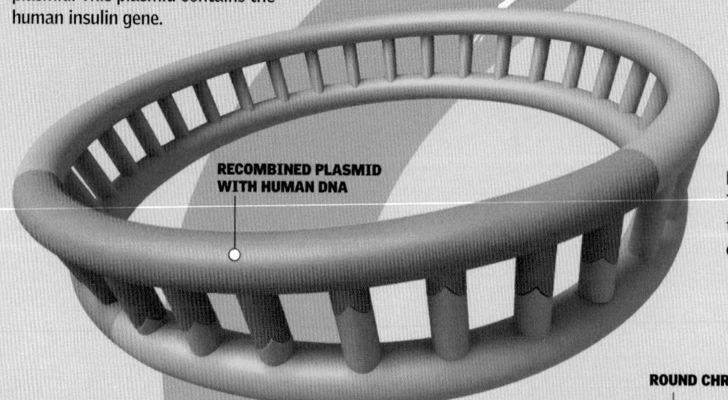


### 1 Extraction

DNA is extracted from a human cell to obtain the gene that codes for producing insulin. The DNA is cut using restriction enzymes that recognize the points where the gene in question begins and ends. These enzymes also cut the bacterial plasmid. The DNA fragments thus obtained have irregular and complementary ends.

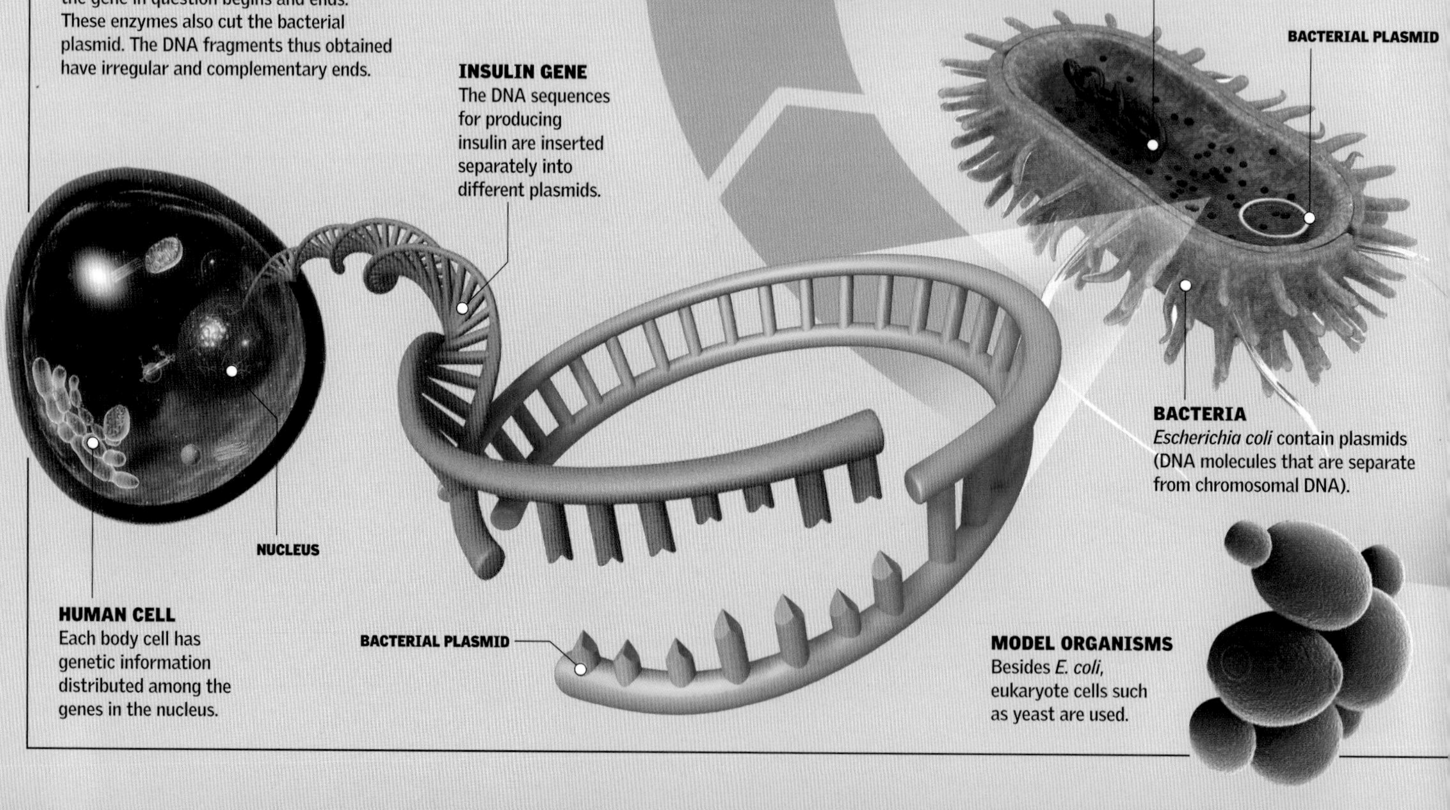

# 10 hours

**ARE NEEDED FOR THE CULTURE POPULATION TO DOUBLE.**

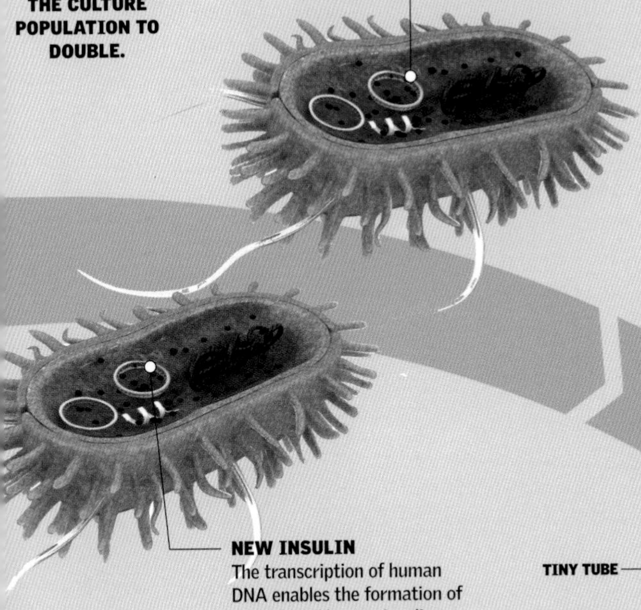

**INSERTION INTO THE CHROMOSOME**
The recombined plasmid is inserted into the bacteria's chromosome.

**NEW INSULIN**
The transcription of human DNA enables the formation of recombined human insulin.

## 4 Reproduction

The bacteria reproduce constantly in fermentation tanks with water and essential nutrients. In these conditions, the recombined bacteria transcribe the information in their chromosomes to produce proteins. The bacteria also read the information from the human DNA that was inserted using the recombined plasmid, and they produce insulin.

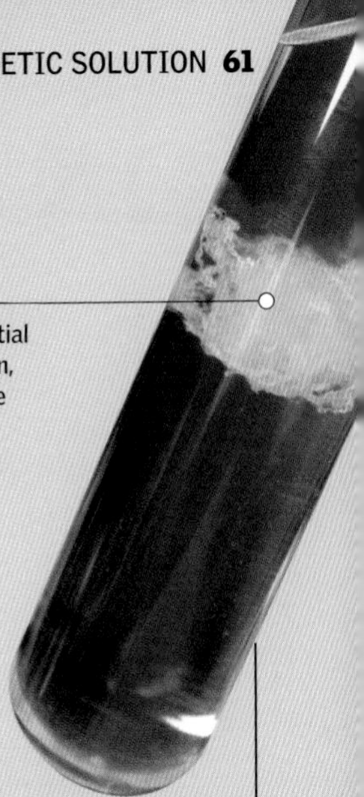

**BACTERIA**
In phase of exponential growth. From now on, they will produce the hormone insulin.

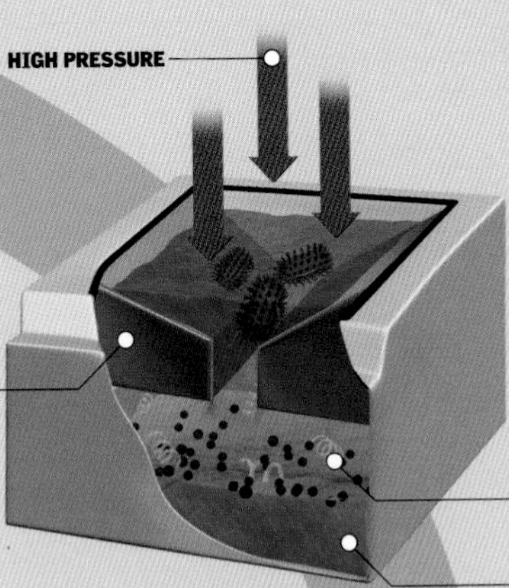


## 5 Purification

The culture is circulated at high pressure through tiny tubes that destroy the bacteria. The solution contains a large amount of insulin that must be separated from the other proteins in the solution.

# First Case

**INSULIN WAS THE FIRST PROTEIN PRODUCED BY GENETIC ENGINEERING. IT WAS APPROVED FOR HUMAN USE IN 1982.**

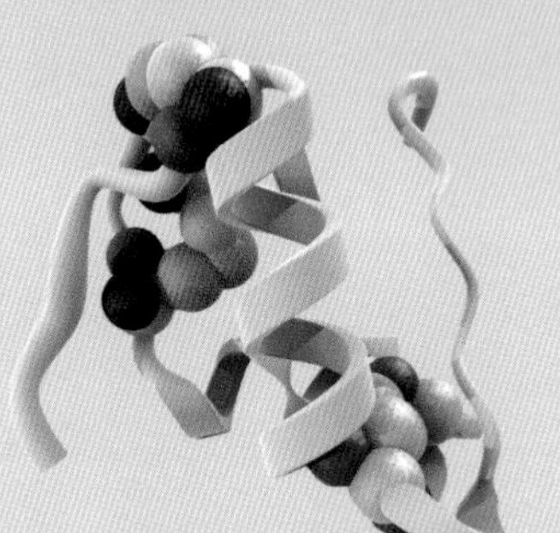

**INSULIN PROTEIN**

**CENTRIFUGAL FORCE**
Centrifugal force accelerates the decantation.

**DECANTATION**
The centrifuges reduce the amount of time necessary to separate the solid matter.

## 6 Centrifugation

Centrifuges separate the various compounds present in the solution from the bacterial remains and the human insulin. The proteins present in the solid matter are separated from the original solution.

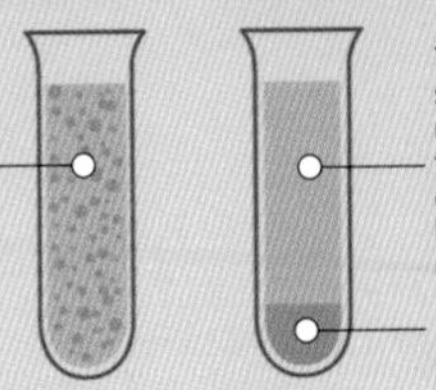


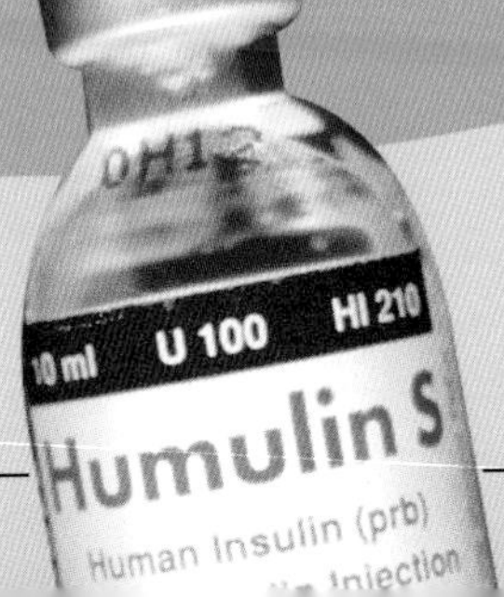


## 7 Formulation

The recombinant human insulin is chemically modified. This produces a stable, aseptic compound that can be administered therapeutically via injection.

# Recombinant antibiotics and vaccines

**ARE ALSO PRODUCED BY GENETIC ENGINEERING.**

# DNA Footprints

Since Sir Alec Jeffreys developed the concept of the DNA profile for the identification of people, this type of forensic technique has taken on significant importance. A practically unmistakable genetic footprint can be established that allows for the correlation of evidence found at the scene of a crime (hair, semen, blood samples) with a suspect. In addition, the use of this technique is a key element to determine the genetic link in kin relationships.

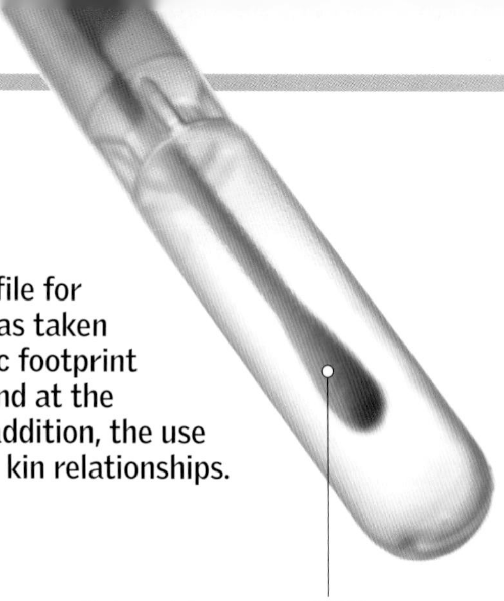

## 1 Sample Collection

Any body fluid, such as urine, blood, semen, sweat, and saliva, or fragments, such as tissues, cells, or hairs, can be analyzed to obtain a person's DNA. There is generally always something left at the scene that can be used as a sample.

Only a very small amount of evidence is needed for sampling. For example, just a small fraction of a drop of blood or sperm is sufficient.

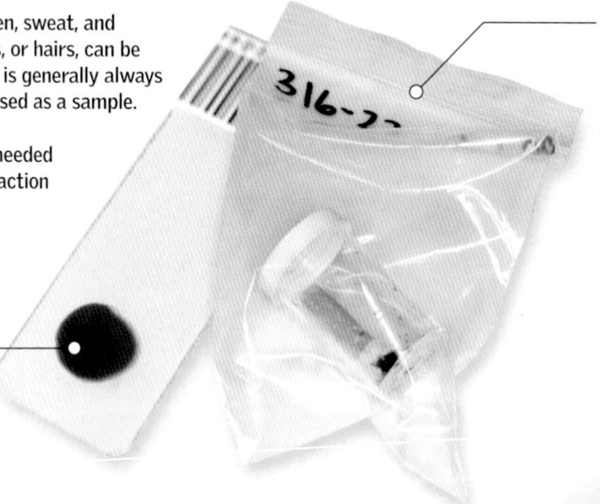


Each sample is placed in separate plastic bags, sealed, and certified to avoid adulterations.

**SWAB**
For saliva samples. Then it is immersed in a solvent solution and the DNA extracted.

**FACTORS THAT ALTER DNA**

Moisture or water will denaturalize a sample faster.

Heat is one of the most destructive factors.

## 2 DNA Separation

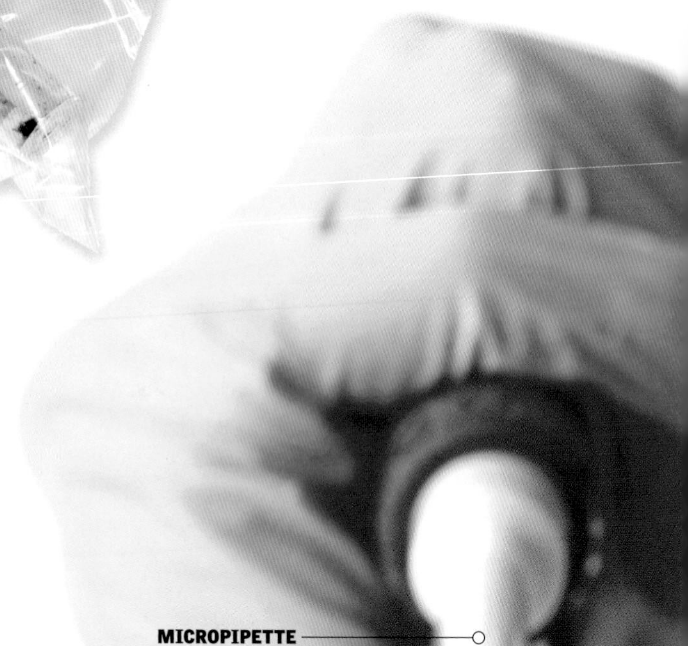

**HAIR FOLLICLE**
A follicle has DNA that is easy to obtain.

**TWEEZERS**
These must be properly sterilized.

**1 HAIR DIGESTION**
The hair is divided into sections. These are then put into a tube, and solvents are applied.

**MICROPIPETTE**
Only the substance floating on the surface is extracted. This is where the DNA is.

**LABELING**
This is absolutely necessary so that the samples are not mixed up.

**2 CENTRIFUGING**
The suspended DNA must be centrifuged to separate it from the rest of the cell material.

**3 PRECIPITATION**
A 95 percent solution of ethanol is added; the sample is shaken and then centrifuged at a higher speed than before.

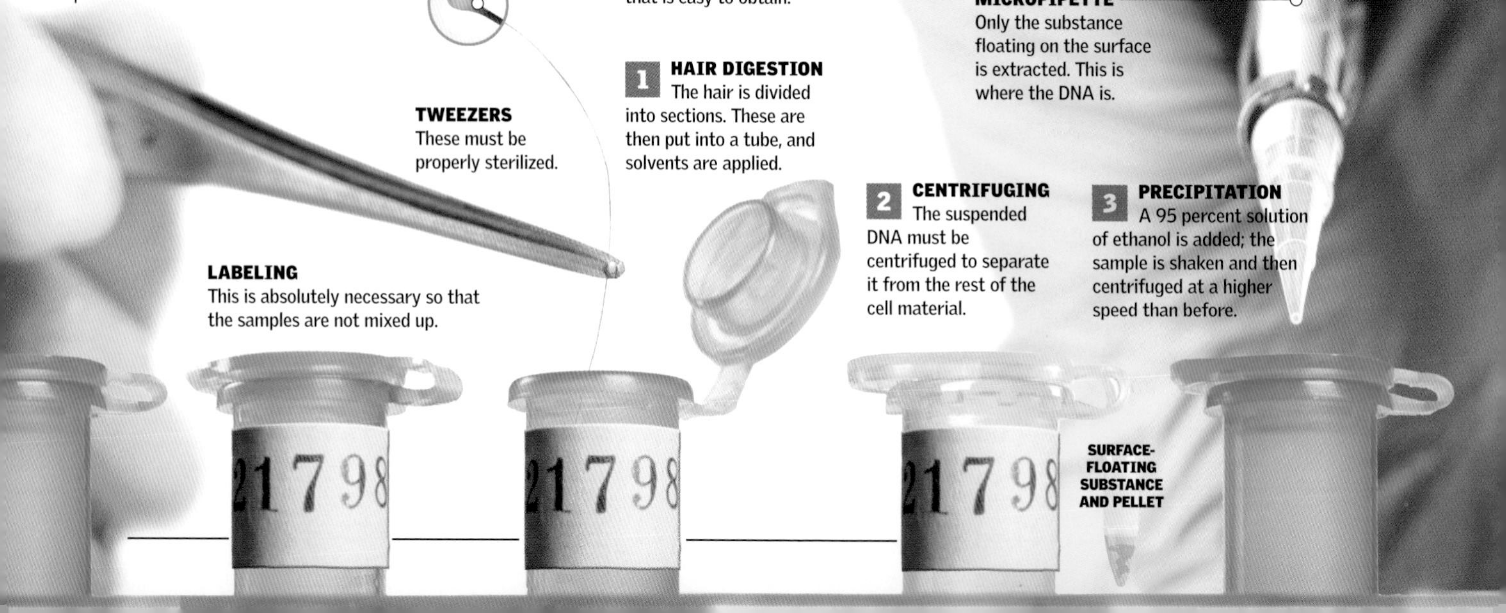


**SURFACE-FLOATING SUBSTANCE AND PELLET**

## 3 DNA Magnification

The polymerase chain reaction (PCR) is carried out by a machine that, using heat, synthetic short nucleotide sequences, and enzymes, copies each fragment of DNA as many times as needed. This amplification makes it possible to conduct a large number of tests while conserving the DNA. Later the DNA fragments are separated by means of capillary electrophoresis.

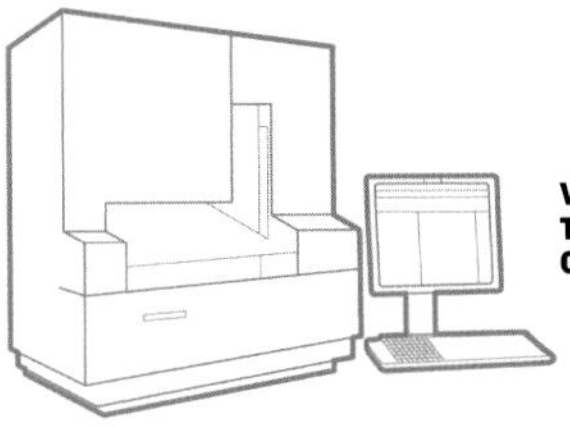


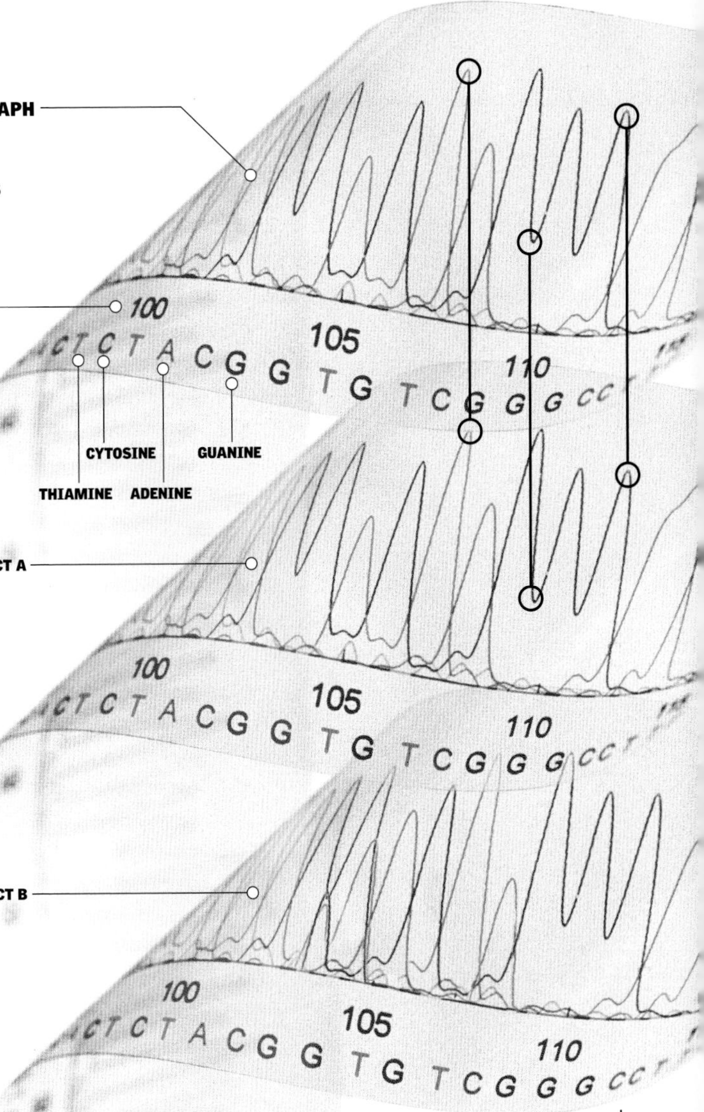


## 4 Impression and Comparison

The machine presents the results as curves, where each base has a specific location according to the height of the curve in the graph sequence. It then compares the sample obtained at the crime scene with those obtained from the crime suspects. If one of them was at the scene of the crime, the curves coincide exactly in at least 13 known positions.

# 13 locations

**IS THE MINIMUM NUMBER OF COINCIDING POINTS THAT NEED TO BE FOUND FOR A SUSPECT TO BE ACCUSED OF A CRIME IN THE UNITED STATES.**

**DISPOSABLE MATERIAL**
All the material that is used must be disposable to avoid contaminating the DNA.

**4 SURFACE-FLOATING SUBSTANCE**
A 70 percent solution of ethanol is added, and the mixture is rinsed with water. The DNA is free of impurities and ready for analysis.

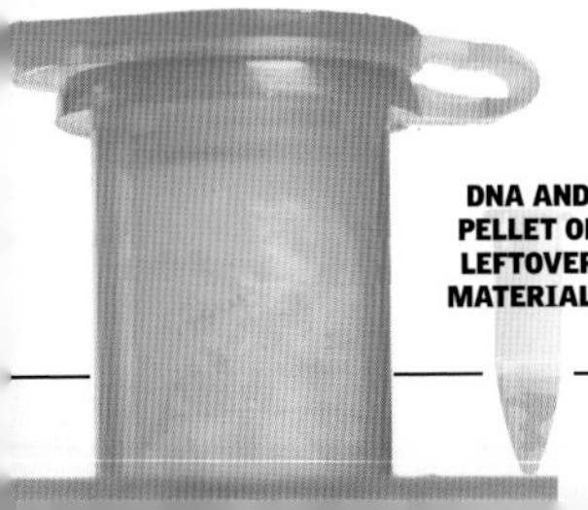


## Power of Exclusion (PE)

Overall, for a DNA test to be considered as valid criminal evidence, at least in theory, it should be able to guarantee a PE with a certainty above 99.9999999 percent. The PE is measured as a percentage but is expressed as the number of people who are excluded as possible bearers of the DNA at the crime scene. Thus, a sample is taken at random from one person, as a type of witness, and it is then compared with the DNA from the evidence and that of the suspect. The detail of the analysis must be so precise that it can, at least theoretically, be able to discriminate one person among one billion people. In practice, the test is valid if it statistically discriminates one person in one billion. All this is done to guarantee the results of the test and so that it can have validity in court. In practice, the suspects are not chosen randomly but fulfill other evidence patterns, among which DNA is used to confirm these patterns.

**1 in 1,000,000,000**
**is the STATISTICAL GUARANTEE.**

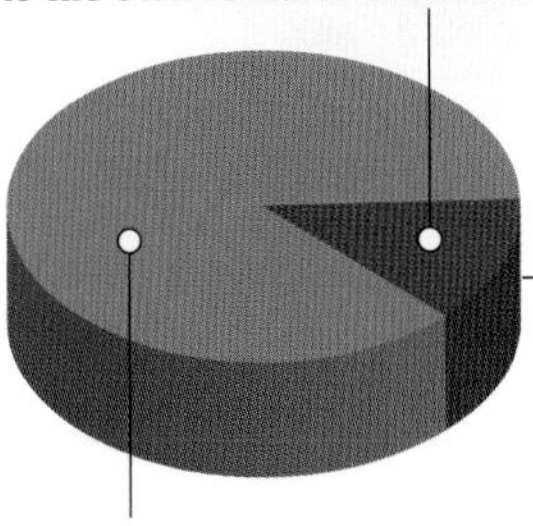

**6,500,000,000**
**is the WORLD'S POPULATION.**

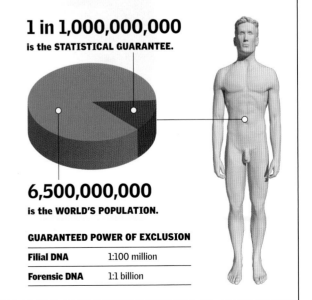

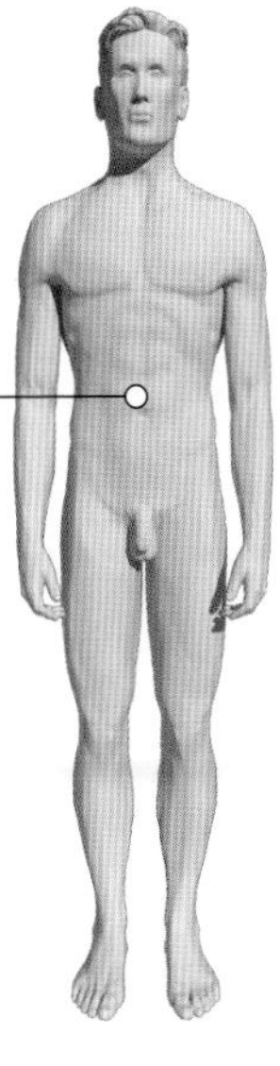

**GUARANTEED POWER OF EXCLUSION**

| | |
|---|---|
| **Filial DNA** | 1:100 million |
| **Forensic DNA** | 1:1 billion |

# The Genetic Ancestor

Ever since Darwin published his theory about the evolution of species, humans have sought to understand their origin in light of a diversity of ideas and theories. With the success of efforts to map the human genome, old evidence is gaining new strength. Many scientific teams used some 100,000 samples of DNA from all over the world to trace the process of human expansion back to a common ancestor—the "Mitochondrial Eve" that lived in sub-Saharan Africa some 150,000 years ago. She was not the only human female of her time, but she was the one that all present-day women recognize as a common genetic ancestor. The key to the trail is in DNA mutations.

## Genetic Material

Each time an organism is conceived, its genetic material is a fusion of equal parts received from its parents. Recovering this material throughout history is impossible because of the large number of combinations, so scientists use mitochondrial DNA from the cells as well as DNA from the chromosomes. Thus, following a single path for each sex, the possible combinations are reduced to a set of hereditary lines that are traceable over time. This method is possible when a cell's DNA, along with the various locations of the genes and recombinant areas, is known.

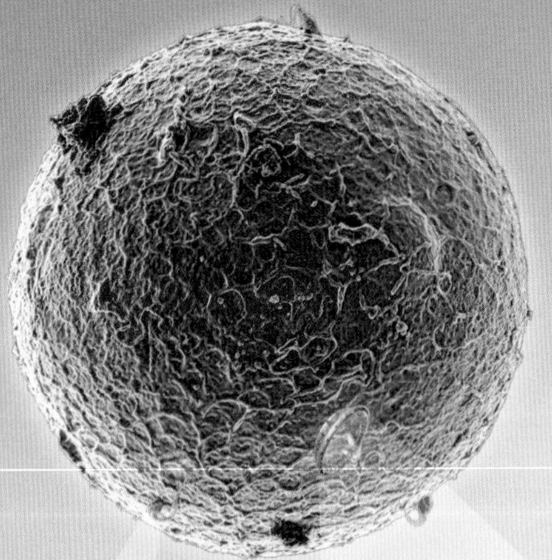

### Ovule

This cell is a haploid cell that at the moment of fertilization provides the cellular organelles as well as half of the chromosomes. Among the organelles, the mitochondria are the most important for genetic studies.

### Spermatozoon

When a spermatozoon fertilizes an ovule, its tail breaks off, along with all cellular material except its nucleus, which contains half of the necessary genetic information for a new individual.

### Mitochondria

These are the organelles that provide energy to the cell through respiration. They contain a portion of DNA.

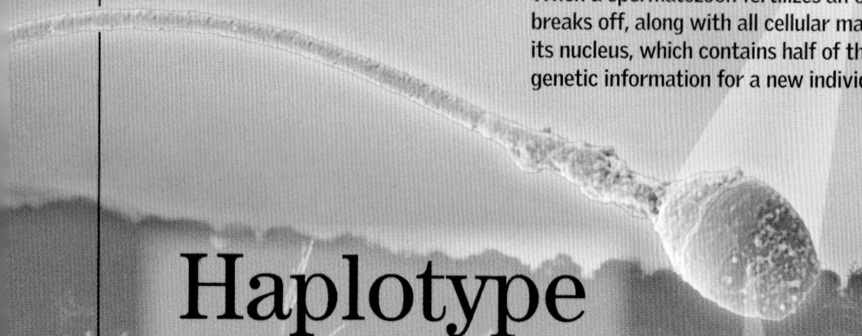

# Haplotype

**IS A SET OF CLOSELY LINKED ALLELES ON A CHROMOSOME.**

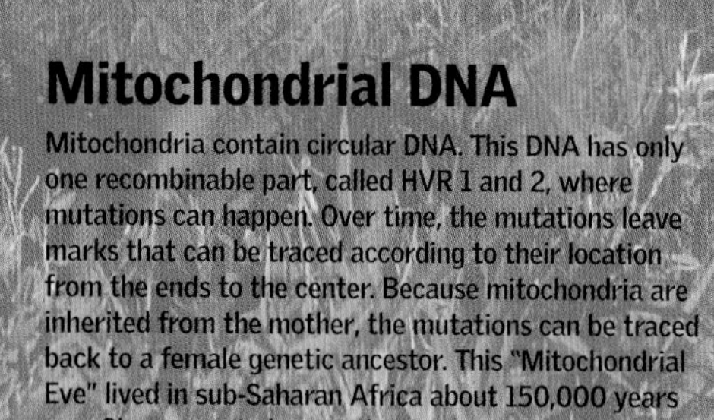


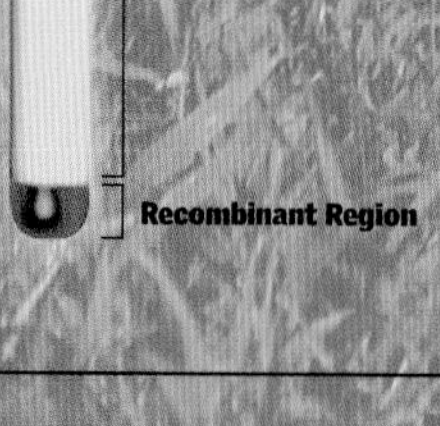


## The Y Chromosome

A baby's sex is determined by the sperm cell that succeeds in fertilizing the ovule. Specifically the male gender is determined by the Y chromosome, which is passed on from father to son. To follow a line of ascendant mutations in the recombinant part, the markers of each mutation must be read from the ends to the center to find a common male ancestor. He is called the chromosomal Adam, and he is estimated to have lived 90,000 years ago in Africa.

## Mitochondrial DNA

Mitochondria contain circular DNA. This DNA has only one recombinable part, called HVR 1 and 2, where mutations can happen. Over time, the mutations leave marks that can be traced according to their location from the ends to the center. Because mitochondria are inherited from the mother, the mutations can be traced back to a female genetic ancestor. This "Mitochondrial Eve" lived in sub-Saharan Africa about 150,000 years ago. She was not alone at the time, nor was she the only one of her species. However, she was the only one of her community whose genetic inheritance survives.

## Genetic Diversity and Phylogenetics

Geneticists have determined statistically that every three generations there is a mutation that will be preserved in the DNA of the descendants. They used this statistic and demographic studies to calculate the age of the "Mitochondrial Eve" and the "Nuclear Adam." If the path of mutations is followed from the present to the past, the line of ascent would lead to these genetic ancestors. However, in reverse, many mutations represent dead ends. That is, they left no descendants for a wide range of reasons. These links are part of the study called phylogenetics and make up well-defined haplogroups. Each haplogroup represents the genetic diversity of a species.

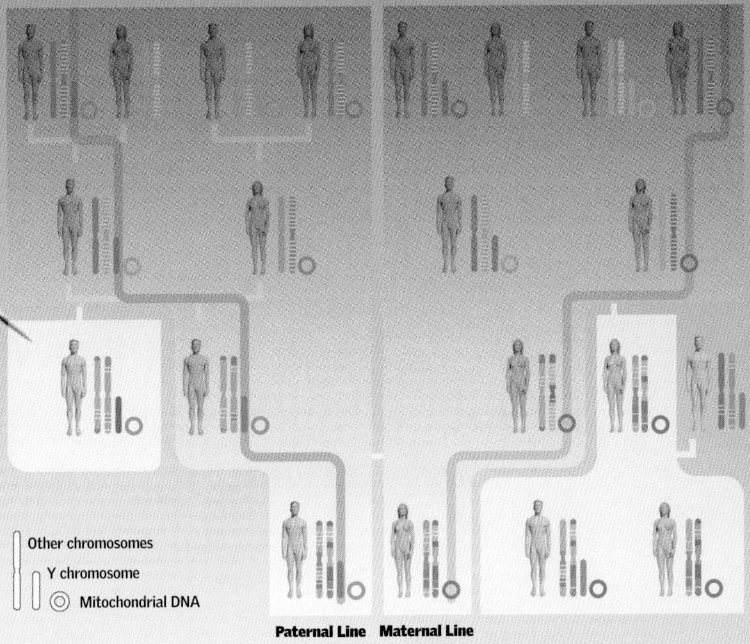


## Genetic Drift

Each time a mutation occurs, it continues as a mark on future generations. Genetic drift explains how this mutation spreads and how the effectiveness of its spread is related to the number of individuals in a group, the time they live in a certain region, and the environment. If the group is small, its chances of success are increased because genetic drift is more effective in changing the genetic pattern. Also, the longer the group remains in one place, the more mutations it will have.

**Africa is where the greatest number of mutations is found. This leads to the supposition that humans have lived there the longest.**

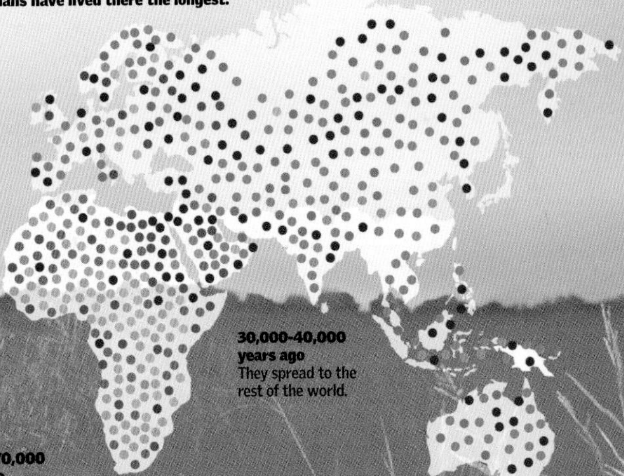

**30,000-40,000 years ago**
They spread to the rest of the world.

# Haplogroup

**IS A HUMAN GROUP WITH THE SAME GENETIC DESCENT, RECOGNIZED BY CHARACTERISTIC MUTATIONS.**

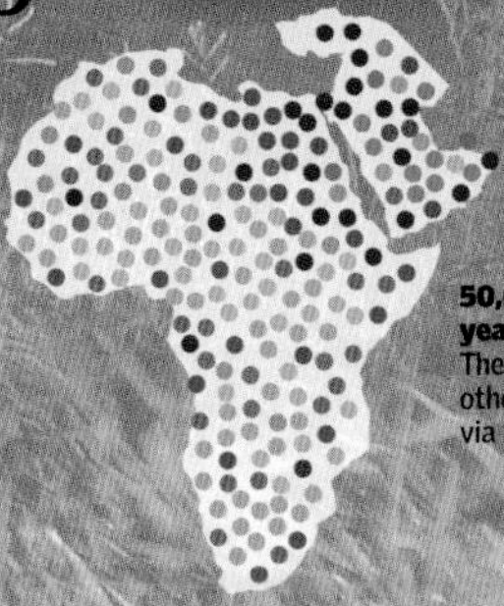

**50,000-70,000 years ago**
They migrate to other continents via the Red Sea.

**L0 and L1, the most ancient**
These haplogroups have the greatest number of mutations in their DNA and are the oldest human groups. They are the San and Khoekhoe peoples in Africa.

**150,000 years ago**
*Homo sapiens* is found only in Africa.

## The Common Relative

In genetic terms, DNA enables us to conceive of a primordial Adam and Eve, our genetic ancestors. However, the common ancestor of all humans alive today is quite a different matter. Several scientific hypotheses estimate that an ancestor to whom we are all related lived between 1,000 and 10,000 years ago.

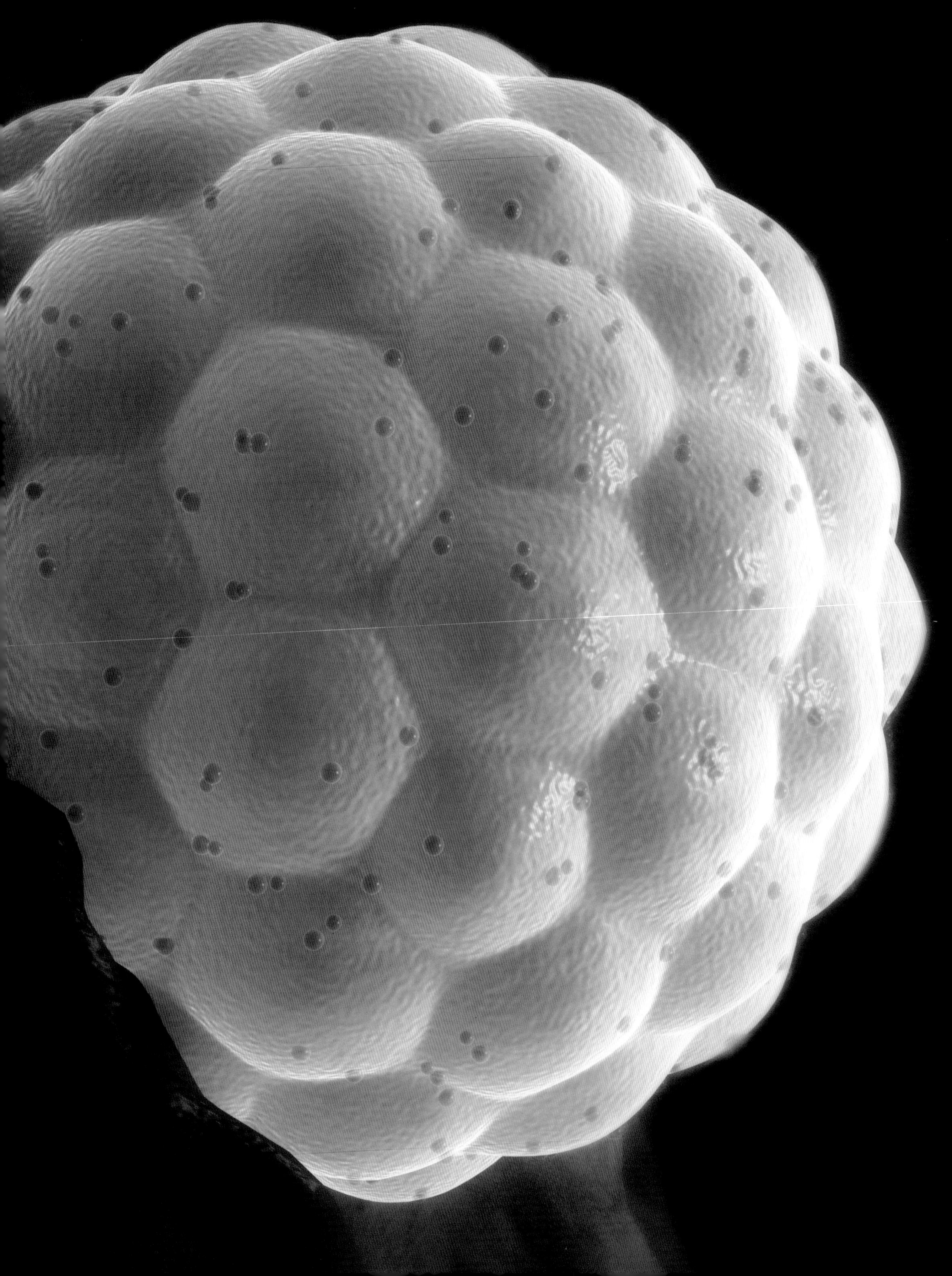

# From Zygote to Embryo

From conception until the third month of pregnancy, what takes place inside the mother's belly? Day by day, during the phases of mitosis, what happens at this embryonic stage, the most critical one during pregnancy? What changes does the embryo go through? Here we present incredible images that show the embryo from its formation to the moment it implants itself in the endometrium and measures about 0.2 inch (5 mm). At what time do the heartbeats begin and the eyes, mouth, and legs begin to form? Also, what is the role of the placenta, the organ that gives the unborn baby the different nutrients and oxygen it needs to continue developing? Turn the page.



**FERTILIZED EGG** (opposite)
After fertilization, the egg implants itself in the endometrium and begins to develop.

# The Origin of Sex and Life

The origin of human reproduction is sexual. Men and women can have sex any time during the year, unlike most other species, which have their specific times of heat. The ability to have sex begins at puberty, the age when the sexual organs develop. Women are fertile from their first menstrual period until menopause at around age 45. Although their sexual activity continues after this age, they no longer produce eggs, the female sexual gametes capable of being fertilized by sperm.

## The Male Sexual Apparatus

The testicles, or male sexual glands, lie below the pelvis within a structure called the scrotum. It is there that sperm—the mobile sex cells—are produced. During sexual intercourse, these cells, if they reach the female vaginal canal, head toward the egg so that one of them may fertilize it. The ductus deferens is the path through which the sperm travels to be joined by materials from the seminal vesicles and the prostate. This combination makes up the semen, which, in the moments of maximum sexual excitation, will move to the urethra to exit the man's body through the penis.

**1 FROM REST TO AN ERECTION**
A physical or mental stimulus causes the cavernous bodies to fill with blood and the penis to swell.

**2 EJACULATION**
If the penis continues being stimulated, the seminal vesicle contracts and expels the semen.

**EJACULATION**
Ejaculation (the exit of the semen from the male body) is produced by the intense excitement of the erect sexual organ.

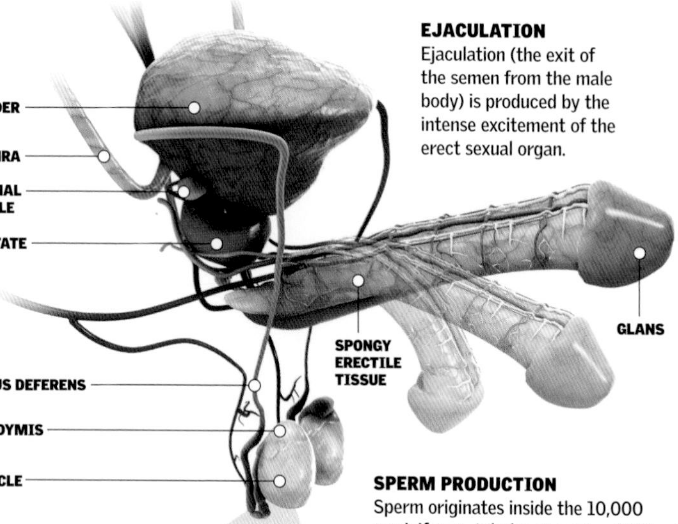


**SPERM PRODUCTION**
Sperm originates inside the 10,000 seminiferous tubules at a rate of 120 million a day and are stored in the epididymis. This process requires a temperature of 93° F (34° C), which the testicles achieve by being outside the abdomen.

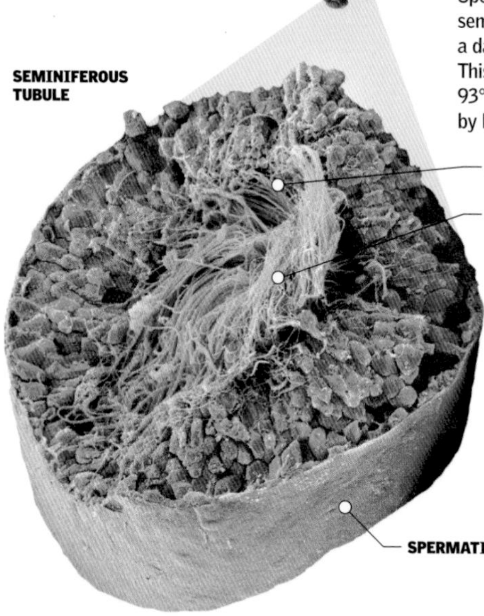


## The Sexual Organs of the Woman

With the exception of the vulva, which is external, the female sexual apparatus (which allows a woman to have an active sexual life, become pregnant, and give birth) lies completely inside the abdominal cavity, where it is supported and protected by the pelvis. Its basic shape is that of a cavity formed by the vagina and the uterus. The ovaries produce the eggs, or sex cells, and hormones. Periodically a mature egg leaves the ovary and installs itself in the uterus (ovulation). There, if it has not been fertilized in the fallopian tube, the body will expel it naturally together with the residues of the endometrium (menstruation).

**FALLOPIAN TUBE**
A tube 4 to 5 inches (10-12 cm) in length and about 0.1 inch (3 mm) in diameter, with internal cilia that propel the egg toward the uterus.

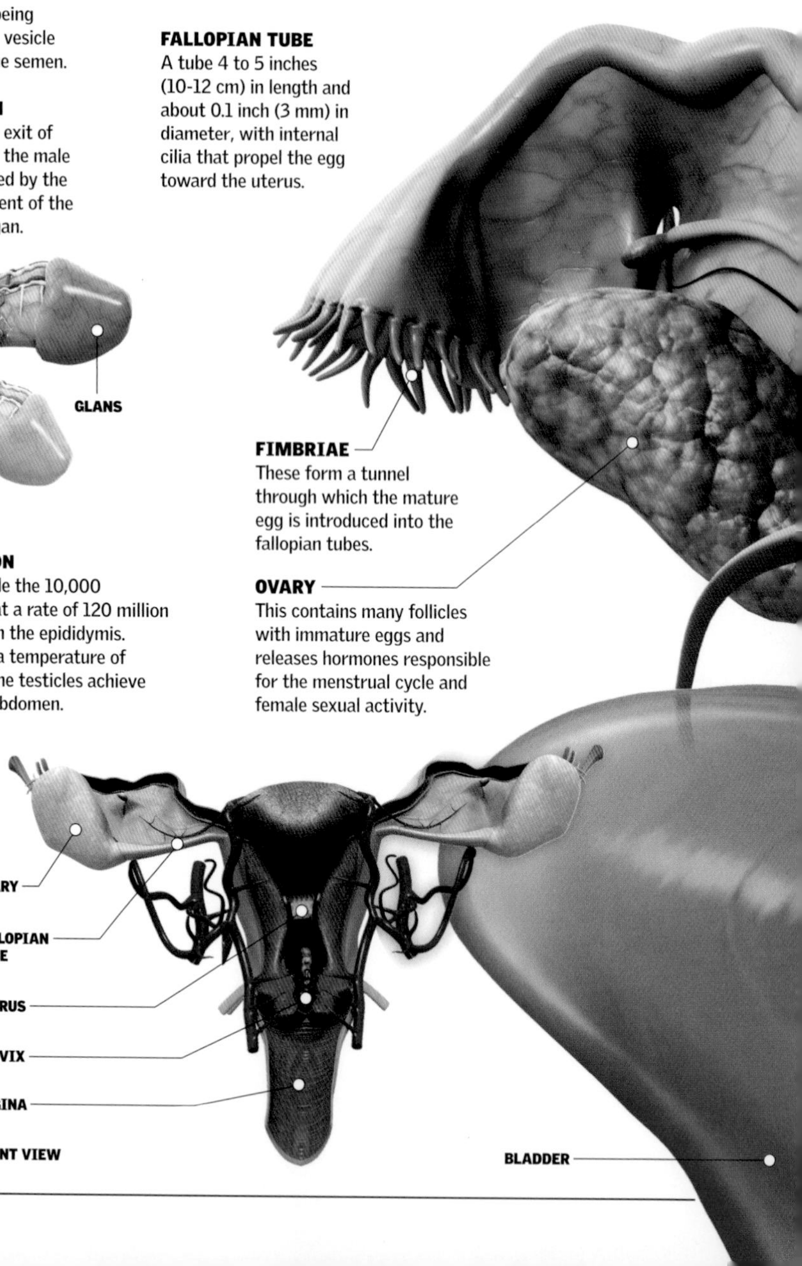


**FIMBRIAE**
These form a tunnel through which the mature egg is introduced into the fallopian tubes.

**OVARY**
This contains many follicles with immature eggs and releases hormones responsible for the menstrual cycle and female sexual activity.

## The Menstrual Cycle

The uterus is prepared for the implantation of the fertilized egg. For this, the woman's hormones have stimulated the uterus to thicken its internal wall (endometrium). If no egg is implanted, the thickened wall breaks down and the waste material is disposed of outside the body, together with the unfertilized egg. This process is synchronized with ovulation and is repeated regularly throughout the woman's fertile life, from puberty until menopause.

**MENSTRUATION**
The female body disposes of the dead cells from the endometrium.

**THICKENING**
The blood vessels of the uterus lengthen, and the wall grows.

**MAXIMUM HORMONE LEVELS**
Estrogen, luteinizing hormone (LH), and follicle-stimulating hormone (FSH).

**OVULATION**
This process occurs around the 14th day after menstruation.

**INCREASE IN PROGESTERONE**
The hormone that prepares the endometrium for implantation.

**ARRIVAL OF THE EGG**
If it is fertilized, it becomes implanted; if not, menstruation occurs.

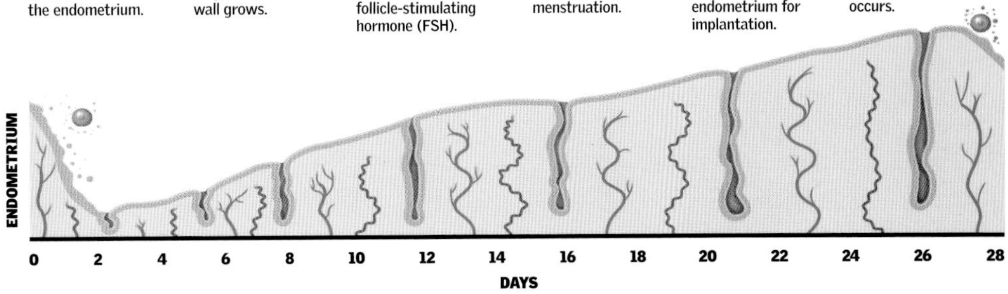


## Gametes and Hormones

Testicles and ovaries are glands that produce the sex cells, or gametes—sperm and eggs, respectively. Gametes are haploid cells. In other words, they possess half the chromosomes of any other human tissue cell, which contains a total of 46. Upon uniting at conception, each gamete contributes half of the genetic load of the new embryo. The sex glands also produce hormones that determine secondary characteristics and, in women, ovulation.

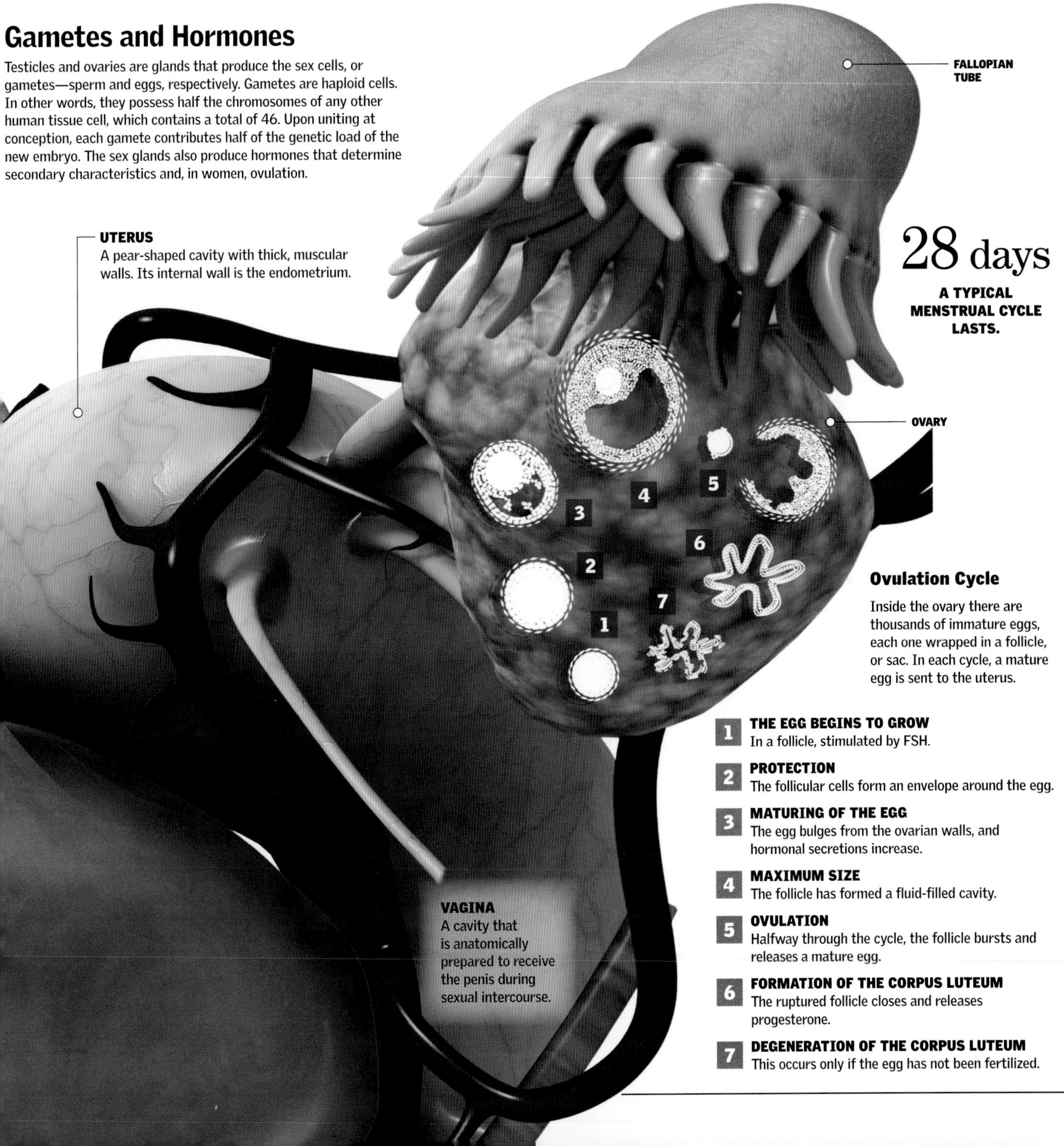

FALLOPIAN TUBE
UTERUS
A pear-shaped cavity with thick, muscular walls. Its internal wall is the endometrium.
OVARY
1
2
3
4
5
6
7
VAGINA
A cavity that is anatomically prepared to receive the penis during sexual intercourse.

**28 days**

**A TYPICAL MENSTRUAL CYCLE LASTS.**

### Ovulation Cycle

Inside the ovary there are thousands of immature eggs, each one wrapped in a follicle, or sac. In each cycle, a mature egg is sent to the uterus.

1. **THE EGG BEGINS TO GROW**
   In a follicle, stimulated by FSH.
2. **PROTECTION**
   The follicular cells form an envelope around the egg.
3. **MATURING OF THE EGG**
   The egg bulges from the ovarian walls, and hormonal secretions increase.
4. **MAXIMUM SIZE**
   The follicle has formed a fluid-filled cavity.
5. **OVULATION**
   Halfway through the cycle, the follicle bursts and releases a mature egg.
6. **FORMATION OF THE CORPUS LUTEUM**
   The ruptured follicle closes and releases progesterone.
7. **DEGENERATION OF THE CORPUS LUTEUM**
   This occurs only if the egg has not been fertilized.

# Fertilization of the Egg

Day 1

Fertilization is the starting point for the development of pregnancy. After intercourse, two sex cells, or gametes, fuse together, giving rise to an ovum, or zygote, in which the chromosomes of the two gametes are united. In humans, these sex cells are the sperm and the egg. For conception of a new life, a sperm must fertilize the egg in a tough competition with hundreds of millions of other sperm.

## The Journey of a Sperm

After ejaculation, millions of sperm begin their journey through the genital tract. Only 200 will reach the egg. The trip toward the fallopian tubes takes anywhere from 15 minutes to several hours. To reach them, sperm use their tails, and they are helped by contractions in the walls of the vagina and the uterus. Inside the egg, the sperm loses its tail and midsection, which dissolve. The head, which contains the genetic material, moves toward the plasma membrane of the egg. The march toward fertilization is underway.

### From Penetration to Fertilization

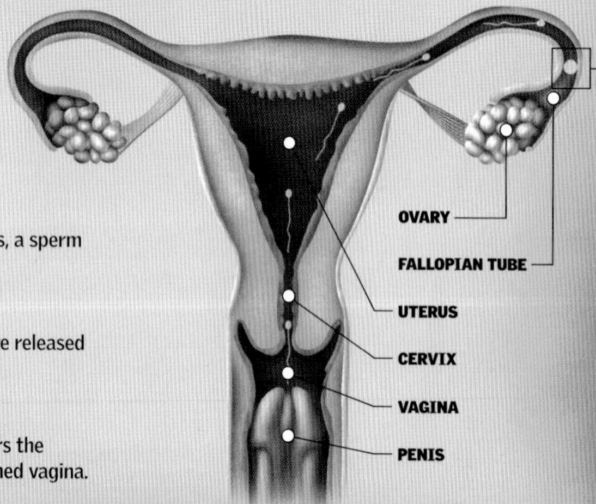


**3 FERTILIZATION**
In the fallopian tubes, a sperm fertilizes the egg.

**2 EJACULATION**
250 million sperm are released into the vagina.

**1 PENETRATION**
The erect penis enters the widened and moistened vagina.

**250 million**

**SPERM BEGIN THEIR JOURNEY THROUGH THE GENITAL TRACT AFTER EJACULATION. ONLY ONE WILL FERTILIZE THE EGG.**

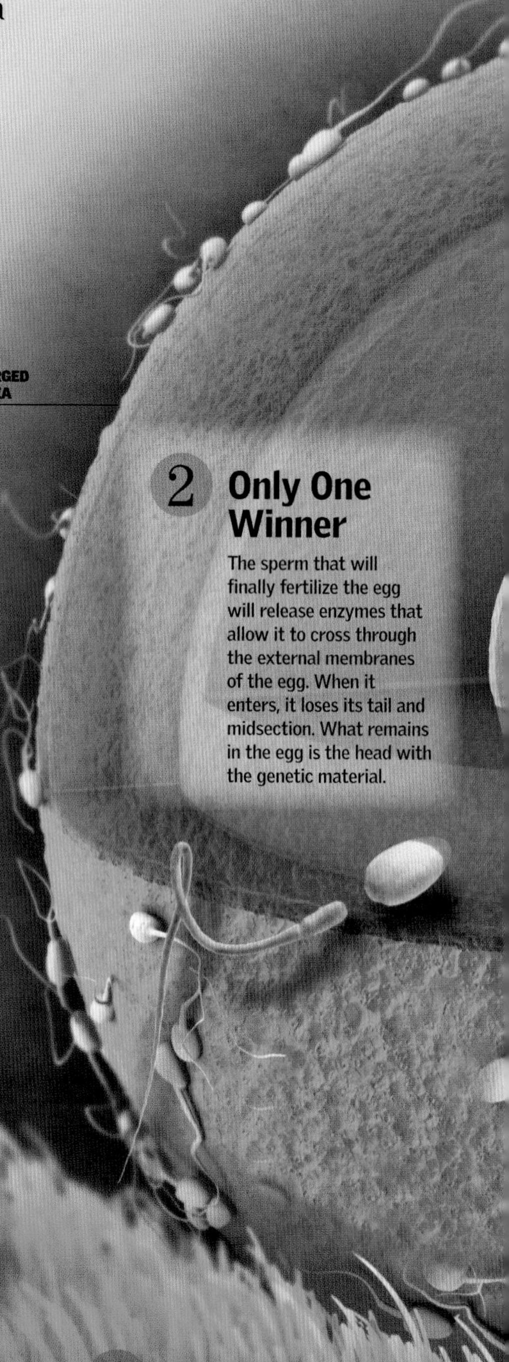

## 2 Only One Winner

The sperm that will finally fertilize the egg will release enzymes that allow it to cross through the external membranes of the egg. When it enters, it loses its tail and midsection. What remains in the egg is the head with the genetic material.

## The Sperm

The male sex cell. With a tail, a head, and a midsection, millions of sperm fight to fertilize the egg, a mission that only one of them will accomplish. It measures 0.002 inch (0.05 mm) in length.

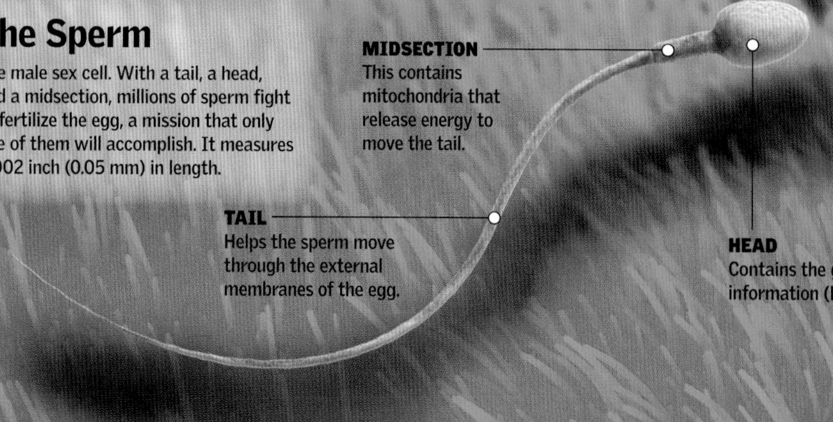

**MIDSECTION**
This contains mitochondria that release energy to move the tail.

**TAIL**
Helps the sperm move through the external membranes of the egg.

**HEAD**
Contains the genetic information (DNA).

## 1 In the Race

Hundreds of millions of sperm go in search of the egg immediately after ejaculation during reproduction.

## 3 Fertilization

A zygote, or ovum, is produced from the union of the egg and the sperm. The cell will begin its cell division through mitosis.

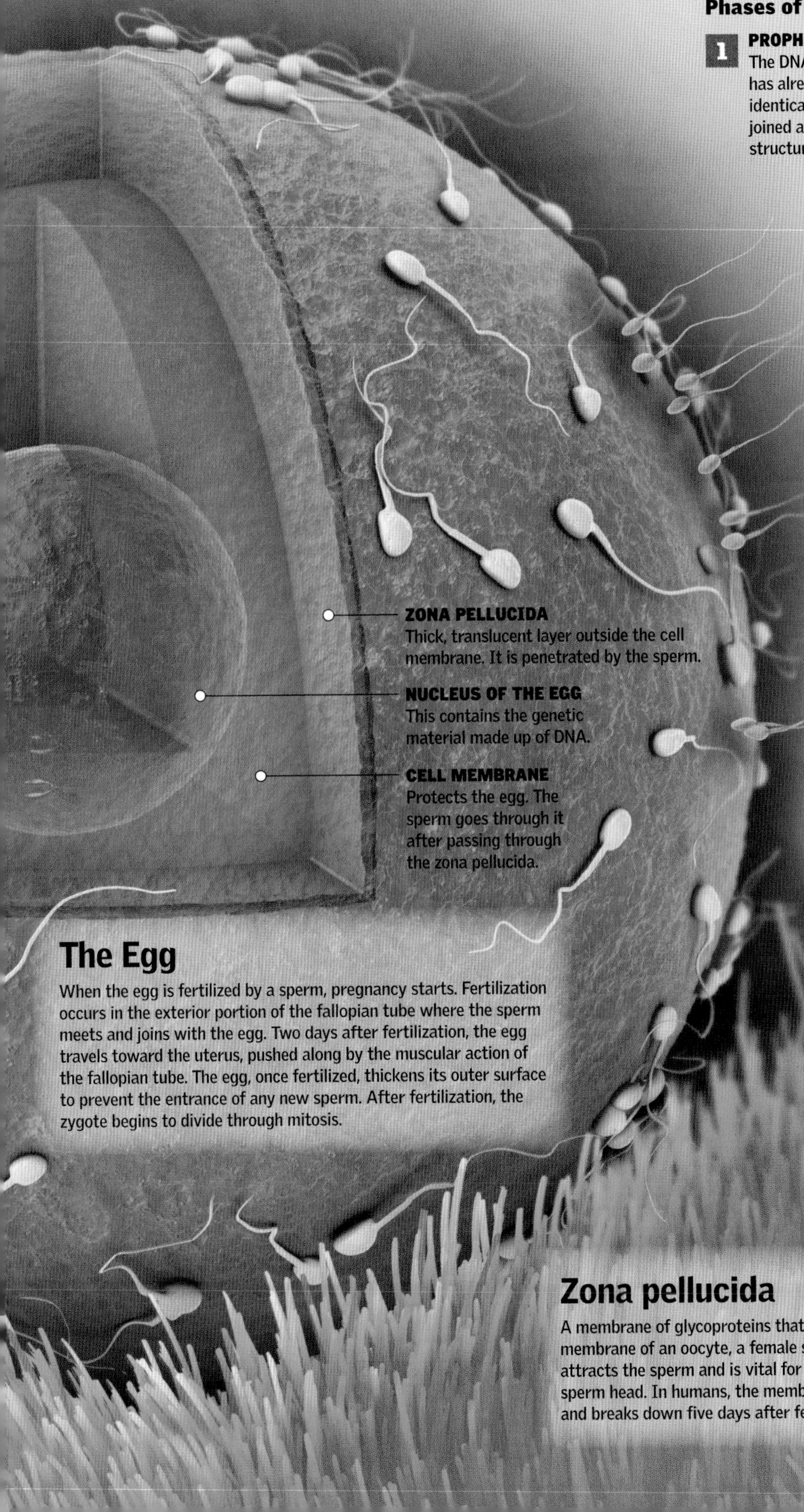


### The Egg

When the egg is fertilized by a sperm, pregnancy starts. Fertilization occurs in the exterior portion of the fallopian tube where the sperm meets and joins with the egg. Two days after fertilization, the egg travels toward the uterus, pushed along by the muscular action of the fallopian tube. The egg, once fertilized, thickens its outer surface to prevent the entrance of any new sperm. After fertilization, the zygote begins to divide through mitosis.

## 4 Mitosis

The process of generating new cells. Cell division begins with the replication of the DNA. In this way, a "mother" cell generates two identical "daughter" cells that contain the same genetic information as the mother cell. The process of replication for cell division occurs for each of the 46 chromosomes in each cell. The cells of the embryo divide through mitosis, just like most adult tissue cells.

### Phases of Mitosis

**1 PROPHASE**
The DNA of the chromosomes has already been copied. Two identical strands are formed, joined at the center by a structure called a centromere.

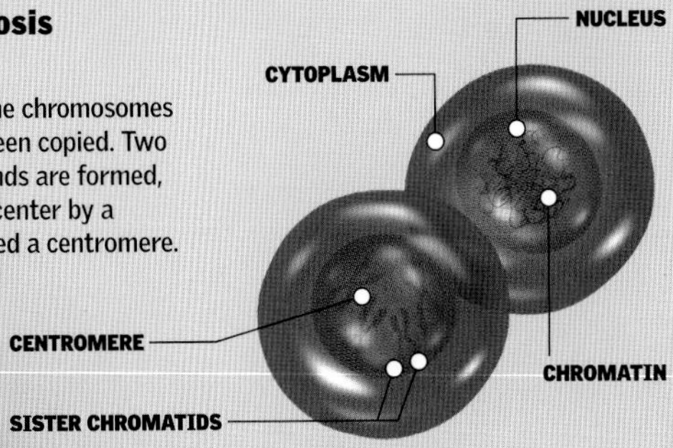


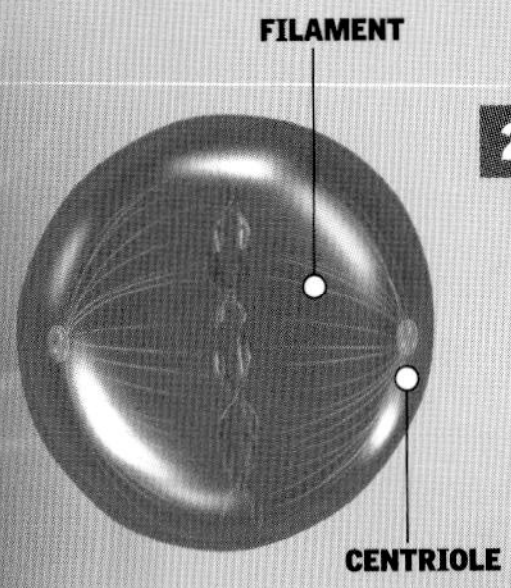


**2 METAPHASE**
The membrane that covers the cell nucleus disappears, and filaments form in the cell. The chromosomes align themselves along these filaments across the middle of the cell.

**3 ANAPHASE**
The filaments "tug" the duplicated chromosomes. The duplicated individual chromosomes move toward the two ends of the cell.

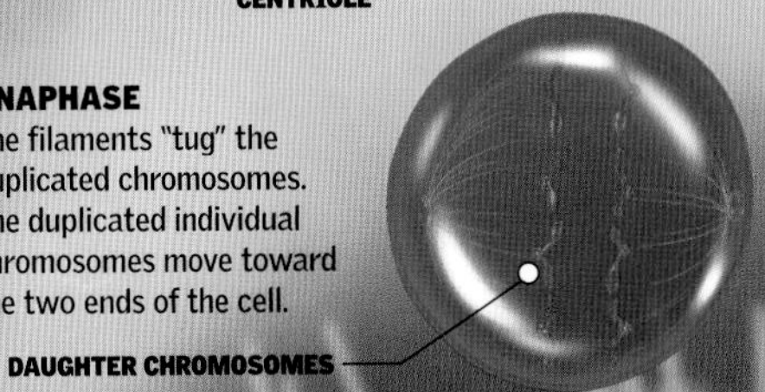

**4 TELOPHASE**
The filaments disappear, and a new nuclear envelope forms around each group of 46 chromosomes. The nucleus has divided in two.

**5 LATE TELOPHASE**
The new cells have been formed. The daughter cells contain the same genetic material as the original cell.

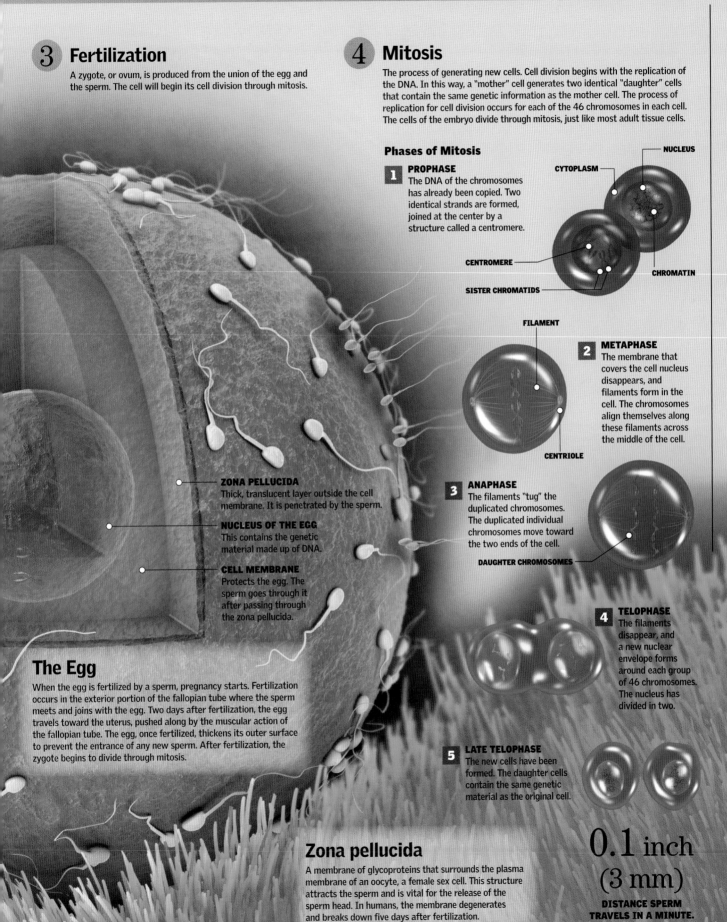

### Zona pellucida

A membrane of glycoproteins that surrounds the plasma membrane of an oocyte, a female sex cell. This structure attracts the sperm and is vital for the release of the sperm head. In humans, the membrane degenerates and breaks down five days after fertilization.

**0.1 inch (3 mm)**
**DISTANCE SPERM TRAVELS IN A MINUTE.**

# Day 2

## Fertilization

Fertilization occurs in the upper part of the fallopian tube. When the head of the sperm penetrates a mature egg, the nuclei of both sex cells, each one with 23 chromosomes, fuse to form the zygote, or ovum. With 46 chromosomes, the zygote will begin the process of successive cell divisions through mitosis. It will begin the journey from the fallopian tubes toward the endometrium, where it will implant itself.

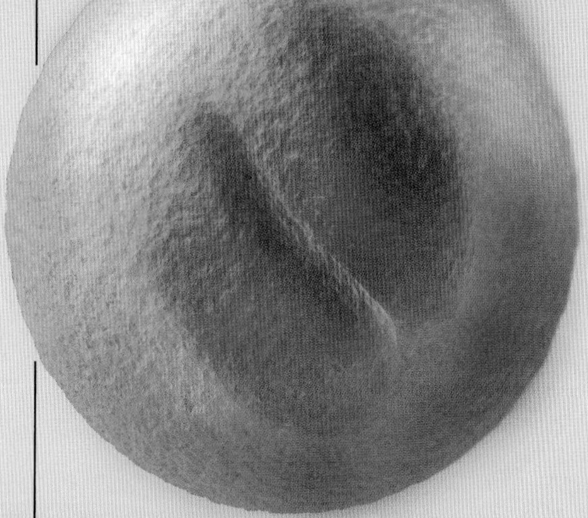

### Zygote

The resultant cell from the union of the male gamete (sperm) with the female gamete (egg) in sexual reproduction is called the zygote. Its cytoplasm and organelles are from the maternal egg. It contains all the necessary genetic material for fetal development.

**0.004 inch (0.1 mm)**

**DIAMETER OF THE ZYGOTE.**

### Morula

The second important stage of development prior to the formation of the blastocyst. It forms from the repeated mitosis of the zygote. Initially its interior contains 16 blastomeres, which are the first cells that develop from the zygote. Inside the morula, these cells are uniform in shape, size, and physiological potential.

# Day 4

## Formation of the Morula

The zygote goes through three stages of cell division. While it travels through the fallopian tube, it divides first into two and then into four identical cells. After 72 hours, it will have reached the stage of 16 cells, at which point a mulberry-shaped cell agglomerate called the morula is formed (the name comes from the Latin word morum, meaning "mulberry"). The morula continues its journey through the fallopian tube until it reaches the uterus. Cell division continues until a more solid ball with 64 cells, the blastocyst, is formed. Once the blastocyst attaches itself to the interior of the uterus, the formation of the embryo begins.

**12 hours**

**HOW LONG IT TAKES THE ZYGOTE, OR OVUM, TO DIVIDE THROUGH MITOSIS. COMPACT MASSES ARE SUCCESSIVELY FORMED IN THESE CELLULAR MULTIPLICATIONS.**

## The Zygote's Journey

Once the sex cells have formed the zygote, it begins the journey toward the uterus through the fallopian tube. During this journey, several cellular divisions will take place. Before entering the uterine cavity, a mulberry-shaped compact cellular mass is formed (the morula). Within the uterus, cellular divisions take place every 12 hours until the blastocyst stage (about 64 cells) has been reached. Once on the uterine lining, the blastocyst adheres to it, and shortly thereafter implantation takes place. From that moment, embryonic growth begins.

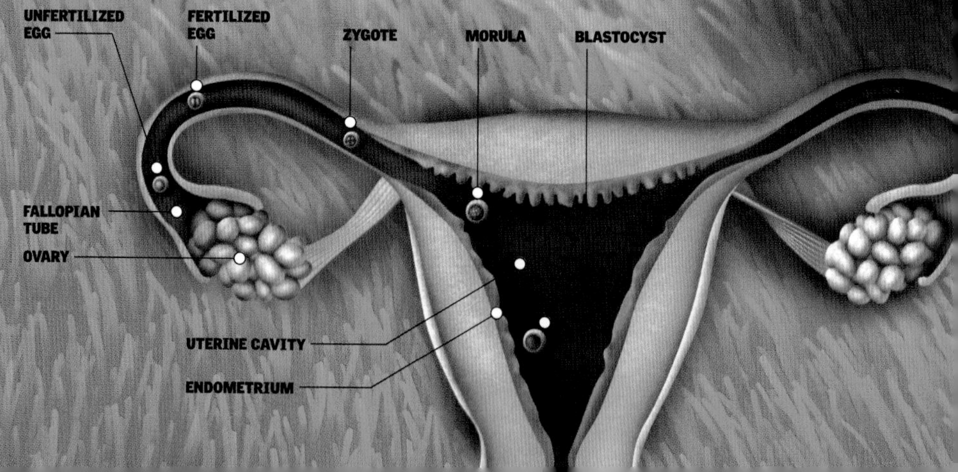

# 9 days

**AFTER FERTILIZATION THE BLASTOCYST, THE STAGE PRIOR TO THE EMBRYONIC STAGE, IMPLANTS ITSELF IN THE UTERINE WALL.**

## Implantation

After cellular division to 64 cells, the morula becomes a blastocyst, a more compact and solid mass. Once formed, the blastocyst moves freely in the uterine cavity for 48 hours before finding a place to implant itself in the endometrium. The endometrium relaxes to ease implantation of the blastocyst. Nine days after fertilization, the embryo will already be in the uterine wall. After implantation, the embryo begins to grow. If the woman has very low levels of estrogen and progesterone, the endometrium can rupture and cause implantation to occur in the wrong place.

Day 9

## X-ray of the Morula

The morula is made up of 16 cells in its initial state. As it divides, it will reach 64 cells, at which time it becomes a blastocyst.

## Blastocyst

The last step before growth of the embryo. The cellular mass is covered with an external layer called the trophoblast. The trophoblast releases enzymes that help the blastocyst adhere to the endometrium.

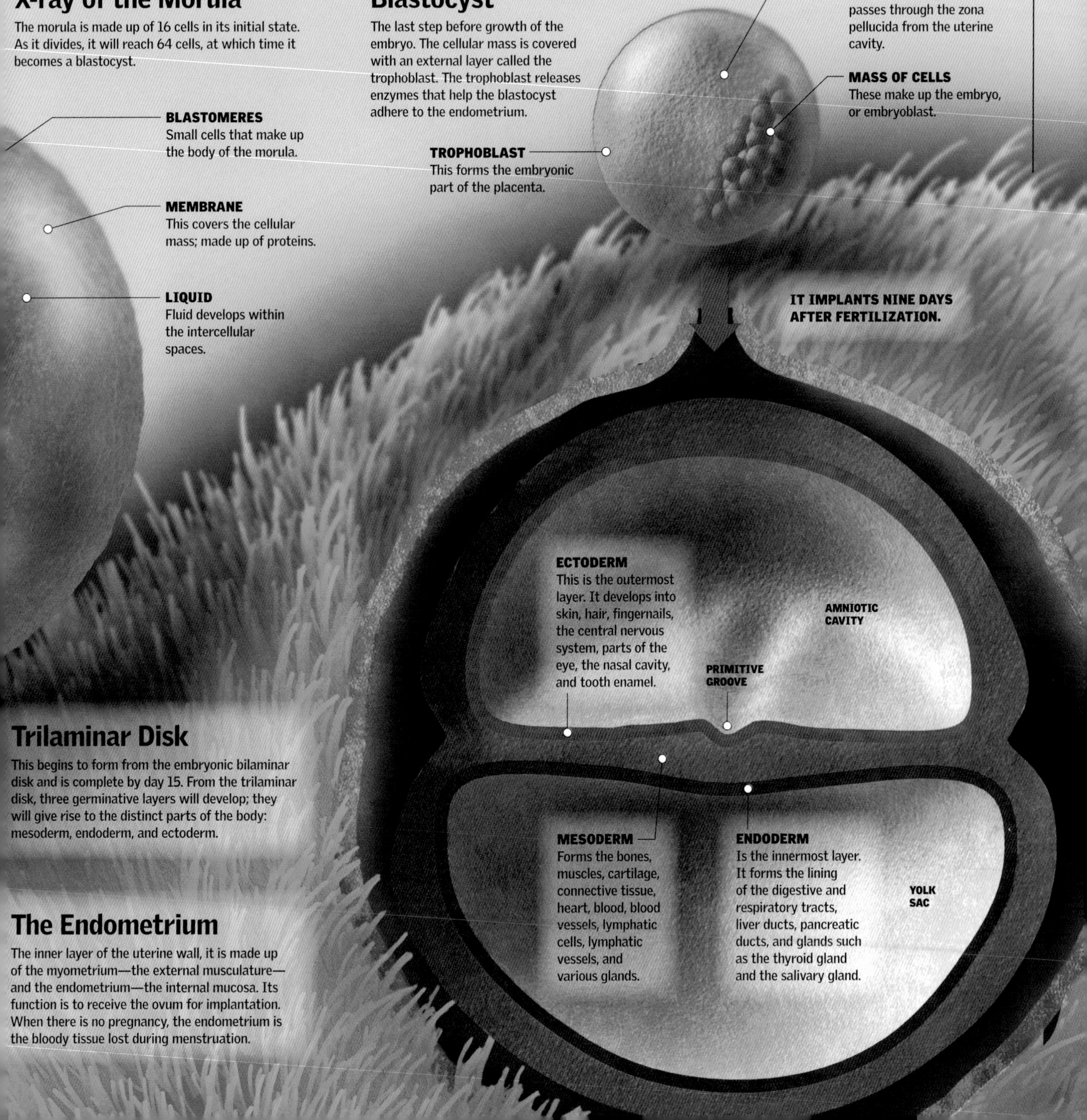
CAVITY, OR BLASTOCOEL
This contains liquid that passes through the zona pellucida from the uterine cavity.
MASS OF CELLS
These make up the embryo, or embryoblast.
BLASTOMERES
Small cells that make up the body of the morula.
TROPHOBLAST
This forms the embryonic part of the placenta.
MEMBRANE
This covers the cellular mass; made up of proteins.
LIQUID
Fluid develops within the intercellular spaces.
IT IMPLANTS NINE DAYS AFTER FERTILIZATION.
ECTODERM
This is the outermost layer. It develops into skin, hair, fingernails, the central nervous system, parts of the eye, the nasal cavity, and tooth enamel.
AMNIOTIC CAVITY
PRIMITIVE GROOVE
MESODERM
Forms the bones, muscles, cartilage, connective tissue, heart, blood, blood vessels, lymphatic cells, lymphatic vessels, and various glands.
ENDODERM
Is the innermost layer. It forms the lining of the digestive and respiratory tracts, liver ducts, pancreatic ducts, and glands such as the thyroid gland and the salivary gland.
YOLK SAC

## Trilaminar Disk

This begins to form from the embryonic bilaminar disk and is complete by day 15. From the trilaminar disk, three germinative layers will develop; they will give rise to the distinct parts of the body: mesoderm, endoderm, and ectoderm.

## The Endometrium

The inner layer of the uterine wall, it is made up of the myometrium—the external musculature—and the endometrium—the internal mucosa. Its function is to receive the ovum for implantation. When there is no pregnancy, the endometrium is the bloody tissue lost during menstruation.

# First Human Forms

Nine days after fertilization, the blastocyst has installed itself in the wall of the uterine endometrium, where it will spend the rest of the nine months of gestation before being born. The blastocyst measures slightly more than 0.004 inch (0.1 mm), and the uterine wall increases in size and attains a spongy consistency, the product of an intense supply of hormones by the ovaries. The uterine wall is where the stages of embryonic development will continue. The formation of the various kinds of tissues begins, and in the third week, the heartbeat starts.

## Protective Membrane

The rubbing of the blastocyst against the zona pellucida of the endometrium (normally in the back of the uterus, the part closest to the spine) leads to the release of enzymes that interact with the embryo. The blastocyst has little trouble penetrating the porous wall. At the same time, a new membrane forms: the chorion, which will protect the embryo.

**MOVEMENTS**
The cells that will form the embryo migrate in and out according to their function.

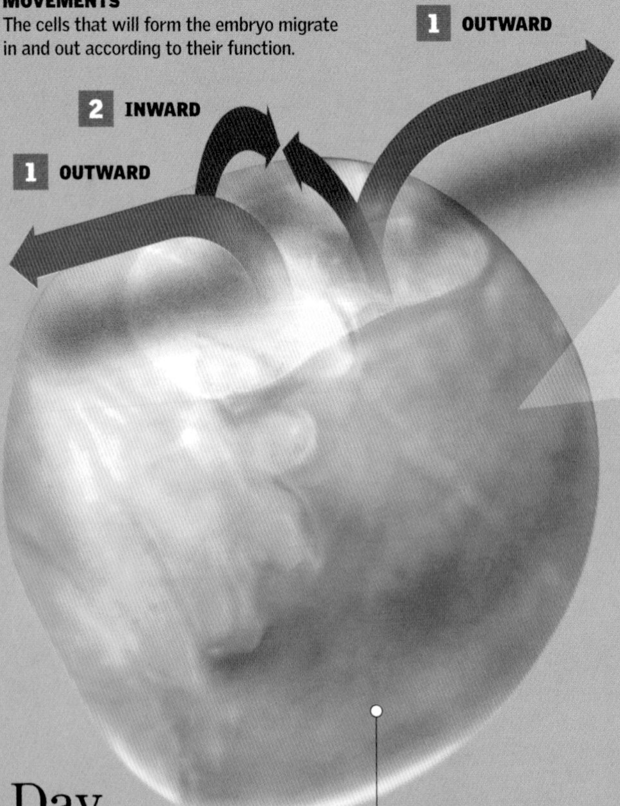


Day 10

**CHORION**
This is a live tissue membrane that surrounds and protects the embryo.

## Cellular Differentiation

Inside the embryo are cells that will form the skeleton as well as cells that will make up the viscera. Originally undifferentiated, they begin to move, seeking their place. Some cells will move outward (those that will form the skeleton) and others inward (those that will make up the viscera). The latest research has shown that some cells release certain chemicals that provoke other cells to do certain tasks. These substances are called morphogens.

1 First, the cells related to skeleton formation migrate toward the outside. They place themselves on the wall of the embryo.

2 Soon after, the cells related to visceral growth begin to migrate toward the inside. The embryonic disk undergoes a transformation.

## Morphogenesis

This includes the formation of the tissues and organs of the embryo. In this process, the cells are distributed along specific sites according to the tissues or organs they will form.

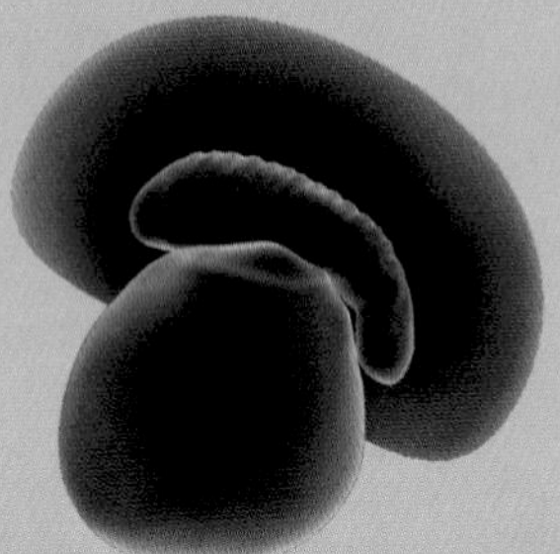

Day 13

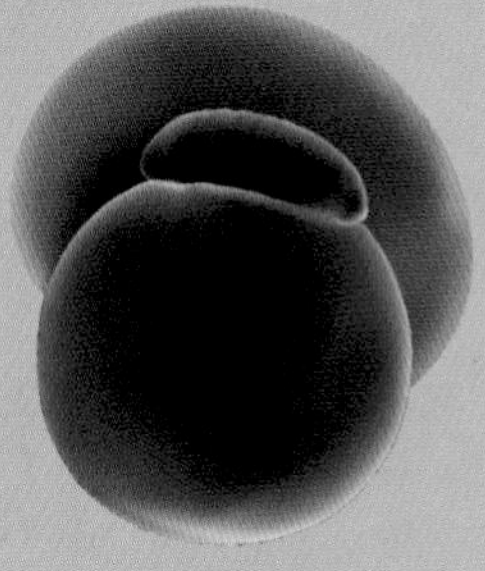

**CHANGES IN SHAPE**
When the cells that will form the viscera find their place, the embryo undergoes a transformation within a few hours. From the disklike appearance of day 13, a tube forms from filaments that are generated by these cells.

## The Placenta Forms

From the implanted blastocyst, new cellular formations begin to branch out over the chorion. These branches (called trophoblasts) are the source of the placenta, a disk-shaped interchange organ that grows between the chorion and the tissues of the endometrium. In the placenta, the blood vessels of the mother intertwine with those of the embryo without joining. The embryoblast, which contains the source of primitive blood for the development of the liver and the marrow, grows under this disk, which serves as a protective and immunological barrier.

1,000 cells

**MAKE UP THE HUMAN EMBRYO BY THE TIME THE PLACENTA IS FORMED (DAY 13) AND GASTRULATION BEGINS.**

# 87 gallons (330 l) a day

**THE AMOUNT OF BLOOD THAT MUST CIRCULATE THROUGH THE UMBILICAL CORD TO SUSTAIN EMBRYONIC GROWTH.**

# 8 inches (20 cm)

**IS THE AVERAGE DIAMETER OF THE PLACENTA RESPONSIBLE FOR BLOOD CIRCULATION BETWEEN THE MOTHER AND THE FETUS AT THE TIME OF BIRTH.**

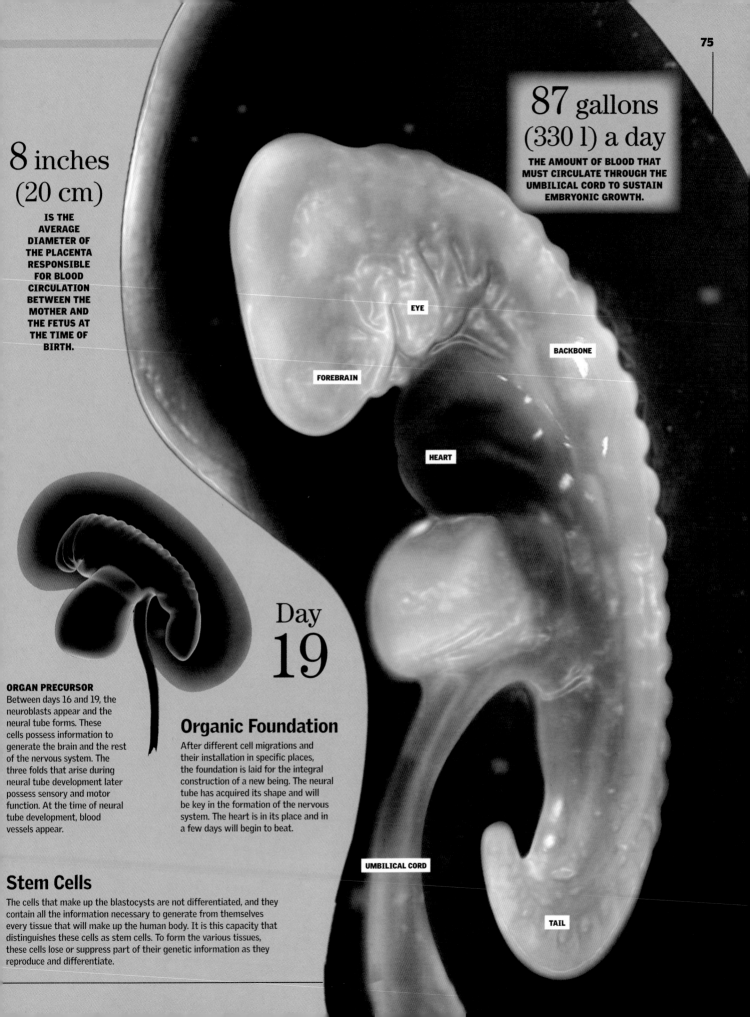

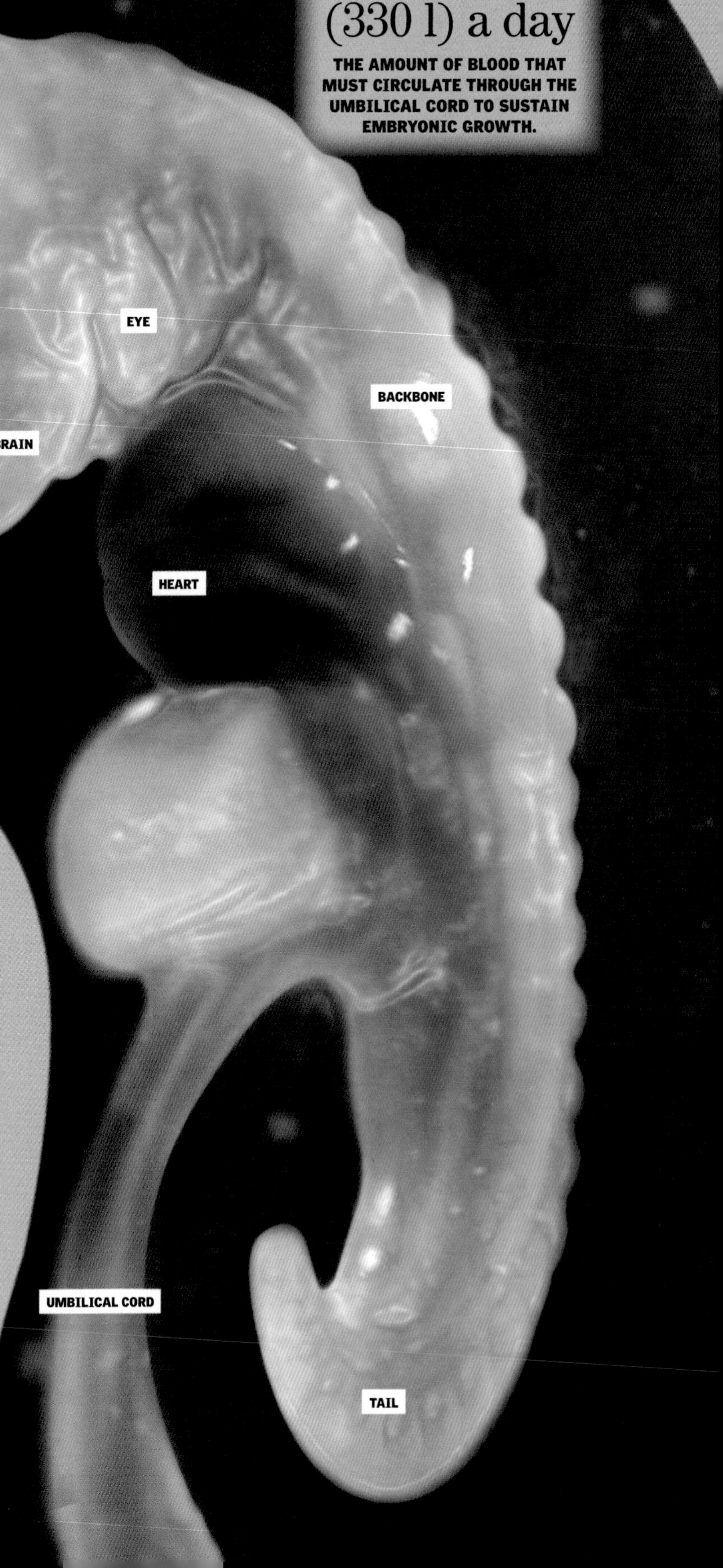


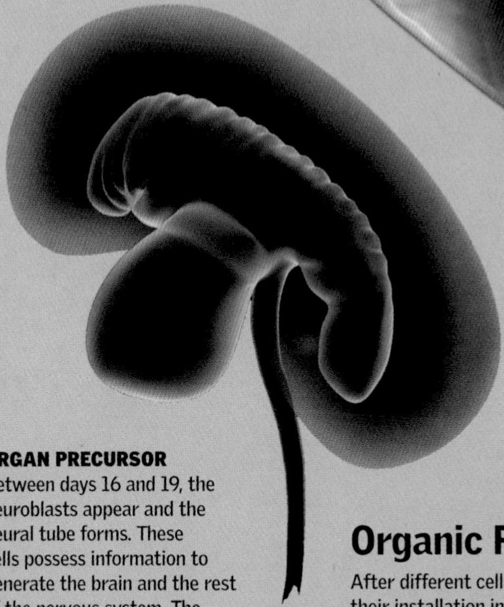

## Day 19

**ORGAN PRECURSOR**
Between days 16 and 19, the neuroblasts appear and the neural tube forms. These cells possess information to generate the brain and the rest of the nervous system. The three folds that arise during neural tube development later possess sensory and motor function. At the time of neural tube development, blood vessels appear.

## Organic Foundation

After different cell migrations and their installation in specific places, the foundation is laid for the integral construction of a new being. The neural tube has acquired its shape and will be key in the formation of the nervous system. The heart is in its place and in a few days will begin to beat.

## Stem Cells

The cells that make up the blastocysts are not differentiated, and they contain all the information necessary to generate from themselves every tissue that will make up the human body. It is this capacity that distinguishes these cells as stem cells. To form the various tissues, these cells lose or suppress part of their genetic information as they reproduce and differentiate.

## Embryonic Stage

It is still impossible to see a human shape at this moment of intrauterine development. The embryo is smaller than a grain of rice and has at one end a type of curved tail that will disappear as development progresses. In the interior and in the folds of the embryo there are groups of various cells, each one with different instructions according to the organs they must form. In this period, the cells of the cardiovascular system initiate the beating of the heart.

# Day 22

**LENGTH:**
**0.2 inch (4 mm)**

**WEIGHT:**
**0.001 ounce (0.03 g)**

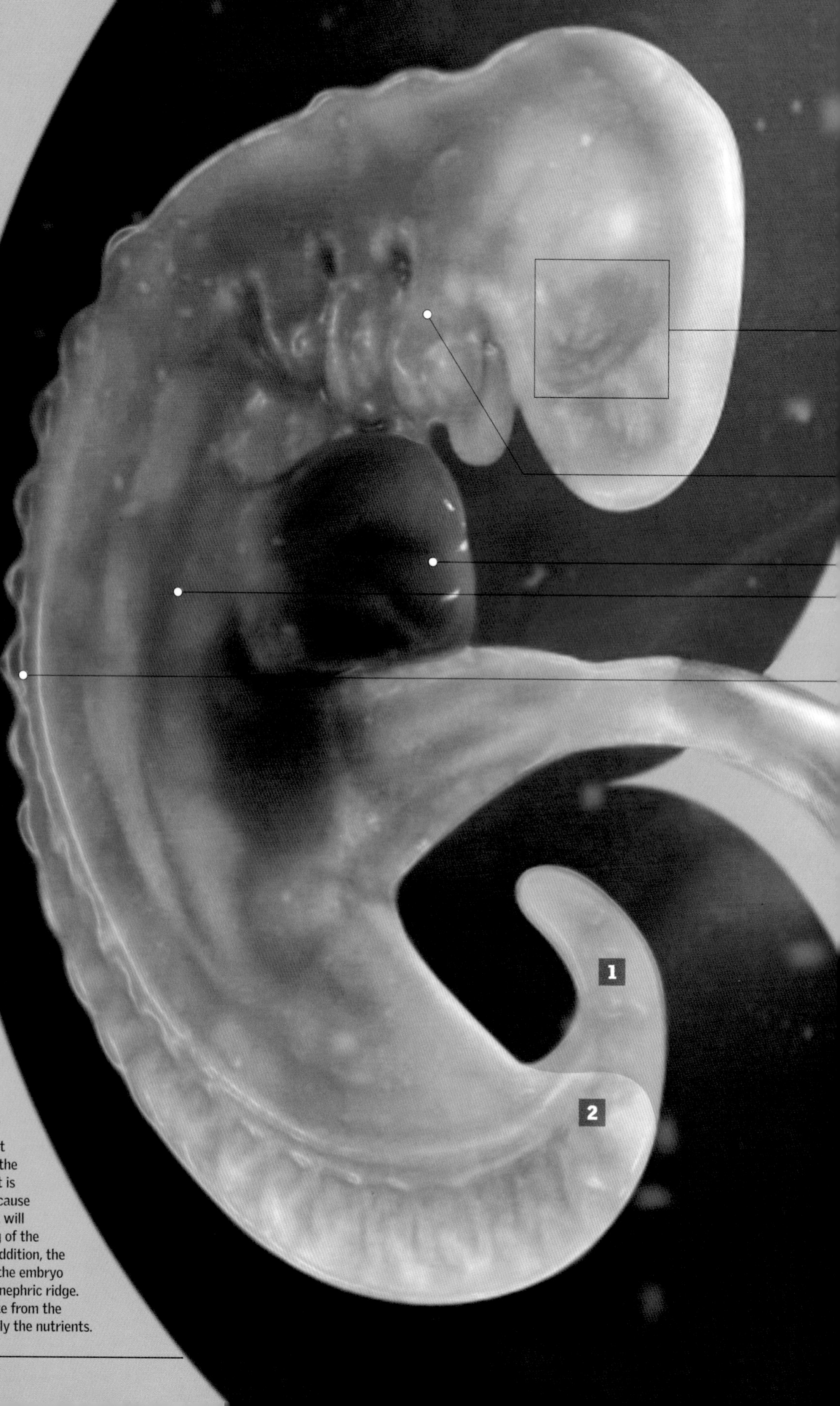


## C-shaped

In most vertebrates, the curved C-shape will disappear as the body slowly grows.

## Tail

1 **FOLD**
The embryonic tail acquires a curved shape before its disappearance.

2 **ABSORPTION**
The tail is absorbed when the embryo begins the road to fetal development.

## Liver and Kidneys

During the embryonic period, the first two months of gestation, the liver is the central organ for blood production. It is in charge of producing blood cells because the bone marrow, the substance that will have this function with the beginning of the fetal period, is not yet complete. In addition, the primitive kidneys begin to appear in the embryo from a protuberance called the mesonephric ridge. The kidneys filter the metabolic waste from the blood so that the embryo receives only the nutrients.

**127 million**

**THE AVERAGE NUMBER OF CELLS IN THE EYE WHEN IT ACQUIRES ITS DEFINITIVE SHAPE.**

**THE EYE WILL BE ABLE TO DISTINGUISH BETWEEN**

**10 million**

**COLORS AND SHADES OF LIGHT AND DARK WHEN IT REACHES MAXIMUM DEVELOPMENT.**

## Formation of the Eye

All vertebrates' eyes develop according to the same process. From certain changes in the ectodermal layer and invagination patterns on the surface of the embryo, the eye develops an "inverted" retina, in which the initial detection of light rays occurs in the outermost portion. In this way, the light-sensitive elements are situated on the outer regions, and the neural connection with the brain is in the inner region. The retina houses in its interior light-sensitive cells that have the function of receiving light and transmitting the correct information to the brain. The final development in the eye's functionality will occur at approximately the seventh month, when the baby will open its eyes for the first time and will react to changes of shade between light and dark.

### Development of the Eye

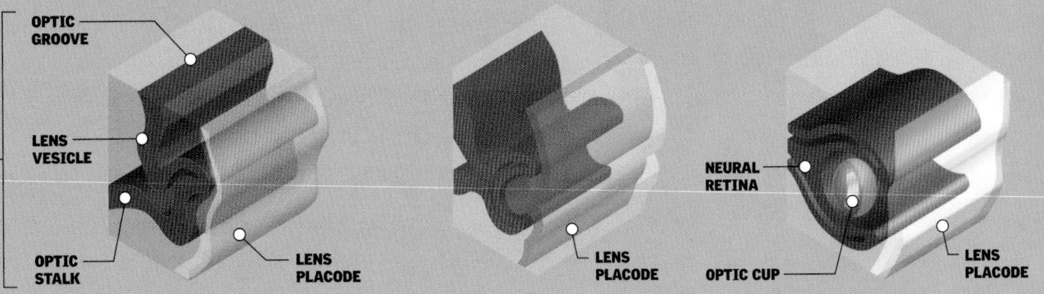


**1 DEVELOPMENT OF THE PLACODE**
By day 30, the lens placode, a region on the embryonic surface, comes in contact with the optic stalk.

**2 FORMATION OF THE VESICLE**
A day later, invagination of the lens placode takes place, and the lens vesicle forms.

**3 DEVELOPMENT OF THE RETINA**
On day 32, the neural retina and the pigment epithelium are formed. The lens vesicle detaches from the placode.

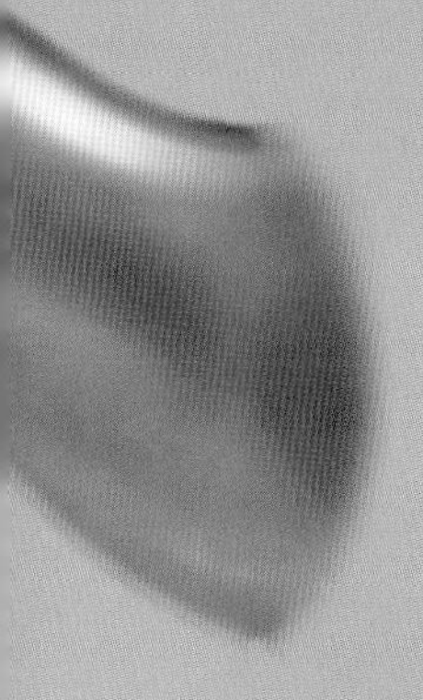

**ESOPHAGUS**
This separates from the breathing tube to allow the appropriate development of the digestive system.

**HEART**

**LUNGS**
These begin to develop. They are the last organs to acquire their shape and be completely functional.

**SPINE**
This has 40 pairs of muscles and 33 pairs of vertebrae. It is the hardest part of the embryo.

## The Heart Begins to Beat

By day 22, the heart is already active, just like the brain. Its division into subregions has begun, and it now makes up, together with the brain, half the size of the fetus. Initially the heart is simply a pump that maintains the flow of blood in the body and toward the placenta. When the four chambers are developed, the heart acquires the ability to gather the blood from the lungs and distribute it toward the organs throughout the body.

## Development of the Heart

After the differentiation of the cells that form the blood vessels, the cardiac muscle appears and begins to pump with the beating of its cells.

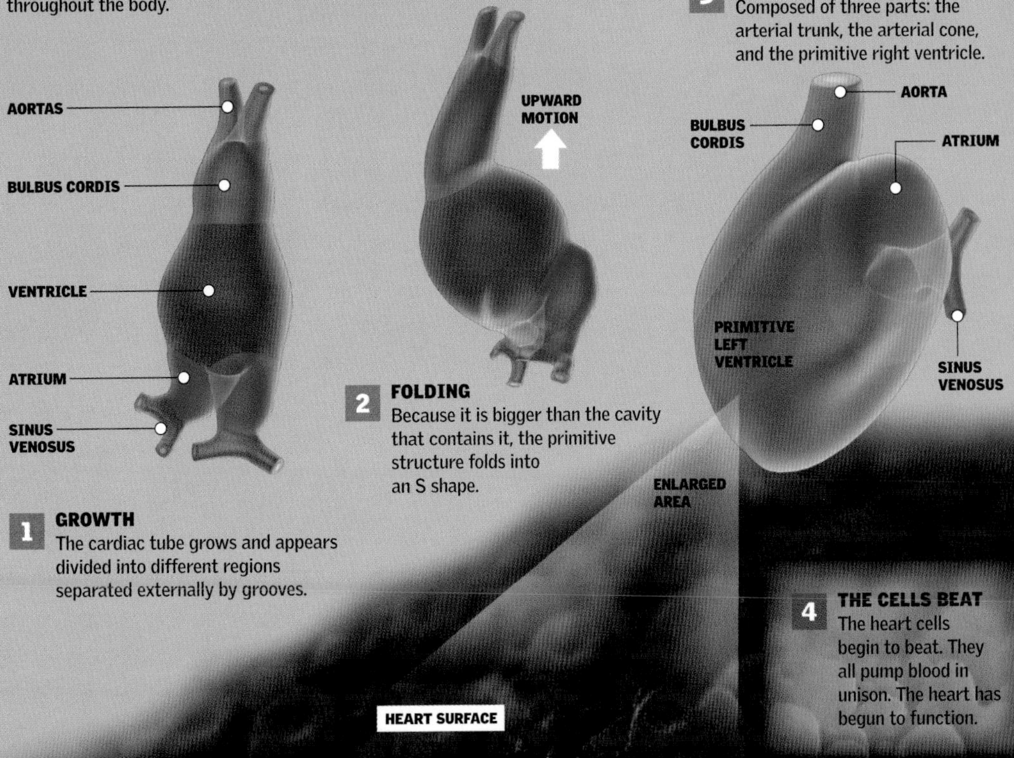


**1 GROWTH**
The cardiac tube grows and appears divided into different regions separated externally by grooves.

**2 FOLDING**
Because it is bigger than the cavity that contains it, the primitive structure folds into an S shape.

**3 BULBUS CORDIS**
Composed of three parts: the arterial trunk, the arterial cone, and the primitive right ventricle.

**4 THE CELLS BEAT**
The heart cells begin to beat. They all pump blood in unison. The heart has begun to function.

**50%**

**OF THE EMBRYO IS COMPOSED OF JUST TWO ORGANS: THE HEART AND THE BRAIN.**

**AMNIOTIC SAC**
This contains the liquid in which the fetus floats and is made up of two membranes that protect the embryo.

# Day 36

**LENGTH:** 0.4 inch (10 mm)

**WEIGHT:** 0.002 ounce (0.05 g)

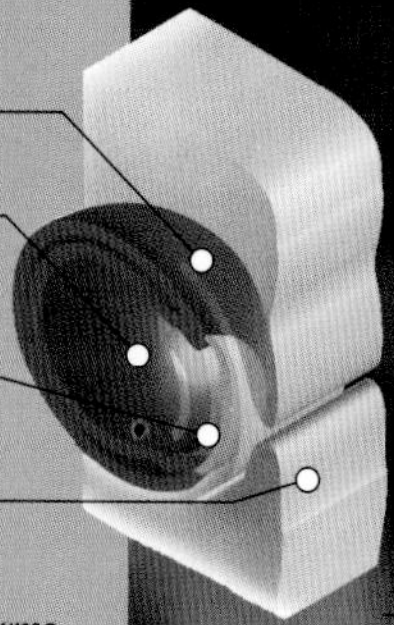


**4 EYE**
The eyelid forms. A membrane prepares the final shape of the pupil.

**CORD**
The umbilical cord begins to develop and, by the time of birth, can measure up to two feet (60 cm) in length.

## Arms and Legs

Small buds begin to appear that will grow until they form the arms and legs. Up to this moment, the arms are in their right place and will remain in proportion to this stage of development. They are only missing the development of the hands. The legs begin to develop, but they take longer than the arms and hands.

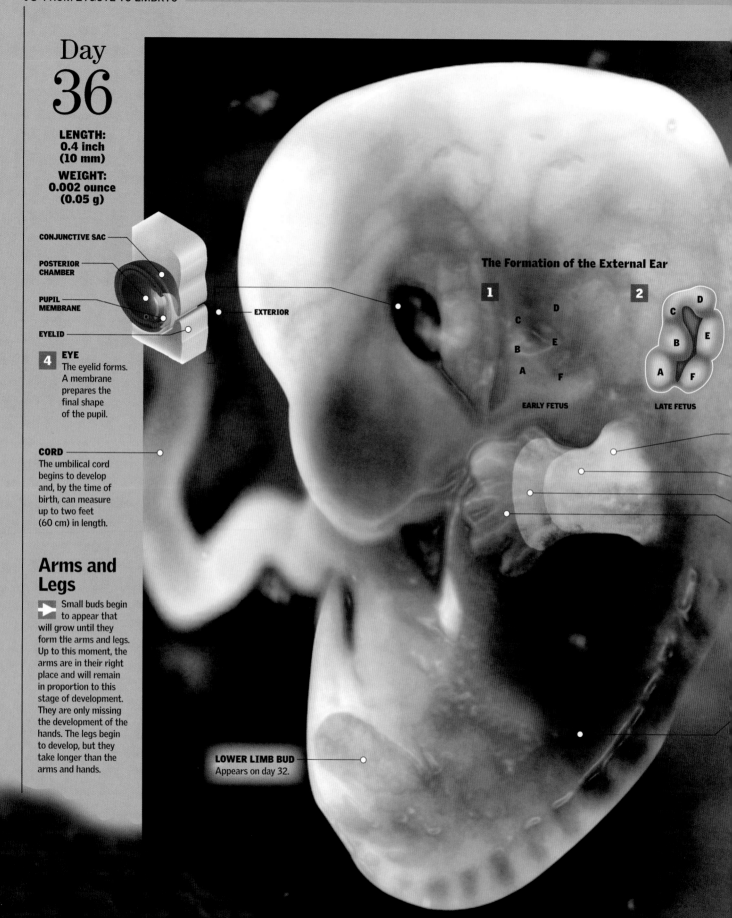

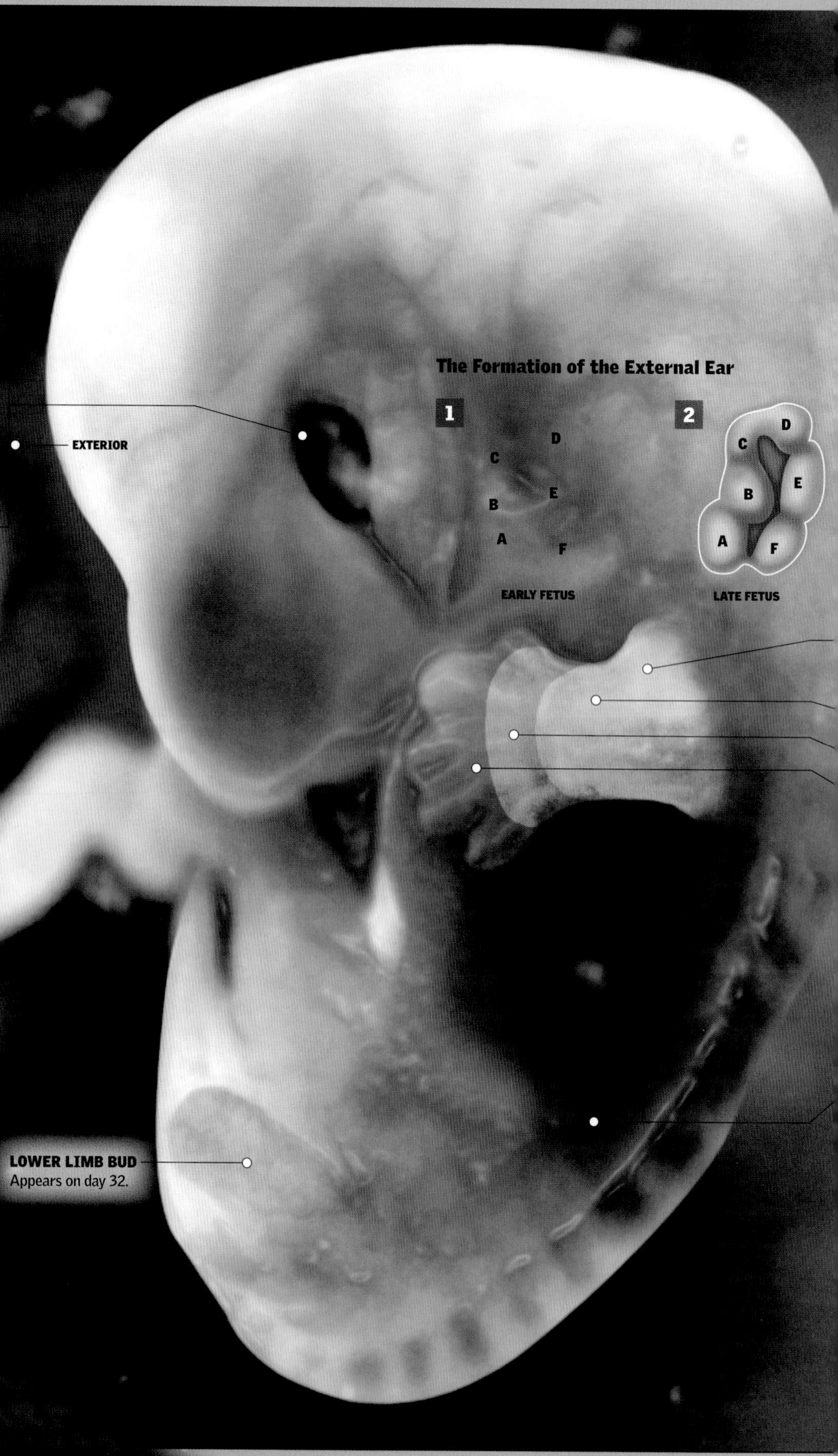

## Changes in the Head

The brain, the organ of the central nervous system that coordinates all muscle movement, begins to develop. Inside it, the pituitary gland (hypophysis) begins to form. It will produce growth hormone and other hormones. The jaw and the facial muscles also begin to develop.

# Day 40

**LENGTH:**
**0.4 inch (10 mm)**

**WEIGHT:**
**0.004 ounce (0.1 g)**

### The Folds of the Brain

These develop gradually as the months of intrauterine life go by.

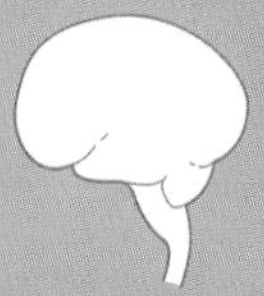

**1 SMOOTH BRAIN**
Initially the embryonic brain has a smooth surface.

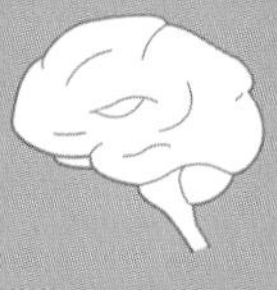

**2 A FEW FOLDS**
By six months, some basic folds can be seen.

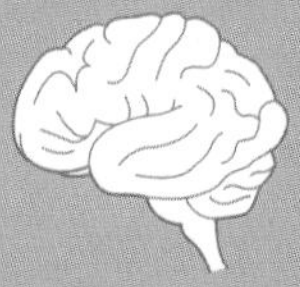

**3 ADULT**
The complete folds allow optimal functioning.

**3**

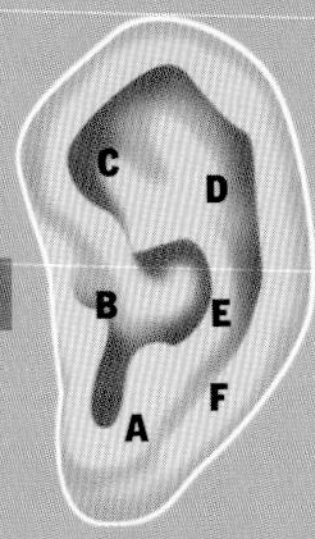


**BIRTH**

## The External Ear

Three auricular hillocks can be found in the first arch and three more in the second. As the jaw and teeth develop, the ears move up from the neck toward the sides of the head. Two ectodermic derivations appear in the cephalic region of the embryo: the otic placode and the lens placode. At birth, the external ear exhibits its typical shape.

## Eyes

The optical vesicles develop on both sides of the head, move toward the center, and form the eyes, as will the ducts that will make up the inner ear.

**THE ARM**
This is developed but is still missing the complete formation of the fingers.

**UPPER LIMB BUD**
Appears at day 26.

**HAND PLATE**
Appears at day 33.

**DIGITAL RAYS**
Appear at day 40.

### Growth of the Fetus

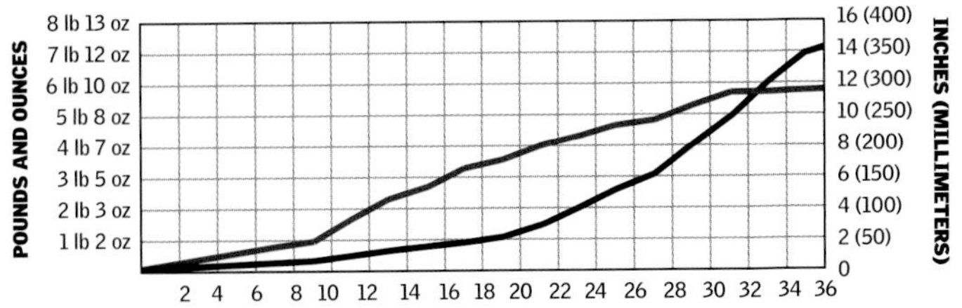


— Weight — Height

1 pound = 450 g; 1 ounce = 30 g

**TISSUE**
Connective tissue forms. It will engender the cells that form the cartilage, bones, and support tissues.

## Formation of the Face

The facial characteristics are quickly delineated. The pharyngeal arches that surround the stomodeum in the center of the face are configured. The mandibular processes and the frontonasal prominence can already be identified. From the pharyngeal arch, the maxillary process will also develop, which will give rise to the premaxillary, the maxillary, the zygomatic bone, and part of the temporal bone. At the roof of the embryo's mouth, the primitive palate is constructed. Through an invagination in the frontonasal prominence, the nose is shaped. The same thing happens with the chin, which acquires normal proportions by day 40 of intrauterine life.

### Features Become Distinct

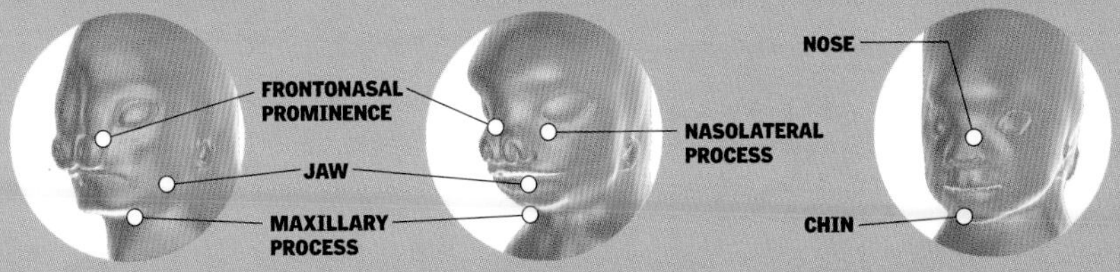


**1 LOWER JAW**
Begins to develop by day 37, together with the lips.

**2 NOSE**
This is roughed out by the invagination of the frontonasal prominence on day 39.

**3 CHIN**
This is already proportional by day 40. The nose has acquired its definitive shape.

**4** The shape of the face has begun to develop and will continue to do so until the third month.

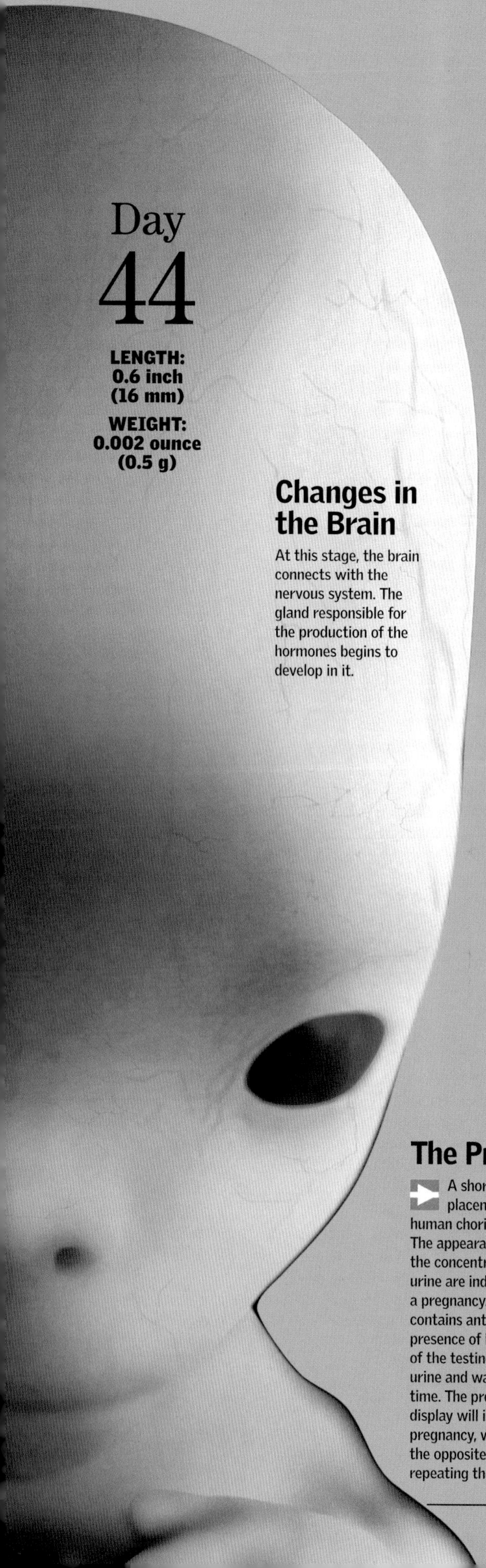

# Day 44

**LENGTH:**
**0.6 inch (16 mm)**

**WEIGHT:**
**0.002 ounce (0.5 g)**

## Changes in the Brain

At this stage, the brain connects with the nervous system. The gland responsible for the production of the hormones begins to develop in it.

## Development of the Placenta

The placenta is a special organ that provides the fetus with different nutrients and oxygen. It also absorbs the waste that the fetus produces and acts as a protective barrier against any harmful substances. The placenta forms from the trophoblast, the external layer of the blastocyst (mass of cells implanted in the uterus after fertilization). It begins to develop after implantation, and by the tenth day, it is complete. The placental hormones help preserve the endometrium.

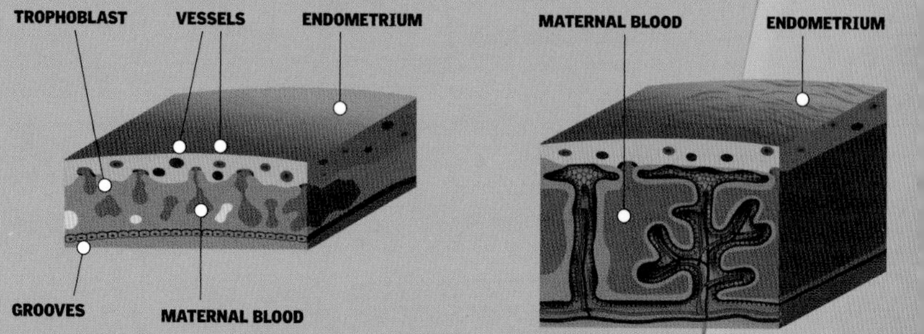


**1 FORMATION OF THE PLACENTA**
The cells of the trophoblast extend inside the blood vessels of the uterus. The blood from the mother flows from these vessels toward empty spaces inside the trophoblast.

**2 THE PLACENTA AS FILTER**
The mother's blood and that of the fetus do not have direct contact inside the placenta. They are separated by a barrier of cells. Oxygen, nutrients, and antibodies travel through the barrier and reach the fetus. The waste is returned to the placenta.

### Internal Ear and Middle Ear

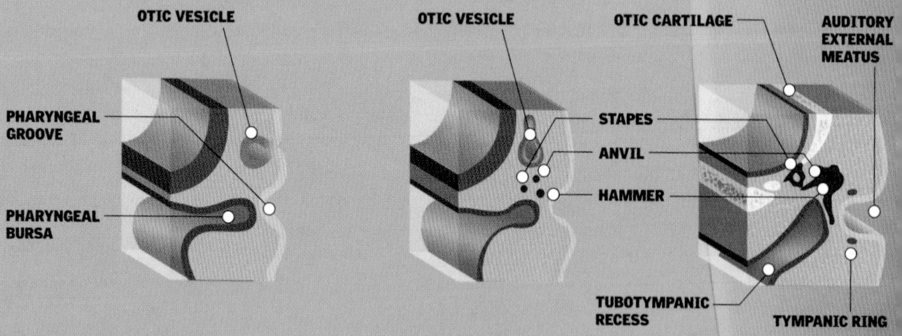


**1 22 DAYS**
A visible groove appears in the place where the ear will be.

**2 28 DAYS**
The structures that will give rise to the bones of the middle ear appear.

**3 32 DAYS**
The middle ear is formed (stapes, anvil, and hammer).

## The Pregnancy Test

A short time after fertilization, the placenta releases a hormone called human chorionic gonadotropin (HCG). The appearance and rapid increase in the concentration of this hormone in the urine are indicators of the existence of a pregnancy. A common pregnancy test contains antibodies that react to the presence of HCG. The user places the tip of the testing device in contact with her urine and waits an indicated amount of time. The presence of two lines in the display will indicate the existence of a pregnancy, while a single line will indicate the opposite. If the test is negative, repeating the test is recommended.

### How it Works

**ABSORPTION OF URINE**
An absorbent tip is placed under the urine stream for six seconds, until it is moist.

**RESULT**
Two lines indicate the presence of pregnancy. If there is only one, it is recommended the test be repeated within 48 to 72 hours.

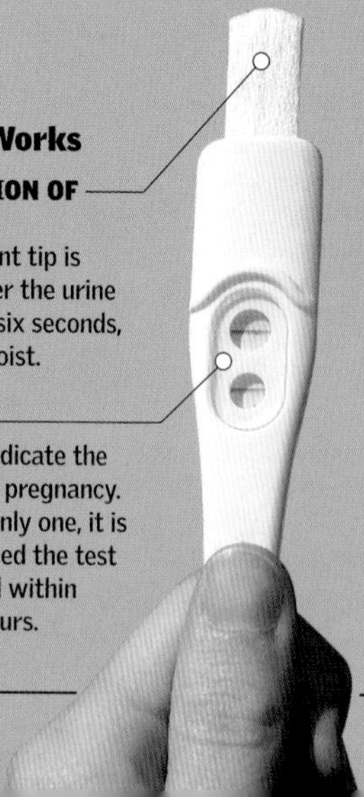

**99%**
**EFFECTIVE IN DETECTING PREGNANCY.**

# 6 weeks

**THE FINGERS APPEAR**
**ALTHOUGH THE HAND HAS A SHAPE RESEMBLING A SMALL PADDLE, THE DIGITS ARE BECOMING DISTINCT.**

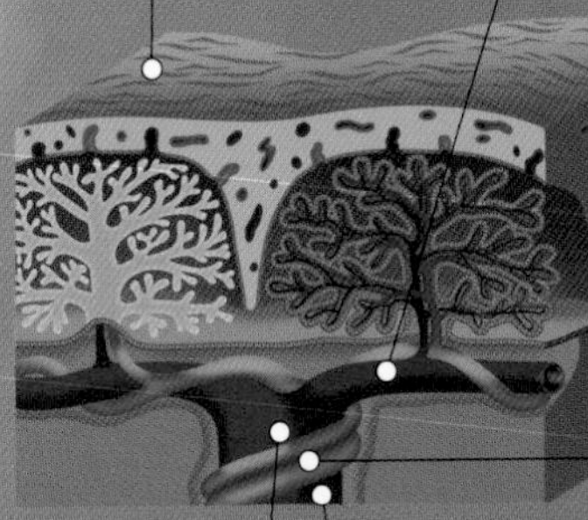


**3 END OF THE PLACENTA**
The placenta continues its development as the fetus grows, such that, by the end of the pregnancy, it measures about eight inches (20 cm). It is connected to the baby by the umbilical cord.

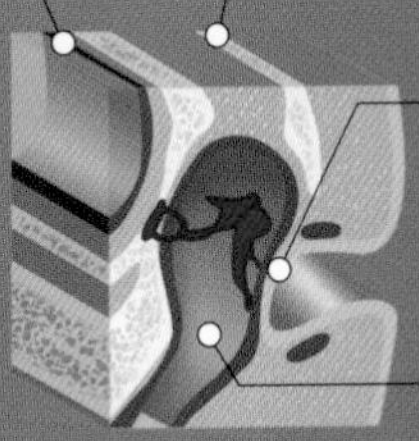


**4 60 DAYS**
The auditory external meatus develops from the pharyngeal groove.

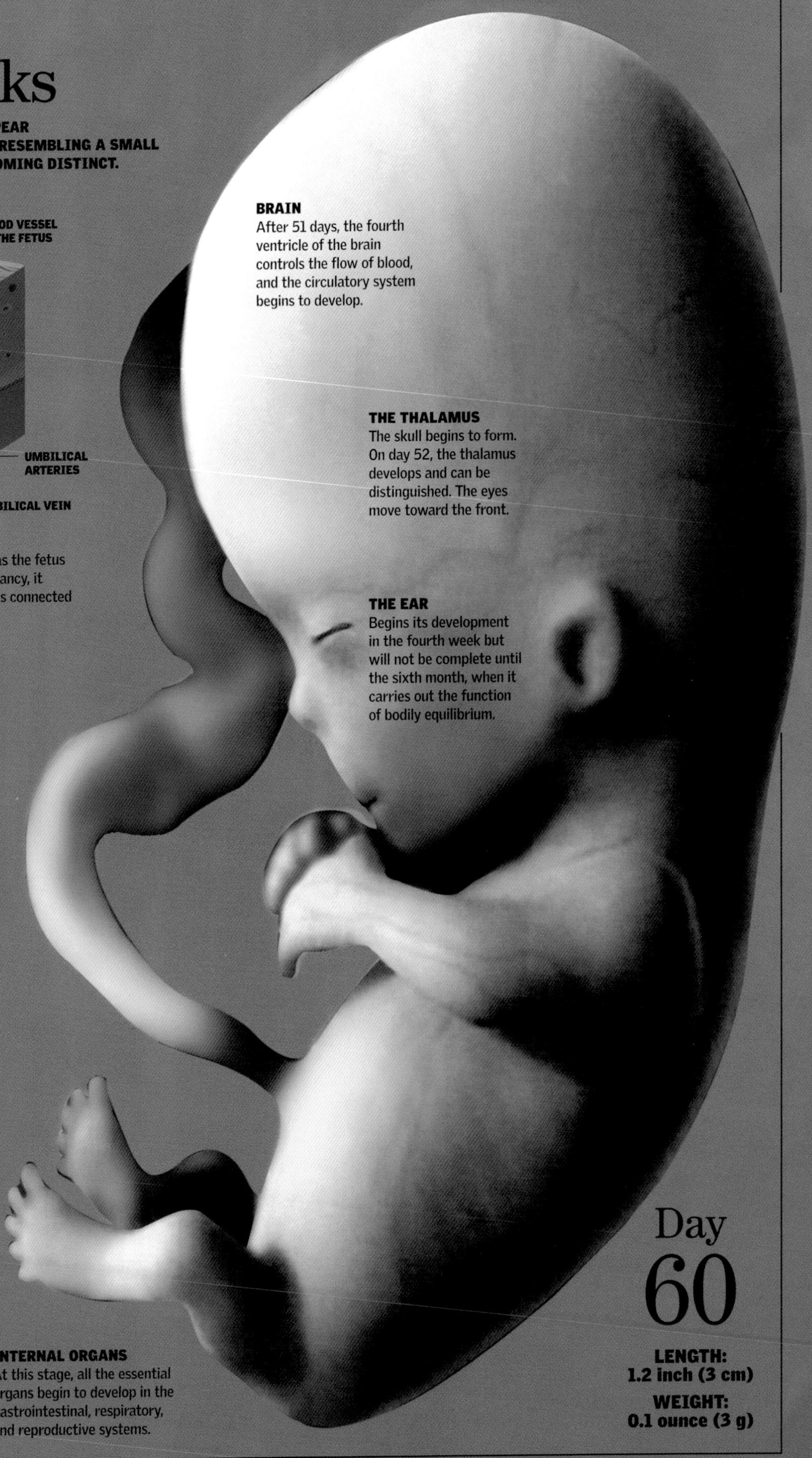

**BRAIN**
After 51 days, the fourth ventricle of the brain controls the flow of blood, and the circulatory system begins to develop.

**THE THALAMUS**
The skull begins to form. On day 52, the thalamus develops and can be distinguished. The eyes move toward the front.

**THE EAR**
Begins its development in the fourth week but will not be complete until the sixth month, when it carries out the function of bodily equilibrium.

## Everything in Place and Working

During this period, the brain and the nervous system develop rapidly. On both sides of the head, the optic vesicles that will make up the eyes have formed, as have the ducts that will make up the inner ear. The heart already beats strongly, and the digestive and respiratory apparatus have begun to take shape. Small buds that will grow to form the arms and legs appear. The fetus, measured from the top of the head to the coccyx, by the sixth week will have reached 0.6 inch (16 mm).

**INTERNAL ORGANS**
At this stage, all the essential organs begin to develop in the gastrointestinal, respiratory, and reproductive systems.

**Day 60**

**LENGTH:**
**1.2 inch (3 cm)**
**WEIGHT:**
**0.1 ounce (3 g)**

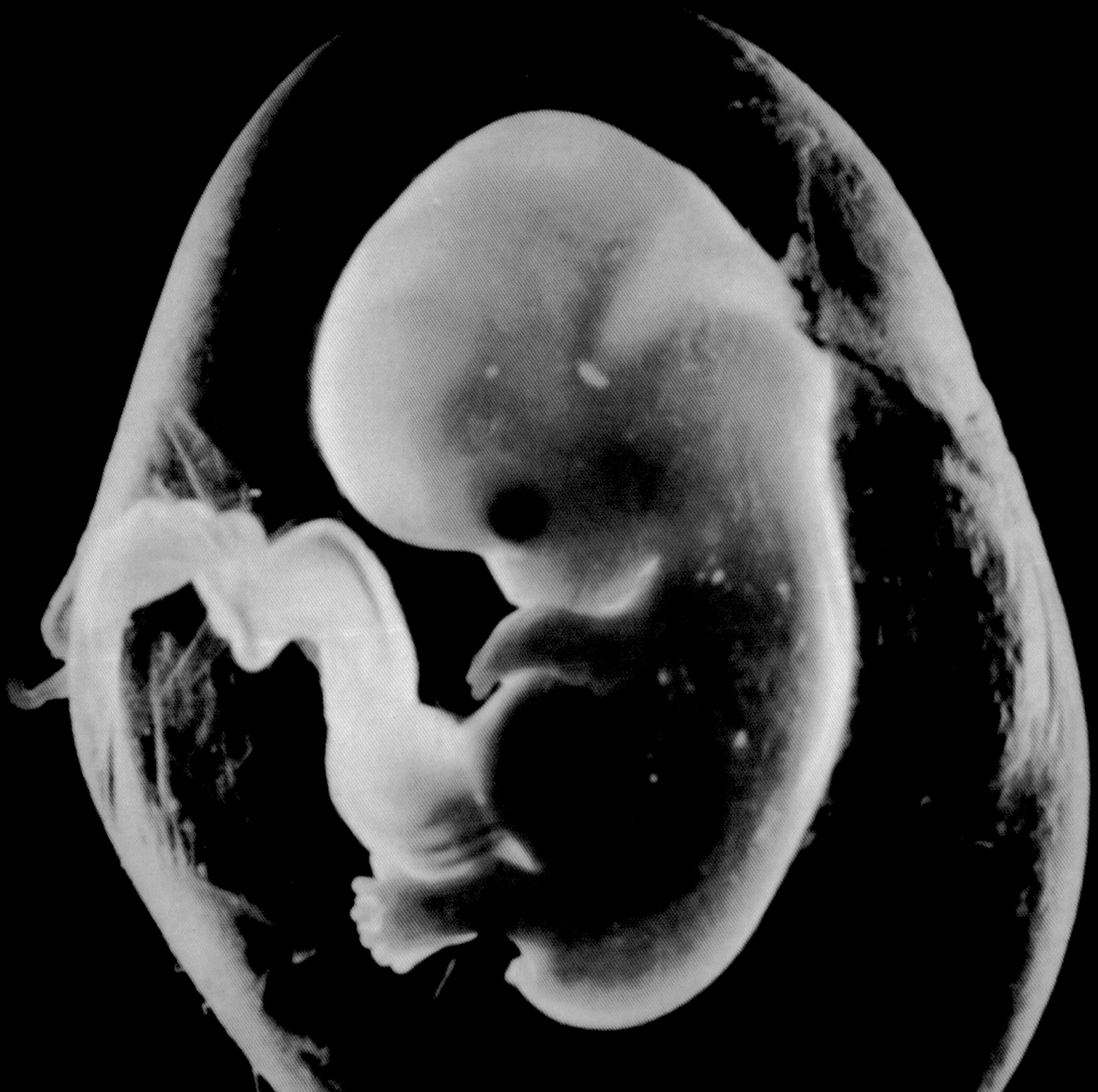

# Fetal Development and Childbirth

The fetus's growth continues progressing day by day, and this chapter illustrates the most notable changes that can be seen. By now, it is possible to distinguish ovaries from testicles, to observe the external parts of the ear, and to see the limbs flex. We will also use pictures to tell you about DNA, the key substance of the body that enables hereditary characteristics to be passed from one generation to the next. Which tests must pregnant women take to find out if fetal development is normal? What are the most notable changes that the woman's body goes through, and what happens once the baby begins to breathe and live outside the womb?



**FIRST TRIMESTER** (opposite)
Picture of an eight-week-old baby. The development of the brain, heart, and extremities can be noted.

# Neuron Development

The third month of fetal development brings distinct changes compared with previous stages. What was once called the embryo is now a fetus. The number of neurons in the brain increases rapidly, and toward the end of the month the fetus has the same number of nerve cells as an adult. However, the interneuron connections have not been established. During this month, through nerve impulses, the network will take shape, which in later months will allow voluntary movements of the joints.

**100 billion**

**NEURONS ARE FORMED BETWEEN THE THIRD AND SEVENTH MONTHS OF FETAL DEVELOPMENT INSIDE THE MOTHER'S WOMB.**

## The Nervous System

In the third month of gestation, the fetus's developing brain changes significantly compared with previous stages. Toward the end of the month, it will have the same number of nerve cells as an adult. The nerves that run from the brain begin to be covered with myelin, a protective lipid layer that insulates the axons of some neurons to speed the transmission of impulses. The nerves and muscles begin to establish connections, setting the foundation for movement controlled by the cerebral cortex. Although the fetus can make a fist and clasp its hands, the movements are still involuntary since the nervous system is not complete.

## Neuron

The neuron is the most important cell in the nervous system. Through the transmission of nerve impulses, neurons establish connections to other neurons to make the brain function.

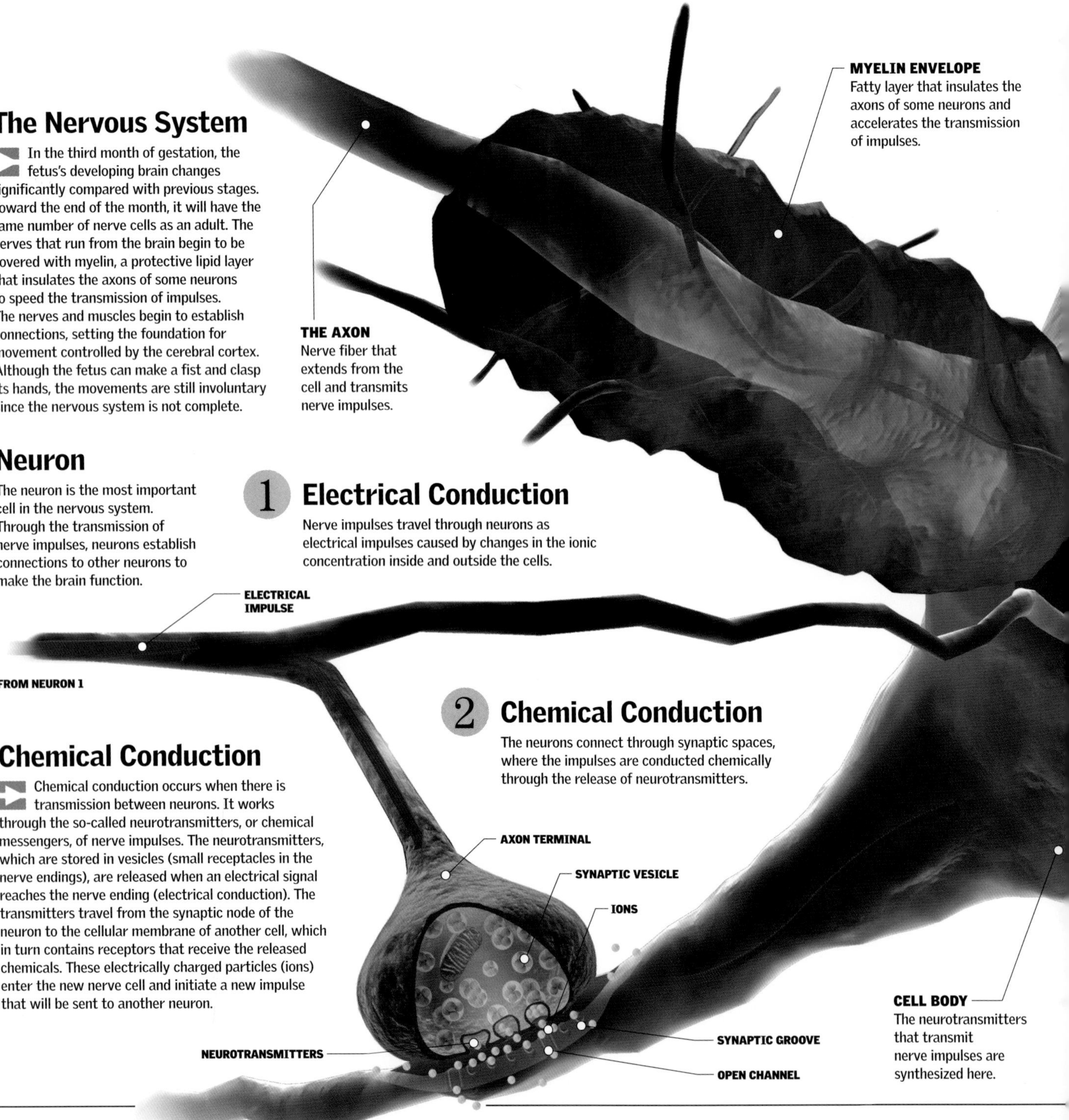
MYELIN ENVELOPE
Fatty layer that insulates the axons of some neurons and accelerates the transmission of impulses.
THE AXON
Nerve fiber that extends from the cell and transmits nerve impulses.
ELECTRICAL IMPULSE
FROM NEURON 1
AXON TERMINAL
SYNAPTIC VESICLE
IONS
NEUROTRANSMITTERS
SYNAPTIC GROOVE
OPEN CHANNEL
CELL BODY
The neurotransmitters that transmit nerve impulses are synthesized here.

### 1 Electrical Conduction

Nerve impulses travel through neurons as electrical impulses caused by changes in the ionic concentration inside and outside the cells.

### 2 Chemical Conduction

The neurons connect through synaptic spaces, where the impulses are conducted chemically through the release of neurotransmitters.

## Chemical Conduction

Chemical conduction occurs when there is transmission between neurons. It works through the so-called neurotransmitters, or chemical messengers, of nerve impulses. The neurotransmitters, which are stored in vesicles (small receptacles in the nerve endings), are released when an electrical signal reaches the nerve ending (electrical conduction). The transmitters travel from the synaptic node of the neuron to the cellular membrane of another cell, which in turn contains receptors that receive the released chemicals. These electrically charged particles (ions) enter the new nerve cell and initiate a new impulse that will be sent to another neuron.

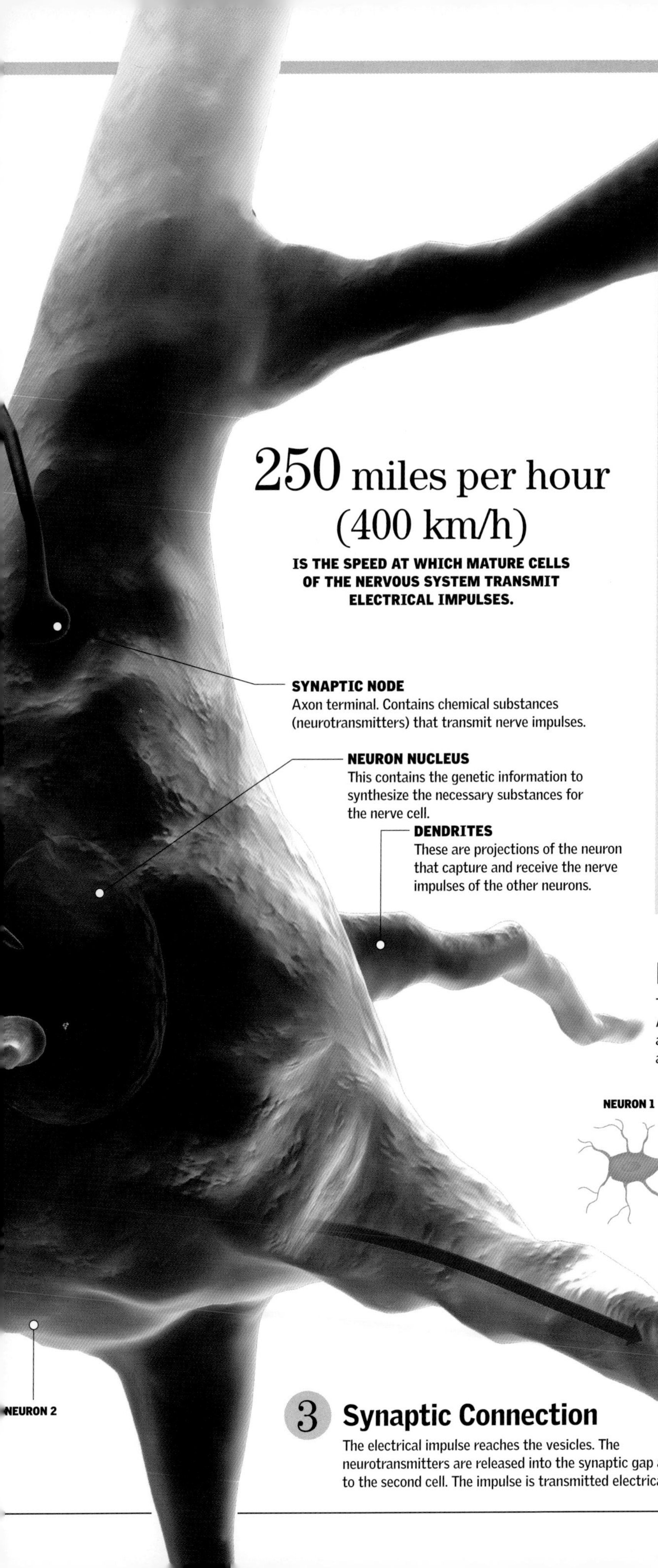


# 250 miles per hour (400 km/h)

**IS THE SPEED AT WHICH MATURE CELLS OF THE NERVOUS SYSTEM TRANSMIT ELECTRICAL IMPULSES.**

**SYNAPTIC NODE**
Axon terminal. Contains chemical substances (neurotransmitters) that transmit nerve impulses.

**NEURON NUCLEUS**
This contains the genetic information to synthesize the necessary substances for the nerve cell.

**DENDRITES**
These are projections of the neuron that capture and receive the nerve impulses of the other neurons.

## Formation of the Brain

Developing from folds in the neural tube, the brain finishes taking shape by the third month. Later it will develop the folds of the cortex.

**1 28 DAYS**
At this stage of development, it is still possible to see the neural tube without any folds.

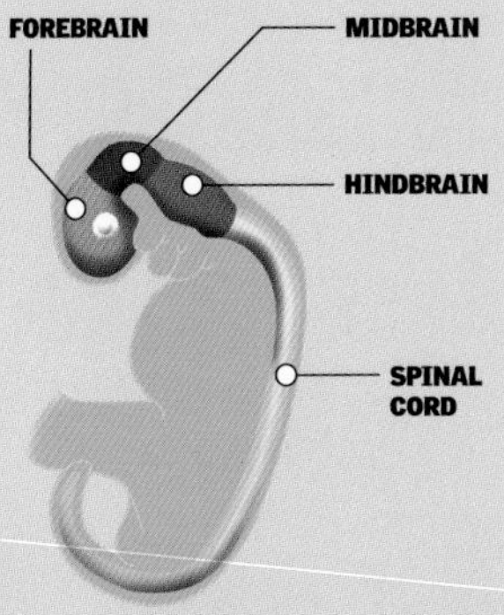


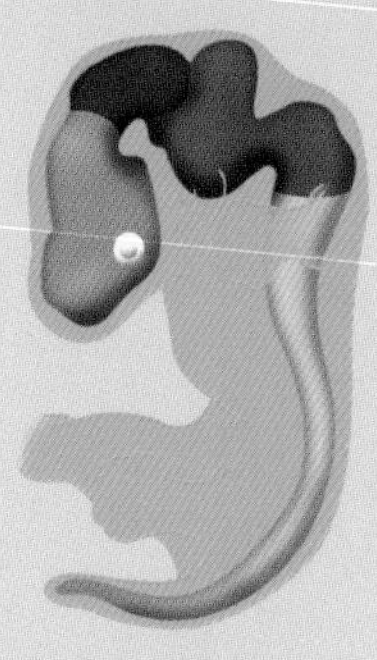

**2 49 DAYS**
An increase in brain size can be seen, and the general structure of the embryo begins to take shape.

**3 3RD MONTH**
We now see a real fetus. The brain attains its definitive shape, but the cortex is still developing.

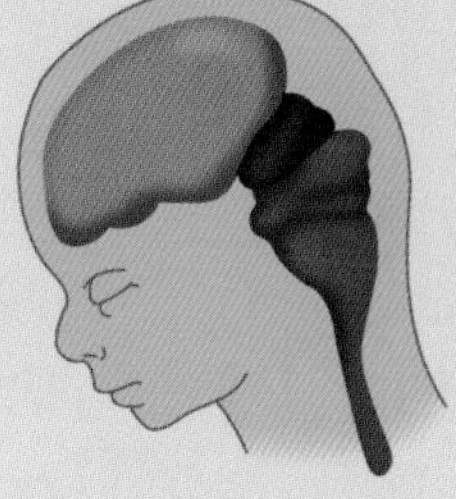

## Electrical Conduction

The nerve impulses inside a neuron travel as electrical impulses. After an electrical impulse is generated, it travels through the axon. This conduction is produced by the exchange of calcium and potassium ions along the length of the nerve cell membrane.

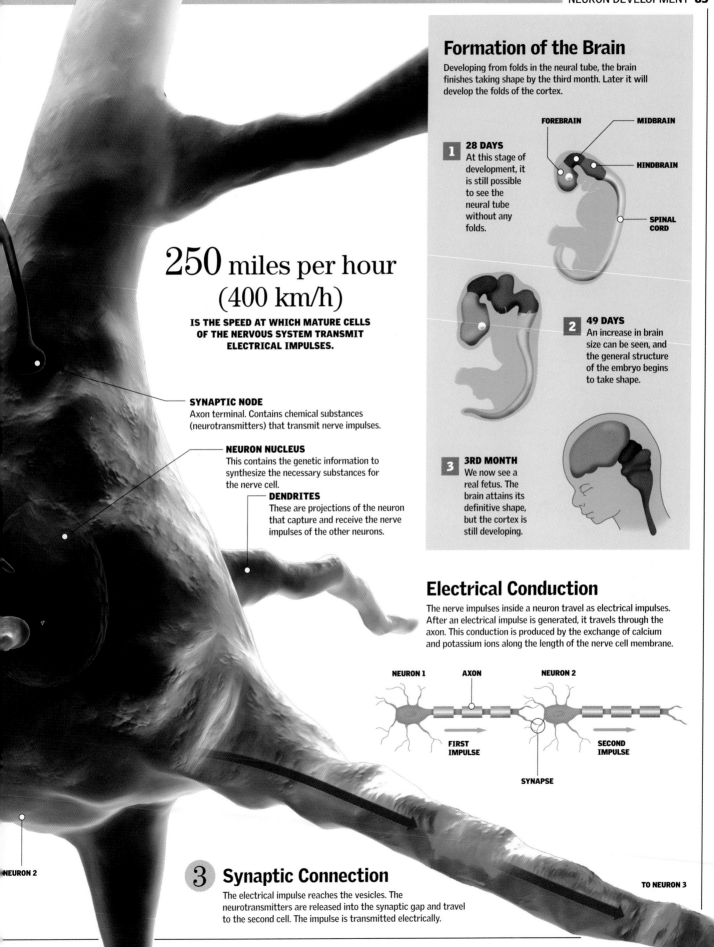

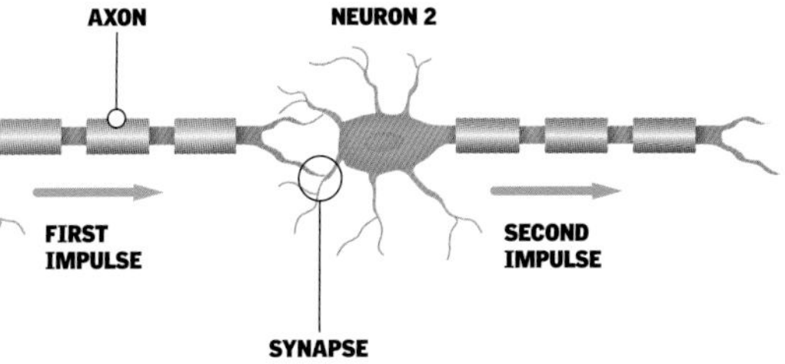


## 3 Synaptic Connection

The electrical impulse reaches the vesicles. The neurotransmitters are released into the synaptic gap and travel to the second cell. The impulse is transmitted electrically.

# Boy or Girl?

By the third month of pregnancy, the mother may be anxious to find out if the unborn baby will be a boy or a girl. Although the sex of the fetus has been determined genetically since fertilization, it still cannot be observed at this early stage of development. By the second trimester, at about 12 weeks, the fetus's genitals begin to appear but still cannot be distinguished as being male or female. The initial undifferentiated bulge has a particular shape that allows it to turn into either a penis or a clitoris.

## Sex is Defined

Until the fifth week after fertilization, the embryonic sex organs of boys and girls are identical. Genetically, sex has already been defined, but under a microscope, the genital regions are indistinguishable. The female and male genitals are not yet differentiated. In the third month, the initial bulge that has developed has a characteristic form and is shaped in such a manner that it can turn into either a penis or a clitoris, with the usual characteristics of either. The genitals have a groove in the urethra that is distinctive. If this groove closes, then a boy is on its way. If it stays open, then the baby will be a girl. The genitals of the fetus, then, begin to grow in the fourth week, becoming visible and external in the eighth week. However, the sex will not be distinguishable until after the 12th week.

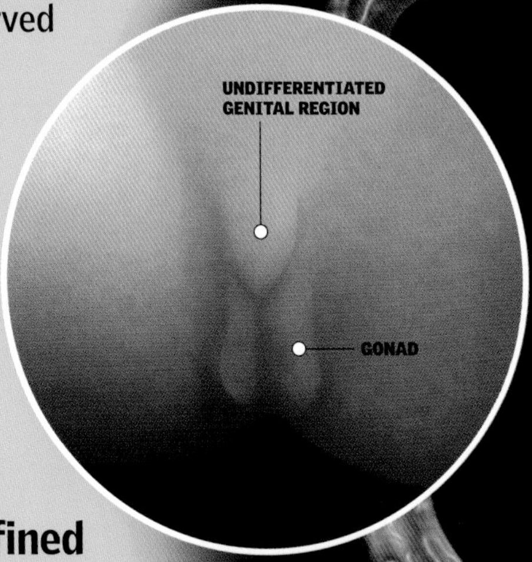


### A Undefined

Each embryo has an undifferentiated genital system and the structures necessary to develop into either sex. The gonads are sexually undefined and have both male and female components.

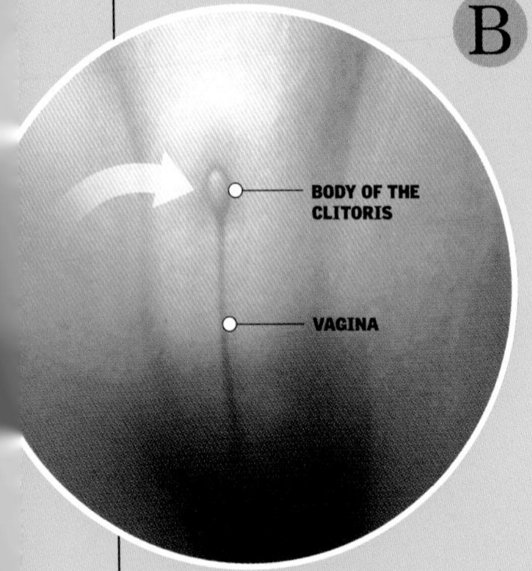


### B If it is a Girl

The vulva (which contains separate openings for the vagina and urethra) and the vagina develop from the same common structures. The clitoris will begin to form from a bulge in the urogenital sinus, the genital tubercle. The intervention of hormones is key in influencing the differential formation of each organ. The evolution might be different, but the origin is exactly the same.

**THE CORD**
This is completely mature and is rolled up so that the baby can move around without any risk.

### C If it is a Boy

By the 11th week, the genital tubercle lengthens rapidly and forms the penis. The components of what will be the genitals are progressively modified and form the elements that define the external genitalia of the male—testicles, scrotum, and penis.

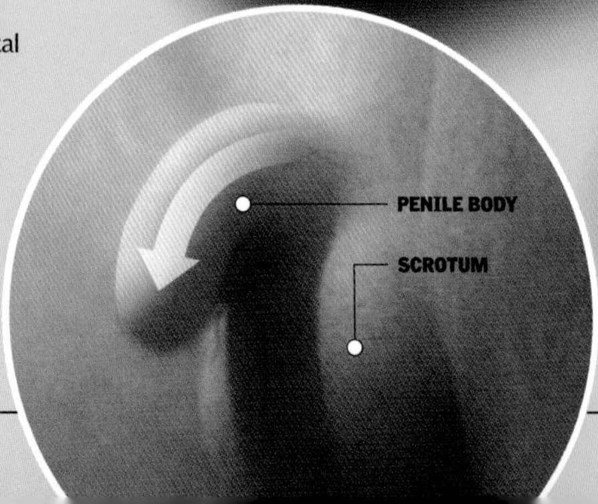

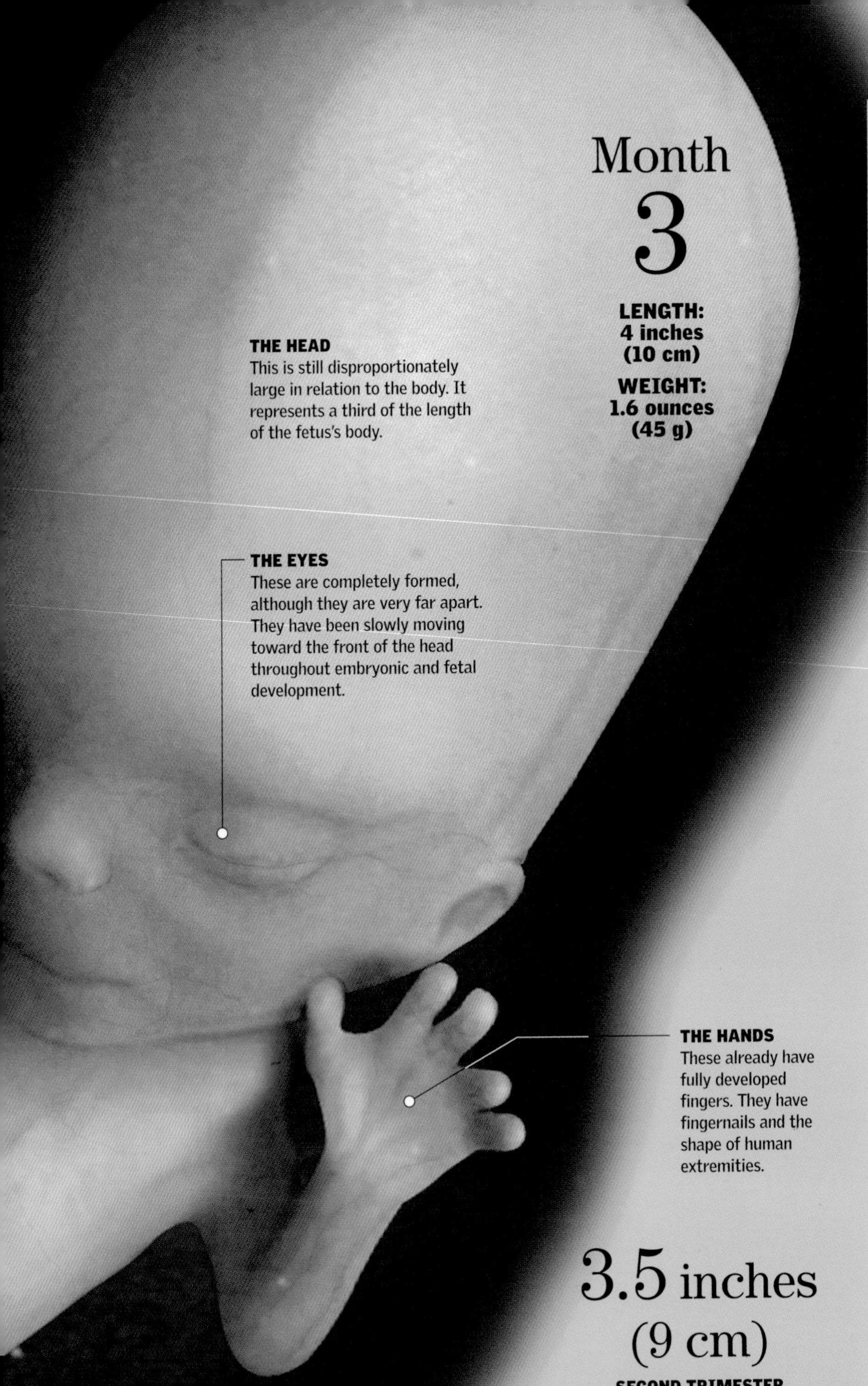

# Month 3

**LENGTH:**
**4 inches**
**(10 cm)**

**WEIGHT:**
**1.6 ounces**
**(45 g)**

**THE HEAD**
This is still disproportionately large in relation to the body. It represents a third of the length of the fetus's body.

**THE EYES**
These are completely formed, although they are very far apart. They have been slowly moving toward the front of the head throughout embryonic and fetal development.

**THE HANDS**
These already have fully developed fingers. They have fingernails and the shape of human extremities.

## 3.5 inches (9 cm)

**SECOND TRIMESTER**
**BY THE BEGINNING OF THE SECOND TRIMESTER, THE FETUS MEASURES 3.5 INCHES (9 CM) LONG.**

## The Sonogram

An ultrasound image, also known as a sonogram, uses inaudible sound waves to produce images of the different structures of the body. During the examination, a small device called a transducer is pressed against the skin. It generates high-frequency sound waves that pass into the body and return as echoes as the sound waves bounces against organs, blood-vessel walls, and tissues. A special computer converts the echoes into an image.

### How a Sonogram is Made

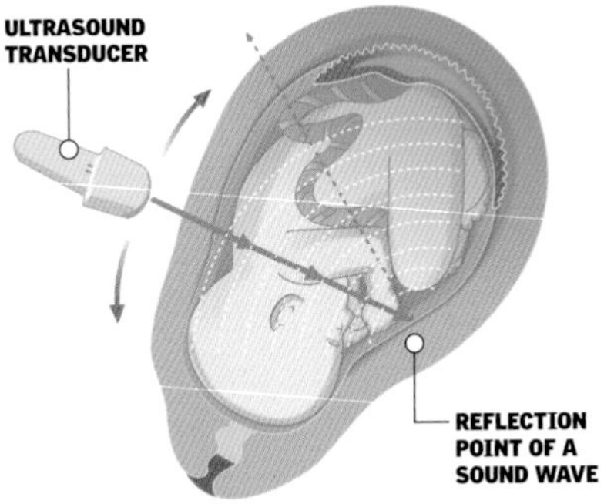


**1 TRANSMISSION OF IMPULSES**
The ultrasound transducer emits high-frequency sound waves.

**2 THE PATH**
The ultrasound impulses pass through the body's tissues and bounce off surfaces.

**3 DETECTING THE ECHO**
Some impulses are reflected as echoes that the wand picks up and sends to the sonograph.

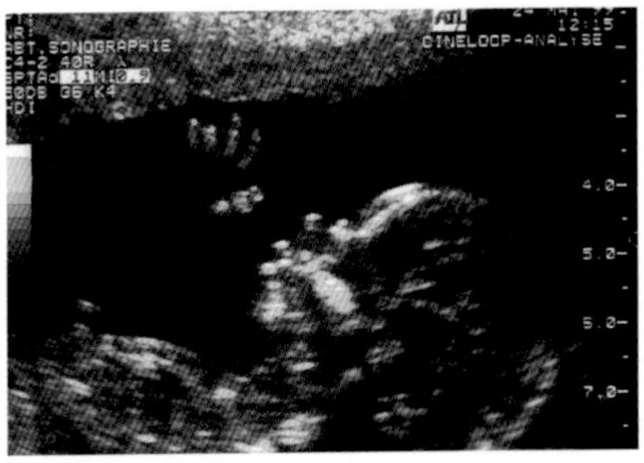

### Forming the Image

The sonograph calculates the distance from the wand to the tissues, the echo intensity, and the return time of each echo in millionths of a second. Some sonographs show three-dimensional images. They show the entire surface of the fetus and help to identify any deformation.

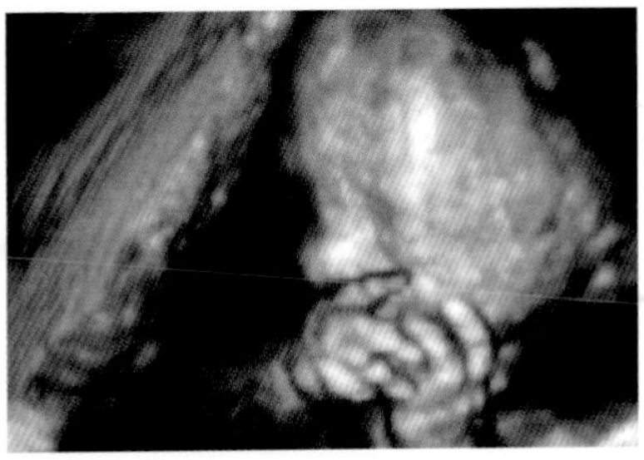

## The Sperm

According to popular belief, to conceive a boy one must have sex on the day the woman is ovulating or the day after, since sperm with a Y chromosome (determinant of the male sex) are quicker than those that contain the X chromosome (female) and reach the egg first. If the desire is to have a girl, it is best to have sex a few days earlier: X sperm are slower, but have more endurance and live longer.

### Life Comparison

The sperm with an X chromosome are slower but have more endurance. They can last up to 72 hours. The sperm with a Y chromosome, on the other hand, are quicker but last only approximately 48 hours.

# Growth Begins

Month
4

**LENGTH:**
**5.9 inches**
**(15 cm)**

**WEIGHT:**
**5.3 ounces**
**(150 g)**

During the fourth month, the mother senses the fetus's first movements. The fetal body changes; its face is completely formed. The skin has a pinkish tone, and the first ribs and cartilage appear. The external sex organs finish forming, and the internal ones differentiate. The first subtle movements of the fetus begin, although they are barely perceptible because of its small size. The fetus now occupies the entire uterine cavity and pushes the abdomen forward. Its extremities can be clearly seen. The little one enters its full growth phase.

## Sex Development

During this period, the fetus begins to reveal the differences in its urogenital system. The undifferentiated gonad, which has male and female components that developed during the embryonic stage, is converted into ovaries in a girl or testicles in a boy. In either case, its presence will determine the development of the individual's sexual characteristics.

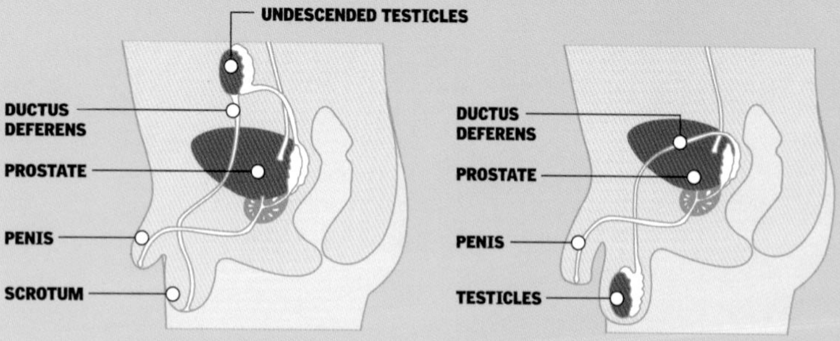


**1 MALE GONAD**
Toward the seventh week, it has already been determined whether the fetus is XY (male) or XX (female). If a gonad is evolving into a testicle, the undifferentiated gonad increases in size as it descends into the scrotum.

**2 DESCENT OF THE TESTICLES**
At about the eighth week, the testicles leave the abdominal cavity and descend toward the scrotum. For males, the presence of the testicles and the actions of their hormones are necessary.

**160**

**IS THE NUMBER OF HEARTBEATS PER MINUTE IN THE EARLY STAGES OF INTRAUTERINE LIFE. TOWARD THE END OF GESTATION, THE NUMBER DROPS TO 120.**

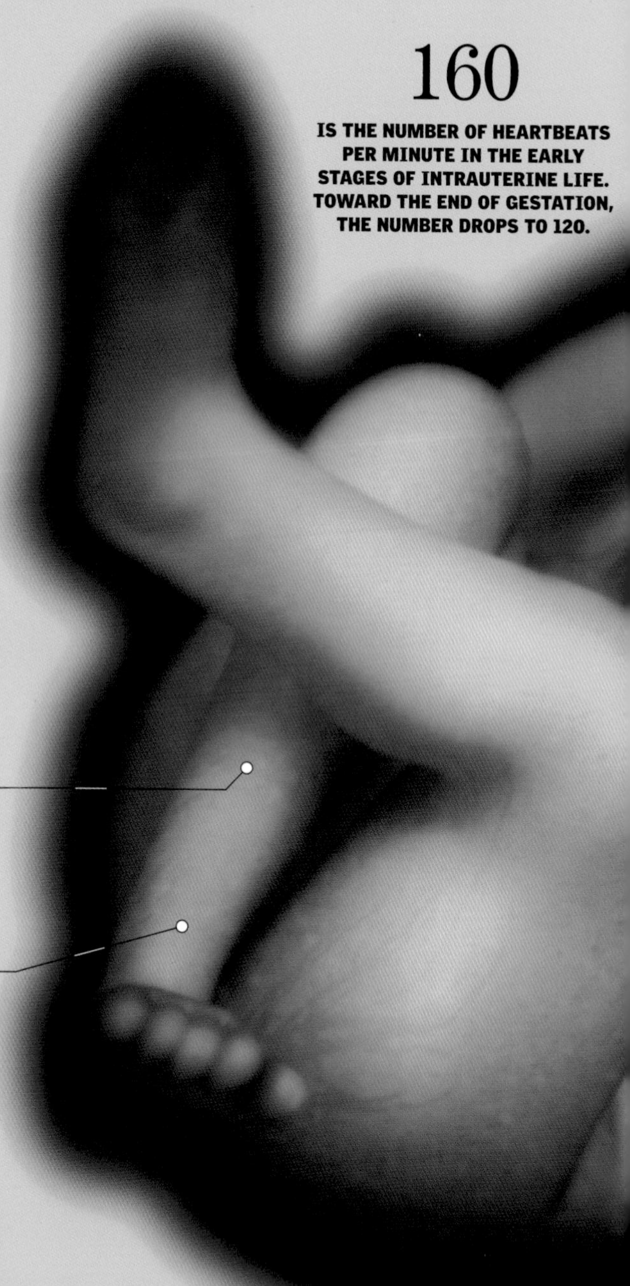

**BONES**
These can be distinguished with X-rays and have begun to change from cartilage into calcified bones.

**LOWER LIMBS**
In this stage, the legs grow rapidly in a proportional manner and are longer than the arms.

## Amniocentesis

This is a test that is performed by studying the amniotic fluid in the sac that surrounds the fetus. After the insertion of a hollow needle into the abdominal wall of the uterus, a small amount of liquid is extracted. It is not a routine test, and it is invasive. It is done when there is suspicion of abnormalities that cannot be detected with other tests (e.g., tests for spina bifida or metabolic diseases).

**COMPONENTS OF THE AMNIOTIC FLUID**

| | |
|---|---|
| **Water** | 98% |
| **Organic solutes**: proteins, lipids, carbohydrates, and nonprotein hydrogenated components. **Inorganic solutes**: zinc, copper, iron, and magnesium. | 2% |

## Chromosomal Study

Amniocentesis gives a cytogenetic map (map of chromosomes) from which different chromosomal disorders can be detected, such as Down's syndrome (an extra chromosome in pair 21) or the existence of an abnormal gene that can cause neurological or metabolic diseases.

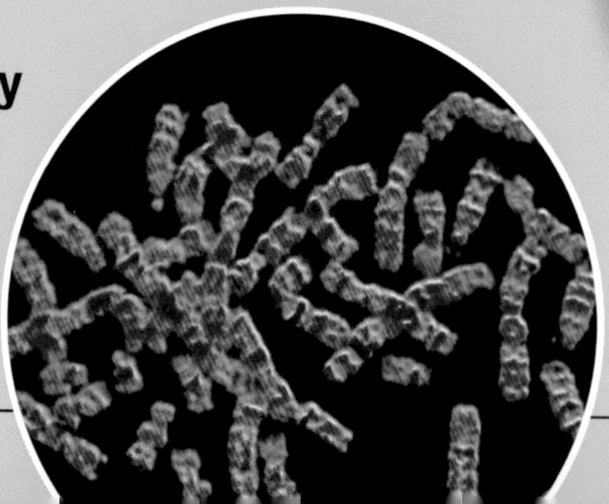

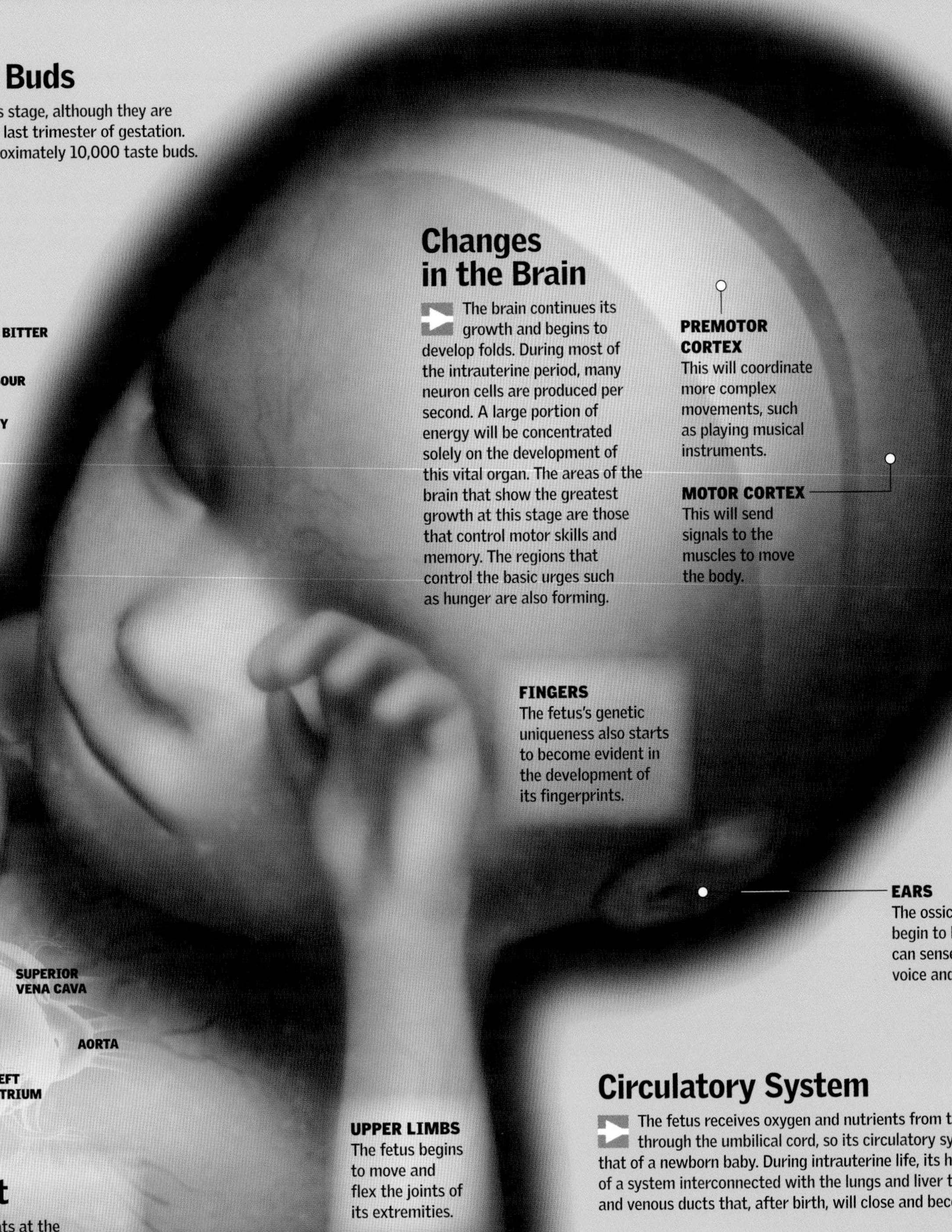

## The Taste Buds

These develop at this stage, although they are activated only in the last trimester of gestation. The tongue has approximately 10,000 taste buds.

**The Tongue**

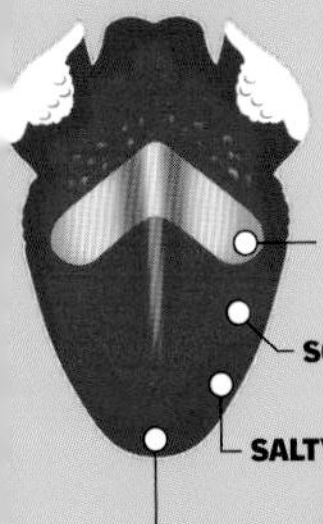


## Changes in the Brain

The brain continues its growth and begins to develop folds. During most of the intrauterine period, many neuron cells are produced per second. A large portion of energy will be concentrated solely on the development of this vital organ. The areas of the brain that show the greatest growth at this stage are those that control motor skills and memory. The regions that control the basic urges such as hunger are also forming.

**PREMOTOR CORTEX**
This will coordinate more complex movements, such as playing musical instruments.

**MOTOR CORTEX**
This will send signals to the muscles to move the body.

**FINGERS**
The fetus's genetic uniqueness also starts to become evident in the development of its fingerprints.

**EARS**
The ossicles (tiny bones) begin to harden. The fetus can sense its mother's voice and heartbeat.

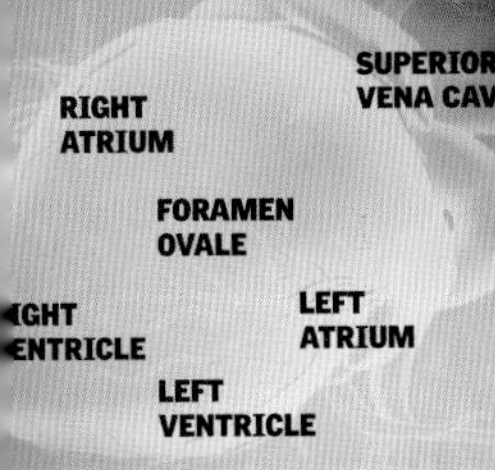


## The Heart

At this stage, it beats at the mother's heart rate and pumps more than 6 gallons (25 l) of blood each day. Its size is large relative to the body. The foramen ovale in the fetal heart is a hole that allows the blood to circulate from the right atrium to the left one. It will close during the first three months after birth.

**UPPER LIMBS**
The fetus begins to move and flex the joints of its extremities.

## Circulatory System

The fetus receives oxygen and nutrients from the placenta through the umbilical cord, so its circulatory system differs from that of a newborn baby. During intrauterine life, its heart is the center of a system interconnected with the lungs and liver through arterial and venous ducts that, after birth, will close and become ligaments.

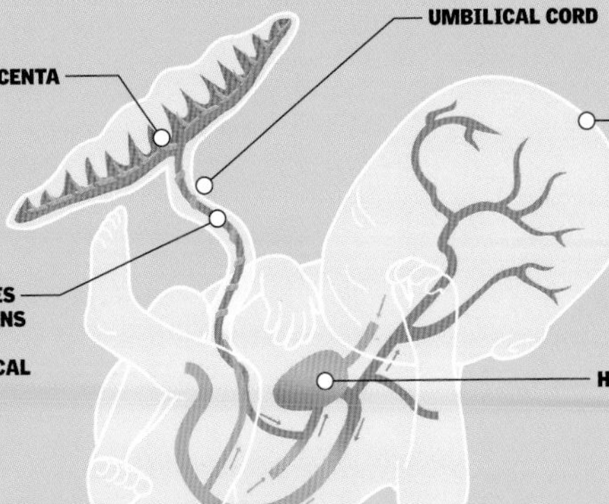


**THE SKIN**
It is still translucent, thin, and wrinkled, and it allows the developing blood vessels and bones to be seen.

# Intense Movements

The fifth month of intrauterine life reveals marked changes: the fetus's movements are more obvious and intense and are perceptible to the touch. During this period, it is important to have ultrasound exams to check for the position of the placenta, the proper circulation between the uterus and placenta, and the risk of premature birth. The future baby's features are clearly visible.

## Energetic Movement

Because of the accelerated growth and development of its internal organs, the fetus is much more active. It turns, moves from side to side, and finds ways to be more comfortable inside the uterus. It is exploring the surroundings where it lives, which makes its movements strong enough to notice. When she least expects it, the mother can receive a kick from the unborn baby. Anyone that gets close to the mother's belly can hear the fetal heartbeat through a special device.

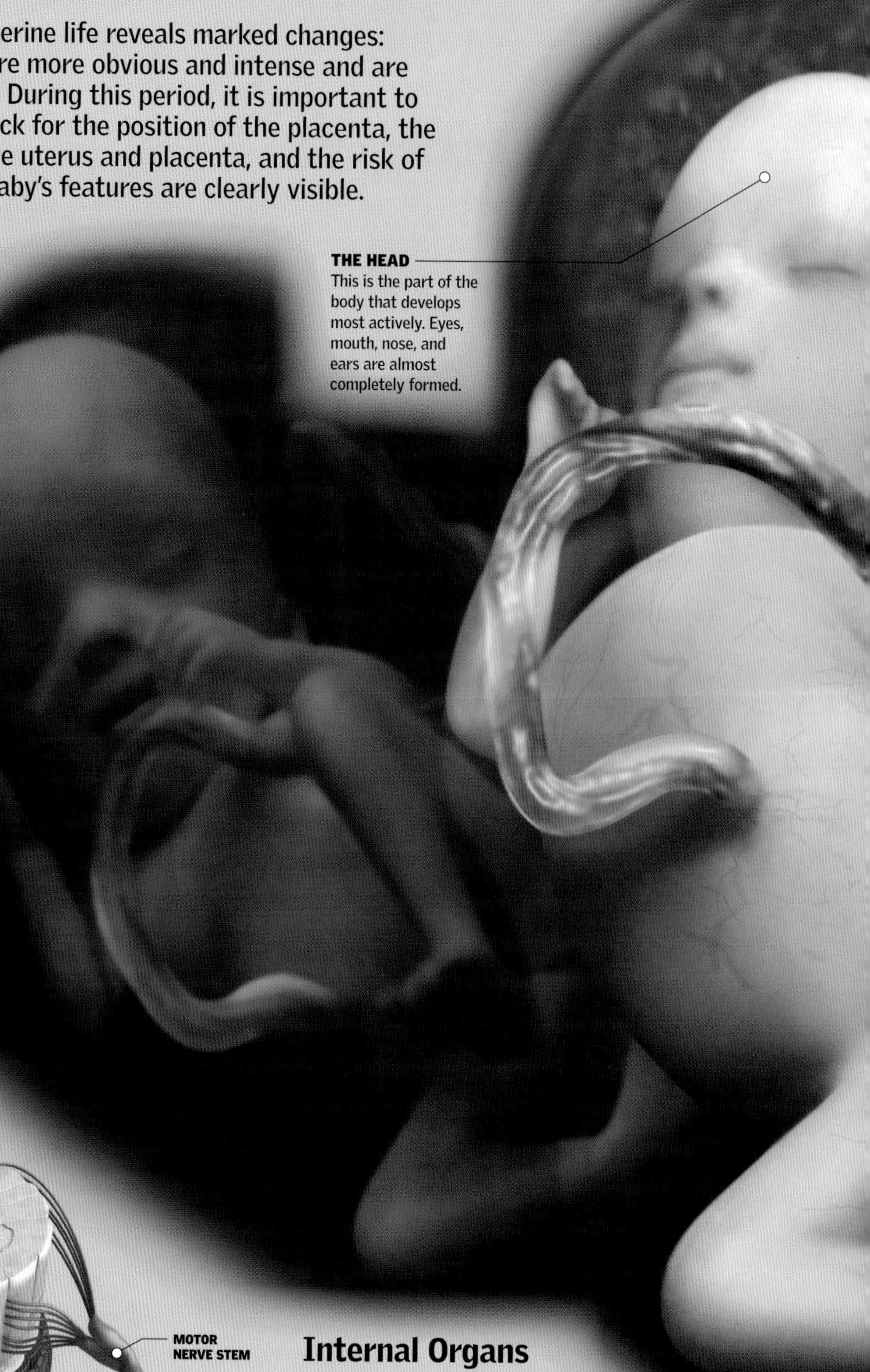

THE HEAD
This is the part of the body that develops most actively. Eyes, mouth, nose, and ears are almost completely formed.

## Spinal Cord

This begins to develop in the fetus. The spinal cord will be the communication link between the brain and the rest of the body. The spinal cord receives and transmits information through the nerves. The nerve impulse stimulates the muscle to move.

### Muscle Movement

**1** The brain processes sensory data and sends information to the spinal cord.

**2** The spinal cord receives the nerve impulse from the brain and sends a response to the muscle.

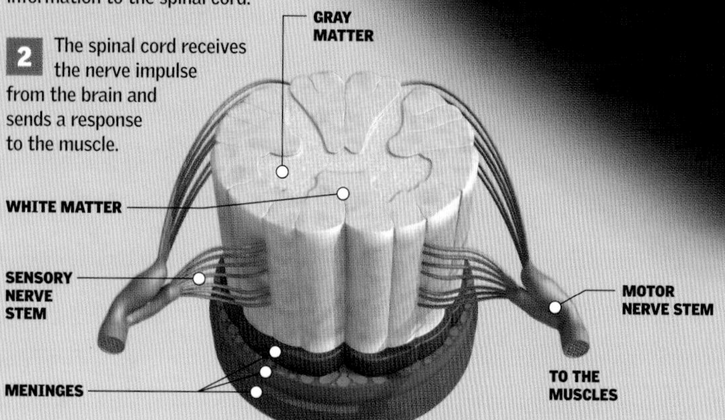


## Internal Organs

These are in a maturing stage and most are already formed. However, the lungs and the digestive system are not yet complete. The fetus cannot maintain its body temperature or survive outside the uterus.

# Month 5

**LENGTH:**
**7.9 inches (20 cm)**

**WEIGHT:**
**17.6 ounces (500 g)**

## Exploration

It is very important for the mother to have prenatal tests done periodically to monitor for possible problems or abnormalities in the fetus. Different techniques can be used to verify the fetal position and the development of its features. Sonographs produce images of the internal organs or masses for diagnostic purposes. Three-dimensional magnetic resonance imaging (MRI) allows the diagnosis of previously undetectable diseases and pathologies. In addition, it is not harmful to the fetus. 4-D ultrasound allows monitoring of the fetus in real time.

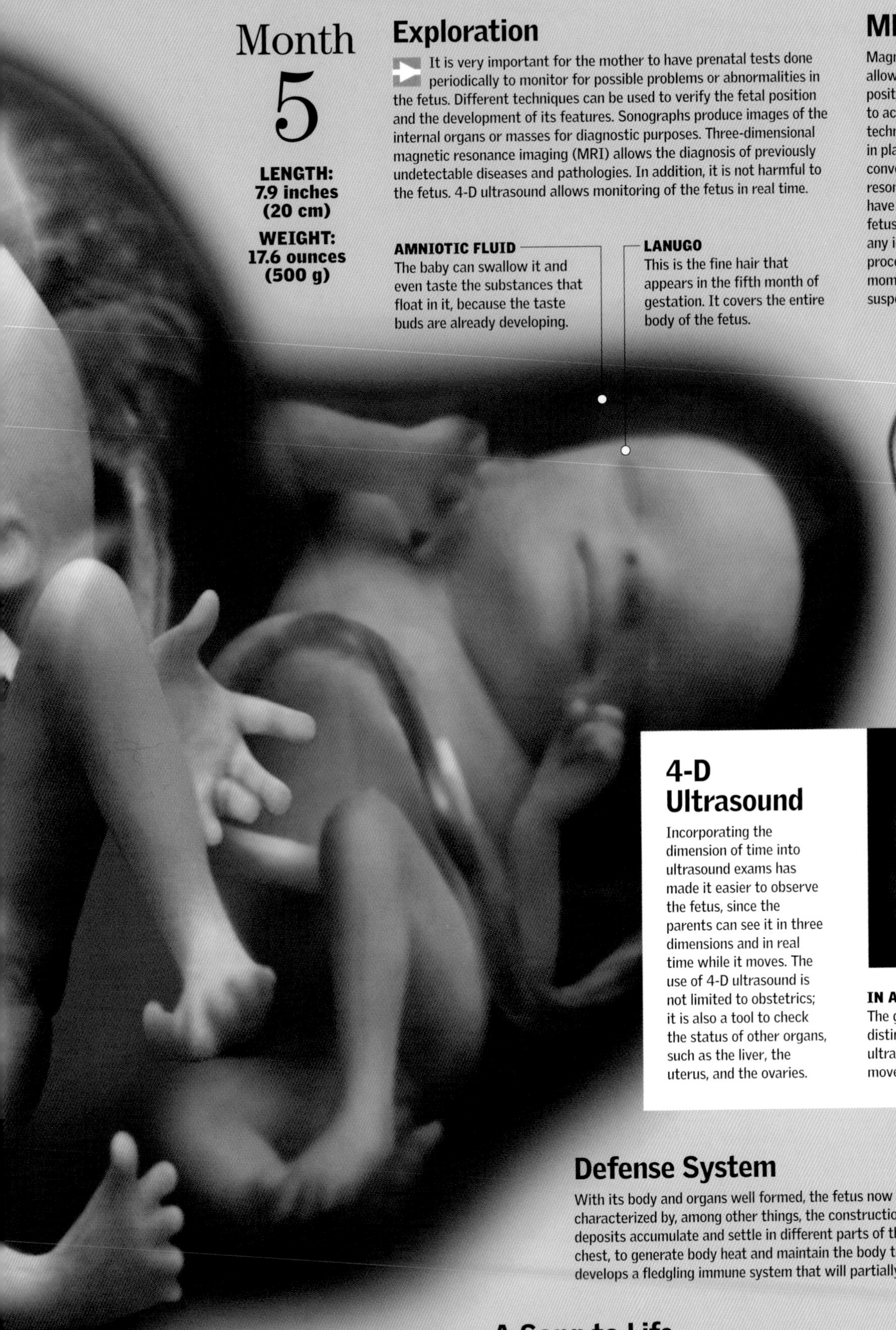

**AMNIOTIC FLUID**
The baby can swallow it and even taste the substances that float in it, because the taste buds are already developing.

**LANUGO**
This is the fine hair that appears in the fifth month of gestation. It covers the entire body of the fetus.

## MRI

Magnetic resonance imaging allows diagnosis of the fetal position when this is difficult to accomplish with other techniques. This can be helpful in planning for the birth. Unlike conventional X-rays, magnetic resonance imaging does not have harmful effects on the fetus because it does not emit any ionizing radiation. Use of the process is recommended from the moment a fetal abnormality is suspected until birth.

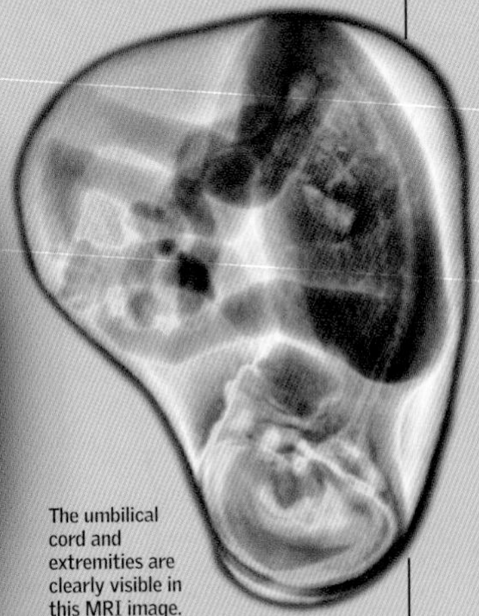

The umbilical cord and extremities are clearly visible in this MRI image.

## 4-D Ultrasound

Incorporating the dimension of time into ultrasound exams has made it easier to observe the fetus, since the parents can see it in three dimensions and in real time while it moves. The use of 4-D ultrasound is not limited to obstetrics; it is also a tool to check the status of other organs, such as the liver, the uterus, and the ovaries.

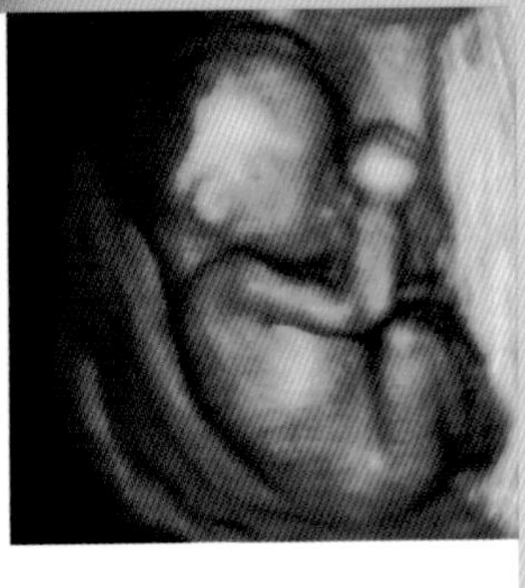

**IN ACTION**
The growing baby can be clearly distinguished in a three-dimensional ultrasound, which also allows fetal movement to be seen.

## Defense System

With its body and organs well formed, the fetus now enters a stage of maturation characterized by, among other things, the construction of a defense system. Fatty deposits accumulate and settle in different parts of the body, such as the neck and chest, to generate body heat and maintain the body temperature. The fetus also develops a fledgling immune system that will partially protect it from some infections.

## A Song to Life

Even though the ear has not reached its peak development, it can already perceive sounds from the outside, besides those coming from the mother (heart, stomach sounds). The mother's physical state and her mood strongly influence the future baby, who can tell at all times if things are right or not.

# Refinement of Hearing

Month

6

**LENGTH:**
**9.8 inches (25 cm)**

**WEIGHT:**
**2 pounds 3 ounces (1 kg)**

During the sixth month of gestation, the ears exhibit their peak development. The fetus is sensitive to sounds outside the uterus and can hear very loud sounds. The cochlea, in the inner ear, is vital for processing sounds and already has attained its characteristic coiled shape. This is the month in which the fetus prepares for life as an independent being.

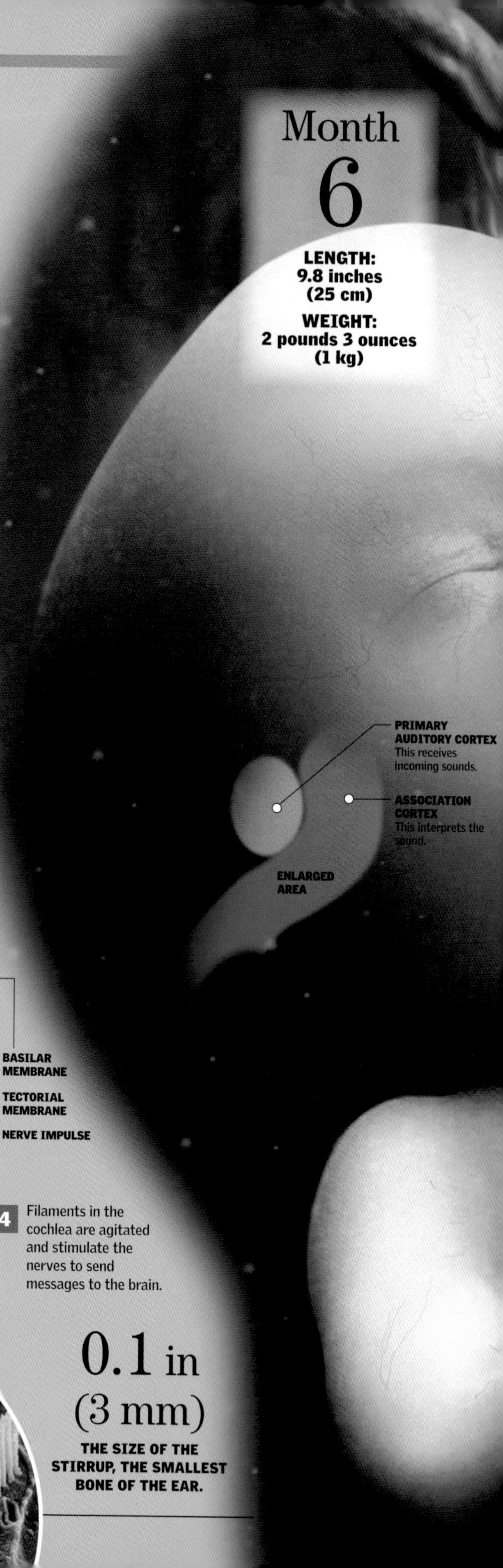


## Recognizing the Parents' Voices

With the perfection of the sense of hearing, the baby not only can hear noises and voices from outside but also can memorize them. It can recognize both the mother's and the father's voices. Since the fetus can respond to external stimuli, the parents are usually recommended to talk and play music. It can also move to the rhythm of the music and show musical preferences. With its eardrums fully developed and fully functional, the fetus can now hear sounds originating from itself, such as its heartbeat.

## Balance

The functioning of the sense of hearing is essential also for understanding the sense of balance. The inner ear has fluid that sends nerve impulses to the brain to update the information about the body's movement and to maintain balance and posture.

### Sound Wave Path

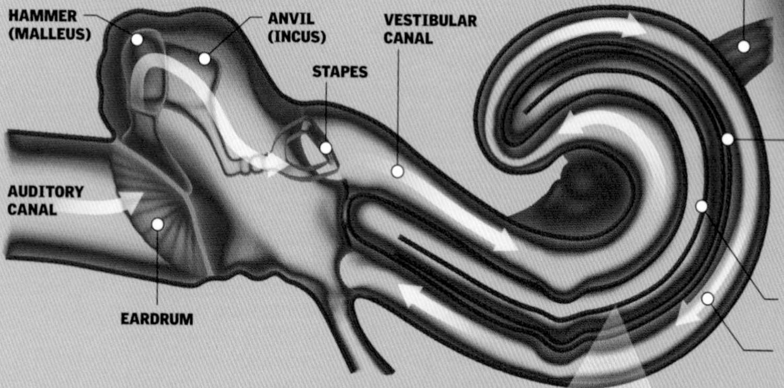


1. Sound waves enter the outer ear canal and are transmitted to the eardrum.
2. The eardrum receives the sound waves as vibrations, which later reach the cochlea.
3. In the cochlea, the organ of Corti gathers the vibrations through hair cells.
4. Filaments in the cochlea are agitated and stimulate the nerves to send messages to the brain.

**15,000**

**HAIR CELLS ARE IN THE ORGAN OF CORTI. THEY CONVERT THE SOUND VIBRATIONS INTO NERVE IMPULSES, WHICH TRAVEL TO THE BRAIN TO BE PROCESSED AS SOUND.**

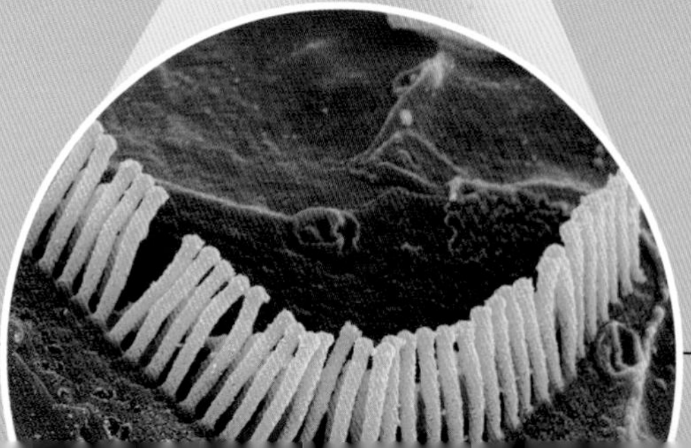

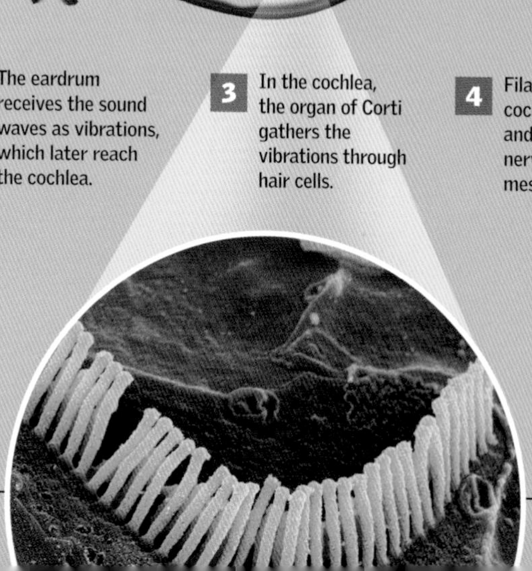

**0.1 in (3 mm)**

**THE SIZE OF THE STIRRUP, THE SMALLEST BONE OF THE EAR.**

## The Umbilical Cord

This is the structure that connects the fetus to the placenta. It constitutes the immunological, nutritional, and hormonal link with the mother. It contains two arteries and a vein that regulate the exchange of nutritional substances and oxygen-rich blood between the embryo and the placenta. It is 12 to 39 inches (30-100 cm) in length, connects the fetus's navel to the placenta, and constitutes the first physical tie between the mother-to-be and the fetus. Usually there are no complications related to the umbilical cord, although there are cases where knots form that block the flow of blood. These knots can be deadly if they are not controlled or corrected.

### Cross Section of the Umbilical Cord

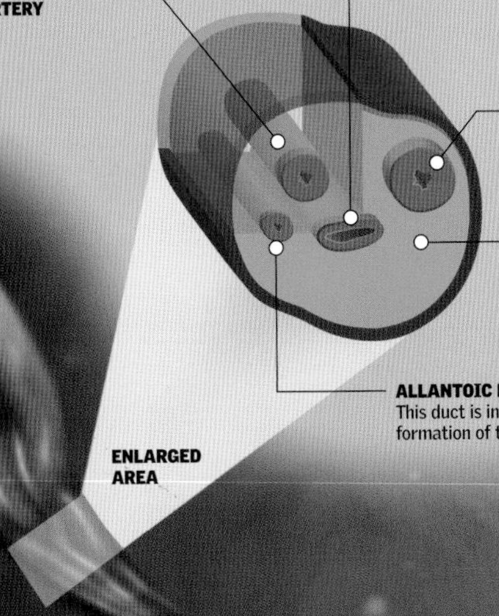

UMBILICAL ARTERY
UMBILICAL VEIN This transports oxygenated blood from the placenta to the fetus.
UMBILICAL ARTERY This takes deoxygenated blood from the fetus to the placenta.
AMNIOTIC EPITHELIUM This produces amniotic fluid and speeds its circulation.
ALLANTOIC DUCT This duct is involved in the formation of the bladder.
ENLARGED AREA

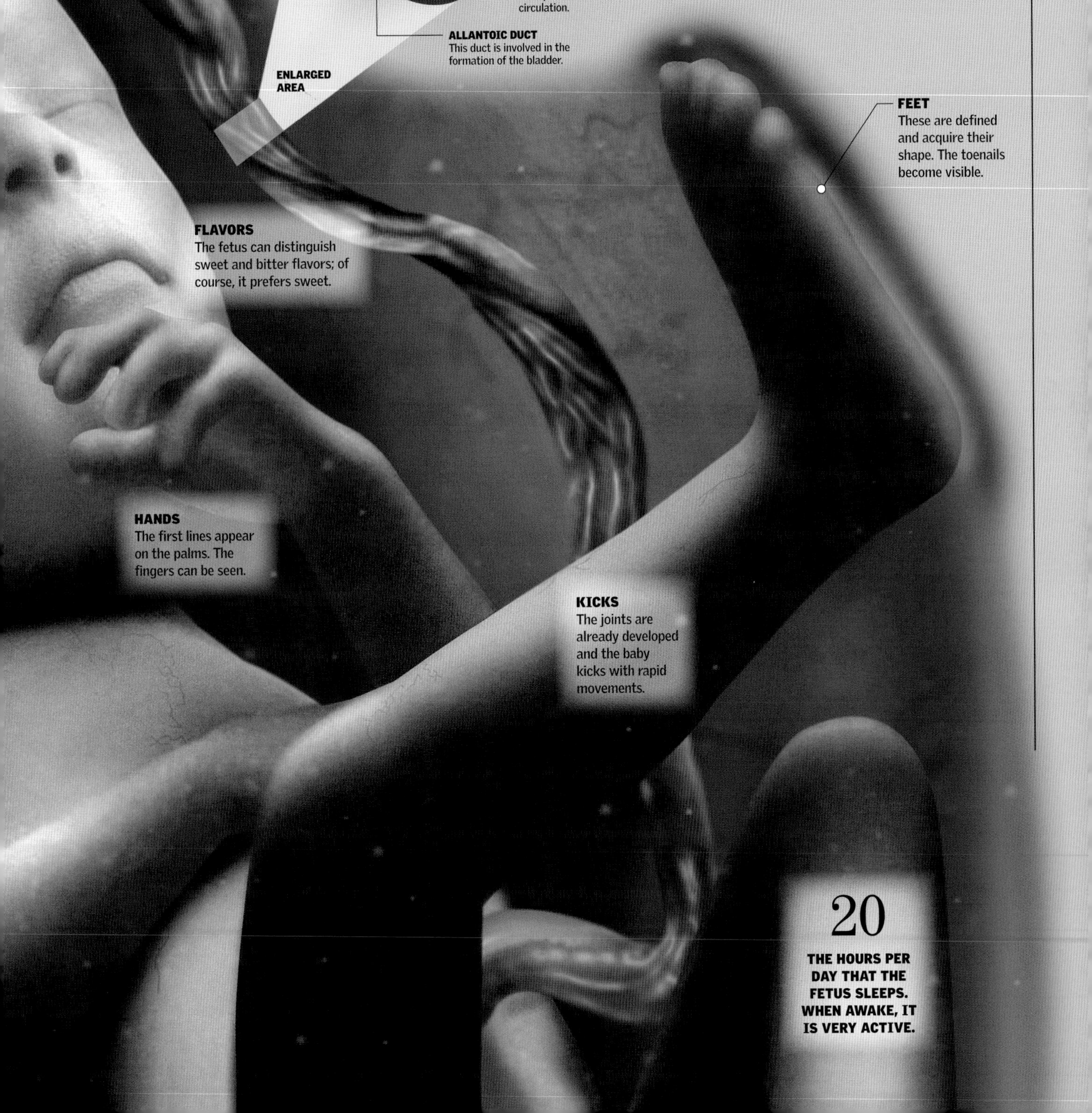

FEET These are defined and acquire their shape. The toenails become visible.
FLAVORS The fetus can distinguish sweet and bitter flavors; of course, it prefers sweet.
HANDS The first lines appear on the palms. The fingers can be seen.
KICKS The joints are already developed and the baby kicks with rapid movements.

**20**

**THE HOURS PER DAY THAT THE FETUS SLEEPS. WHEN AWAKE, IT IS VERY ACTIVE.**

# Closer with Every Moment

The beginning of the third trimester of pregnancy marks a key point in gestation. The process of strengthening and calcification of the bones of the fetus begins. Its body needs nutrients, such as calcium, folic acid, and iron. The baby can already open and close its hands (which will soon have defined fingerprints) and also opens and closes its mouth, sticks out its tongue, and can even suck its thumb. Its skin is still very thin but has begun to turn opaque. The bones and muscles begin to have more consistency. The organs are completely formed.

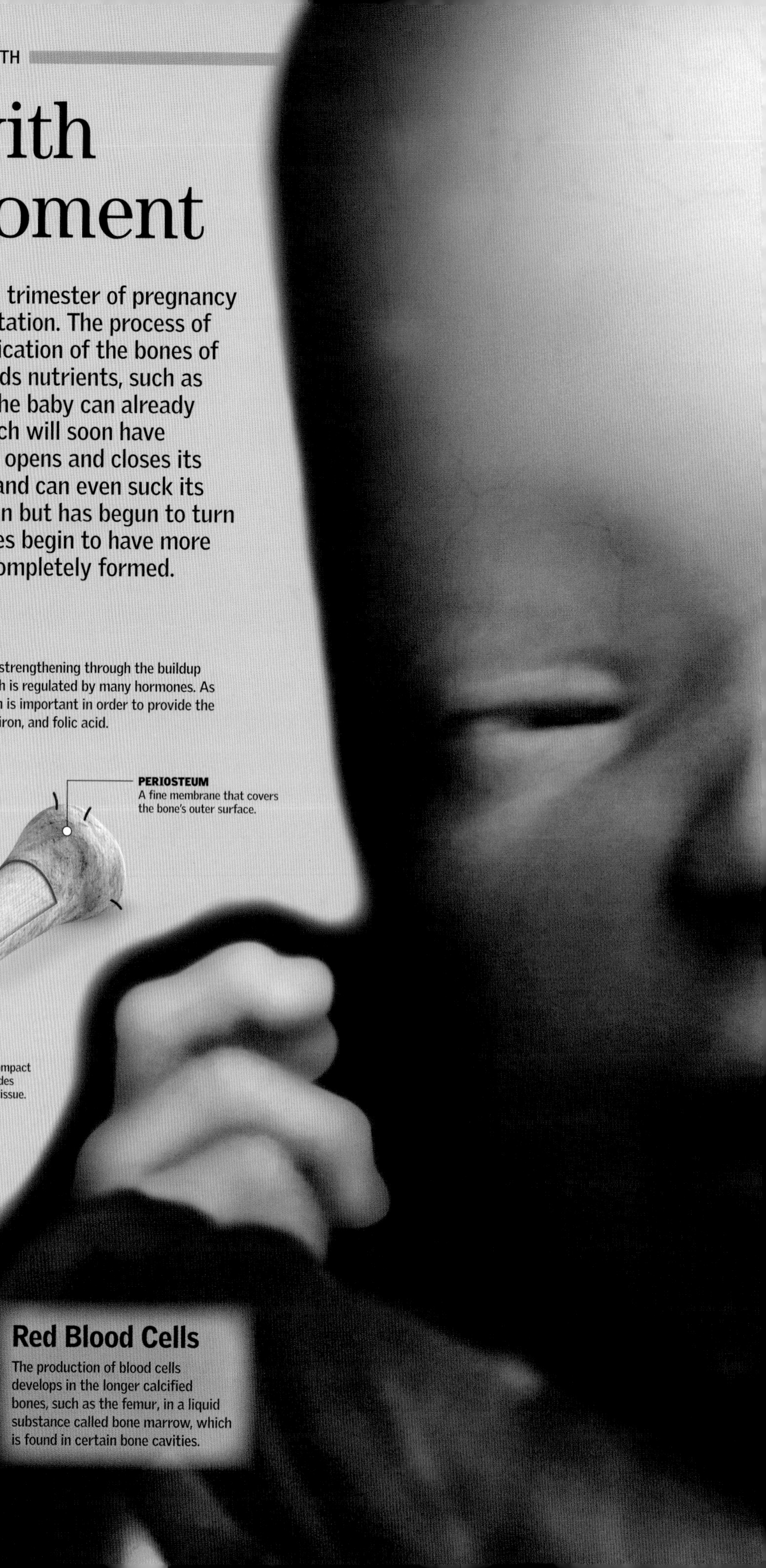

## Bone Calcification

The baby's bones have begun the process of strengthening through the buildup of calcium and also phosphorus. Bone growth is regulated by many hormones. As the bones are getting harder, appropriate nutrition is important in order to provide the necessary amounts of calcium, vitamin D, protein, iron, and folic acid.

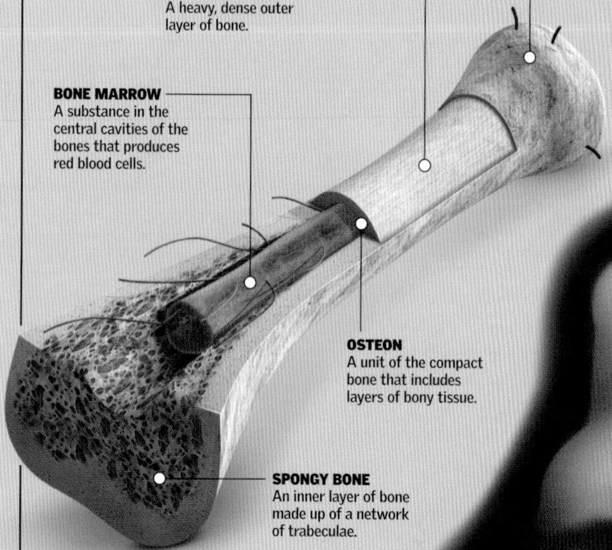

COMPACT BONE
A heavy, dense outer layer of bone.
PERIOSTEUM
A fine membrane that covers the bone's outer surface.
BONE MARROW
A substance in the central cavities of the bones that produces red blood cells.
OSTEON
A unit of the compact bone that includes layers of bony tissue.
SPONGY BONE
An inner layer of bone made up of a network of trabeculae.

## Red Blood Cells

The production of blood cells develops in the longer calcified bones, such as the femur, in a liquid substance called bone marrow, which is found in certain bone cavities.

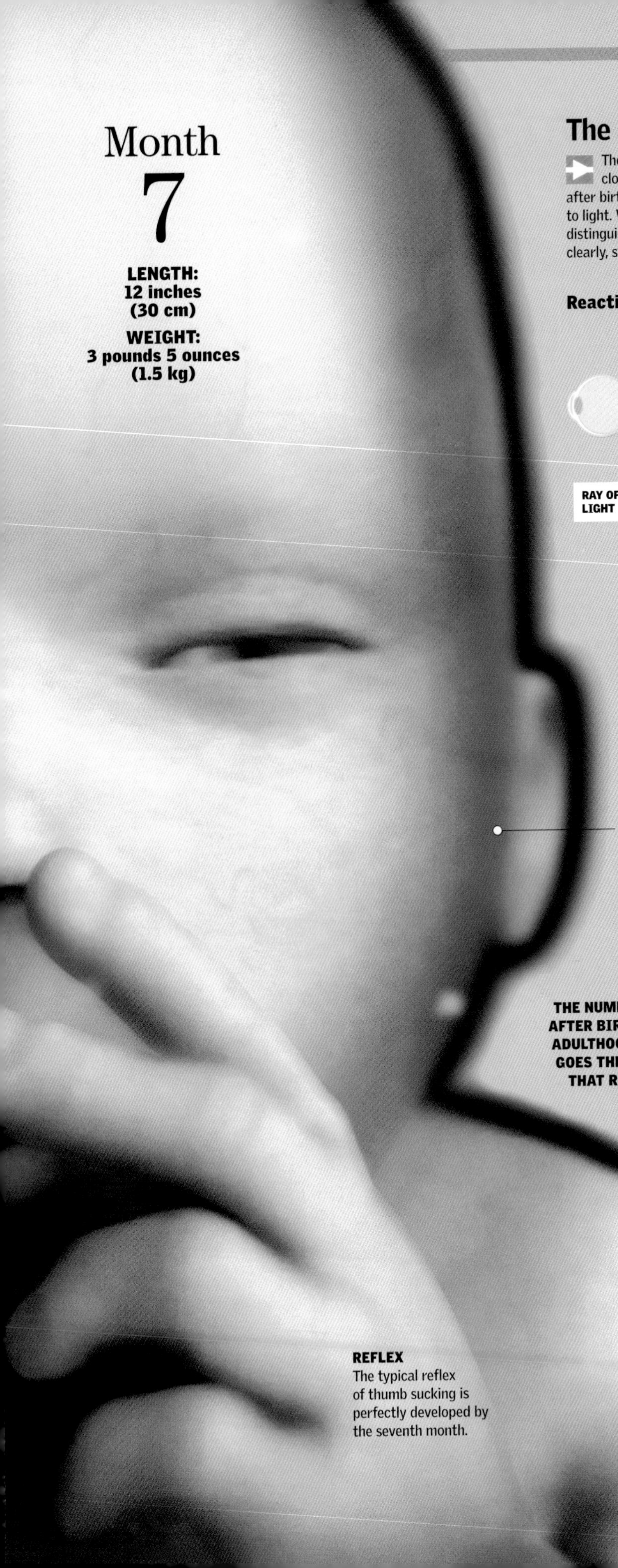

# Month 7

**LENGTH:**
**12 inches (30 cm)**

**WEIGHT:**
**3 pounds 5 ounces (1.5 kg)**

## The Fetus Opens its Eyes

The optical structure is practically fully formed. The fetus can open and close its eyes, which will keep their sky blue color until the second week after birth, since the definitive pigmentation is attained through exposure to light. With the general development of the eyes, the fetus can already distinguish changes from light to dark. It might also be able to see its hand clearly, since it puts it into its mouth with ease.

### Reaction to Light

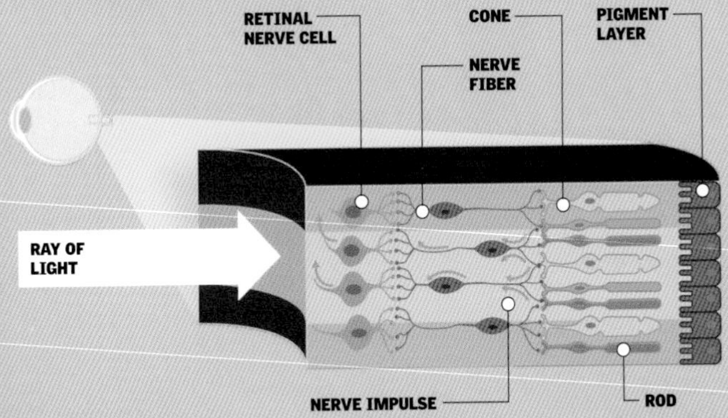


**1 LIGHT ENTERS**
It goes through the pupil and reaches the pigment layer in the retina.

**2 NERVE IMPULSE**
The cone and rod cells, when stimulated, transmit impulses to the fibers.

**3 RECEPTION**
The retinal nerve cells receive the impulse and relay the information to the brain.

**THE SKIN**
This is no longer transparent and takes on a more opaque color. Layers of fat begin to accumulate under the epidermis, which makes the skin smoother.

# 300

**THE NUMBER OF BONES IN A FETUS. AFTER BIRTH AND BEFORE REACHING ADULTHOOD, THE SKELETAL SYSTEM GOES THROUGH A FUSION PROCESS THAT REDUCES THE NUMBER OF BONES TO 206.**

## Central Nervous System

The folds in the cerebral cortex undergo rapid development that is more noticeable toward the end of the month. The body temperature and breathing are already controlled by the central nervous system, which controls the inhalation of air.

## Glucose Tolerance Test

In the seventh month of pregnancy, a crucial test is performed to detect the possible presence of gestational diabetes (diabetes that develops during pregnancy). In this test, called the Glucose Tolerance Test, or the O'Sullivan Test, a glucose load (about 1.8 ounces [50 g]) is administered orally to the fasting woman. An hour later, blood is drawn and the glucose level is measured.

**REFLEX**
The typical reflex of thumb sucking is perfectly developed by the seventh month.

# Crucial Moments

## Month 8

**LENGTH:**
**13.8 inches**
**(35 cm)**

**WEIGHT:**
**5 pounds 8 ounces**
**(2.5 kg)**

The eighth month of pregnancy brings many obvious changes to the fetus. The lanugo disappears from the fetus's face, and the limbs become chubby. Birth is imminent, and before the month is finished, most fetuses assume a head-down position. The space in which the fetus has to move within the uterus is minimal, so during this time the fetus remains almost still. Except for the lungs, the organs are fully functional. That is why birth at this stage entails many risks.

### Final Preparations

At the beginning of the eighth month, the unborn baby's kicks become increasingly forceful and frequent. The shifting toward its final position begins, which in most cases is cephalic (the head toward the pelvis), although sometimes it is breech (with the buttocks toward the pelvis). If the baby is in a breech position, a cesarean section might be necessary. It is common to do ultrasound exams at this stage to verify that the baby's weight is adequate.

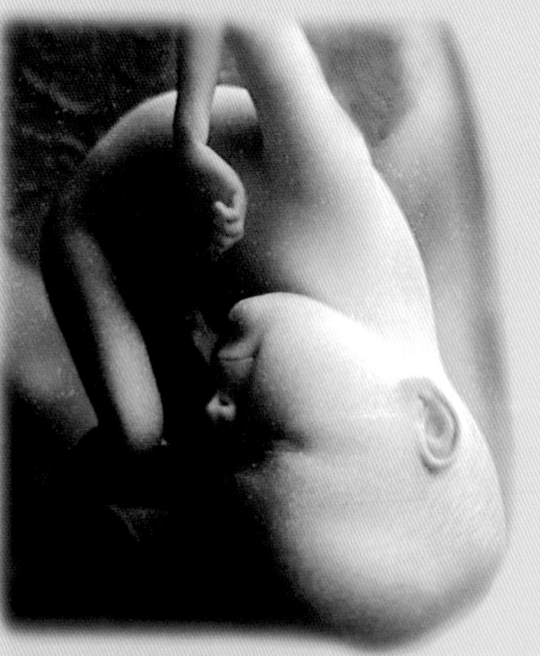

### 1 Reduced Space

Since the fetus has reached a considerable size, it now has little room in which to move. Hence, it begins to turn and kick forcefully.

### 2 Cephalic Presentation

In 90 percent of cases, the fetus is positioned so that the head will come out first during delivery.

### 3 Final Position

The fetus has assumed its final position before delivery. Its buttocks will start pressing against the mother's diaphragm.

### Appearance of the Pulmonary Surfactant

In the eighth month of pregnancy, a substance called surfactant appears in the alveoli. This liquid covers the alveoli, which are surrounded by blood vessels and provide the surface for gaseous exchange. The surfactant maintains equilibrium in the lungs and keeps them from completely collapsing after every breath. With the presence of proteins and lipids with hydrophobic and hydrophilic regions, water is absorbed by the former regions and air is absorbed by the latter. A baby born at eight months can have problems because it lacks surfactant.

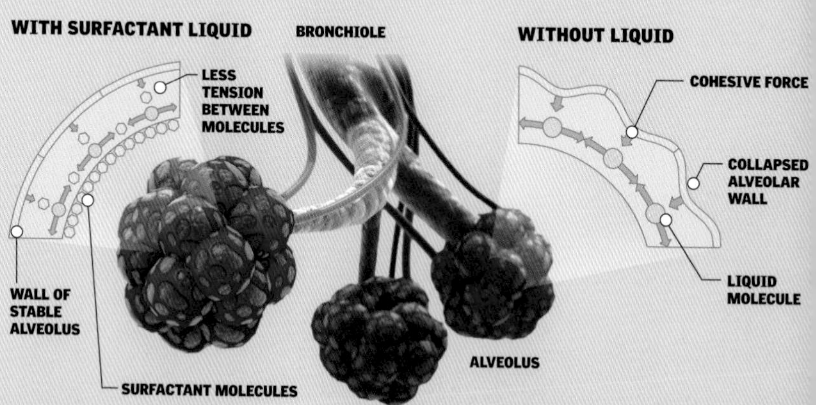

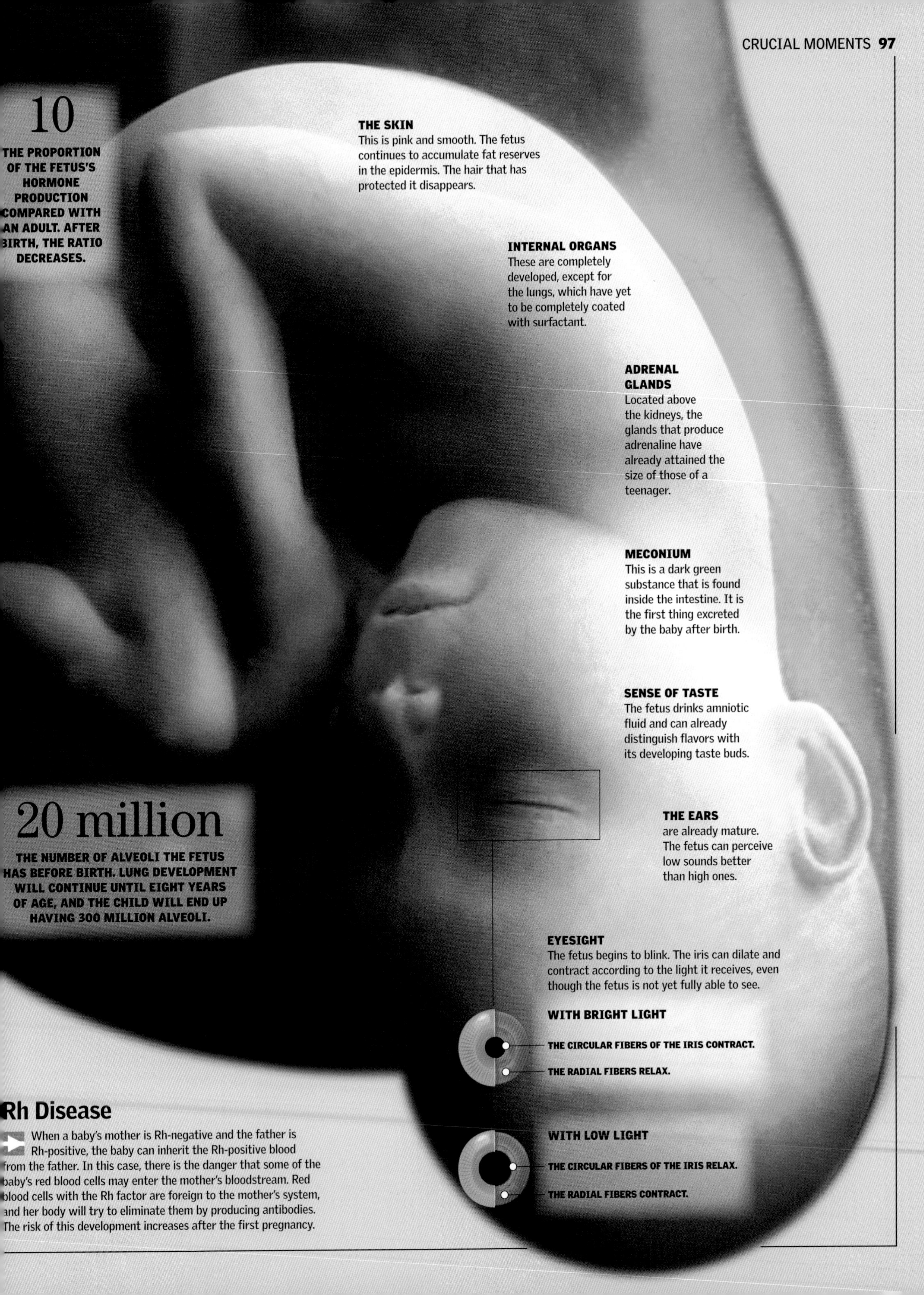

10
THE PROPORTION OF THE FETUS'S HORMONE PRODUCTION COMPARED WITH AN ADULT. AFTER BIRTH, THE RATIO DECREASES.
THE SKIN
This is pink and smooth. The fetus continues to accumulate fat reserves in the epidermis. The hair that has protected it disappears.
INTERNAL ORGANS
These are completely developed, except for the lungs, which have yet to be completely coated with surfactant.
ADRENAL GLANDS
Located above the kidneys, the glands that produce adrenaline have already attained the size of those of a teenager.
MECONIUM
This is a dark green substance that is found inside the intestine. It is the first thing excreted by the baby after birth.
SENSE OF TASTE
The fetus drinks amniotic fluid and can already distinguish flavors with its developing taste buds.
THE EARS
are already mature. The fetus can perceive low sounds better than high ones.
20 million
THE NUMBER OF ALVEOLI THE FETUS HAS BEFORE BIRTH. LUNG DEVELOPMENT WILL CONTINUE UNTIL EIGHT YEARS OF AGE, AND THE CHILD WILL END UP HAVING 300 MILLION ALVEOLI.
EYESIGHT
The fetus begins to blink. The iris can dilate and contract according to the light it receives, even though the fetus is not yet fully able to see.
WITH BRIGHT LIGHT
THE CIRCULAR FIBERS OF THE IRIS CONTRACT.
THE RADIAL FIBERS RELAX.
WITH LOW LIGHT
THE CIRCULAR FIBERS OF THE IRIS RELAX.
THE RADIAL FIBERS CONTRACT.

## Rh Disease

When a baby's mother is Rh-negative and the father is Rh-positive, the baby can inherit the Rh-positive blood from the father. In this case, there is the danger that some of the baby's red blood cells may enter the mother's bloodstream. Red blood cells with the Rh factor are foreign to the mother's system, and her body will try to eliminate them by producing antibodies. The risk of this development increases after the first pregnancy.

# Forty Weeks of Sweet Anticipation

The pregnancy is reaching its end. During the last few months, besides the enlargement of the belly and breasts, the mother has undergone many psychological and emotional changes because of altered hormone levels. Now, a step away from birth, it is possible that she might not sleep well and may tire easily. Moreover, in this situation, certain fears and anxieties are common to every woman, so it is best to be well informed.

## The alveolus

This is the functional unit that produces milk.

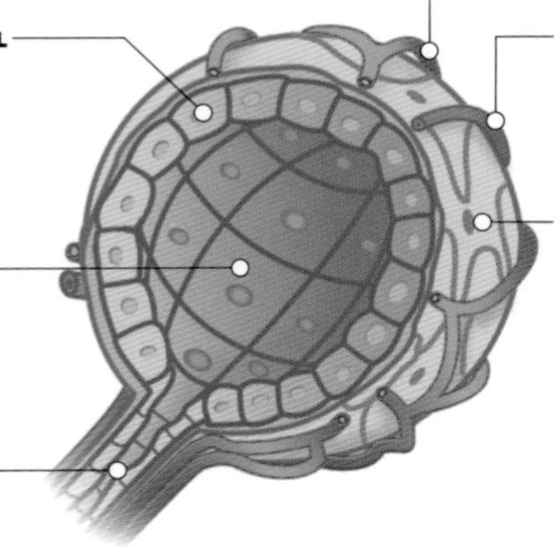


## Milk Ejection

When the ducts contract as a response to oxytocin (letdown reflex), the milk flows inside the galactophorous, or lactiferous, ducts toward the reservoir of the mammary gland.

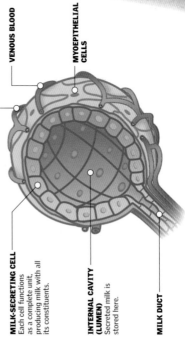

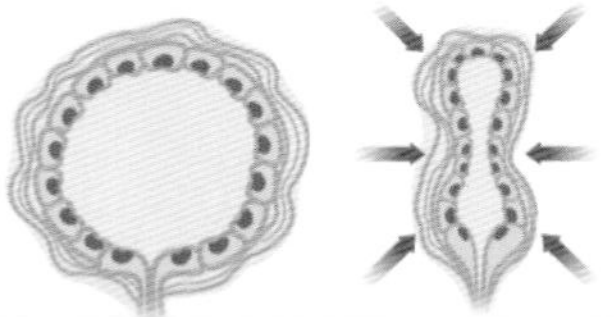

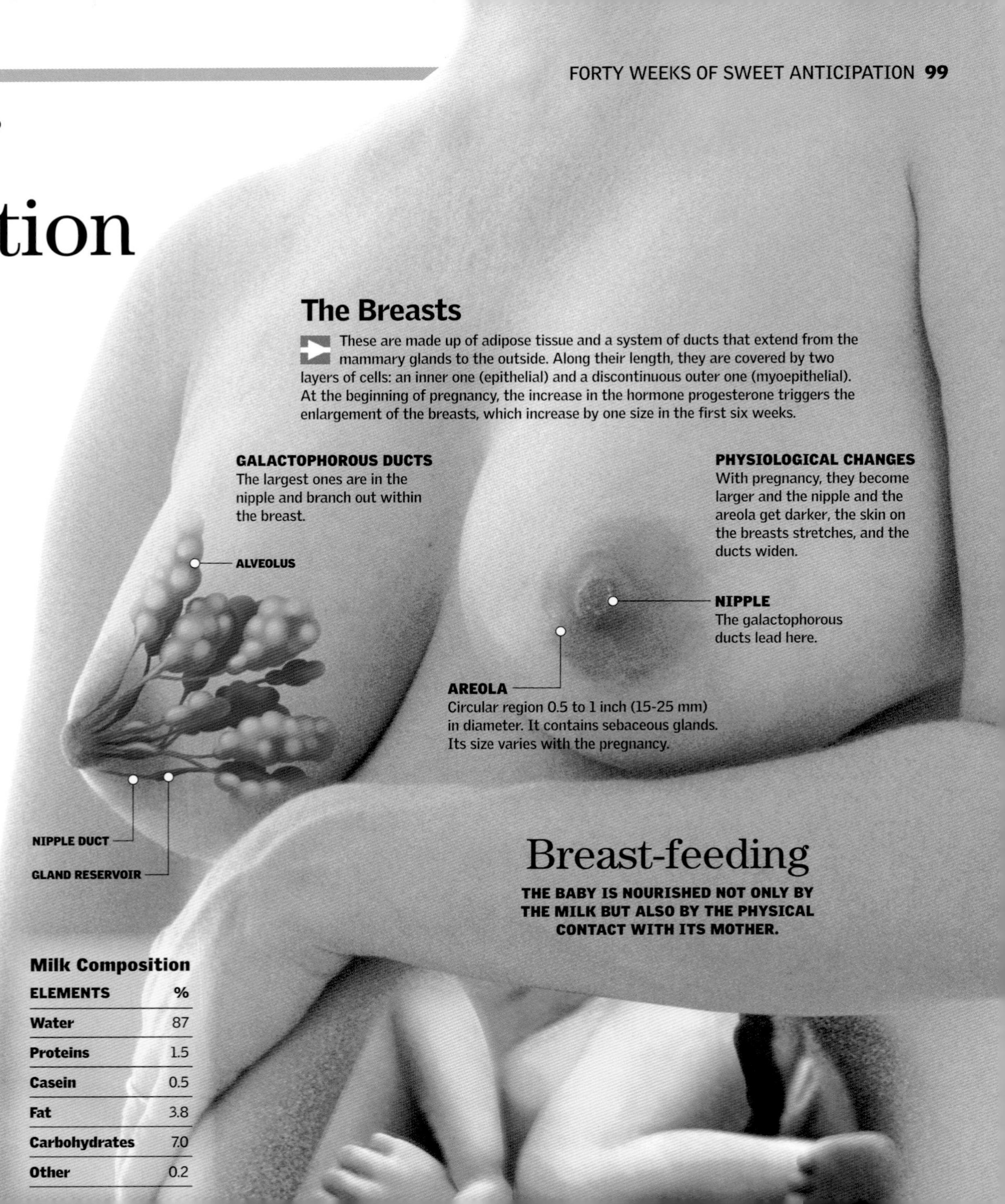

GALACTOPHOROUS DUCTS
The largest ones are in the nipple and branch out within the breast.
ALVEOLUS
NIPPLE DUCT
GLAND RESERVOIR
PHYSIOLOGICAL CHANGES
With pregnancy, they become larger and the nipple and the areola get darker, the skin on the breasts stretches, and the ducts widen.
NIPPLE
The galactophorous ducts lead here.
AREOLA
Circular region 0.5 to 1 inch (15-25 mm) in diameter. It contains sebaceous glands. Its size varies with the pregnancy.

## The Breasts

These are made up of adipose tissue and a system of ducts that extend from the mammary glands to the outside. Along their length, they are covered by two layers of cells: an inner one (epithelial) and a discontinuous outer one (myoepithelial). At the beginning of pregnancy, the increase in the hormone progesterone triggers the enlargement of the breasts, which increase by one size in the first six weeks.

### Milk Composition

| ELEMENTS | % |
|---|---|
| Water | 87 |
| Proteins | 1.5 |
| Casein | 0.5 |
| Fat | 3.8 |
| Carbohydrates | 7.0 |
| Other | 0.2 |

# Breast-feeding

**THE BABY IS NOURISHED NOT ONLY BY THE MILK BUT ALSO BY THE PHYSICAL CONTACT WITH ITS MOTHER.**

## Changes per Trimester

A pregnancy lasts 40 weeks, which by convention are divided into three trimesters. Each trimester corresponds to a series of more or less specific changes that come from the fetus's different developmental stages. Many of these transformations are painful, such as the pressure of the enlarged womb against the spine. There is also an increase in weight, dizziness, mood swings, and changes in heart rate.

**NEW LIFE**
The baby has grown from a mere embryo, and the woman's entire belly grows to accommodate this increase in size.

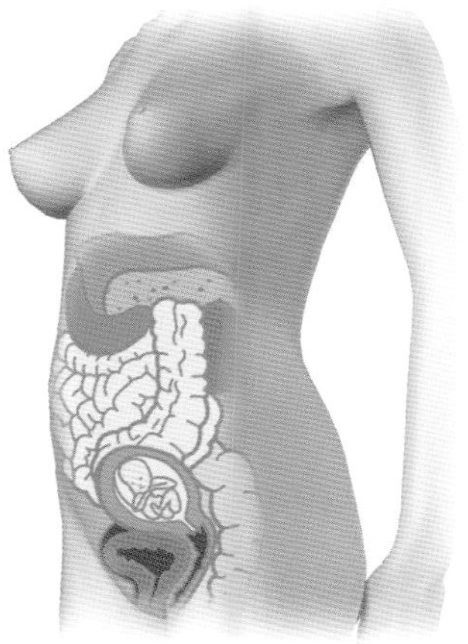

### 1 First Trimester

During the first trimester of pregnancy, the body prepares to carry the fetus. The woman's breasts grow and their conditioning for breastfeeding begins. Dizziness and nausea are frequent during this period, the cause of which is not precisely known. Also normal is an increased need to urinate due to the activity of certain hormones that generate a need to empty the bladder repeatedly. It may also be apparent that the waistline is beginning to fade.

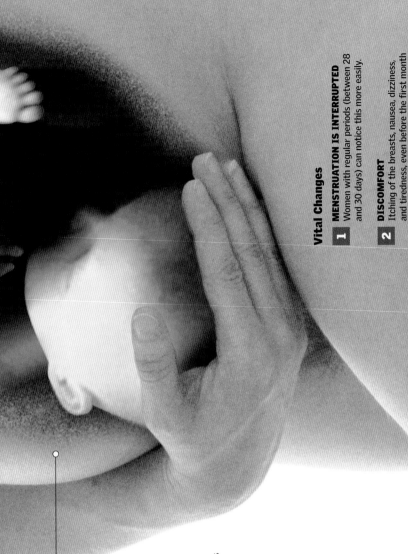

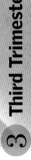

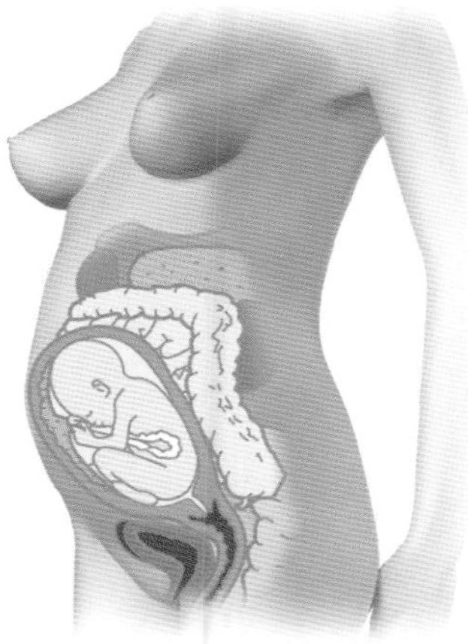

### 2 Second Trimester

This is the period when it first becomes noticeable that the woman is pregnant. The uterus now extends from the pubic bone to the navel, and the belly is noticeable. The heart rate is altered by changes in the circulatory system. Varicose veins can also form in the legs due to the difficulty blood has returning through the veins from the lower limbs.

### 3 Third Trimester

The skin stretches over the belly and very soft contractions begin to be felt. The uterus has grown and pushes on the bladder, which in some cases causes incontinence. In this period, back pains become more recurrent. The large volume of the belly can often cause deformations of the spine. In some women, breathing difficulties and repeated fatigue can develop. It is also normal to develop hemorrhoids.

**40%**

**MORE BLOOD IS PUMPED BY A PREGNANT WOMAN'S HEART.**

### Vital Changes

**1 MENSTRUATION IS INTERRUPTED**
Women with regular periods (between 28 and 30 days) can notice this more easily.

**2 DISCOMFORT**
Itching of the breasts, nausea, dizziness, and tiredness, even before the first month is finished.

**3 THE UTERUS EXPANDS**
At eight weeks, this is perceptible through a gynecological exam.

**4 MOVEMENTS ARE FELT**
Beginning in the fourth month, it is possible to perceive the movements of the fetus's hands and feet by ultrasound.

# Childbirth, One More Step

Finally, the long-awaited day has arrived—the end of gestation and the moment of delivery. Labor is said to begin with the onset of regular uterine contractions. Labor has four stages: dilation, expulsion, delivery proper, and delivery of the placenta. With each contraction, a little more of the baby's head appears, and after about 15 minutes, the rest of the body comes out by itself and the umbilical cord is cut.

## Labor

The labor process of birth is a joint effort between the fetus and the mother. Labor is divided into four stages: dilation, which starts with the contractions; expulsion, in which the fetus travels down the birth canal; delivery; and delivery of the placenta. Once the umbilical cord is cut, the newborn begins to breathe independently with its own respiratory system.

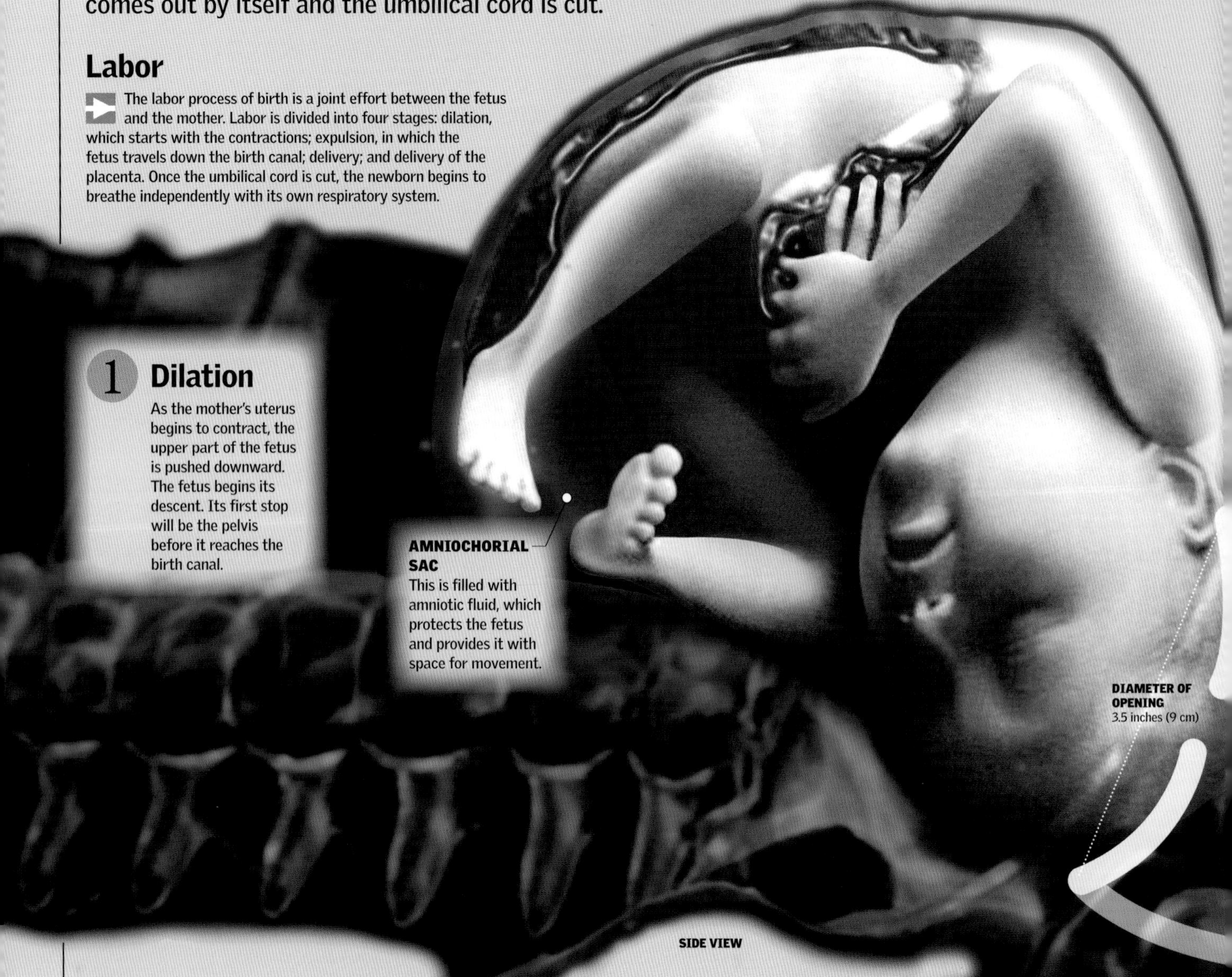


### 1 Dilation

As the mother's uterus begins to contract, the upper part of the fetus is pushed downward. The fetus begins its descent. Its first stop will be the pelvis before it reaches the birth canal.

**AMNIOCHORIAL SAC**
This is filled with amniotic fluid, which protects the fetus and provides it with space for movement.

## Fetal Monitoring

During labor, the fetus's heart rate, between 120 and 160 beats per minute, is monitored. The rate decreases with each contraction and then returns to normal. If this does not happen, it could be problematic.

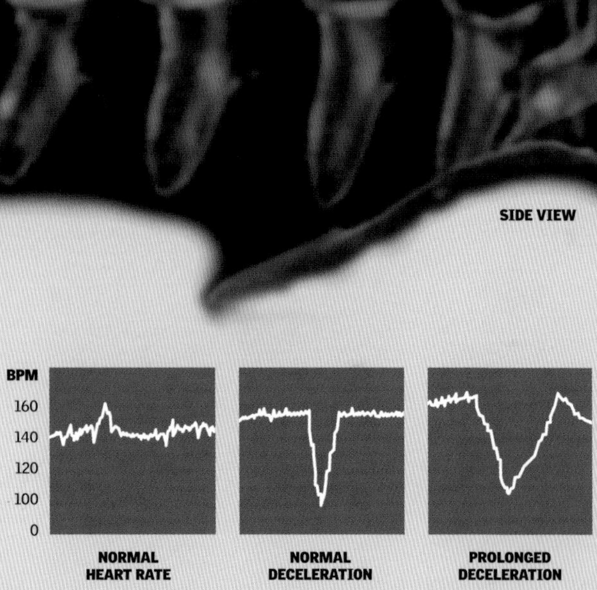


### 2 First Obstacle

The pelvis is the first obstacle that the fetus must face. To overcome it, the fetus adjusts its head according to the largest diameter, the oblique one, which is normally 4.3 inches (11 cm).

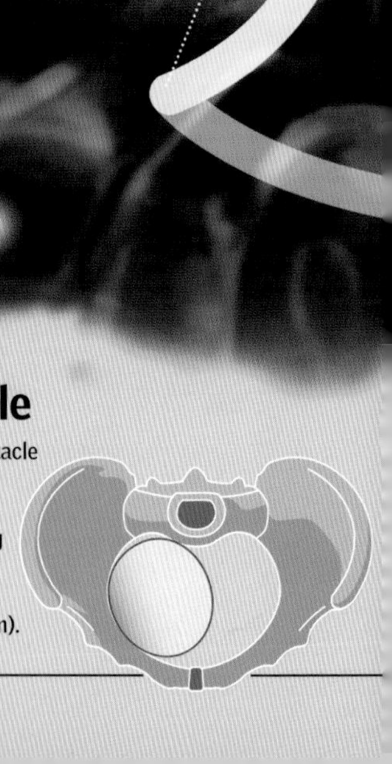

## Contractions

The regular and frequent contractions of the uterus generally appear on the date of delivery. They are indispensable for childbirth to be natural and spontaneous. The uterus is a muscle, and each contraction shortens the muscle fibers of the cervix and contracts it to open it. The stage of contractions is the first phase of labor and the most important. If it proceeds normally, the baby will come out of the uterus naturally and begin its journey to the outside. Without contractions, the mother will not be able to push the baby, and it will be necessary to resort to assisted-labor techniques.

### Pushing the Fetus

1. In preparation for delivery, the mother's uterus begins to contract at short intervals.
2. The mother's uterus contracts, putting more pressure on the upper part and pushing the fetus, which begins its descent.
3. The opening of the cervix dilates gradually with each contraction. The dilation is complete when it reaches 4 inches (10 cm).

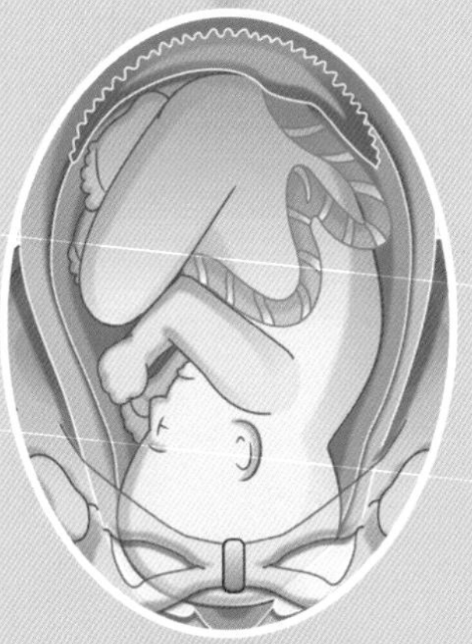

4 INCHES (10 CM)

### The Cervix

The contractions of the uterus cause the gradual dilation of the cervix. It dilates completely when the opening is 4 inches (10 cm) in diameter. From this moment, labor passes to the second stage. The amniotic membranes can rupture at any time.

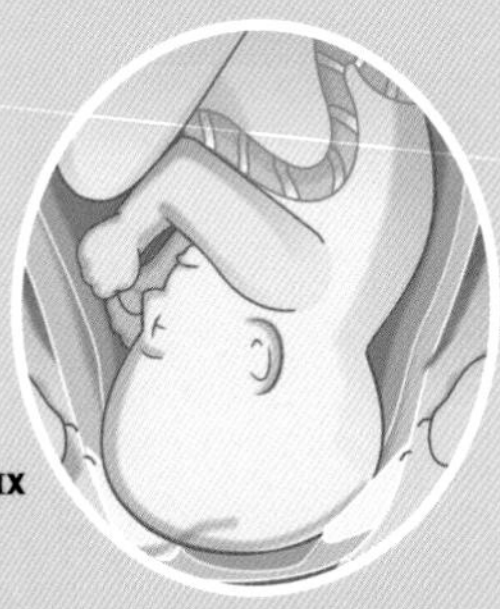

CERVIX

**Month 9**

**LENGTH:** 19.7 inches (50 cm)

**WEIGHT:** 6 pounds 10 ounces (3kg)

**0.4 inch (1 cm) per hour**

**IS THE RATE OF CERVIX DILATION FOR FIRST-TIME MOTHERS. THE RATE INCREASES WITH SUBSEQUENT BIRTHS.**

## The Pelvis

It is important to know the shape and size of the future mother's pelvis to determine how difficult delivery will be. Any difference between the dimensions of the mother's pelvis and the unborn baby's head could obstruct normal delivery.

**PELVIC ENTRANCE**
5.1 inches (13 cm)

**ROUND PELVIS**
This is the most common pelvis shape. Sometimes it may be oval-shaped. The pelvic exit usually has a diamond shape.

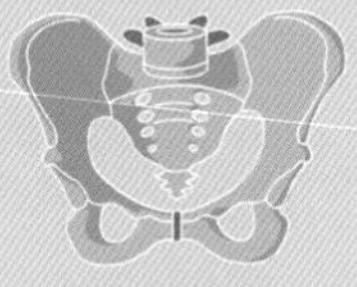

**PELVIC EXIT**
4.3 inches (11 cm)

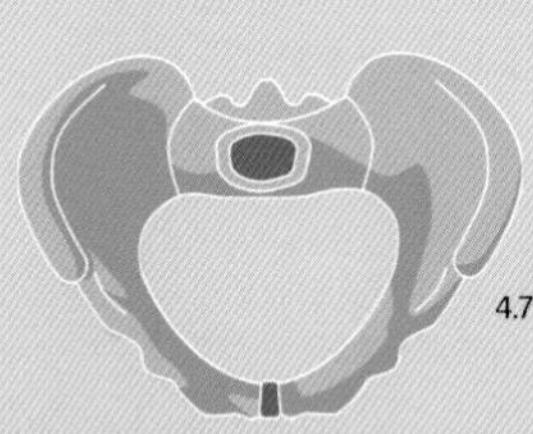

**PELVIC ENTRANCE**
4.7 inches (12 cm)

**TRIANGULAR PELVIS**
In some cases, the pelvic entrance is triangular and the exit is narrower. Delivery is more complicated in these cases.

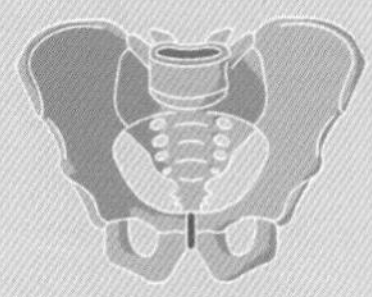

**PELVIC EXIT**
4 inches (10 cm)

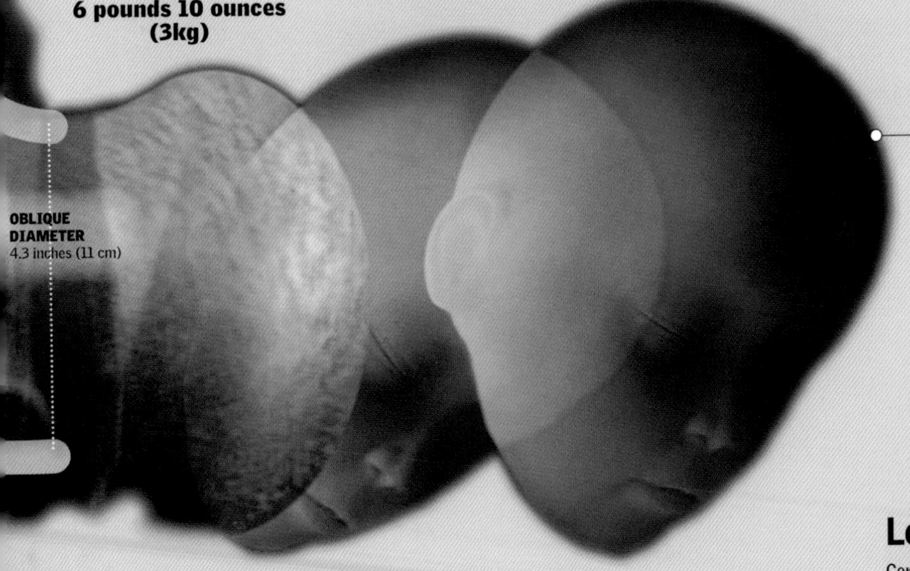


**THE SKULL**
Until 18 months after birth, the skull will have cracks between its bones that will later fuse.

## Relaxation

After each contraction, the mother should be able to relax the uterus so that the fetus gets enough oxygen. Without relaxation, the amount of blood reaching the fetus is reduced because the uterus flattens the blood vessels as it contracts.

## Less Pain

Certain natural techniques, such as relaxation and deep breathing, can help the mother experience less pain during childbirth. In other cases, a mixture of half air, half nitrous oxide can be administered by the doctor through a mask at the beginning of each contraction. Another option is the use of epidural anesthesia to relieve pelvic pain. This anesthesia is inserted through a needle into the spinal canal. Epidural anesthesia numbs the nerves that feed the pelvis and lower abdomen. This type of injection reduces the possibility of the mother feeling the contractions.

## 3 Birth Canal

The fetus finds that the birth canal has stretched. It rests its head on the pelvis and pushes against it. It pushes on the coccyx and is able to get its head out.

## 4 Exit to the Outside

Once the head passes through the birth canal, the baby passes its shoulders, one at a time. The rest of the body comes out without difficulty. Finally the umbilical cord is cut.

# After Childbirth

Once the baby is born, many changes take place in the child and in the mother. After the umbilical cord is cut, the baby begins to breathe on its own, and its circulatory system is autonomous. For the mother—in pain, with breasts full of milk, and a crying baby—the situation can be stressful. At this new stage, the best thing for the brand-new mother is to rely on her intuition to understand what it is that this much-anticipated baby needs. At the same time, the presence of an involved father will favor the development of a deeper and more intense bond with the child.

## Changes in Circulation

The fetus's circulatory system, which receives oxygen and nutrients from the placenta, is different from that of the baby after its umbilical cord is cut. The fetus's heart, which receives blood from the mother through the cord, has an oval opening called the foramen ovale. This hole, which allows blood to flow from the left atrium to the right one, closes after birth. The arterial duct, a tube that takes blood from the lungs to the aorta, also closes. The same happens with the umbilical blood vessels. When these ducts close, those that remain in the newborn's circulatory system become ligaments.

### Before the Umbilical Cord is Cut

1. The oxygenated blood enters the right atrium through the umbilical cord.
2. Since the lungs are contracted, they exert pressure in the opposite direction to that of the blood and force it to change direction.
3. The blood reaches the aorta mostly through the foramen ovale and, to a lesser extent, through the arterial duct. Once in the aorta, the blood is distributed throughout the body. This brings oxygen and nourishes the fetus.

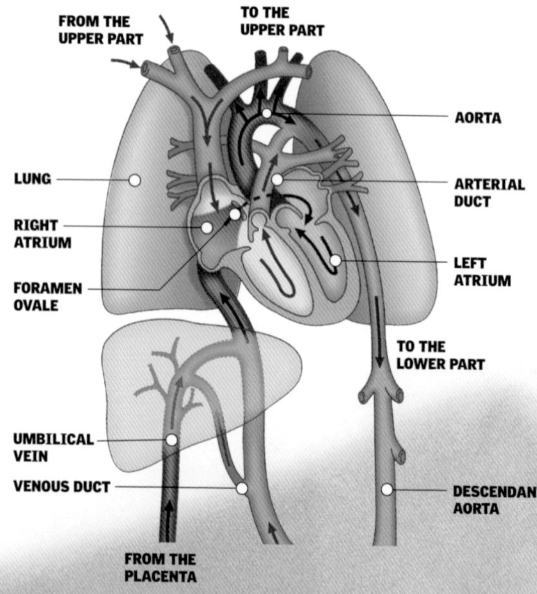


### After the Umbilical Cord is Cut

4. The newborn takes its first breath and fills its lungs with air for the first time. The direction of blood flow reverses.
5. Upon the cutting of the umbilical cord, the baby stops receiving blood from the mother.

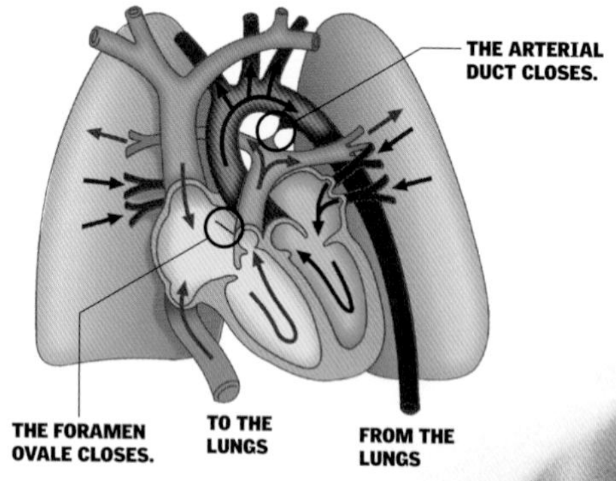


6. Blood is oxygenated in the lungs and reaches the aorta through the pulmonary veins. At the same time, the foramen ovale closes and a ligament forms in its place.

## Sexual Disorders

The months after childbirth are usually traumatic for the sex life of the couple. In the beginning, sexual desire may be diminished due to the place assumed by the baby as the new center of attention. Moreover, in the first three months after childbirth, there may be vaginal dryness stemming from a lack of lubrication caused by hormonal changes. It is also normal for intercourse to be painful because of the scarring of wounds caused by the delivery. It is all a matter of time—time to readjust to the new situation, to give oneself permission to experience new sensations.

## Hormonal Changes

During pregnancy, the levels of prolactin, a hormone produced in the anterior lobe of the pituitary gland (hypophysis), increase. This hormone remains at high levels while the mother breast-feeds. Prolactin is the hormone that causes milk production in the mammary glands. Another hormone released after pregnancy is oxytocin. Oxytocin brings on a reflex that causes milk to come out of the nipple. It is produced in the posterior lobe of the pituitary gland. The secretion of both prolactin and oxytocin, vital hormones in breast-feeding, is stimulated when the baby sucks on the breast. Milk production increases as the baby grows and requires more milk for feeding.

# 7.9 gallons (30 liters)

**IS THE AVERAGE AMOUNT OF MILK PRODUCED BY THE MOTHER IN A MONTH. THE BREAST MILK CONTAINS LACTOSE (A TYPE OF SUGAR), PROTEINS, AND FATS.**

## Everything Returns to Normal

During the postpartum period, the genital tract gradually returns to its state prior to the pregnancy. The uterus expels the remaining placental tissue in the form of a liquid called lochia, which is red at first, but later takes on a whitish color. The vagina gradually returns to its original size.

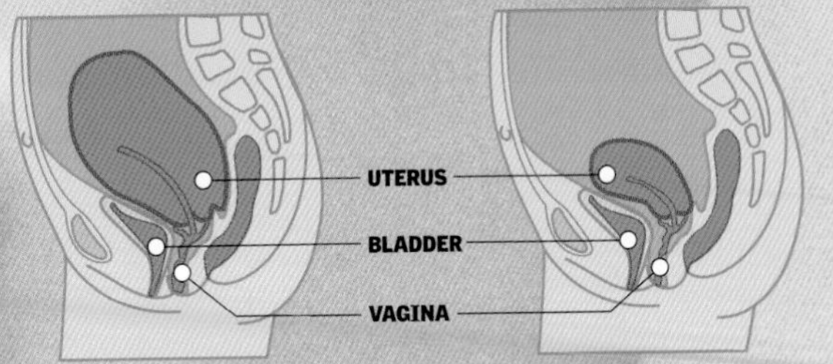


**1** After birth, within a month, the placental site has healed, but the uterus is still distended.

**2** Until the uterus returns to its original size, contractions continue and can be painful.

# 2 A PERFECT MACHINE

108

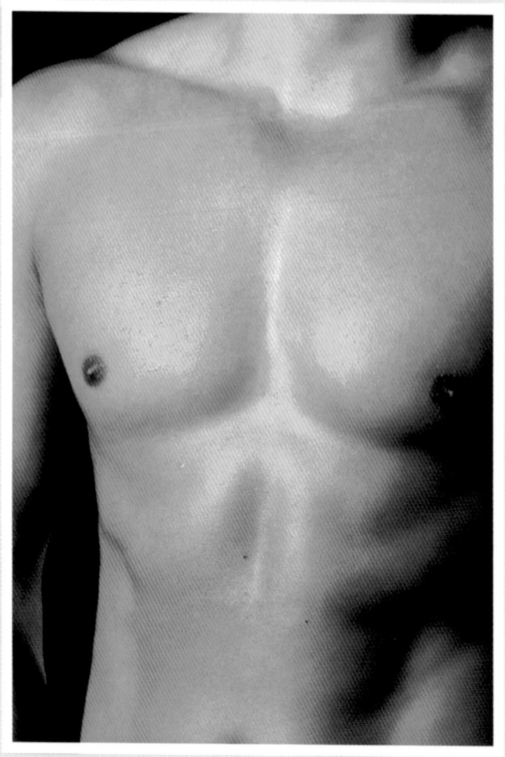

120

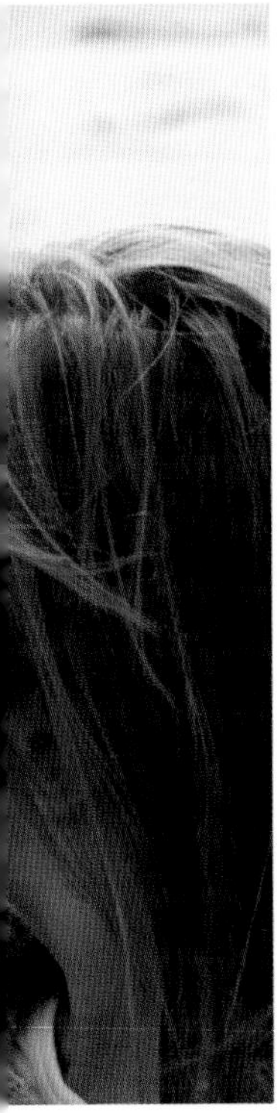
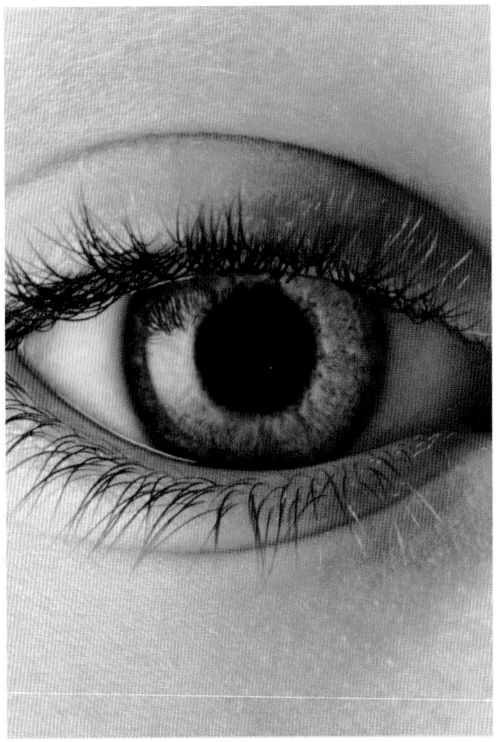
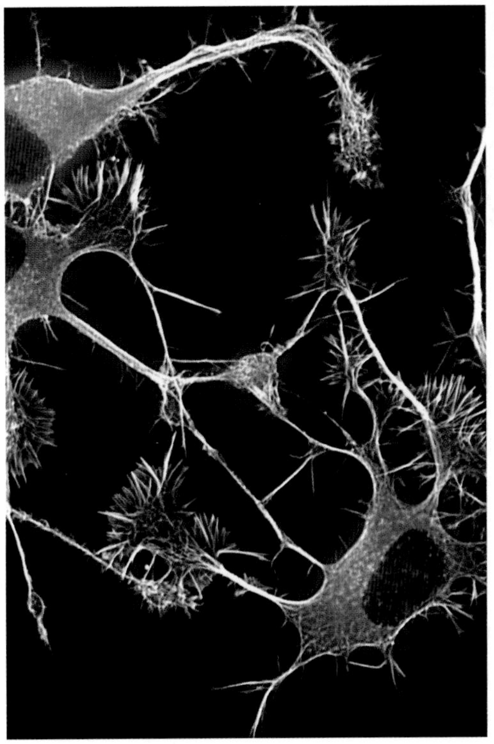

# A PERFECT MACHINE

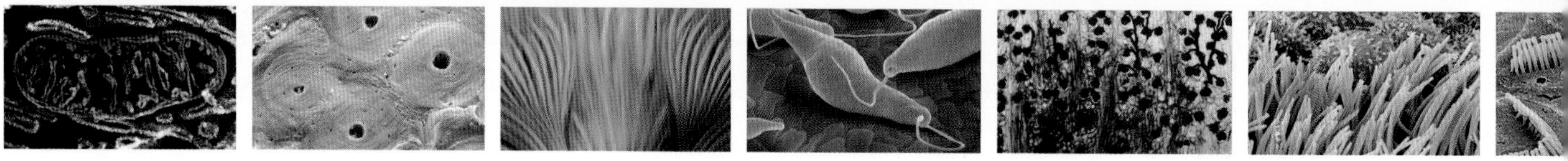

**How can we understand what we are? What are we made of? Are we aware that all that we do—including reading this book—is the work of a marvelous machine?**

We know very little about how we are able to be conscious of our own actions; nevertheless, even though we are usually not very aware of it, this community of organs that is the body—an integrated system that includes the brain, heart, lungs, liver, kidneys, muscles, bones, skin, and endocrine glands—acts together in exquisitely regulated harmony. It is interesting that various mechanisms work together to keep the temperature of the body at 98.6° F (37° C); thanks to the dynamic structure of bones and cartilage, the body is maintained in perfect balance. The body also has a fantastic ability to transform the food it ingests into living tissues, bones, and teeth, all of which contribute to its growth. By this same process, we obtain the energy for working and playing. It is hard to imagine that not long ago the cells of the body of the person reading this book were autonomous and were duplicating themselves freely within the walls of a mother's uterus. Certainly no one reading this book could recognize herself or himself in those cells. Nevertheless, each cell carried within it the information necessary for the development of that person. Everything that happens inside us is truly fascinating.

What are cells like, and how do they form tissue? What is blood, and why are proteins so important? The heart, usually thought of as the wellspring of love and the emotions, is actually the engine of the circulatory system. It is because of the heart that all the cells of the body receive a constant supply of nutrients, oxygen, and other essential substances. The nervous system is the most intricate of all the body's systems. It works every second of every day, gathering information about the organism and its surroundings and issuing instructions so that the organism can react. It is this computer that permits us to think and remember and that makes us who we are.

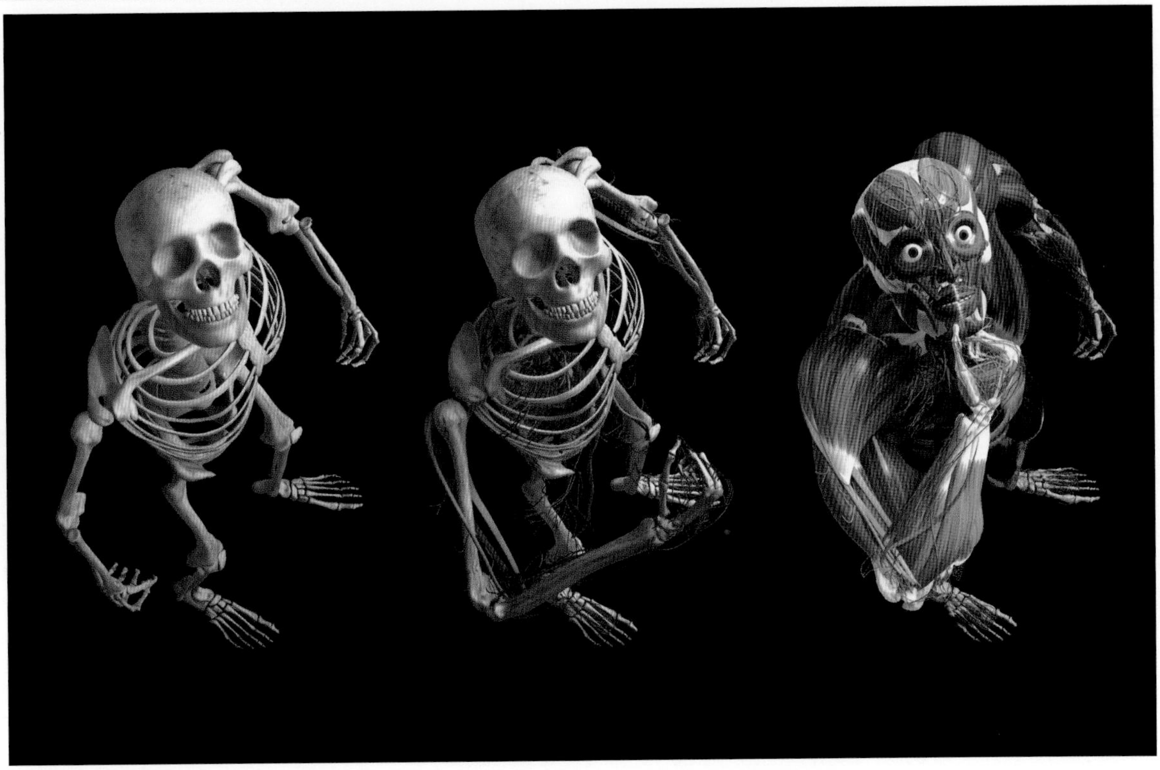

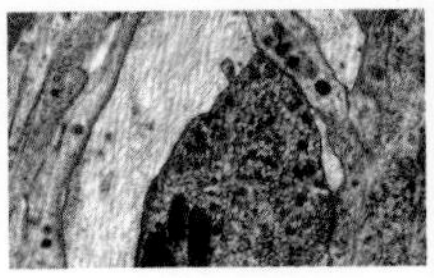

**A LIVING STRUCTURE**
The skeleton consists of 206 separate bones, which differ in form, size, and name. It supports and shapes the body, protects the internal organs, and—in the bone marrow of certain bones—manufactures various types of blood cells.

The nervous system is a complex network of sensory cells, originating in the brain and spinal cord, that transmits signals throughout the body, using a variety of chemical messengers to make sense of this marvelous complex that we describe as touch, taste, smell, hearing, and vision. Modern cameras are designed on the same basic principles as our eye, but they have never been able to equal the visual power of the eye. The focus and the automatic aperture of the human eye are perfect. Our ears share a similar complexity and allow us to have excellent hearing. The external ear operates by receiving sound waves in the air. Sound waves travel through the auditory canal and are transmitted by the bones of the intermediate ear toward the cochlea, which contains liquid and is spiraled like the shell of a small sea snail. The cochlea converts waves of air into vibrations of liquid, which are detected by special filaments in the ear that are of many lengths and that detect sound waves of different lengths. These filaments then transmit nerve impulses to the brain and provide us with our ability to interpret what we hear. This section of the book will also tell you about the function of our skin, the largest organ of the body, which serves as an elastic barrier covering and protecting everything inside our bodies. Captivating images will show you how each of our extraordinary body systems function, and incredible facts will help you understand why the human body is so amazing.

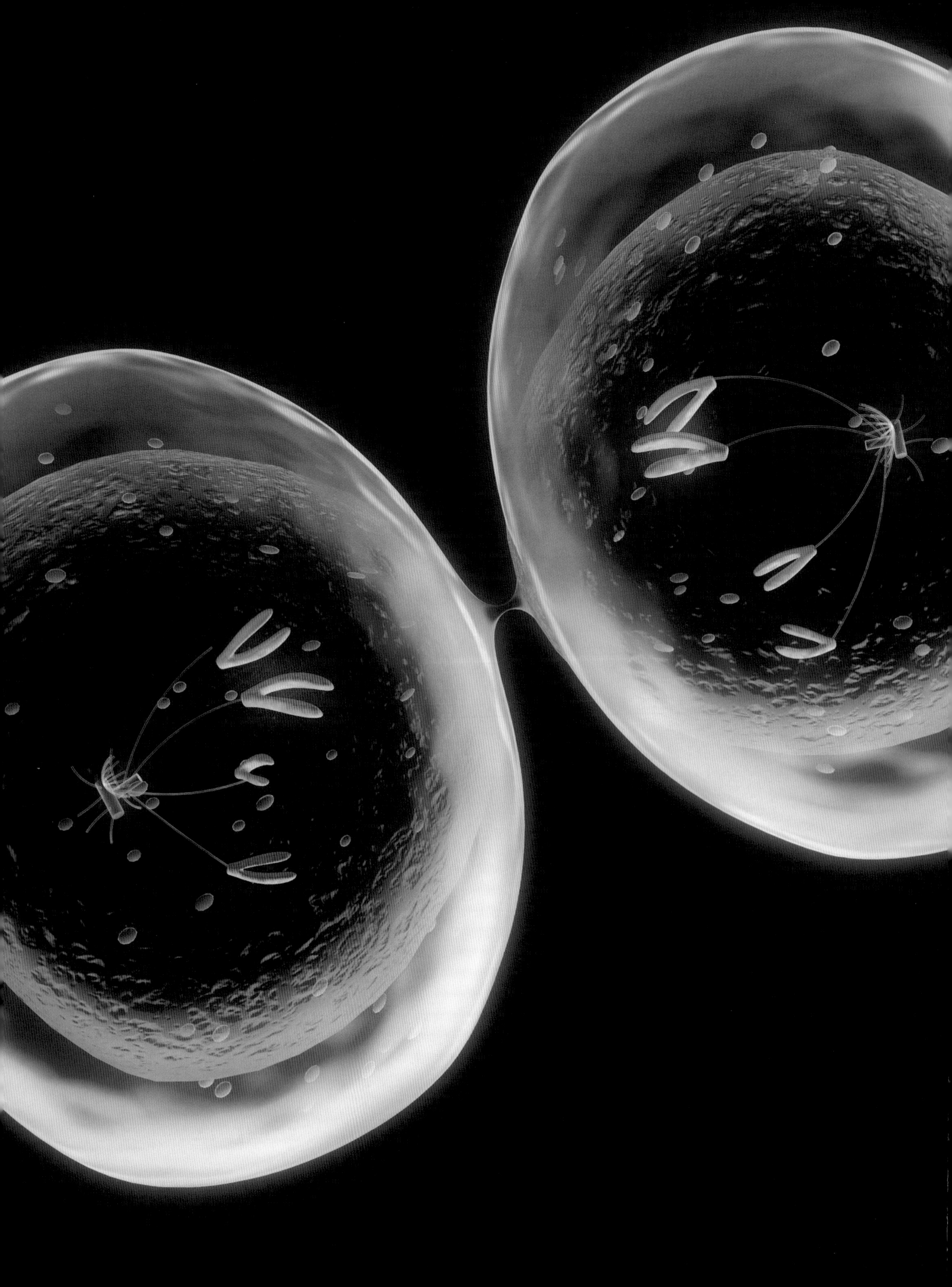

# What Are We Made Of?

To understand the truest and most elementary characteristics of life, we must begin with the cell—the tiny organizing structure of life in all its forms. Most cells are too small to be observed with the naked eye, but they can be distinguished easily through an ordinary microscope. Human body tissues are groups of cells whose size and shape depend on the specific tissue to which they belong. Did you know that an embryo is a mass of rapidly dividing cells that continue to develop during infancy? We invite you to turn the page and discover many surprising things in this fascinating and complex world.



**MITOSIS** (opposite)
An enlarged view that shows the process of mitosis, the most common form of cellular division.

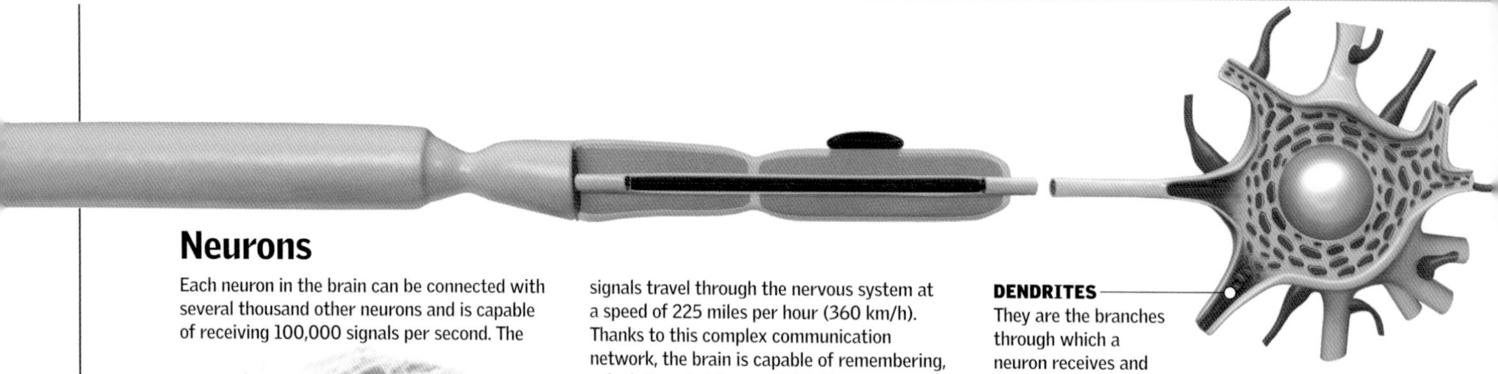

## Neurons

Each neuron in the brain can be connected with several thousand other neurons and is capable of receiving 100,000 signals per second. The signals travel through the nervous system at a speed of 225 miles per hour (360 km/h). Thanks to this complex communication network, the brain is capable of remembering, calculating, deciding, and thinking.

**DENDRITES**
They are the branches through which a neuron receives and sends messages. With this system each neuron can be stimulated by thousands of other neurons, which in turn can stimulate other neurons, and so forth.

# Undivided Attention

From birth the infant's brain cells develop rapidly, making connections that can shape all of life's experiences. The first three years are crucial. When neurons receive visual, auditory, or gustatory stimuli, they send messages that generate new physical connections with neighboring cells. The signals are sent through a gap called a synapse by means of a complex electrochemical process. What determines the formation of a person's synapses and neural networks? One key factor is believed to be the undivided attention and mental effort exerted by the person.

## Learning

Each child has his or her own intellectual filter; the quality of the filter depends on undivided attention and on how the child responds to a broad variety of stimuli.

225 miles per hour (360 km/h)

**THE VELOCITY OF THE NERVOUS SYSTEM'S SIGNALS.**

## Brain

At birth the infant brain contains 100 billion neurons. That is about as many nerve cells as there are stars in the entire Milky Way Galaxy! Then as the infant receives messages from the senses, the cerebral cortex begins its dynamic development.

**3 pounds (1.4 kg)**

**IS THE WEIGHT OF A HUMAN BRAIN.**

## Respiration

Respiration is usually an involuntary, automatic action that allows us to take in the oxygen we need from the air and exhale carbon dioxide. These gases are exchanged in the pulmonary alveoli.

**A WORLD OF SENSATIONS**

The tongue recognizes four tastes (sweet, salty, sour, and bitter), and the nasal fossas contain cells that have more than 200 million filaments, called cilia, which are capable of detecting thousands of odors.

**THE SENSE OF TOUCH**

It is predominant in the fingers and hands. The information is transmitted along nerves, which carry impulses to the brain and serve to detect sensations such as cold, heat, pressure, and pain.

**SKIN**

The skin is one of the most important organs of the body. It contains approximately five million tiny nerve endings that transmit sensations.

# Water and Fluids

Water is of such great importance that it makes up almost two thirds of the human body by weight. Water is present in all the tissues of the body. It plays a fundamental role in digestion and absorption and in the elimination of indigestible metabolic waste. Water also serves as the basis of the circulatory system, which uses blood to distribute nutrients to the entire body. Moreover, water helps maintain body temperature by expelling excess heat through the skin via perspiration and evaporation. Perspiration and evaporation of water account for most of the weight a person loses while exercising.

## Water Balance and Food

In its continuous process of taking in and eliminating water, one of the most important functions of the body is to maintain a continuous equilibrium between the water that enters and the water that leaves the body. Because the body does not have an organ or other place for storing water, quantities that are lost must be continuously replenished. The human body can survive for several weeks without taking in food, but going without water for the same length of time would have tragic consequences. The human being takes in about 2.5 to 3 quarts (2.5-3 l) of water per day. About half is taken in by drinking, and the rest comes from eating solid food. Some foods, such as fruits and vegetables, consist of 95 percent water. Eggs are 90 percent water, and red meat and fish are 60 to 70 percent water.

**HOW WATER IS ABSORBED**
Water for the body is obtained primarily by drinking and ingesting food and through internal chemical reactions.

**50%**
of the water comes from ingesting fluids.

**35%**
of the water is obtained from food.

**15%**
comes from metabolic activities.

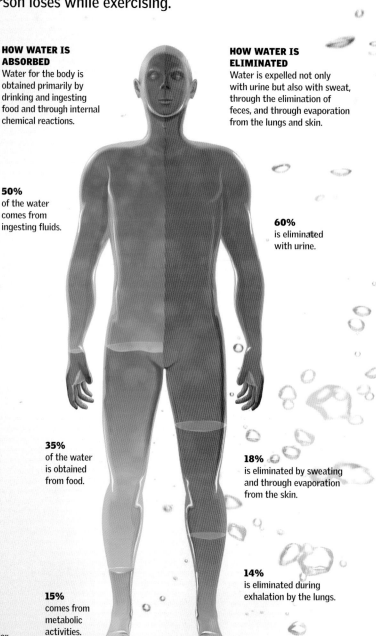

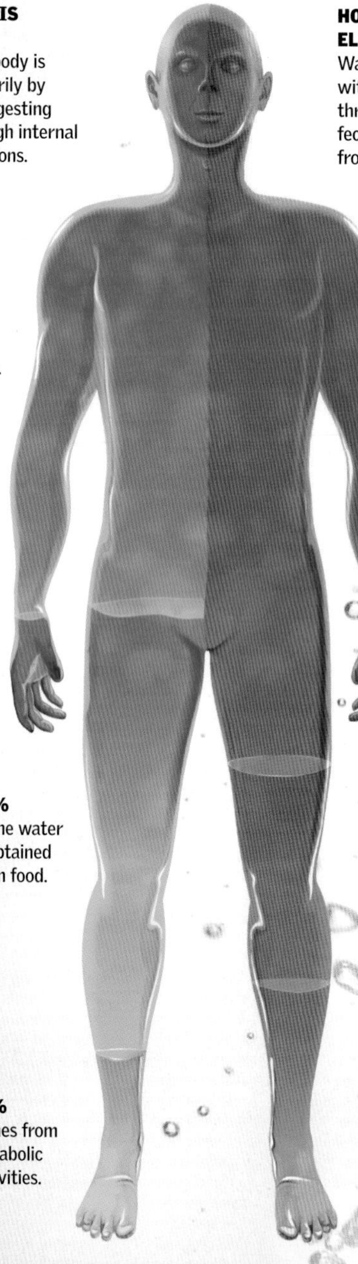

**HOW WATER IS ELIMINATED**
Water is expelled not only with urine but also with sweat, through the elimination of feces, and through evaporation from the lungs and skin.

**60%**
is eliminated with urine.

**18%**
is eliminated by sweating and through evaporation from the skin.

**14%**
is eliminated during exhalation by the lungs.

**8%**
is eliminated in excrement.

# 60%

**THE PERCENTAGE OF A PERSON'S WEIGHT THAT IS DUE TO WATER. IN GENERAL, A 10 PERCENT LOSS OF WATER LEADS TO SERIOUS DISORDERS, AND A LOSS OF 20 PERCENT RESULTS IN DEATH.**

**HOW THIRST IS CONTROLLED**
Thirst is the sensation through which the nervous system informs its major organ, the brain, that the body needs water. The control center is the hypothalamus. If the concentration of plasma in the blood increases, it means the body is losing water. Dry mouth and a lack of saliva are also indications that the body needs water.

## Chemical Elements

The body contains many chemical elements. The most common are oxygen, hydrogen, carbon, and nitrogen, which are found mainly in proteins. Nine chemical elements are present in moderate amounts, and the rest (such as zinc) are present only in very small amounts, so they are called trace elements.

**MAGNESIUM 0.05%**
Lungs, kidneys, liver, thyroid, brain, muscles, heart.

**SODIUM 0.15%**
Fluids and tissues, in the form of salt.

**POTASSIUM 0.3%**
Nerves and muscles; inside the cell.

**SULFUR 0.3%**
Contained in numerous proteins, especially in the contractile proteins.

**CALCIUM 1.5%**
Bones, lungs, kidneys, liver, thyroid, brain, muscles, heart.

**CHLORINE 0.2%**
maintains the equilibrium of water in the blood.

**PHOSPHORUS 1%**
Urine, bones.

**0.004% IRON**
Fluids and tissues, bones, proteins. An iron deficiency causes anemia, whose symptoms include fatigue and paleness. Iron is essential for the formation of hemoglobin in the blood.

**0.0004% IODINE**
Urine, bones. When consumed, iodine passes into the blood and from there into the thyroid gland. Among its other functions, iodine is used by the thyroid to produce growth hormones for most of the organs and for brain development.

## Proteins

Proteins are formed through the combination of the four most common chemical elements found in the body. Proteins include insulin, which is secreted by the pancreas to regulate the amount of sugar in the blood.

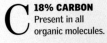

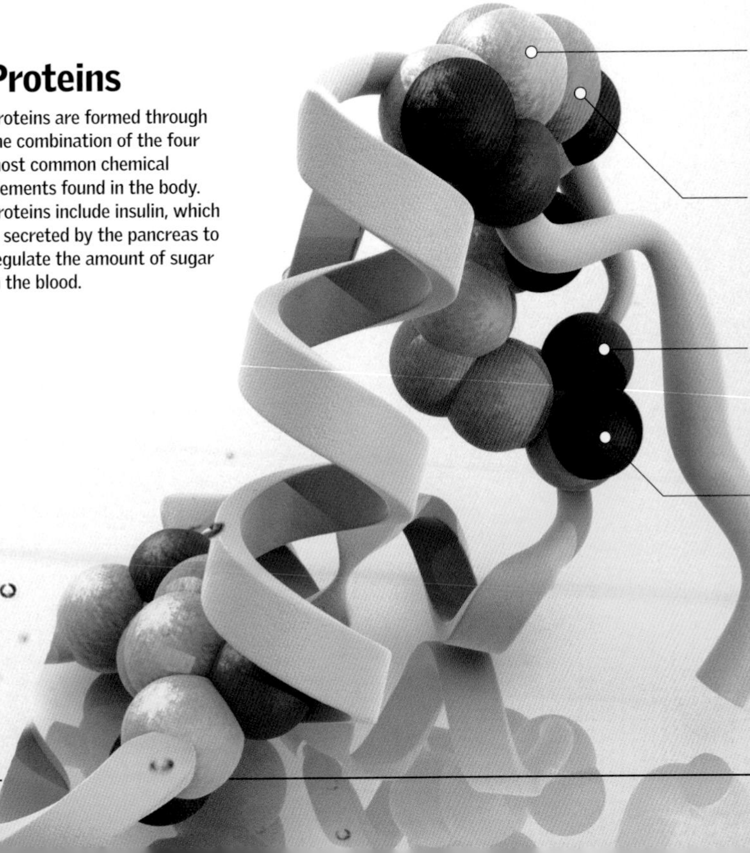

C **18% CARBON**
Present in all organic molecules.

H **10% HYDROGEN**
Present in water, nutrients, and organic molecules.

N **3% NITROGEN**
Present in proteins and nucleic acids.

O **65% OXYGEN**
Present in water and in almost all organic molecules.

# The Cell

It is the smallest unit of the human body—and of all living organisms—able to function autonomously. It is so small that it can be seen only with a microscope. Its essential parts are the nucleus and cytoplasm, which are surrounded by a membrane. Each cell reproduces independently through a process called mitosis. The animal kingdom does have single-celled organisms, but in a body such as that of a human being millions of cells are organized into tissues and organs. The word "cell" comes from Latin; it is the diminutive of cella, which means "hollow." The science of studying cells is called cytology.

## Cell Theory

Before the invention of the microscope, it was impossible to see cells. Some biological theories were therefore based on logical speculations rather than on observation. People believed in "spontaneous generation" because it was inconceivable that cells would regenerate. The development of the microscope, including that of an electronic version in the 20th century, made detailed observation of the internal structure of the cell possible. Robert Hooke was the first to see dead cells in 1665. In 1838 Mathias Schleiden observed living cells, and in 1839, in collaboration with Theodor Schwann, he developed the first theory of cells: that all living organisms consist of cells.

**THEODOR SCHWANN**

**MATHIAS SCHLEIDEN**

**UNDER THE MICROSCOPE**
This cell has been magnified 4,000 times with an electron microscope. The nucleus is clearly visible, along with some typical organelles in the green-colored cytoplasm.

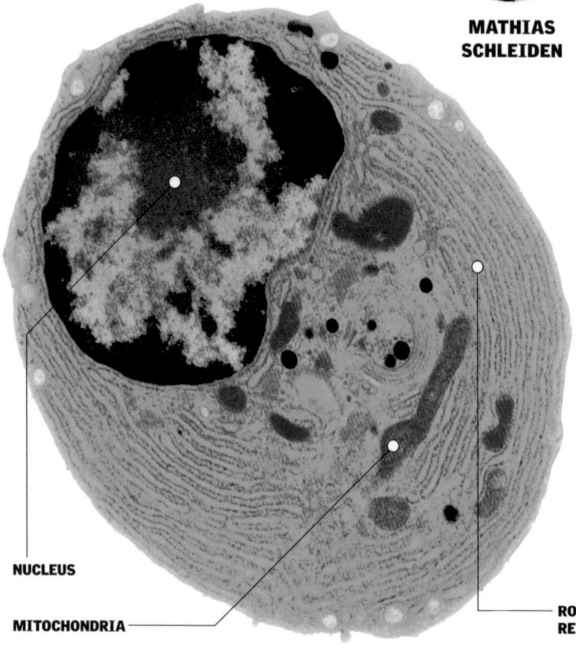

NUCLEUS
MITOCHONDRIA
ROUGH ENDOPLASMIC RETICULUM

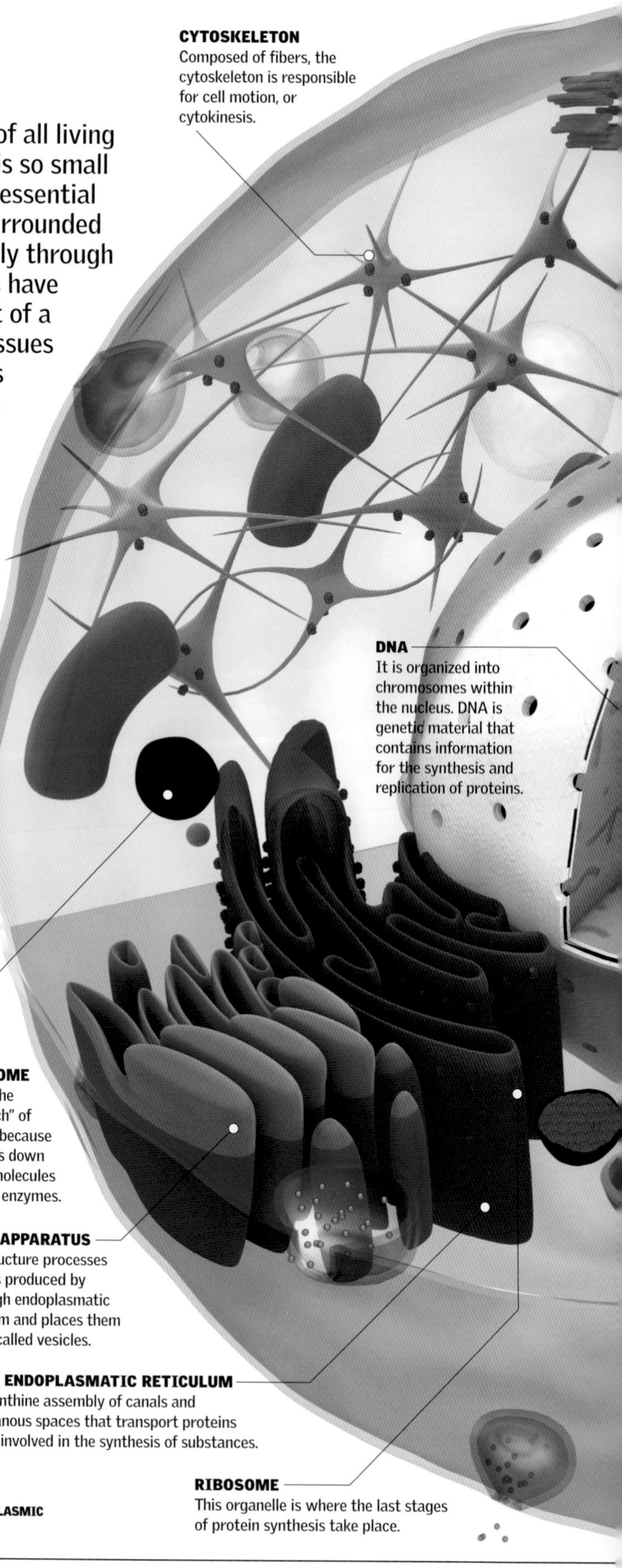

CYTOSKELETON
Composed of fibers, the cytoskeleton is responsible for cell motion, or cytokinesis.
DNA
It is organized into chromosomes within the nucleus. DNA is genetic material that contains information for the synthesis and replication of proteins.
LYSOSOME
This is the "stomach" of the cell because it breaks down waste molecules with its enzymes.
GOLGI APPARATUS
This structure processes proteins produced by the rough endoplasmatic reticulum and places them in sacs called vesicles.
ROUGH ENDOPLASMATIC RETICULUM
A labyrinthine assembly of canals and membranous spaces that transport proteins and are involved in the synthesis of substances.
RIBOSOME
This organelle is where the last stages of protein synthesis take place.

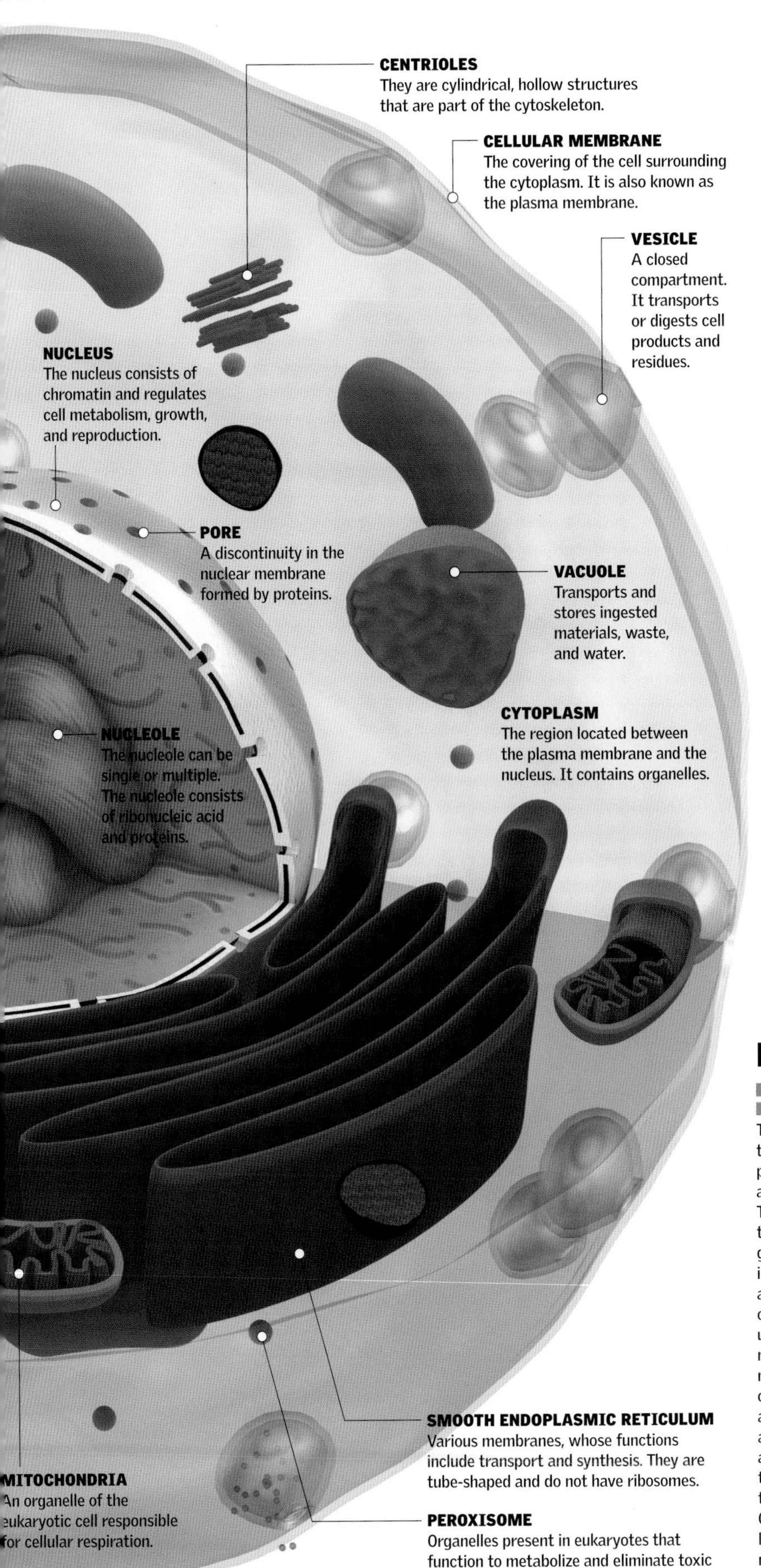
CENTRIOLES
They are cylindrical, hollow structures that are part of the cytoskeleton.
CELLULAR MEMBRANE
The covering of the cell surrounding the cytoplasm. It is also known as the plasma membrane.
VESICLE
A closed compartment. It transports or digests cell products and residues.
NUCLEUS
The nucleus consists of chromatin and regulates cell metabolism, growth, and reproduction.
PORE
A discontinuity in the nuclear membrane formed by proteins.
VACUOLE
Transports and stores ingested materials, waste, and water.
NUCLEOLE
The nucleole can be single or multiple. The nucleole consists of ribonucleic acid and proteins.
CYTOPLASM
The region located between the plasma membrane and the nucleus. It contains organelles.
SMOOTH ENDOPLASMIC RETICULUM
Various membranes, whose functions include transport and synthesis. They are tube-shaped and do not have ribosomes.
MITOCHONDRIA
An organelle of the eukaryotic cell responsible for cellular respiration.
PEROXISOME
Organelles present in eukaryotes that function to metabolize and eliminate toxic substances from cells.

**TRANSPORT MECHANISMS**

The cell membrane is a semipermeable barrier. The cell exchanges nutrients and waste between its cytoplasm and the extracellular medium via passive and active transport mechanisms.

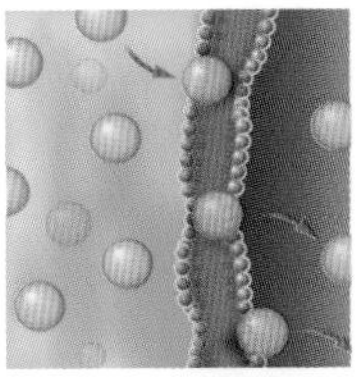

**DIFFUSION** It is a passive transport mechanism in which the cell does not use energy. The particles that cross the cell membrane do so because of a concentration gradient. For example, water, oxygen, and carbon dioxide circulate by diffusion.

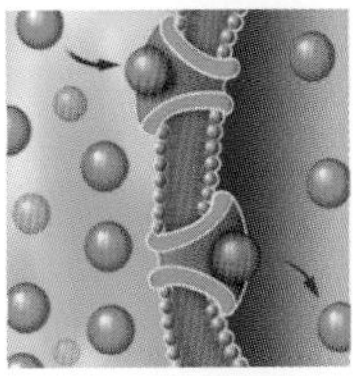

**FACILITATED DIFFUSION** Passive transport in which substances, typically ions (electrically charged particles), that because of their size could not otherwise penetrate the cell's bilayer, can do so through a pore consisting of proteins. Glucose enters the cell in this way.

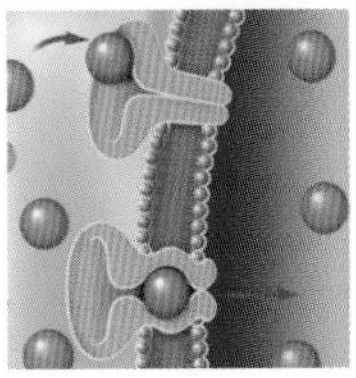

**ACTIVE TRANSPORT** It occurs by means of proteins and requires energy consumption by the cell because the direction of ion transport is against the concentration gradient. In some cells, such as neurons, the Na+/K+ pump uses active transport to move ions into or out of the cell.

# 1 billion

**THE AVERAGE NUMBER OF CELLS IN THE BODY OF AN ADULT. ONE CELL ALONE CAN DIVIDE UP TO 50 TIMES BEFORE DYING.**

## Mitochondria

The mitochondria provide large amounts of energy to the cell. They contain a variety of enzymes that, together with oxygen, degrade products derived from glycolysis and carry out cellular respiration. The amount of energy obtained in this process is almost 20 times as great as that released by glycolysis in the cytoplasm. Mitochondria are very different from other organelles because they have a unique structure: an external membrane enclosing an internal membrane with a great number of folds that delimit the internal area, or mitochondrial matrix. In addition, the mitochondria have a circular chromosome similar to that of bacteria that allows the mitochondria to replicate. Cells that need a relatively large amount of energy have many mitochondria because the cells reproduce frequently.

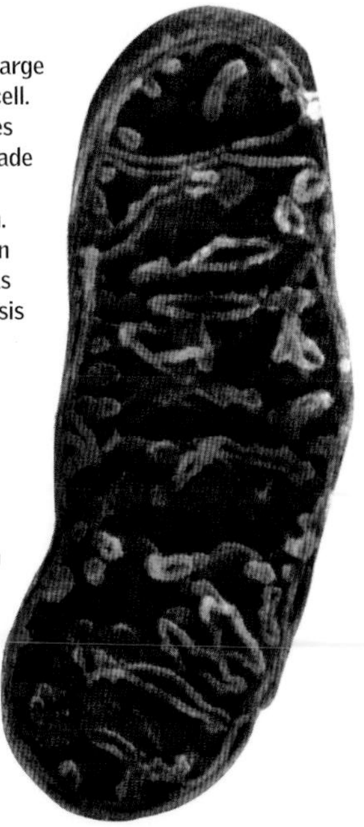

# Mitosis

It is the cell-division process that results in the formation of cells that are genetically identical to the original (or mother) cell and to each other. The copies arise through replication and division of the chromosomes, or genetic material, in such a way that each of the daughter cells receives a similar inheritance of chromosomes. Mitosis is characteristic of eukaryotic cells. It ensures that the genetic information of the species and the individual is conserved. It also permits the multiplication of cells, which is necessary for the development, growth, and regeneration of the organism. The word "mitosis" comes from the Greek *mitos*, which means "thread," or "weave."

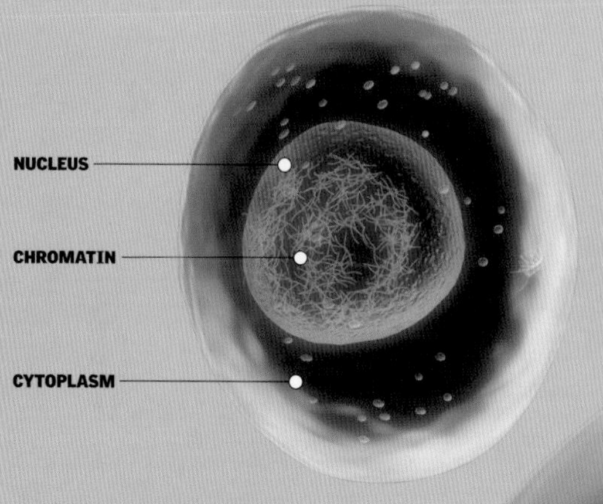


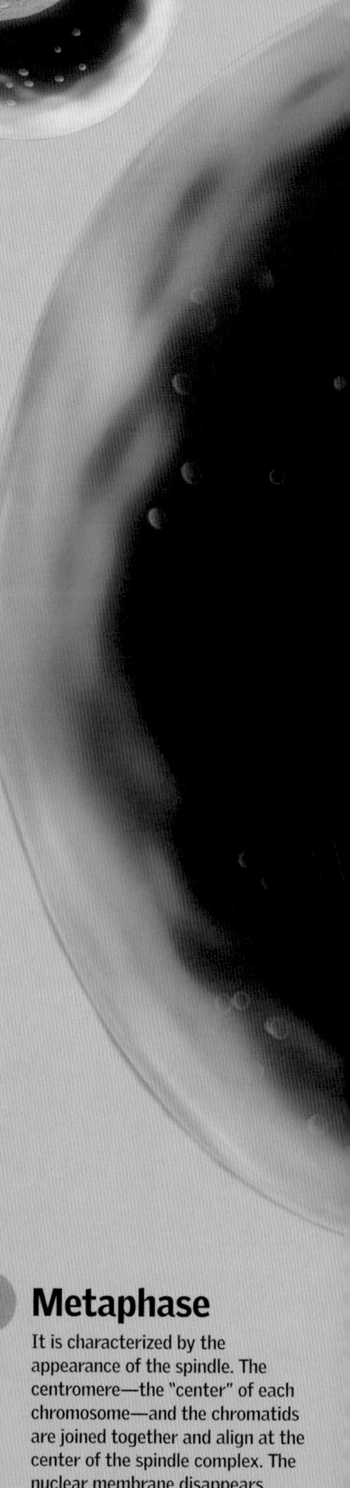

## 1 Interphase

An independent stage that precedes mitosis. The chromatin consists of DNA.

## Antioxidants

Antioxidants are various types of substances (vitamins, enzymes, minerals, etc.) that combat the pernicious effects of free radicals—molecules that are highly reactive and form as a result of oxidation (when an atom loses an electron), which is often caused by coming into contact with oxygen. A consequence of this oxidative action is the aging of the body. One action of antioxidants is the regulation of mitosis. Preventive geriatrics has focused on using antioxidants to prevent disease and to slow aging, in part because properly regulated mitosis is fundamental to these processes.

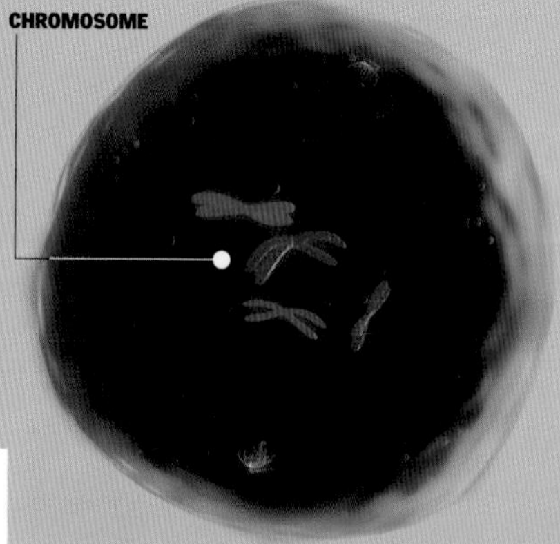


## 2 Prophase

In prophase the chromatin condenses to form chromosomes. The karyotheca (nuclear envelope) begins to disappear. Chromosomes are formed by two chromatids that are joined together by a centromere.

### The Ever-Changing Skin

Mitosis, or cellular division, occurs intensely within the skin, a fundamental organ of the sense of touch. The dead cells on the surface are continuously being replaced by new cells, which are produced by mitosis in the lowest, or basal, layer. From there the cells move upward until they reach the epidermis, the outer layer of the skin. A person typically sheds 30,000 dead skin cells every minute.

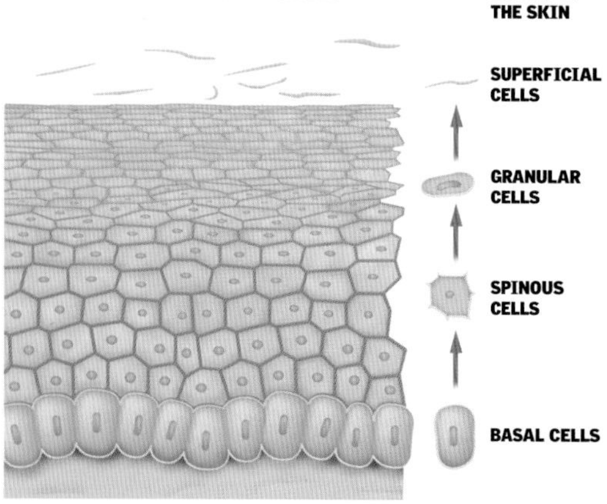


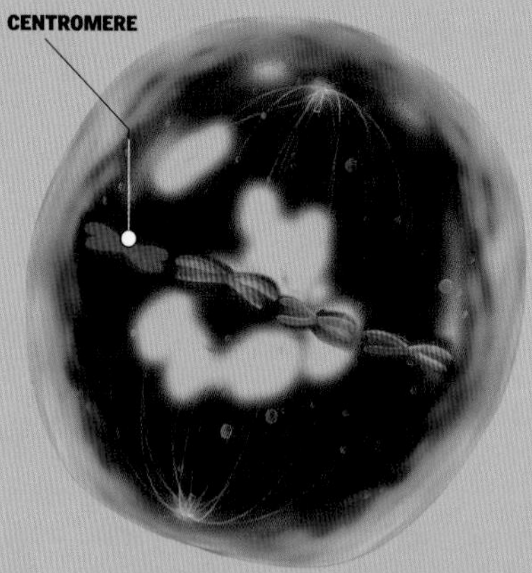


## 3 Metaphase

It is characterized by the appearance of the spindle. The centromere—the "center" of each chromosome—and the chromatids are joined together and align at the center of the spindle complex. The nuclear membrane disappears.

# 50,000

**THE ESTIMATED NUMBER OF CELLS REPLACED EVERY SECOND IN THE HUMAN BODY THROUGH CELLULAR DIVISION.**

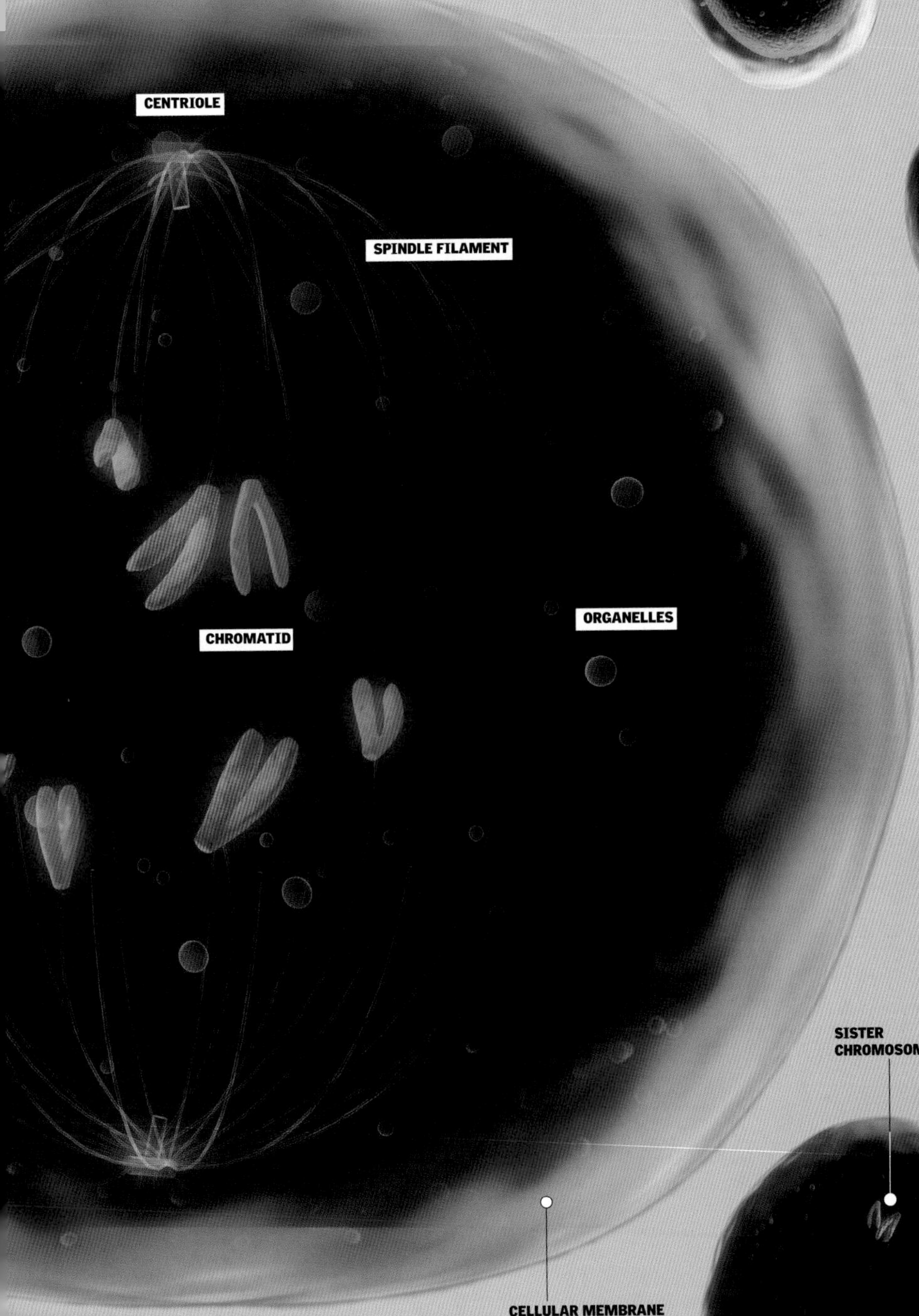


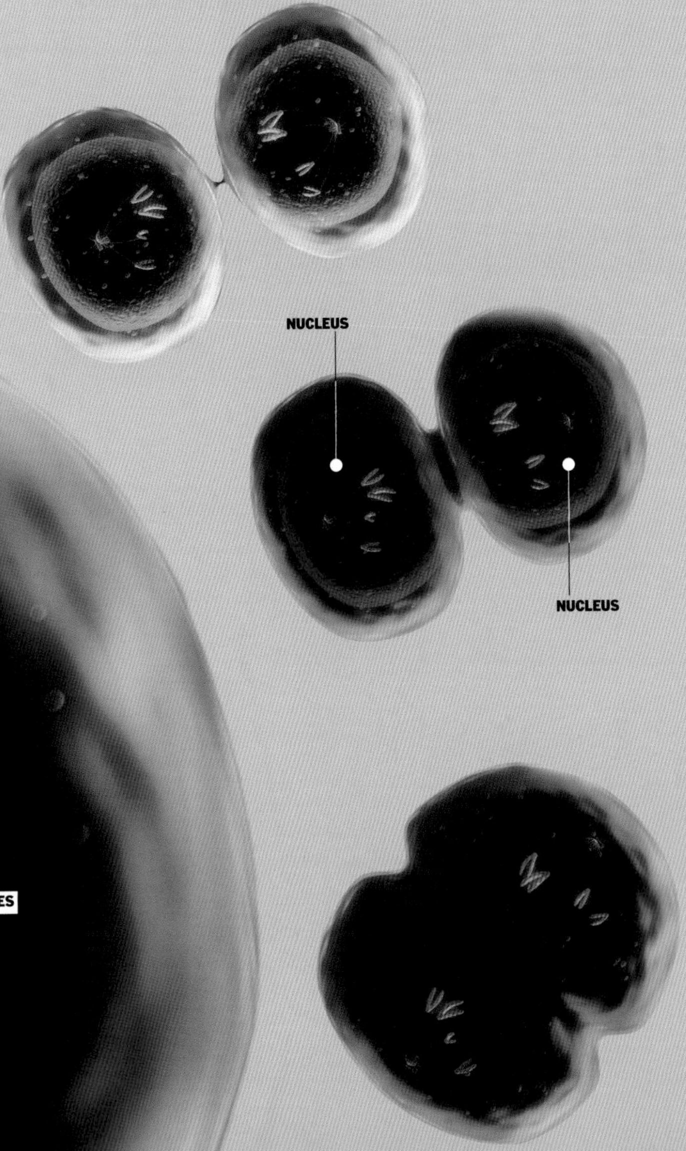


## 5 Telophase

The spindle disappears, and a new nuclear membrane begins to form around each new set of chromosomes. The membrane divides, resulting in two new cells that are identical daughters of the original cell.

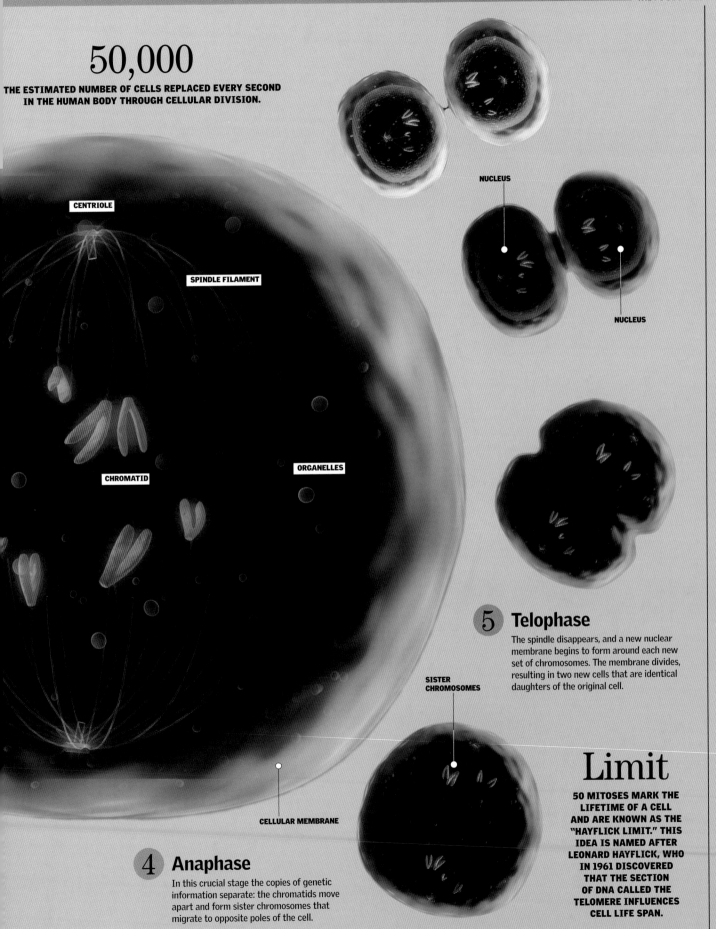

## 4 Anaphase

In this crucial stage the copies of genetic information separate: the chromatids move apart and form sister chromosomes that migrate to opposite poles of the cell.

# Limit

**50 MITOSES MARK THE LIFETIME OF A CELL AND ARE KNOWN AS THE "HAYFLICK LIMIT." THIS IDEA IS NAMED AFTER LEONARD HAYFLICK, WHO IN 1961 DISCOVERED THAT THE SECTION OF DNA CALLED THE TELOMERE INFLUENCES CELL LIFE SPAN.**

# Systems of the Body

The body has various systems with different functions. These functions range from reproducing a cell to developing a new human being, from circulating the blood to capturing oxygen from the air, and from processing food through grinding and chemical transformations to absorbing nutrients and discarding waste. These functions act in harmony, and their interaction is surprisingly efficient.

## Reproductive System

### MALE

The various male organs contribute one of the two cells needed to create a new human being. Two testicles (or gonads) and a penis are the principal organs of the system. The system is continuously active, producing millions of tiny cells called spermatozoa. **See page 166**.

## Circulatory System

This system carries blood to and from the heart and reaches the organs and cells in every part of the body. The supreme pump—the heart—drives the vital fluid—blood—through the arteries and collects it by means of the veins, with a continuous driving impulse that makes the heart the central engine of the body. **See page 138**.

## Skeletal System

The skeleton, or skeletal system, is a solid structure consisting of bones that are supported by ligaments and cartilage. The main functions of the system are to give the body form and to support it, to cover and protect the internal organs, and to allow motion to occur. The skeleton also generates red blood cells (called erythrocytes). **See page 122**.

## Nervous System

The central nervous system consists of the brain, which is the principal organ of the body, along with the spinal cord. The peripheral nervous system consists of the cranial and spinal nerves. Together they send external and internal sensations to the brain, where the sensations are processed and responded to whether the person is asleep or awake. **See page 184**.

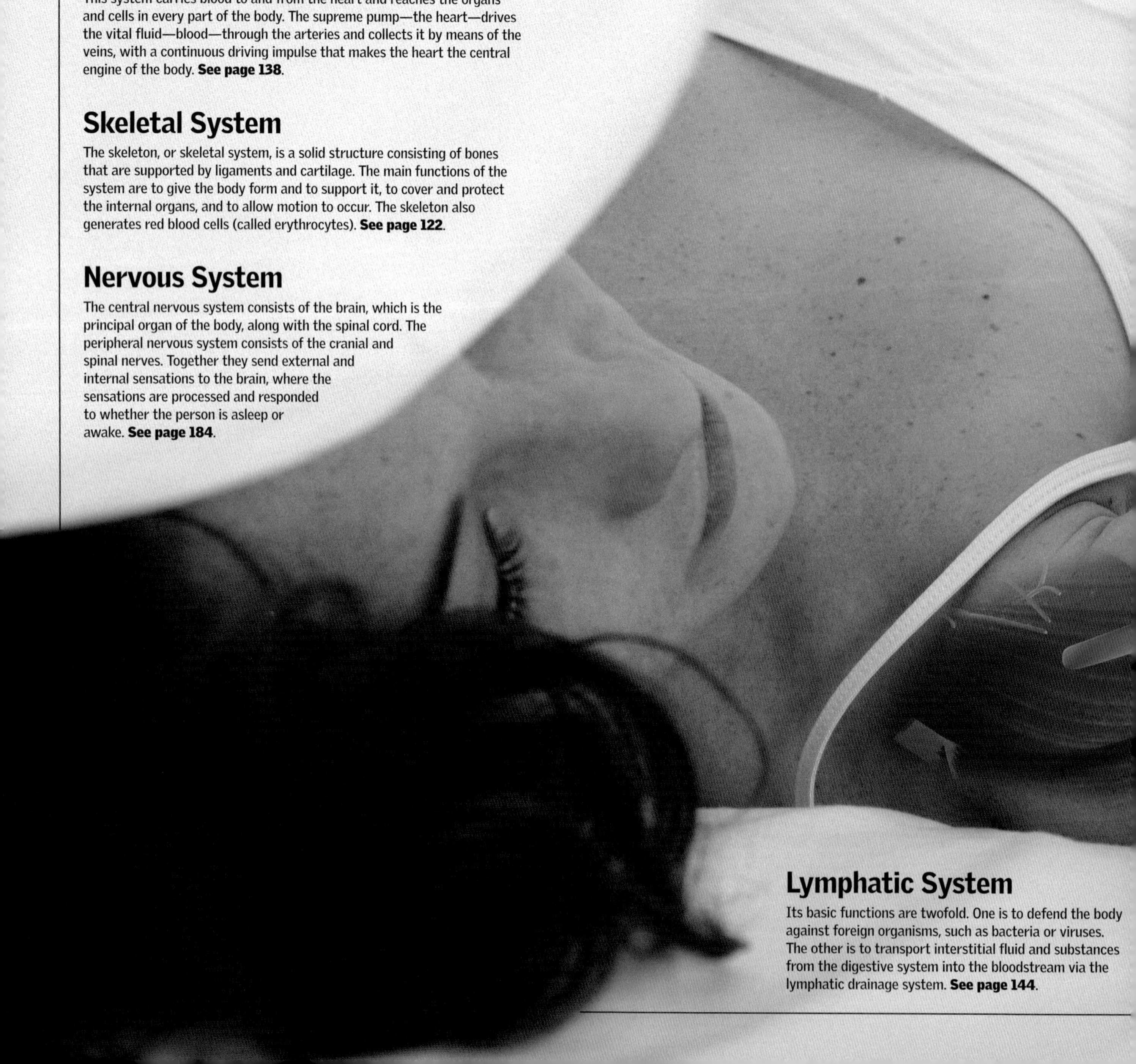

## Lymphatic System

Its basic functions are twofold. One is to defend the body against foreign organisms, such as bacteria or viruses. The other is to transport interstitial fluid and substances from the digestive system into the bloodstream via the lymphatic drainage system. **See page 144**.

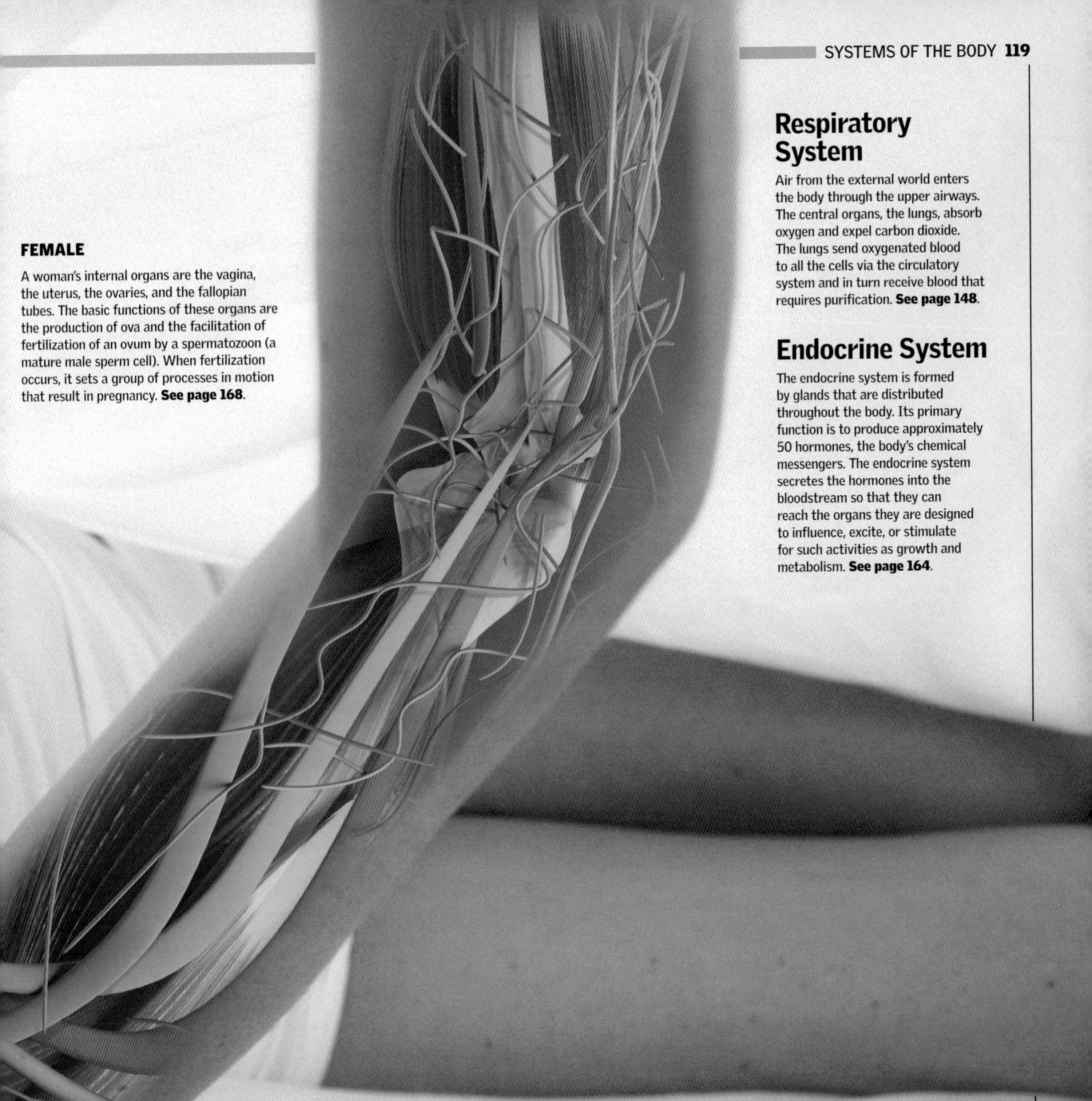

**FEMALE**

A woman's internal organs are the vagina, the uterus, the ovaries, and the fallopian tubes. The basic functions of these organs are the production of ova and the facilitation of fertilization of an ovum by a spermatozoon (a mature male sperm cell). When fertilization occurs, it sets a group of processes in motion that result in pregnancy. **See page 168**.

## Respiratory System

Air from the external world enters the body through the upper airways. The central organs, the lungs, absorb oxygen and expel carbon dioxide. The lungs send oxygenated blood to all the cells via the circulatory system and in turn receive blood that requires purification. **See page 148**.

## Endocrine System

The endocrine system is formed by glands that are distributed throughout the body. Its primary function is to produce approximately 50 hormones, the body's chemical messengers. The endocrine system secretes the hormones into the bloodstream so that they can reach the organs they are designed to influence, excite, or stimulate for such activities as growth and metabolism. **See page 164**.

## Muscular System

Its function is to define the shape of the organism and protect it. The muscular system is essential for producing movement. It consists of muscles, organs made of fleshy tissue, and contractile cells. There are two types of muscles: striated and smooth. Striated muscles are attached to the bones and govern voluntary movement. Smooth muscles also obey the brain, but their movement is not under voluntary control. The myocardium, the muscle tissue of the heart, is unique and is in a class by itself. **See page 132**.

## Digestive System

This system is a large tract that changes form and function as it goes from the mouth to the rectum and anus, passing through the pharynx, the esophagus, the stomach, and the small and large intestines. The liver and pancreas help process ingested food to extract its chemical components. Some of these components are welcome nutrients that are absorbed by the system, but others are useless substances that are discarded and eliminated. **See page 152**.

## Urinary System

This system is a key system for homeostasis—that is, the equilibrium of the body's internal conditions. Its specific function is to regulate the amount of water and other substances in the body, discarding any that are toxic or that form an unnecessary surplus. The kidneys and the bladder are the urinary system's principal organs. The ureters transport the urine from the kidneys to the bladder, and the urethra carries the urine out of the body. **See page 160**.

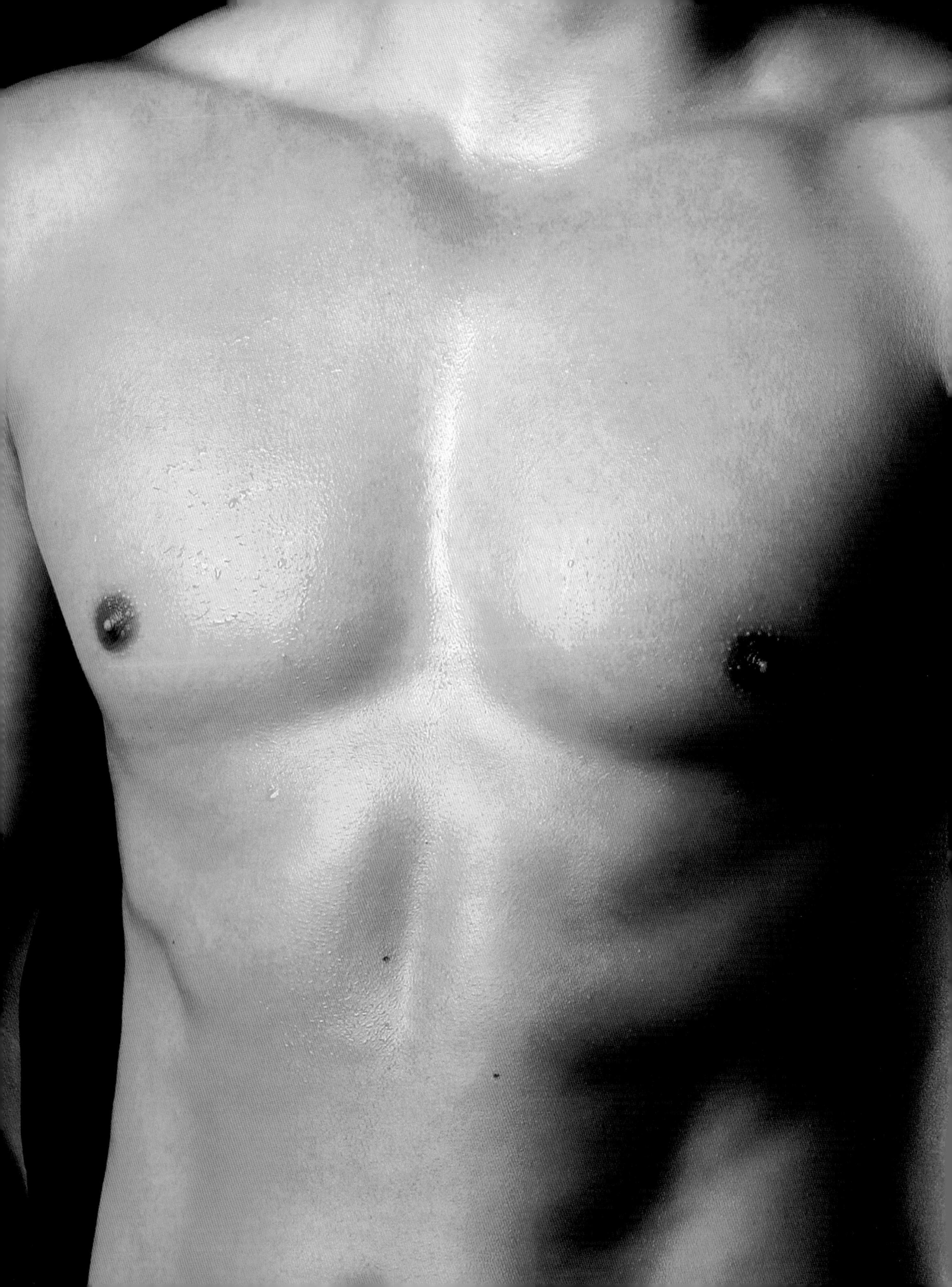

# Bones and Muscles

The musculoskeletal system consists of the skeletal system of bones, attached to each other by ligaments to form joints, and the skeletal muscles, which use tendons to attach muscles to bone. The skeleton gives resistance and stability to the body and serves as a support structure for the muscles to work and produce movement. The bones also serve as a shield to protect the internal organs. In this chapter you will see in detail—even down to the inside of a muscle fiber—how each part works. Did you know that bones are constantly being regenerated and that, besides supporting the body, they are charged with producing red blood cells? In this chapter you will find incredible images, curiosities, and other information.



**MUSCLES OF THE THORAX** (opposite)
They play an important role in breathing by facilitating the contraction and expansion of the thoracic cavity.

# Skeleton

The skeleton, or the skeletal system, is a strong, resistant structure made up of bones and their supporting ligaments and cartilage. The skeleton gives the body form and structure, covers and protects the internal organs, and makes movement possible. The bones store minerals and produce blood cells in the bone marrow.

## Well-Defined Form

The structure of the skeleton can be described as a vertical column of chained vertebrae with a pair of limbs at each end and topped off by the cranium. The upper limbs, or arms, are connected to the shoulder blades and clavicles in what is called the scapular belt, and the lower limbs, or legs, are connected at the hips, or pelvic belt. The joints reach such a level of perfection that modern engineering often uses them as a model in the study of levers when designing such objects as cranes or desk lamps. Although the bones that make up the skeleton are solid, they have a flexible structure and to a large degree consist of spongy tissue. Nevertheless, a small bone is capable of supporting up to 9 tons without breaking. A comparable weight would crush a block of concrete. For a long time anatomists thought that bones themselves were not alive and that their strength merely provided support for the other organs. Modern medicine recognizes that bones are actively living, furnished with nerves and supplied with blood.

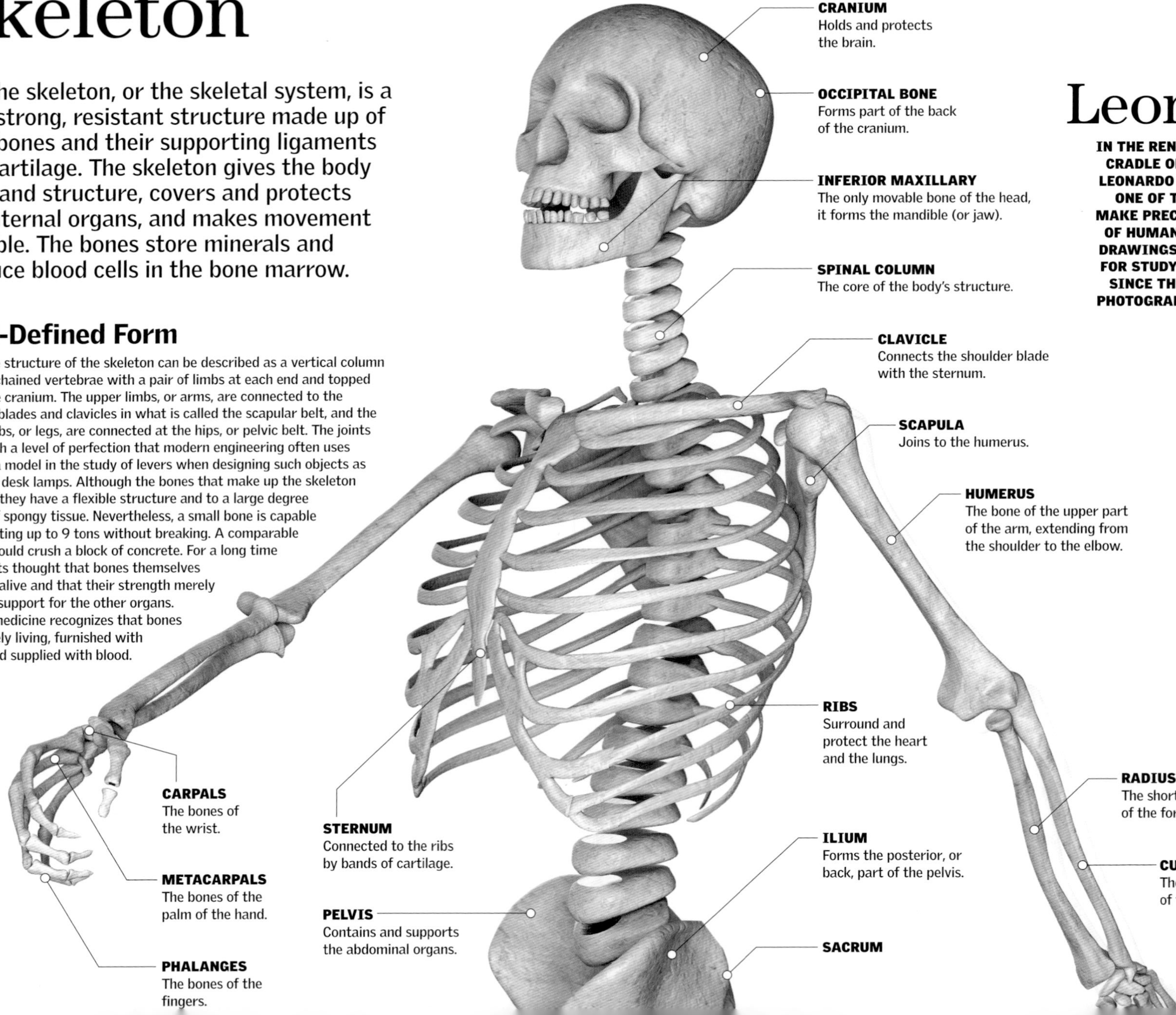
**CRANIUM**
Holds and protects the brain.
**OCCIPITAL BONE**
Forms part of the back of the cranium.
**INFERIOR MAXILLARY**
The only movable bone of the head, it forms the mandible (or jaw).
**SPINAL COLUMN**
The core of the body's structure.
**CLAVICLE**
Connects the shoulder blade with the sternum.
**SCAPULA**
Joins to the humerus.
**HUMERUS**
The bone of the upper part of the arm, extending from the shoulder to the elbow.
**RIBS**
Surround and protect the heart and the lungs.
**RADIUS**
The shorter bone of the forearm.
**CUBITUM**
The inside bone of the forearm.
**ILIUM**
Forms the posterior, or back, part of the pelvis.
**SACRUM**
**CARPALS**
The bones of the wrist.
**STERNUM**
Connected to the ribs by bands of cartilage.
**METACARPALS**
The bones of the palm of the hand.
**PELVIS**
Contains and supports the abdominal organs.
**PHALANGES**
The bones of the fingers.

## Leonardo

**IN THE RENAISSANCE, THE CRADLE OF MODERNITY, LEONARDO DA VINCI WAS ONE OF THE FIRST TO MAKE PRECISE DRAWINGS OF HUMAN BONES. SUCH DRAWINGS WERE NEEDED FOR STUDYING ANATOMY SINCE THERE WERE NO PHOTOGRAPHS OR X-RAYS.**

## Types of Bones

Depending on their characteristics, such as size or shape, the bones of the human body are generally classified as follows:

**SHORT BONES:** have a spherical or conical shape. The heel bone is a short bone.

**LONG BONES:** have a central section that lies between two end points, or epiphyses. The femur is a long bone.

**FLAT BONES:** form thin bony plates. Most bones of the cranium are flat bones.

**IRREGULAR BONES:** take various shapes. The sphenoids ("wedgelike" bones) in the skull are irregular bones.

**SESAMOID BONES:** are small and round. The patella and the bones between tendons and in the joints of the hands and feet are sesamoid bones.

# Axial Bones

**THE BODY HAS 80 OF THESE BONES, WHICH BELONG TO THE PART OF THE SKELETON FORMED BY THE SPINAL COLUMN, THE RIBS, AND THE CRANIUM.**

## 208 bones

**THE TOTAL NUMBER OF BONES IN THE ADULT BODY IS APPROXIMATELY 206. SOME ADULTS, HOWEVER, MAY ACTUALLY HAVE AS MANY AS 208 BONES. THIS VARIATION OCCURS WHEN THE INDIVIDUAL BONES OF EITHER THE STERNUM OR THE COCCYX REMAIN SEPARATE.**

# Appendicular Bones

**THESE COMPRISE THE OTHER 126 BONES: THOSE OF THE ARMS, SHOULDERS, HIPS, AND LEGS. THESE BONES PERMIT A GREAT RANGE OF MOTION.**

## 17 inches (43 cm)

**THE SIZE OF THE LARGEST BONE OF THE BODY, THE FEMUR**

## 0.12 inches (3 mm)

**THE LENGTH OF THE SHORTEST BONE OF THE BODY. IT IS THE STIRRUP, A BONE IN THE EAR.**

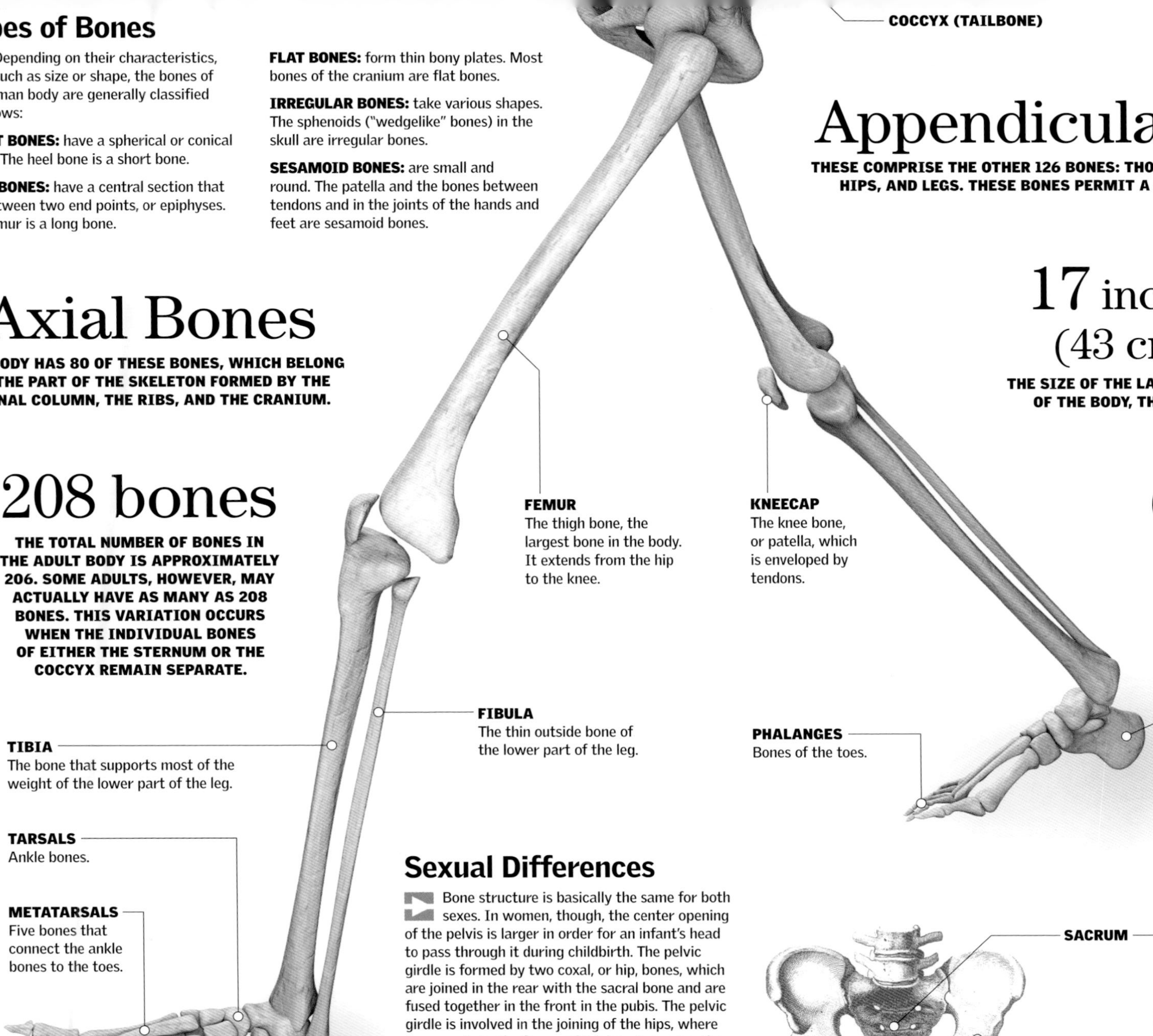


## Sexual Differences

Bone structure is basically the same for both sexes. In women, though, the center opening of the pelvis is larger in order for an infant's head to pass through it during childbirth. The pelvic girdle is formed by two coxal, or hip, bones, which are joined in the rear with the sacral bone and are fused together in the front in the pubis. The pelvic girdle is involved in the joining of the hips, where it connects to the femur (thigh bone), serving the function of transmitting weight downward from the upper part of the body. The pelvic girdle and sacrum form the pelvis, which contains the organs of the digestive, reproductive, and urinary systems.

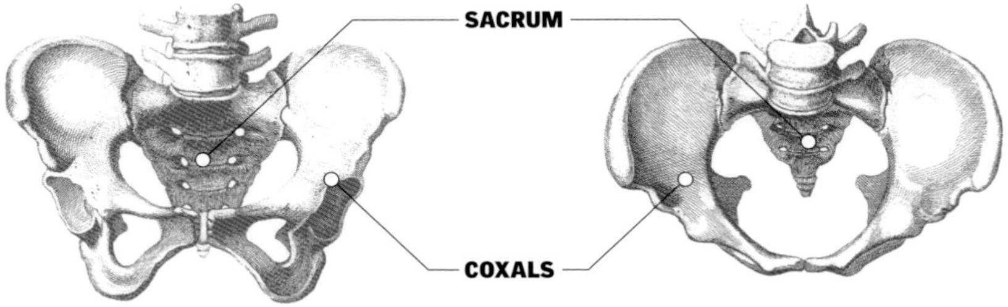

SACROILIAC
The joint that transmits the weight of the body from the spinal column to the pelvis.
SACRUM
COXALS

# Bony Tissue

The primary mission of the bones is to protect the organs of the body. Bones are solid and resilient, which allows them to endure blows and prevent damage to the internal organs. The hard exterior is balanced by the internal spongy part. Over a person's lifetime bones are continuously regenerated; this process continues even after a person reaches maturity. Besides supporting the body and enabling movement, the bones are charged with producing red globules: thousands of millions of new cells are produced daily in the bone marrow, in a never-ending process of replacing old cells.

## Calcium and Marrow

All the hard parts that form the skeleton in vertebrates, such as the human being, are called bones. They may be hard, but they are nevertheless formed by a structure of living cells, nerves, and blood vessels, and they are capable of withstanding pressure of up to 1,000 pounds (450 kg). Because of their constitution and characteristics, they can mend themselves when fractured.
A resistant exterior layer called the periosteum covers the outside of the compact bone. The endosteum, a thin layer of connective tissue lining the interior cavity of bone, contains the trabecular, or spongy mass, which is characterized by innumerable pores. The bone marrow, located in the center of the large bones, acts as a virtual red blood-cell factory and is also known as the medulla ossea. Minerals such as calcium go into making the bones. The fact that calcium is found in foods such as milk explains why healthy bones are usually associated with drinking a lot of milk. Calcium and phosphorous, among other chemical substances, give bones strength and rigidity. Proteins such as collagen provide flexibility and elasticity.

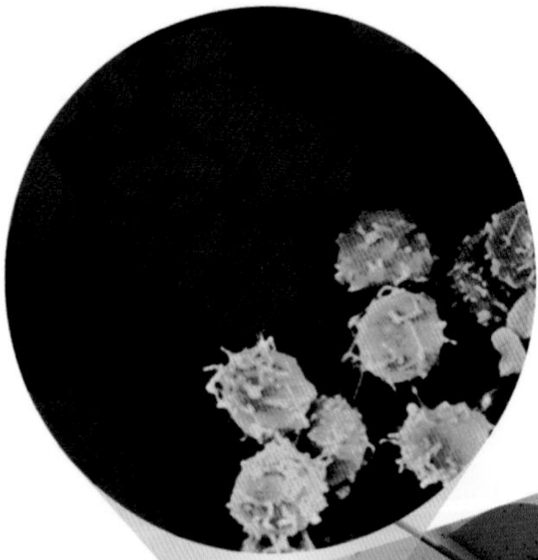

**Bone Marrow**

A soft, fatty substance that fills the central cavities and produces red blood cells. Over time bone marrow in the large bones loses its ability to produce red blood cells.

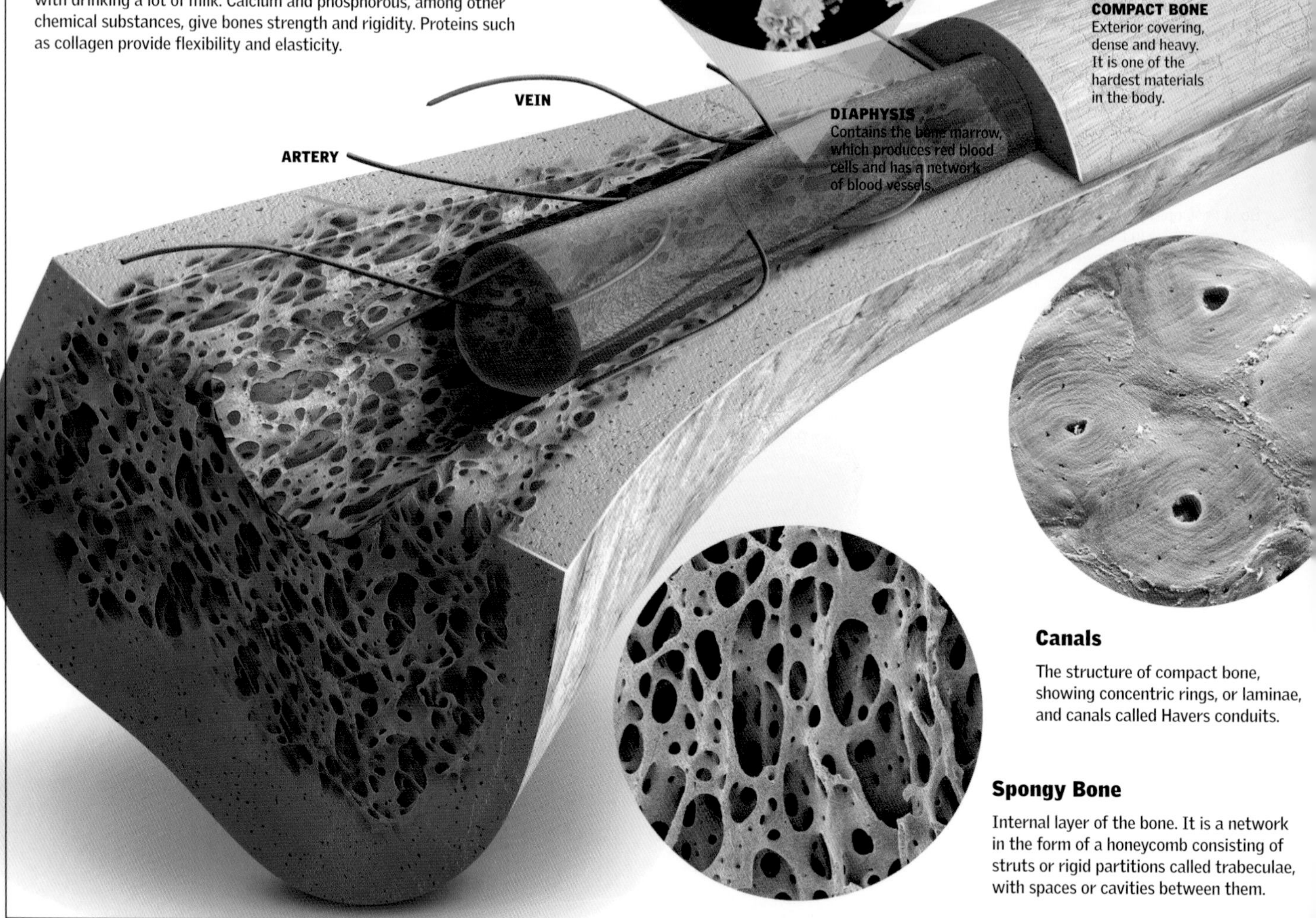

COMPACT BONE
Exterior covering, dense and heavy. It is one of the hardest materials in the body.
VEIN
ARTERY
DIAPHYSIS
Contains the bone marrow, which produces red blood cells and has a network of blood vessels.

**Canals**

The structure of compact bone, showing concentric rings, or laminae, and canals called Havers conduits.

**Spongy Bone**

Internal layer of the bone. It is a network in the form of a honeycomb consisting of struts or rigid partitions called trabeculae, with spaces or cavities between them.

## TWO TYPES OF BONE CELLS

The osseous tissue consists of two types of cells, osteoblasts and osteoclasts. Both are produced by the bone marrow, and their interaction and equilibrium ensure the integrity and continuous renewal of the bone. An osteoclast reabsorbs bone tissue, leaving empty spaces, and an osteoblast fills them. The function of the osteocytes, a variant of the osteoblasts, is to maintain the shape of the bone.

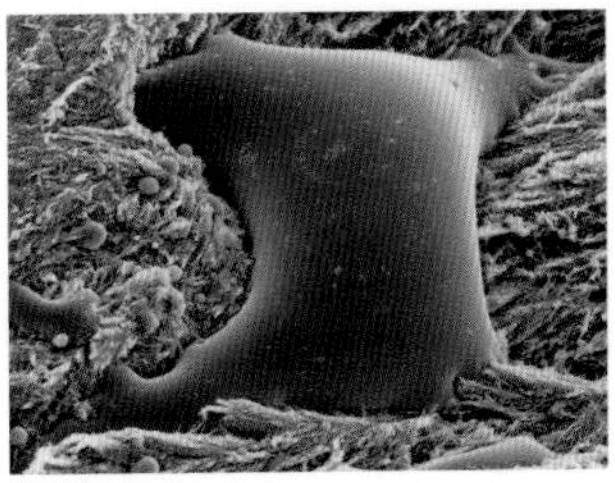

**OSTEOBLAST**
This produces osseous, or bone, tissue, which maintains the strength of the bone.

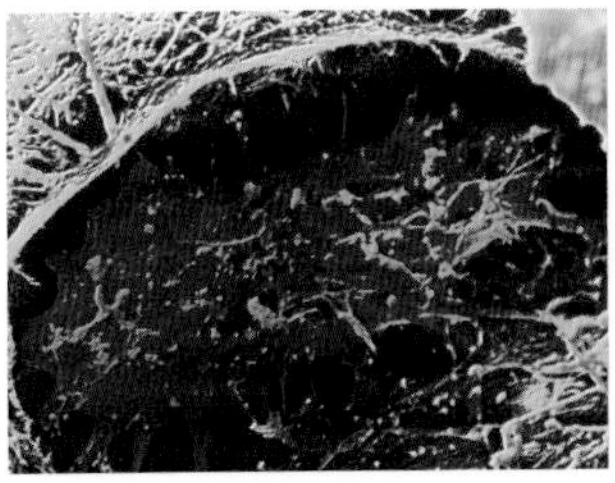

**OSTEOCLAST**
This breaks down the tissue so that it can be replaced with newer tissue.

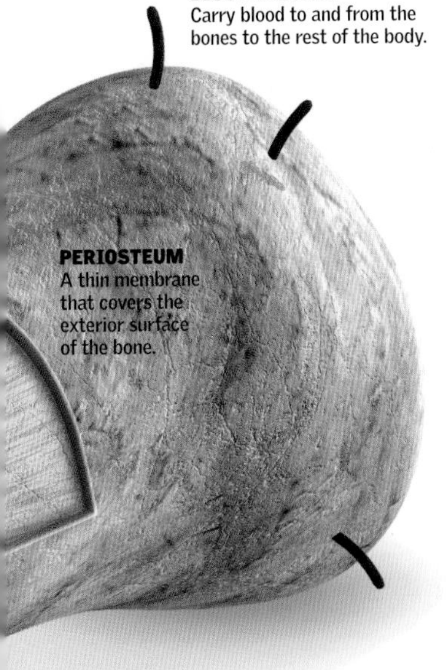

BLOOD VESSELS
Carry blood to and from the bones to the rest of the body.
PERIOSTEUM
A thin membrane that covers the exterior surface of the bone.

## WHY FRACTURES HEAL

Bone has great regenerative capacity. Bone tissue has an extraordinary ability to repair itself after a fracture through processes that include the relatively rapid generation of cells. Medicine can guide these processes to cure other lesions, deformities, etc.

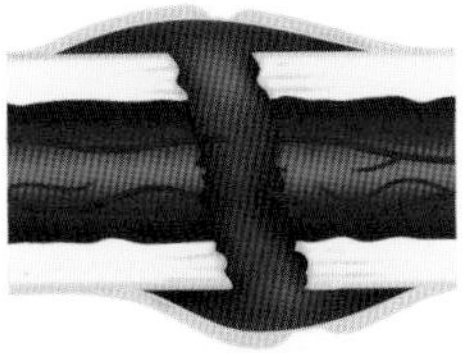

**A** A fracture occurs, and the blood cells coagulate to seal the broken blood vessels.

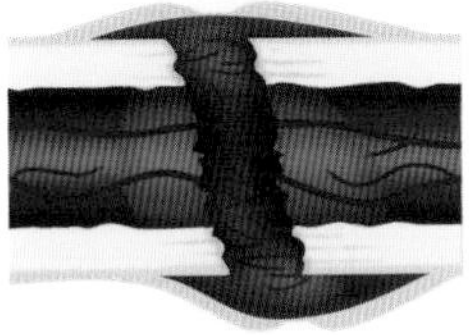

**B** Over a few days a fibrous mesh forms, which closes the ends of the bone and replaces the coagulate.

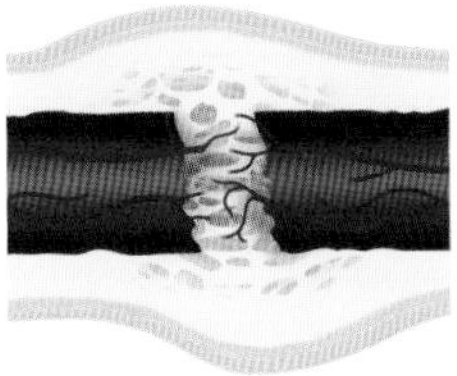

**C** Within one to two weeks new spongy bone develops on a base of fibrous tissue. The spaces created by the fracture are filled, and, finally, the ends are fused.

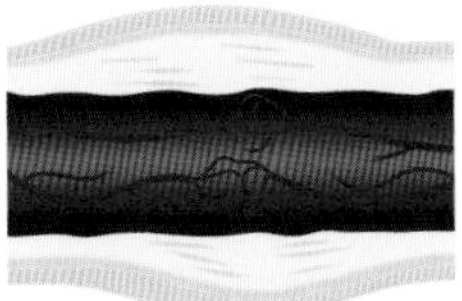

**D** Within two to three months, new blood vessels have developed. Compact bone forms on the bony callous.

# Evolution of Bone

Bone development is completed at about 18 or 20 years of age in a process that begins with an infant's bones, which are largely cartilage, and continues with the ongoing generation of bone in the person as an adult. Calcium is an indispensable element for the healthy development of bones through this process. Until the age of six months, an intake of 0.007 ounce (210 mg) of calcium per day is recommended.

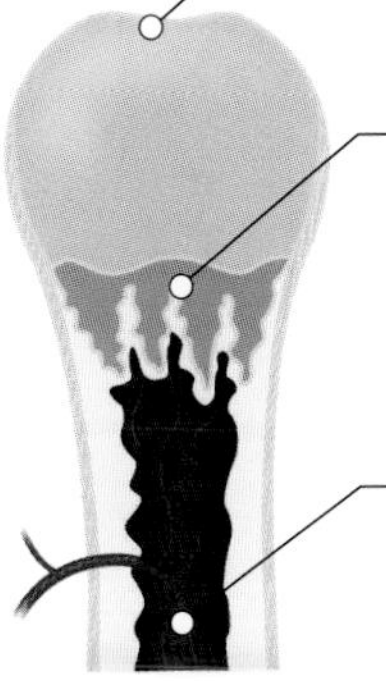

EPIPHYSIS
The end of a long bone, which at birth consists of cartilage.
GROWTH PLATE
Consists of cartilage. It deposits new bone on the diaphysis face of the growth plate so the bone will grow.
DIAPHYSIS
Water is deposited in the new bone.

**1 IN AN INFANT**
In a newborn infant the ends of the long bone (epiphyses) are made of cartilage. Between the bone shaft and an epiphysis, an area called a "growth plate" produces cartilage in order to lengthen the bone.

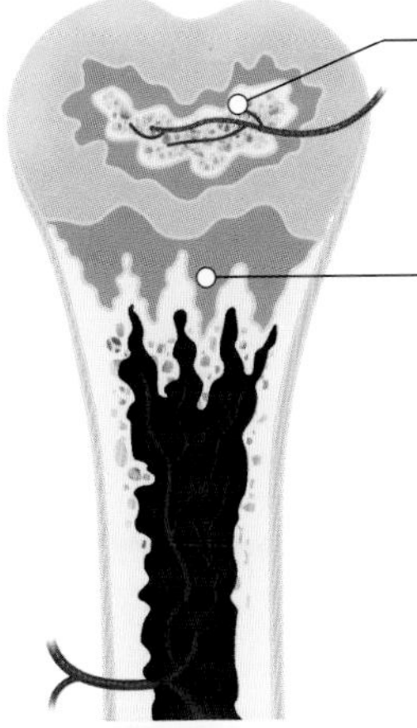

EPIPHYSIS
Secondary ossification centers, to aid in long-term bone growth and to shape the bones.
GROWTH PLATE
Continues to act, depositing bone on the diaphysis face of the growth plate.

**2 IN A CHILD**
In a child ossification continues with the formation of secondary ossification centers in the epiphyses, which enable the bones to grow in length.

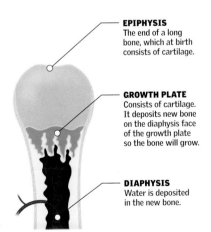

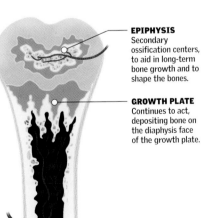

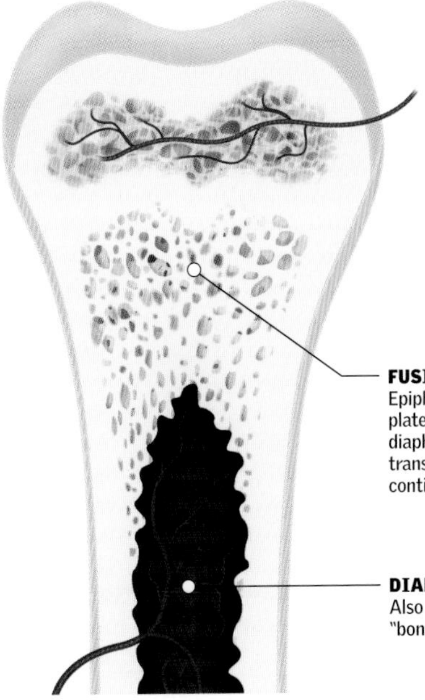


**3 IN AN ADULT**
The process is complete when a person reaches about 18 years of age. The epiphysis, growth plates, and bone shaft fuse and become ossified into a continuous bone.

# Cranium and Face

The cranium surrounds and protects the brain, cerebellum, and cerebral trunk (sometimes called the encephalus). In an adult the cranium consists of eight bones that form the skull and the base of the cranium. The face is the anterior part of the skull. It consists of 14 bones, all of which are fixed except the lower maxillary, which makes up the mandible. The total number of bones in the head as a whole exceeds the total of the face and cranium (22) because it includes the little bones of the middle ear.

## Sutures and Fontanels

The cranium can be compared to a sphere, which consists of separate bones at birth and closes completely at maturity. The narrow separations between the bones, which appear as lines in the fetus for the first months of its life, are called sutures. Spaces called fontanels form where the sutures meet. Their separation has the functional purpose of allowing the brain to grow. Therefore, when brain growth is complete, the sphere closes tightly, because its function is to protect the brain.

## Vibration

When a person speaks, the bones of the cranium vibrate. In Japan a technology was developed based on this vibration. In 2006 the firefighters of the Madrid municipality in Spain adopted this technology. A helmet, furnished with a cranial contact microphone, amplifies the vibrations produced in the bones of the cranium during speech and sends them to radio equipment.

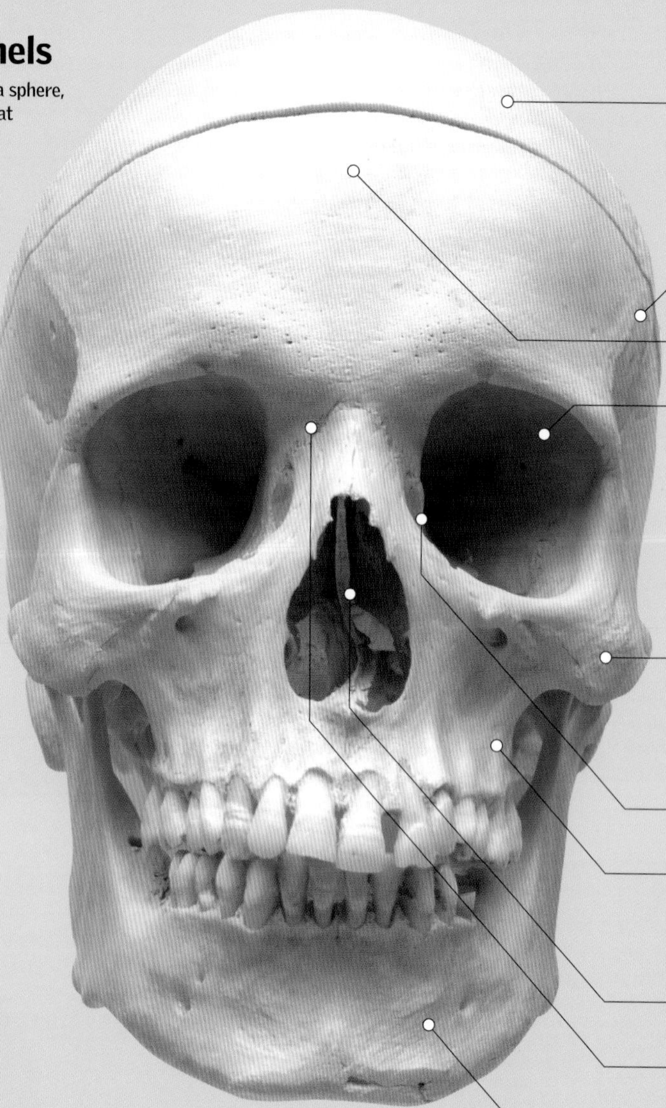

### Cranial Bones (8)

**PARIETAL (2)**
The superior and lateral parts of the cranium.

**OCCIPITAL (1)**
Together with the temporals, it forms the base of the cranium.

**TEMPORAL (2)**
The lateral part of the cranium.

**FRONTAL (1)**
It makes up the forehead.

**SPHENOID (1)**
The front part of the base of the cranium and part of the orbital bone (eye socket).

**ETHMOID (1)**
Upper part of the nasal cavity.

### Facial Bones (14)

**ZYGOMATIC (2)**
The cheekbones.

**PALATINES (2)**
Internal bones that form the roof of the mouth.

**LACHRYMAL BONES (2)**
These form the eye socket.

**SUPERIOR MAXILLARIES (2)**
The upper mandible.

**NASAL CONCHAS (2)**
Independent of the ethmoid conchas.

**VOMER (1)**
Divides the nasal cavity into two halves.

**NASAL BONE (2)**
This forms the bridge of the nose (the rest of the nose is cartilage).

**INFERIOR MAXILLARY (1)**
This constitutes the mandible and is the only facial bone that can move freely.

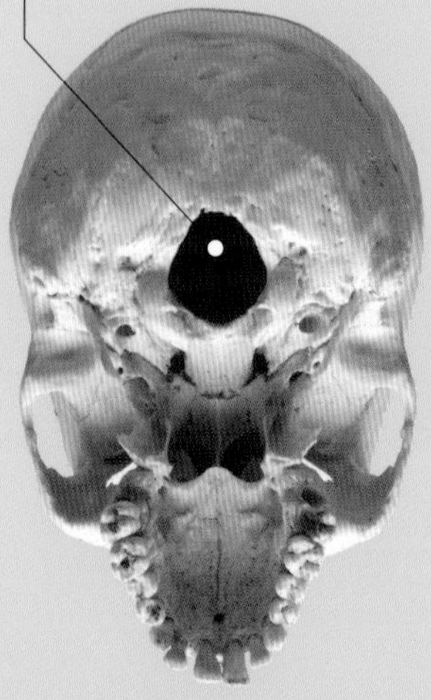


## Foramen Magnum

In Latin this term means "big hole." It is a circular opening, also called the occipital orifice, which is located at the base of the cranium. The foramen magnum allows for the passage of the spinal column, the medulla oblongata, the vertebral arteries, and the spinal nerve. The placement of the foramen magnum toward the bottom of the skull is associated with more highly evolved species.

**83 cubic inches (1,360 cu cm)**

**THE TYPICAL VOLUME OF THE CRANIUM.**

22

**THE TOTAL NUMBER OF BONES IN THE CRANIUM.**

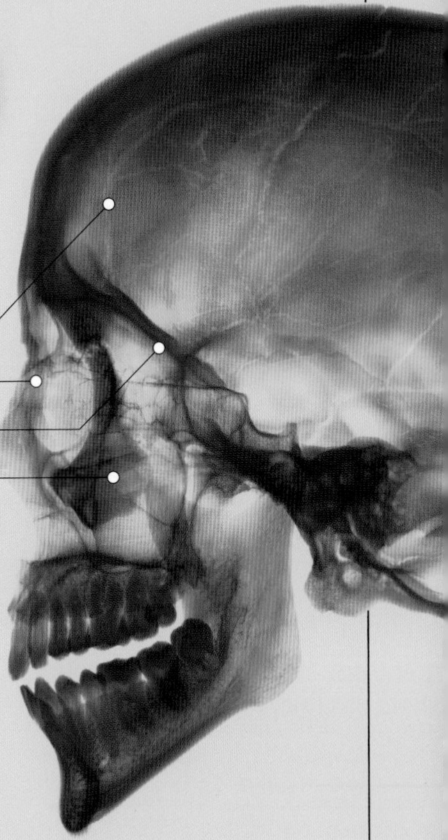


9 pounds (4 kg)

**THE WEIGHT OF AN ADULT HUMAN HEAD.**

## Cranial Sinuses

The sinuses are air-filled cavities whose principal known function is to humidify and heat the air that enters the respiratory tract via the nose. The sinuses reduce the weight of the head, and they also act as resonance cavities, giving the voice its timbre. The sinuses are covered by a moist membrane and are connected via small openings with the interior of the nasal cavity. When the sinuses become inflamed or filled with mucus, there is a risk of infection.

# The Great Axis of the Body

The vertebral, or spinal, column is the flexible axis that lends support to the body. It consists of a series of bones jointed together in a line, or chain, called the vertebrae. The spinal column forms a protective inner channel through which the spinal cord runs. The ribs perform a similar function, wrapping and shielding the vital internal organs, which include the heart and lungs.

## Stability and Motion

The vertebrae have a centrum that allows them to support the body's weight, each vertebra upon the next, as well as the weight of the rest of the body. The vertebrae also have extensions that allow them to articulate with other vertebrae or act as supports for the ligaments and the muscles. This system gives the axis of the body both strength and flexibility. In addition, most of the nerves of the peripheral system (that is, those responsible for voluntary movement, for pain, and for the sense of touch) are connected to the spinal cord inside the spinal column. In the centrum the vertebrae are separated from each other by intervertebral disks that are made of cartilage and have a gelatinous interior. When an intervertebral disk is damaged, some of this material can escape and pinch a nerve. This condition, called a herniated disk, can be very painful.

## The Ribs and the Rib Cage

The 12 pairs of ribs, which also extend from the spinal column, protect the heart, lungs, major arteries, and liver. These bones are flat and curved. The seven upper pairs are called "true ribs," and they are connected to the sternum (a flat bone consisting of fused segments) by cartilage. The next two or three pairs (called "false ribs") are connected indirectly. The remaining pairs ("floating ribs") are not attached to the sternum. The rib cage, formed by the ribs and its muscles, is flexible: it expands and contracts during breathing.

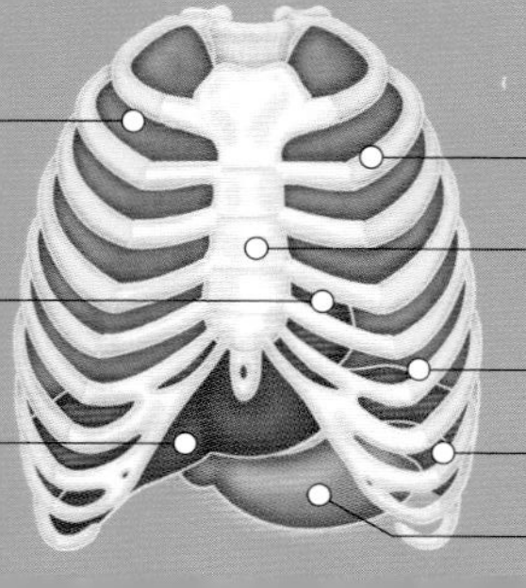


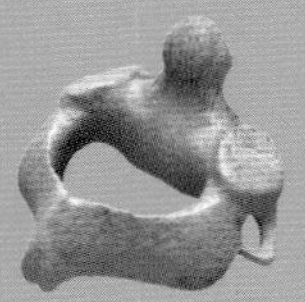

**ATLAS**
This bone is the first of the seven cervical bones; it unites the spinal column with the head.

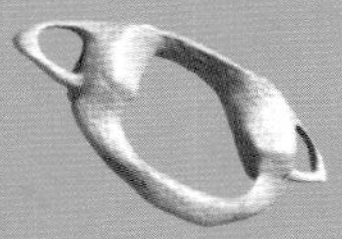

**AXIS**
The second cervical vertebra. Together with the atlas, it permits the movement of the head.

**CERVICAL VERTEBRAE**
These seven vertebrae (including the atlas and the axis) support the head and the neck.

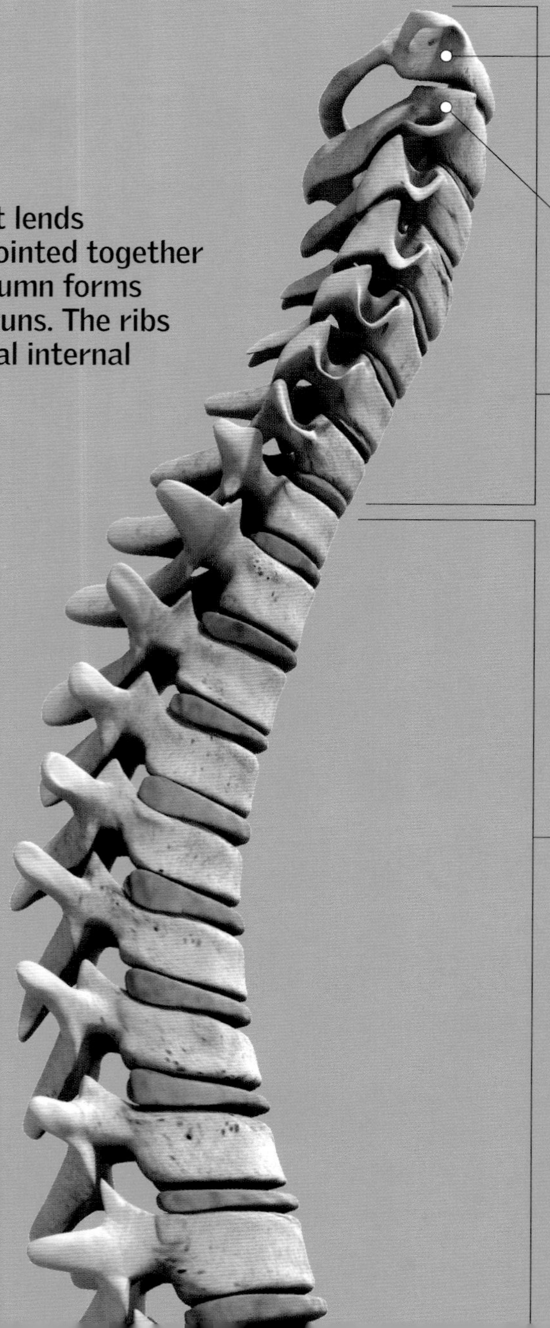

### Downwards

All the vertebrae except the cervical axis and atlas have a cylindrical body, which gives them a particular characteristic: as they approach the pelvis they tend to be longer and stronger.

**THORACIC, OR DORSAL, VERTEBRAE**
There are 12, and they are joined to the ribs.

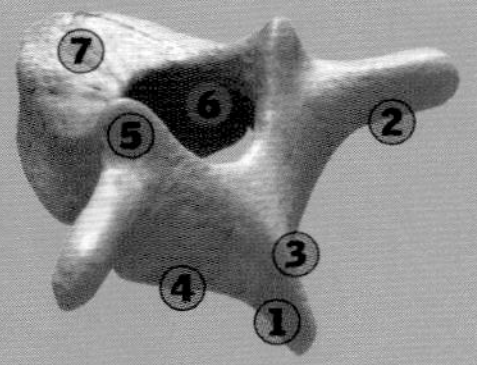


**PARTS OF THE VERTEBRAE**

1. SPINAL APOPHYSIS
2. TRANSVERSE APOPHYSIS (2)
3. ARTICULAR APOPHYSIS (4) (2 SUPERIOR AND 2 INFERIOR)
4. LAMINAE (2)
5. PEDICULAE (2)
6. FORAMEN MAGNUM
7. BODY

**TARSUS (7)**
1. MEDIAL CUNEIFORM
2. INTERMEDIATE CUNEIFORM
3. LATERAL CUNEIFORM
4. TALUS
5. TARSAL SCAPHOIDS
6. CALCANEUS
7. CUBOIDS

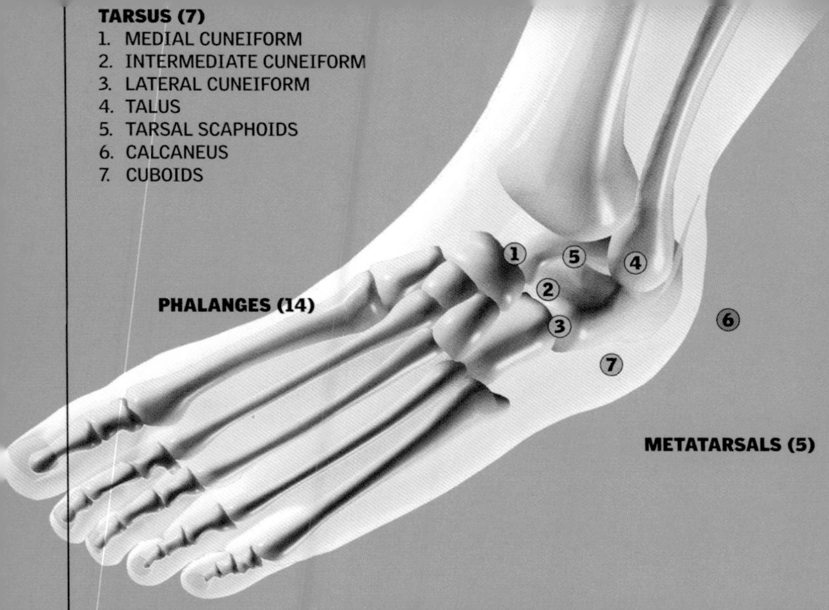


# 33 bones

**OR VERTEBRAE, MAKE UP THE SPINAL COLUMN. DEPENDING ON THE INDIVIDUAL, SOMETIMES THERE ARE 34. THEY ARE CONNECTED BY DISKS OF CARTILAGE THAT ACT AS SHOCK ABSORBERS. THE SACRUM AND THE COCCYX ARE A RUDIMENTARY TAIL LOST DURING EVOLUTION.**

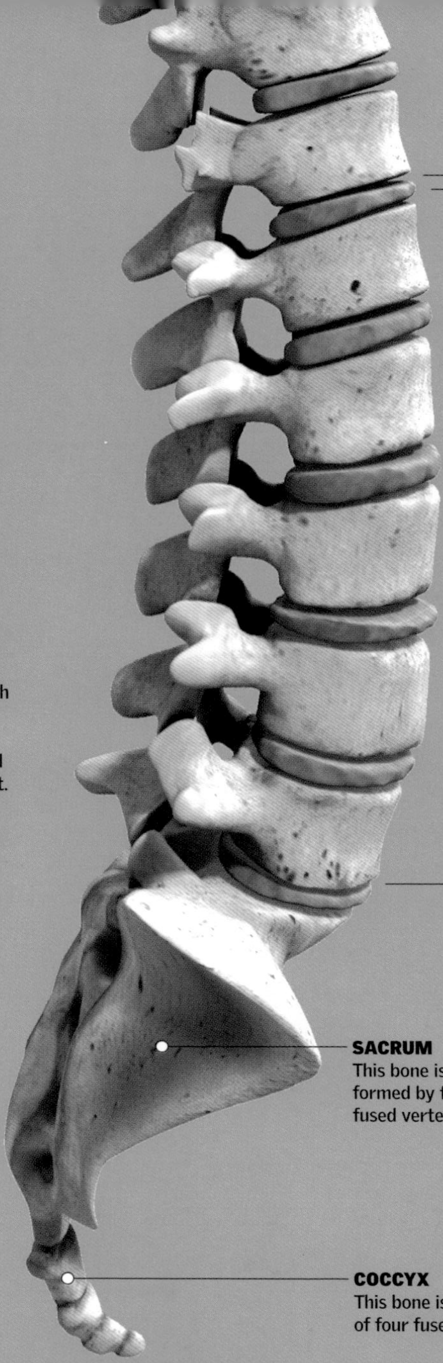

**LUMBAR VERTEBRAE**
There are five of them, and they bear the weight of the upper part of the body.

**SACRUM**
This bone is formed by five fused vertebrae.

**COCCYX**
This bone is composed of four fused vertebrae.

## Bones of the Hands and Feet

Each hand (see the drawing below) has 27 bones, and each foot (see above) has 26. The hand has great mobility, and each of its fingers (five in all) has three phalanges (distal, medial, and proximal), except for the thumb, which has two. The complex of carpal bones makes up the wrist and is connected to the forearm. The metacarpal bone sustains the medial part. The feet function in a similar manner; the toes have first, second, and third phalanges, except for the big toe.

**CARPALS (8)**
1. LUNATE
2. PISIFORM
3. TRIQUETRUM
4. TRAPEZIUM
5. TRAPEZOID
6. CAPITATE
7. SCAPHOID
8. HAMATE

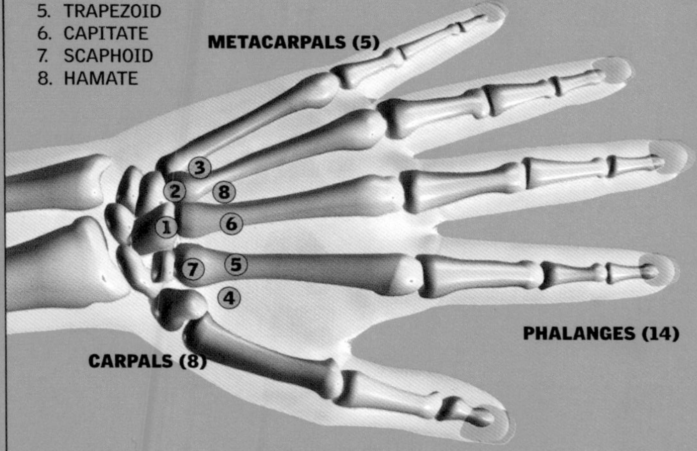


## The Three Curves

The three types of natural curvature in the spinal column include cervical lordosis (forward, or inward, bending in the cervical region of the spine), kyphosis (outward bending of the thoracic region of the spine), and lumbar lordosis (forward bending of the lower back). Shown here is the right side of the spinal column.

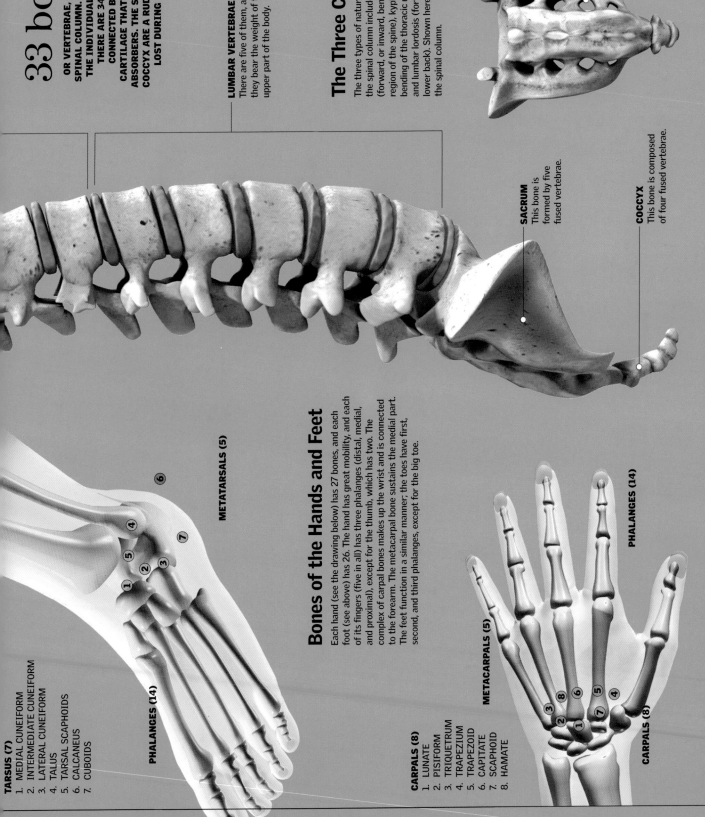

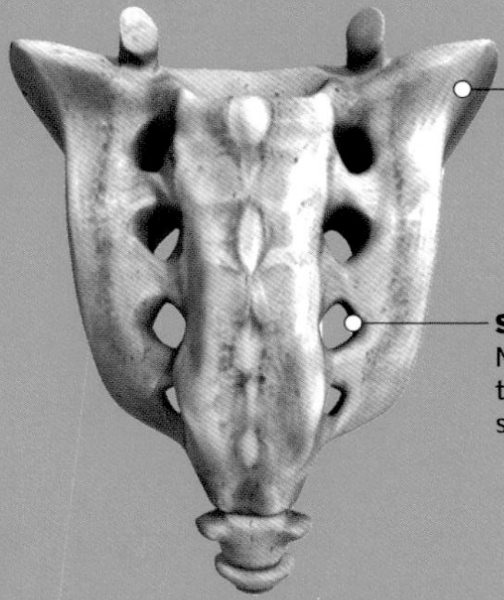

**BLADE**

**SACRAL CANAL**
Nerves pass through the sacral canal.

# Joints

These are the structures where two or more bones come together, either directly or by means of strong fibrous cords called ligaments. The skeleton has movement thanks to its joints. Most joints, like the knee, are synovial joints. They are characterized by mobility, versatility, and lubrication. The muscles that surround them contract to cause movement. When they work as a whole, the bones, muscles, and joints—together with the tendons, ligaments, and cartilage—constitute a grand system that governs the motor activity of the body and allows us to carry out our daily physical activities.

## Hypermobility

The versatility of the joints refers to their characteristic range of motion. Just as there are mobile, semimobile, and fixed joints, there is also a group of joints that are hypermobile. Such joints are less common but are easily recognizable, especially in children and adults who have not lost the flexibility of their joints. The elbows, wrists, fingers, and knees can at an early age and in certain individuals have a greater-than-normal range of motion. For people with hypermobile joints this extra range of motion can be accomplished without difficulty or risk of dislocation.

### Mobile

These are also called diarthroses; they are the joints with the greatest range of motion. The ends of the bones linked together are structured in various ways that facilitate their movement relative to each other, while ensuring the stability of the joint. Most joints in the body are of this type.

### Semimobile

Also known as amphiarthroses. The surfaces of the bone that make contact have cartilaginous tissue. One example is the vertebral joints: they have little individual movement, but as a whole they have ample flexion, extension, and rotation.

### Fixed

Also known as synarthroses. Most fixed joints are found in the cranium and have no need for motion because their primary function is to protect internal organs. They are connected by bone growth or fibrous cartilage and are extremely rigid and very tough.

**MOVEMENTS**

The complex of joints, together with the muscles and bones, allows the body to perform numerous actions, with movements that include turns and twists.

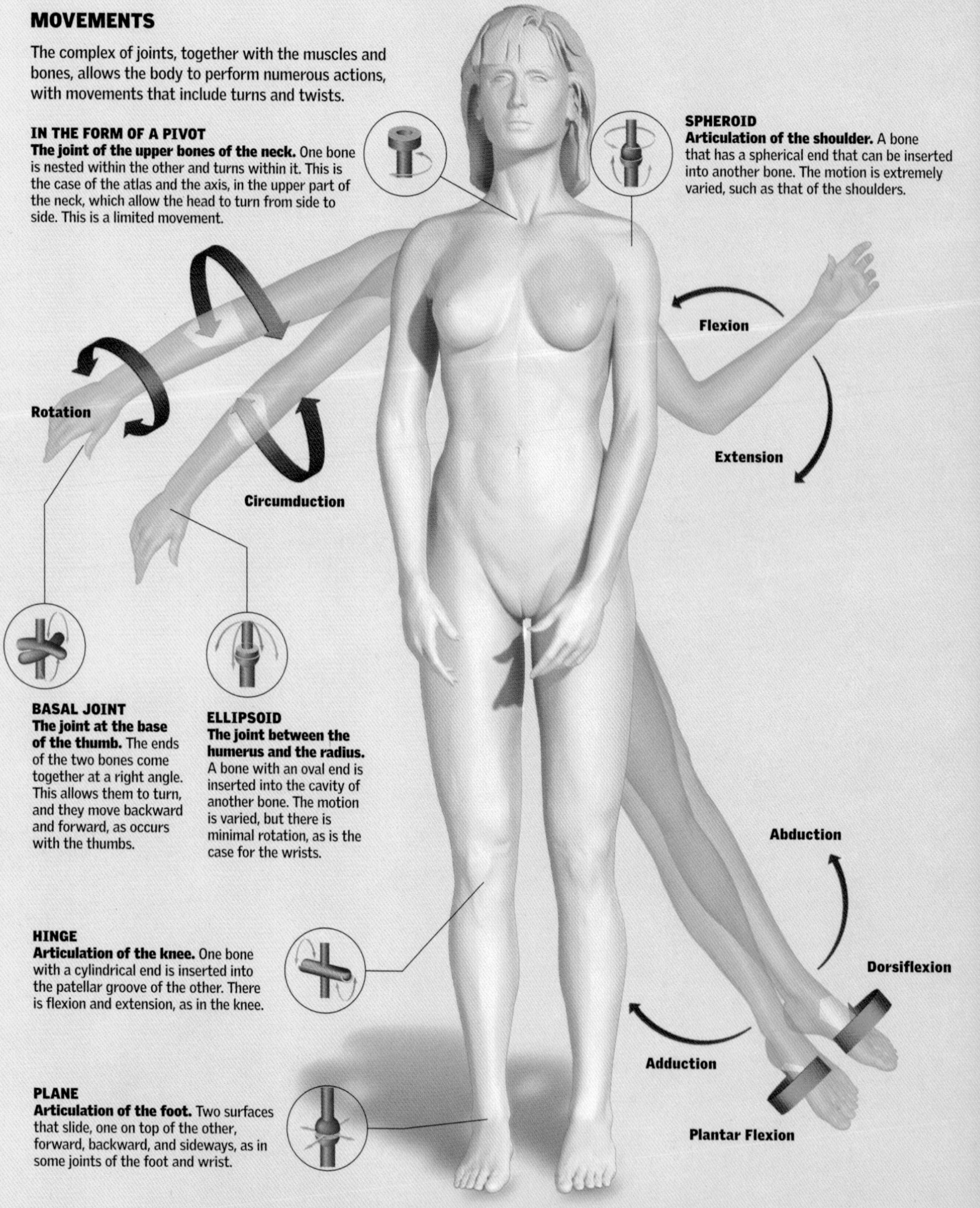


**IN THE FORM OF A PIVOT**
**The joint of the upper bones of the neck.** One bone is nested within the other and turns within it. This is the case of the atlas and the axis, in the upper part of the neck, which allow the head to turn from side to side. This is a limited movement.

**SPHEROID**
**Articulation of the shoulder.** A bone that has a spherical end that can be inserted into another bone. The motion is extremely varied, such as that of the shoulders.

**BASAL JOINT**
**The joint at the base of the thumb.** The ends of the two bones come together at a right angle. This allows them to turn, and they move backward and forward, as occurs with the thumbs.

**ELLIPSOID**
**The joint between the humerus and the radius.** A bone with an oval end is inserted into the cavity of another bone. The motion is varied, but there is minimal rotation, as is the case for the wrists.

**HINGE**
**Articulation of the knee.** One bone with a cylindrical end is inserted into the patellar groove of the other. There is flexion and extension, as in the knee.

**PLANE**
**Articulation of the foot.** Two surfaces that slide, one on top of the other, forward, backward, and sideways, as in some joints of the foot and wrist.

# 1918

**IN THIS YEAR PROFESSOR KENJI TAKAGI OF JAPAN USED A CYSTOSCOPE FOR THE FIRST INTERNAL OBSERVATION OF THE KNEE. TECHNOLOGICAL ADVANCES NOW PERMIT ARTHROSCOPY TO MAKE PRECISE OBSERVATIONS FOR DIAGNOSIS.**

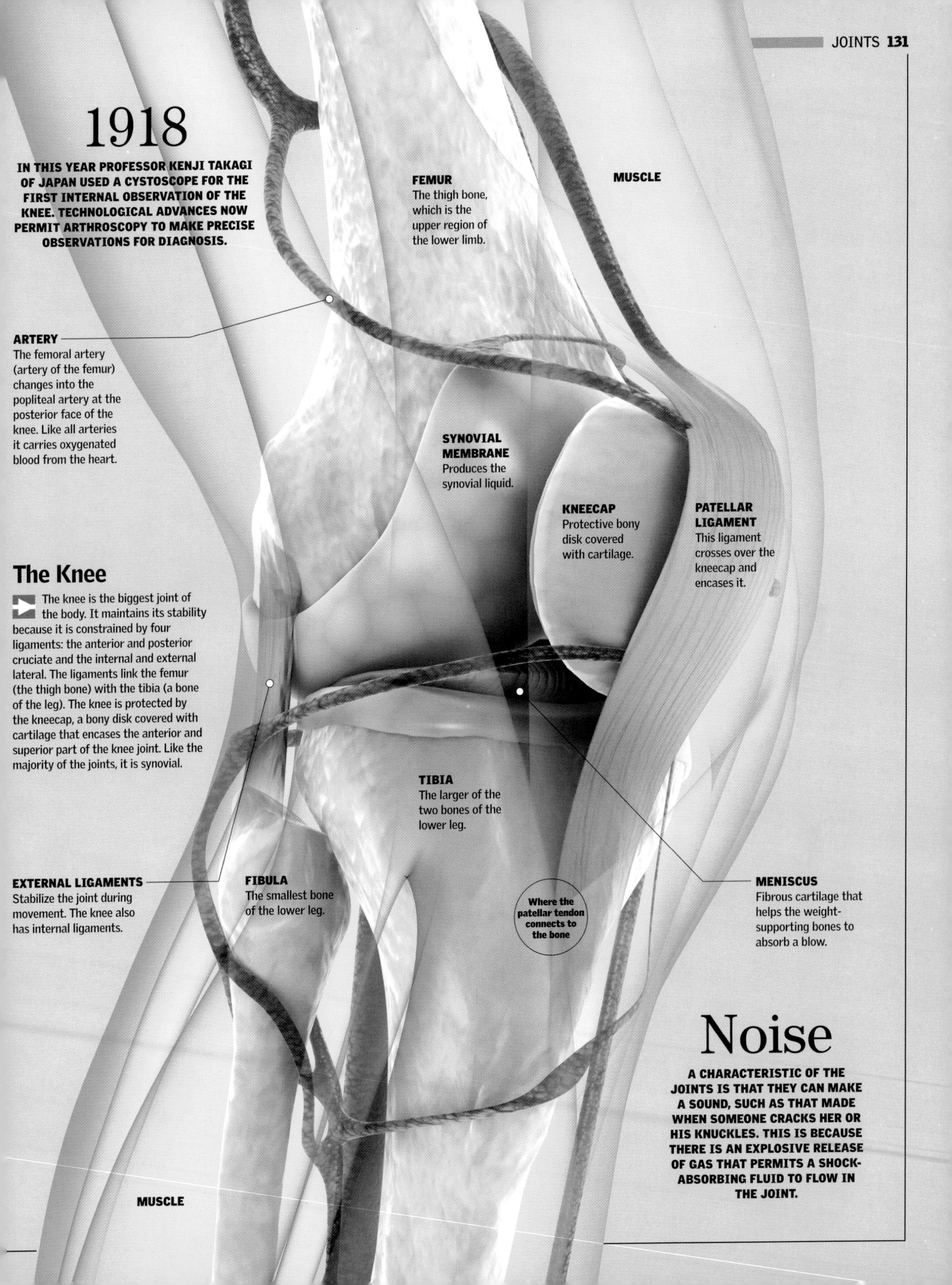
FEMUR
The thigh bone, which is the upper region of the lower limb.
MUSCLE
ARTERY
The femoral artery (artery of the femur) changes into the popliteal artery at the posterior face of the knee. Like all arteries it carries oxygenated blood from the heart.
SYNOVIAL MEMBRANE
Produces the synovial liquid.
KNEECAP
Protective bony disk covered with cartilage.
PATELLAR LIGAMENT
This ligament crosses over the kneecap and encases it.
TIBIA
The larger of the two bones of the lower leg.
EXTERNAL LIGAMENTS
Stabilize the joint during movement. The knee also has internal ligaments.
FIBULA
The smallest bone of the lower leg.
Where the patellar tendon connects to the bone
MENISCUS
Fibrous cartilage that helps the weight-supporting bones to absorb a blow.
MUSCLE

## The Knee

The knee is the biggest joint of the body. It maintains its stability because it is constrained by four ligaments: the anterior and posterior cruciate and the internal and external lateral. The ligaments link the femur (the thigh bone) with the tibia (a bone of the leg). The knee is protected by the kneecap, a bony disk covered with cartilage that encases the anterior and superior part of the knee joint. Like the majority of the joints, it is synovial.

# Noise

**A CHARACTERISTIC OF THE JOINTS IS THAT THEY CAN MAKE A SOUND, SUCH AS THAT MADE WHEN SOMEONE CRACKS HER OR HIS KNUCKLES. THIS IS BECAUSE THERE IS AN EXPLOSIVE RELEASE OF GAS THAT PERMITS A SHOCK-ABSORBING FLUID TO FLOW IN THE JOINT.**

# Muscular System

The muscles are organs formed by fleshy tissue consisting of contractile cells. They are divided into striated, smooth, and, in a unique case, cardiac (the myocardium is the muscular tissue of the heart). Muscles shape and protect the organism. The muscles of the skeleton are attached to the bones to permit voluntary movement, which is consciously directed by the brain. The smooth muscles are also directed by the brain, but their motion is not voluntary, as in the case of digestion. These muscles get most of their energy from alimentary carbohydrates, which can be stored in the liver and muscles in the form of glycogen and can later pass into the blood and be used as glucose. When a person makes a physical effort, there is an increased demand for both oxygen and glucose, as well as an increase in blood circulation. A lack of glucose leads to fatigue.

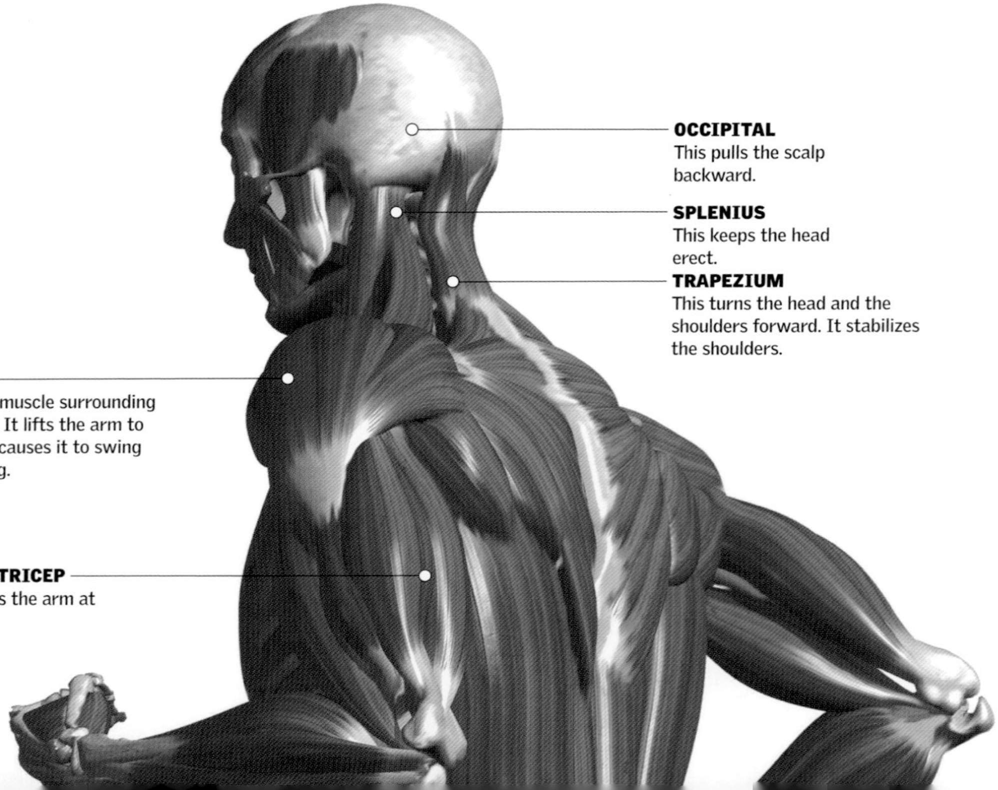
**OCCIPITAL**
This pulls the scalp backward.
**SPLENIUS**
This keeps the head erect.
**TRAPEZIUM**
This turns the head and the shoulders forward. It stabilizes the shoulders.
**DELTOID**
A triangular muscle surrounding the shoulder. It lifts the arm to the side and causes it to swing when walking.
**BRACHIAL TRICEP**
This stretches the arm at the elbow.

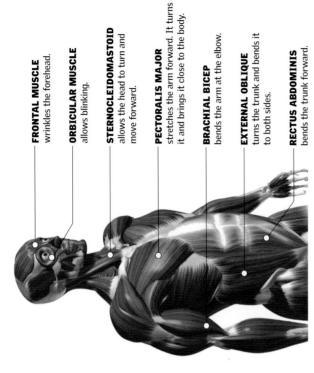

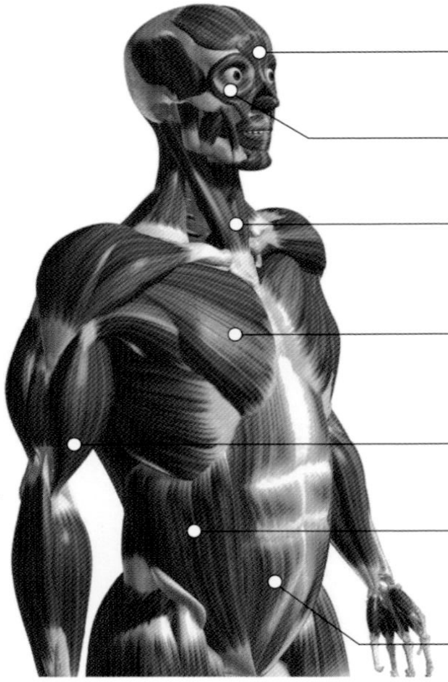
**FRONTAL MUSCLE**
wrinkles the forehead.
**ORBICULAR MUSCLE**
allows blinking.
**STERNOCLEIDOMASTOID**
allows the head to turn and move forward.
**PECTORALIS MAJOR**
stretches the arm forward. It turns it and brings it close to the body.
**BRACHIAL BICEP**
bends the arm at the elbow.
**EXTERNAL OBLIQUE**
turns the trunk and bends it to both sides.
**RECTUS ABDOMINIS**
bends the trunk forward.

## When the Skeleton Moves

The great number of muscles of voluntary action available to the human body makes possible thousands of distinct movements. Actions from the simple blink of an eyelid to the twisting of a belt are accomplished by muscular action. The eye muscles involve the most activity because they carry out 100,000 movements per day. Some 30 muscles control all the movements of the face and define an infinite possible combination of facial expressions. It is calculated that to pronounce one word, the organs for speech and respiration move some 70 muscles. The stirrup muscle, which controls the stirrup of the ear, is one of the smallest in the body. It measures approximately 0.05 inch (1.2 mm). There are other muscles that are very large, including the latissimus dorsi of the shoulder. The foot has 40 muscles and more than 200 ligaments. Because the muscles are connected by a great number of nerves, a lesion or blow causes the brain to react, producing pain. Approximately 40 percent of the total weight of the body consists of the muscular system. When the organism reduces the quantity of calories it normally ingests (for example, when a person goes on a diet), the first thing the body loses is water, which is reflected in a rapid weight loss. Then the metabolism adapts to the diet, and the body resorts to using up muscle tissue before drawing on the fats stored for burning calories. For this reason, when the diet begins this second phase, the consequences can be lack of vigor and loss of muscle tone, which is recovered when the diet returns to normal.

# 650 skeletal muscles

**OR VOLUNTARY MUSCLES ARE IN THE TYPICAL HUMAN BODY.**

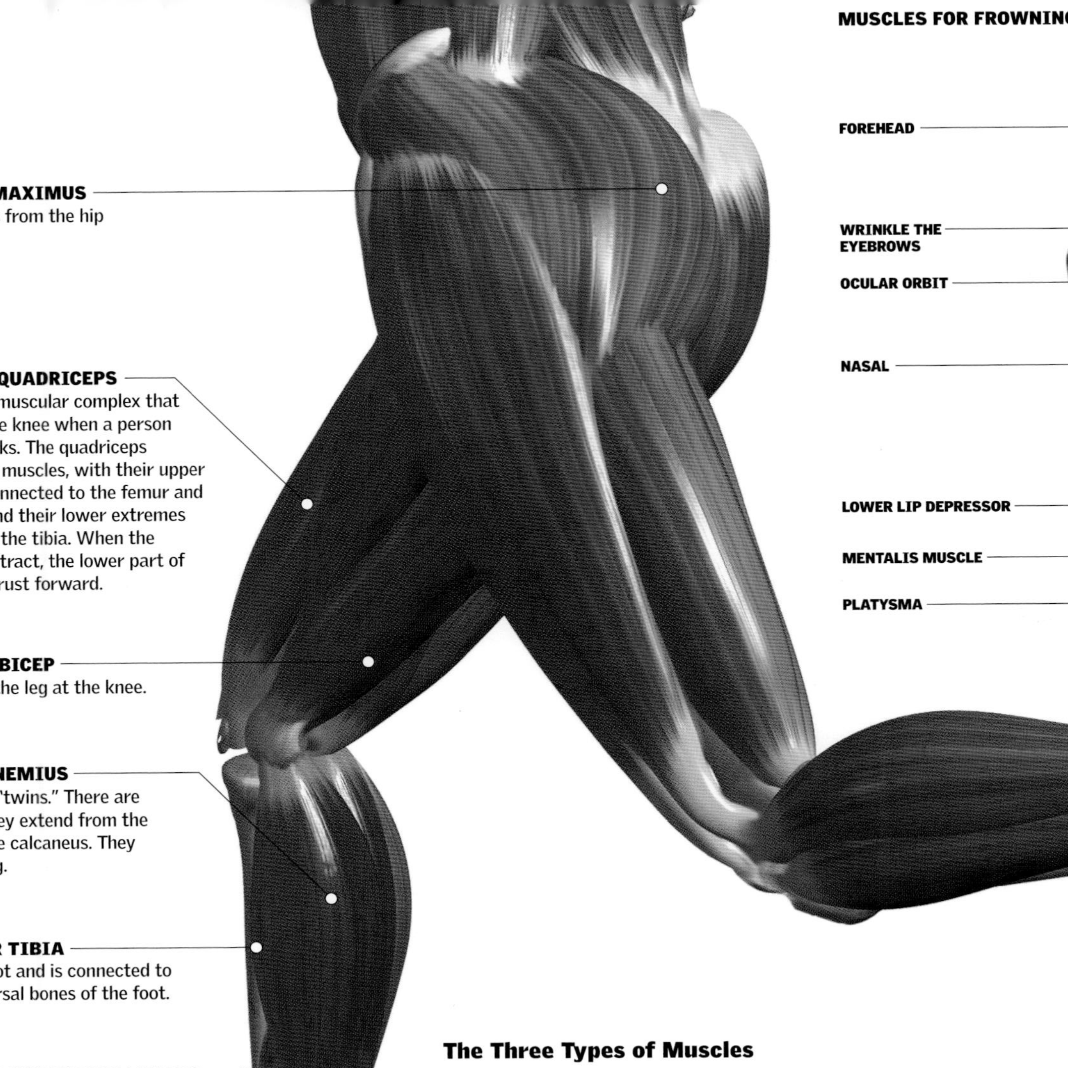

**GLUTEUS MAXIMUS**
This extends from the hip to the thigh.

**FEMORAL QUADRICEPS**
A powerful muscular complex that stretches the knee when a person runs and kicks. The quadriceps include four muscles, with their upper extremes connected to the femur and the pelvis and their lower extremes anchored in the tibia. When the muscles contract, the lower part of the leg is thrust forward.

**FEMORAL BICEP**
This bends the leg at the knee.

**GASTROCNEMIUS**
Also called "twins." There are two, and they extend from the femur to the calcaneus. They bend the leg.

**ANTERIOR TIBIA**
Lifts the foot and is connected to the metatarsal bones of the foot.

**EXTENSOR DIGITORUM LONGUS**
Called the "pedis," it connects to the dorsal part of the foot.

**ACHILLES TENDON**
Connects the gastrocnemius to the calcaneus bone (talus bone).

**MUSCLES FOR FROWNING**

**MUSCLES FOR SMILING**

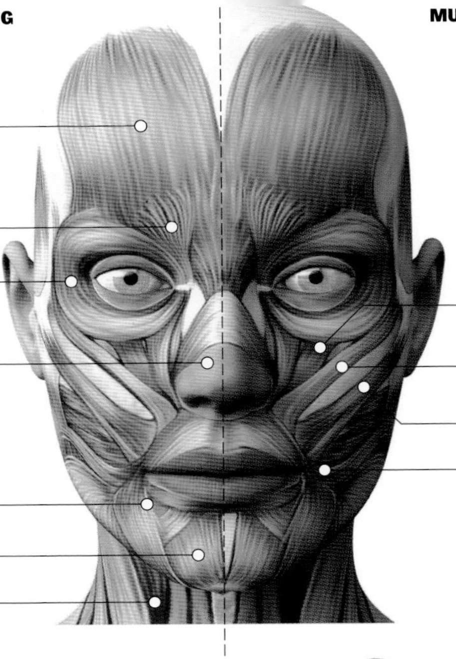


Clearly, far fewer muscles are needed to smile than to frown.

## The Three Types of Muscles

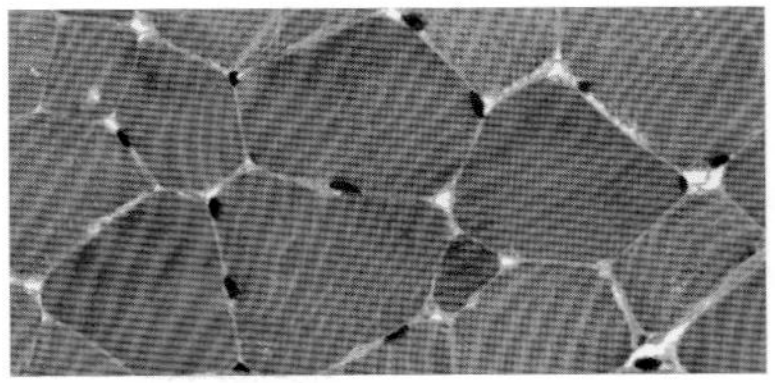

**STRIATED**
They are also called "skeletal" (because they cover the skeleton) and "voluntary." They are composed of cells and fibers that contract rapidly.

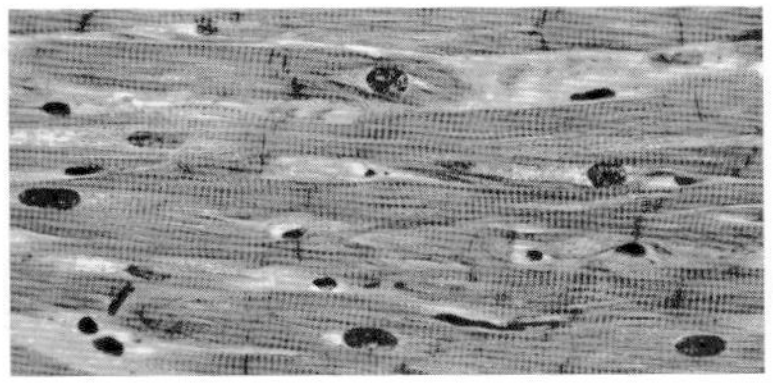

**CARDIAC**
Composed of small interconnected fibers, which maintain the rhythmic and continuous pumping of the heart.

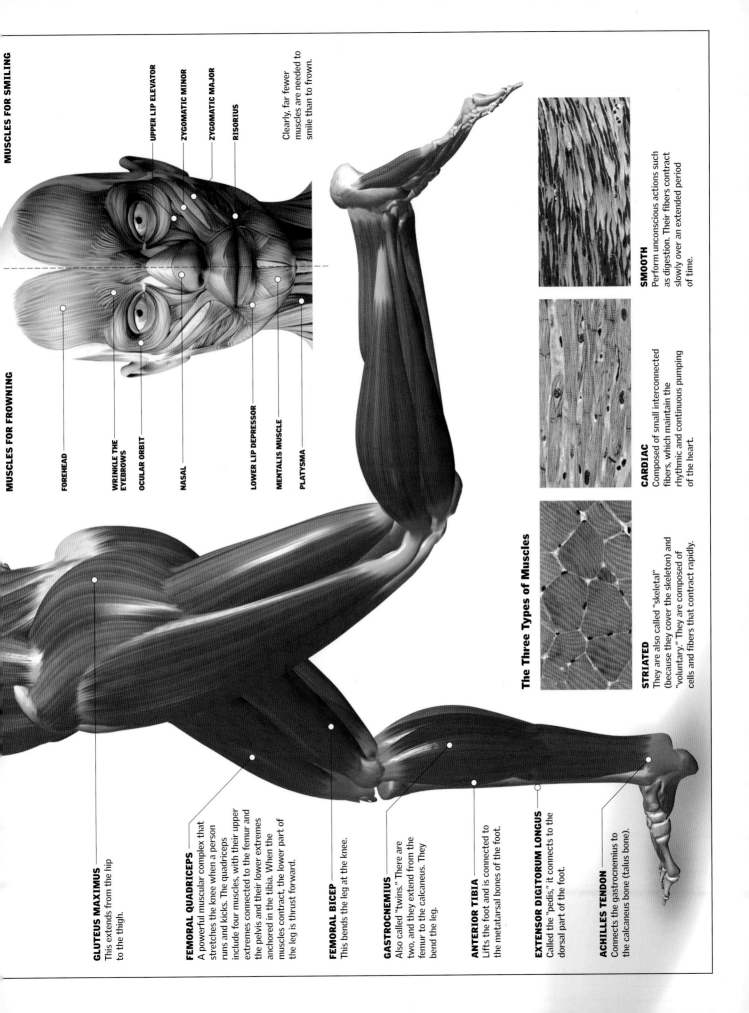

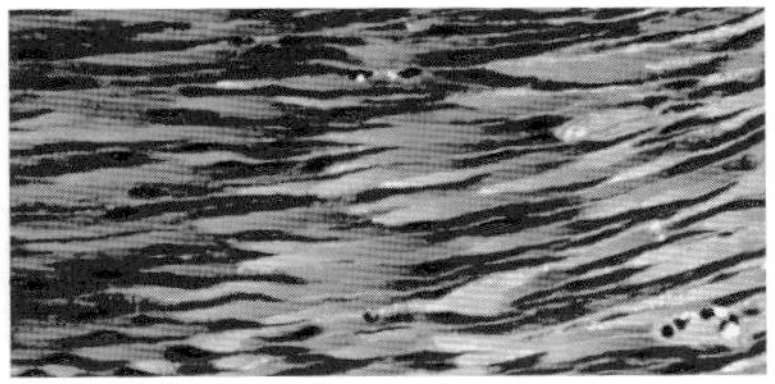

**SMOOTH**
Perform unconscious actions such as digestion. Their fibers contract slowly over an extended period of time.

# Muscular Fiber

A fiber is the long, thin cell that, when organized by the hundreds into groups called fascicles, constitutes the muscles. It is shaped like an elongated cylinder. The amount of fiber present varies according to the function accomplished by each muscle. Fibers are classified as white, which contract readily for actions that require force and power, and red, which perform slow contractions in movements of force and sustained traction. Each muscle fiber contains in its structure numerous filaments called myofibers. Myofibers, in turn, have two classes of protein filaments: myosin, also called thick filaments, and actin, or thin filaments. Both kinds of fibers are arranged in tiny matrices called sarcomeres.

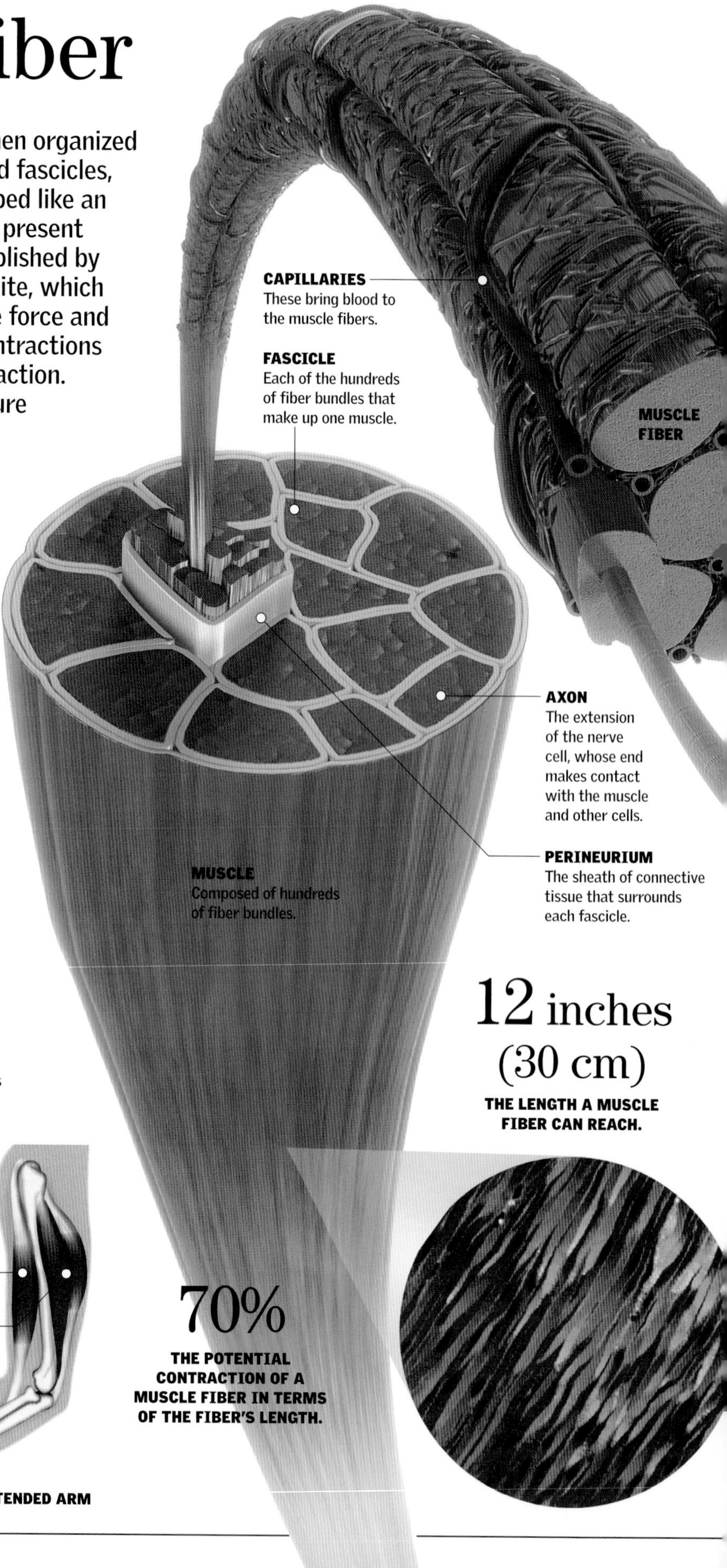

CAPILLARIES
These bring blood to the muscle fibers.
FASCICLE
Each of the hundreds of fiber bundles that make up one muscle.
MUSCLE FIBER
AXON
The extension of the nerve cell, whose end makes contact with the muscle and other cells.
PERINEURIUM
The sheath of connective tissue that surrounds each fascicle.
MUSCLE
Composed of hundreds of fiber bundles.

## Specialization

The quantity of muscle fiber varies according to the size and function of the muscle. Also, the same muscle can combine white fibers (rapid contracters) and red fibers (slow contracters). Even though their percentages differ from one person to the next, the composition of the muscles of the upper limbs tends to be the same as that of the lower in the same person. In other words, the relation between motor neurons and muscle fibers is inscribed in a person's genes. Depending on the type of neuron that stimulates them, the fibers are differentiated into slow fibers (when the neuron or motor neuron innervates between five and 180 fibers) and rapid fibers (when the neuron innervates between 200 and 800 fibers). The neurons and the fiber constitute what is called a motor unit.

**12 inches (30 cm)**

**THE LENGTH A MUSCLE FIBER CAN REACH.**

## Opposites

The muscles contract or relax according to the movement to be accomplished. To make the brain's directive take effect, the muscles involved carry out opposing actions.

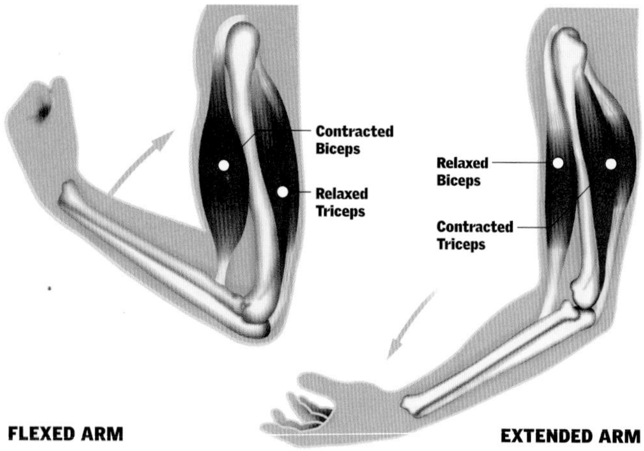


**FLEXED ARM**

**EXTENDED ARM**

**70%**

**THE POTENTIAL CONTRACTION OF A MUSCLE FIBER IN TERMS OF THE FIBER'S LENGTH.**

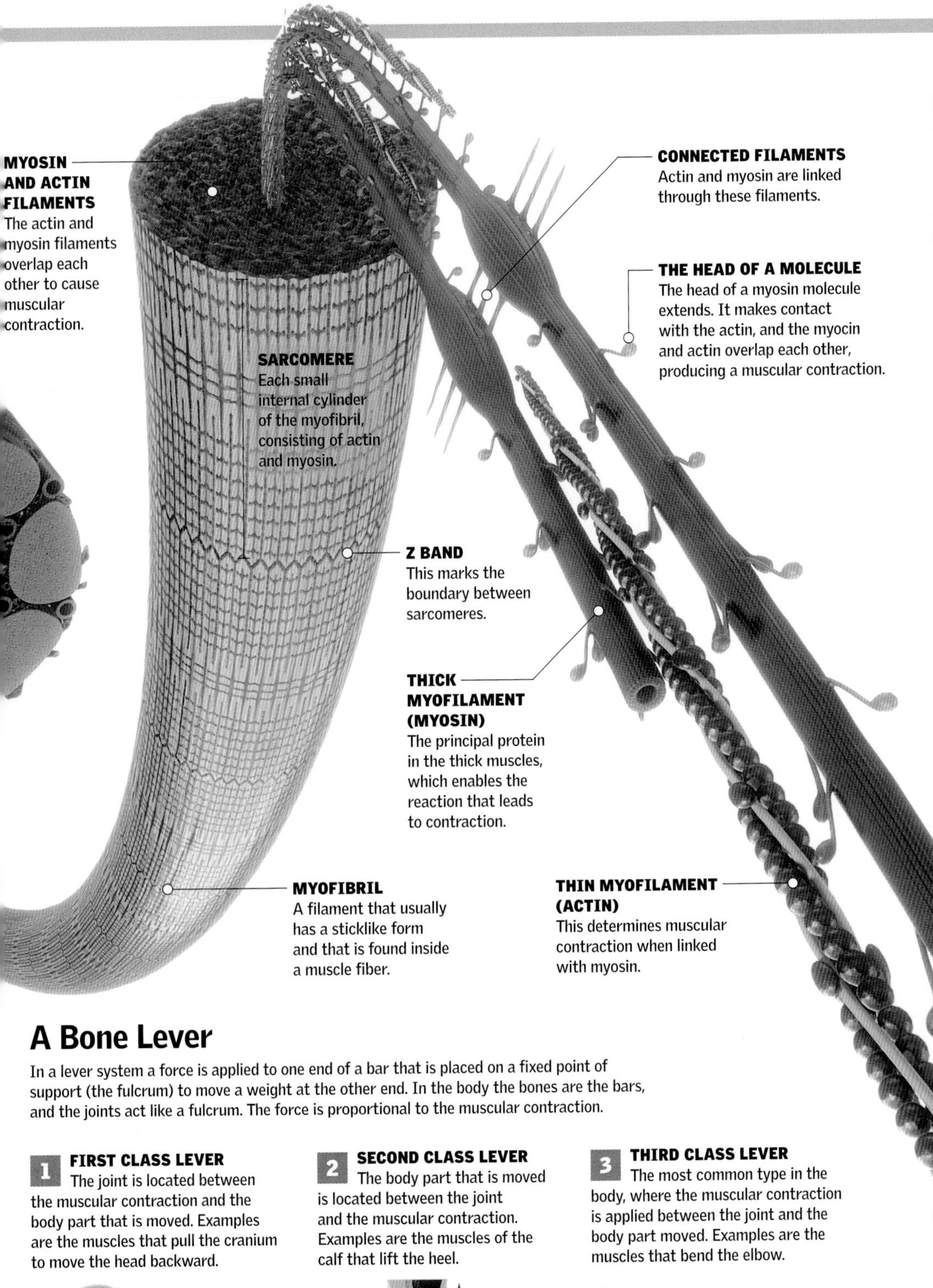

MYOSIN AND ACTIN FILAMENTS
The actin and myosin filaments overlap each other to cause muscular contraction.
SARCOMERE
Each small internal cylinder of the myofibril, consisting of actin and myosin.
CONNECTED FILAMENTS
Actin and myosin are linked through these filaments.
THE HEAD OF A MOLECULE
The head of a myosin molecule extends. It makes contact with the actin, and the myocin and actin overlap each other, producing a muscular contraction.
Z BAND
This marks the boundary between sarcomeres.
THICK MYOFILAMENT (MYOSIN)
The principal protein in the thick muscles, which enables the reaction that leads to contraction.
MYOFIBRIL
A filament that usually has a sticklike form and that is found inside a muscle fiber.
THIN MYOFILAMENT (ACTIN)
This determines muscular contraction when linked with myosin.

### Relaxation

The order to contract given by the nervous system ceases, and the muscle fibers return to a position of rest. This happens to all muscles, regardless of the duration of contraction.

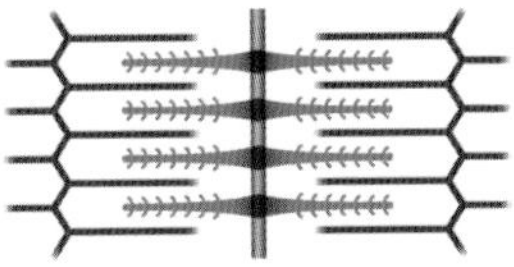

### Contraction

The nervous system orders the muscle fibers, no matter which type, to shorten. In order to create muscle contraction, calcium is released within the muscle cell, which allows the actin and the myosin to come together and overlap each other.

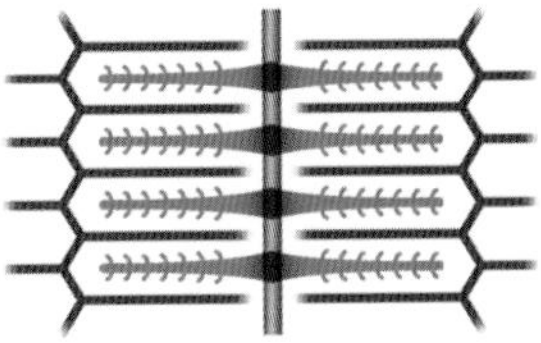

## A Bone Lever

In a lever system a force is applied to one end of a bar that is placed on a fixed point of support (the fulcrum) to move a weight at the other end. In the body the bones are the bars, and the joints act like a fulcrum. The force is proportional to the muscular contraction.

**1 FIRST CLASS LEVER**
The joint is located between the muscular contraction and the body part that is moved. Examples are the muscles that pull the cranium to move the head backward.

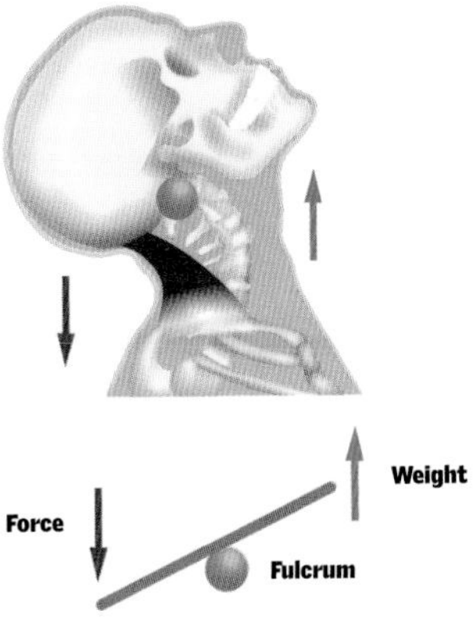


**2 SECOND CLASS LEVER**
The body part that is moved is located between the joint and the muscular contraction. Examples are the muscles of the calf that lift the heel.

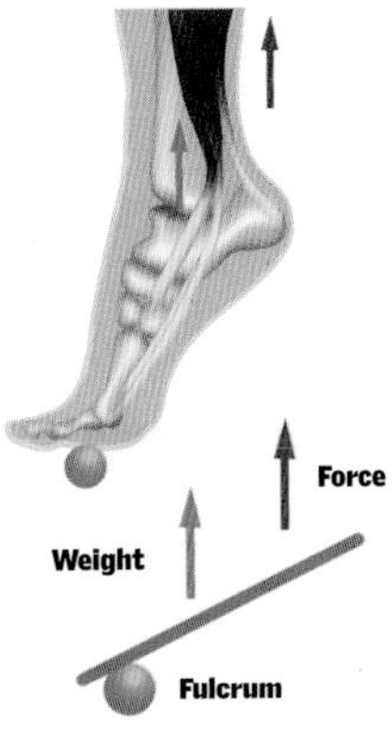


**3 THIRD CLASS LEVER**
The most common type in the body, where the muscular contraction is applied between the joint and the body part moved. Examples are the muscles that bend the elbow.

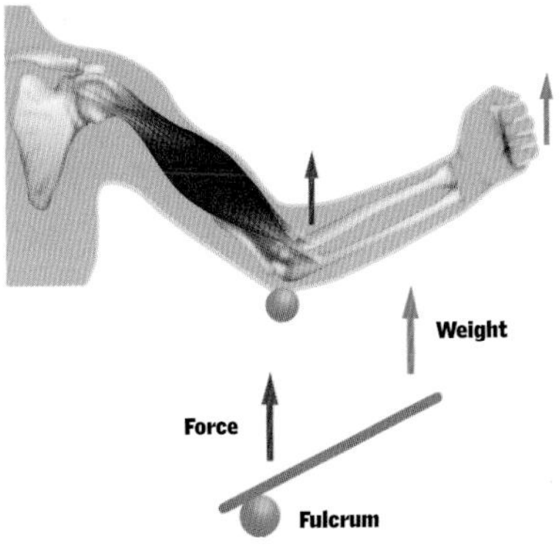


# Running

**MARATHON RUNNERS MAY HAVE AS MUCH AS 90 PERCENT RED, OR SLOW, FIBERS IN THEIR TWIN MUSCLES. CHAMPIONS IN THE 100-METER DASH HAVE ONLY 20 TO 25 PERCENT.**

# Internal Systems and Organs

It is difficult to explain that the sexual attraction between a man and woman—something that appears to be so natural and intimate—is actually a chemical phenomenon. What is certain is that when a couple feels they are in love, it is because hormones have gone into action. Without them, amorous thoughts and sexual fantasies would be drab and dull. We invite you to find out to what extent hormones determine many of our actions and also to investigate in detail, one by one, how the body's systems function. You will learn to understand how various organs of the body work as a team. Although each organ accomplishes specific tasks on its own, they all communicate with each other, and together they form a complete human being.



**THE CHEMISTRY OF LOVE** (opposite)
Even a brief kiss results in the release of adrenaline, causing a sensation of euphoria and joy.

# Circulatory System

Its function is to carry blood to and from all the organs of the body. To drive the constant movement of the blood, the system uses the pumping of the heart, the organ that acts as the system's engine. The arteries bring oxygen-rich blood to all the cells, and the veins retrieve the blood so that it can be oxygenated once again and so that wastes can be removed.

## A System That Goes Around

The center of the system is the heart, which, together with a network of vessels, forms the cardiovascular machinery. This vital engine beats more than 30 million times a year—approximately 2 billion times in a person's lifetime. With each beat it pumps about 5 cubic inches (82 ml) of blood. This means that an adult heart could fill a 2,000-gallon (8,000-l) tank in just one day. Beginning at the heart, the circulatory system completes two circuits: the main, or systemic, circulation via the aortic artery and the minor, or pulmonary, circulation. The main circulation brings oxygenated blood to the capillary system, where the veins are formed; the minor circulation brings oxygen-poor blood through the pulmonary artery to be enriched with oxygen and to have carbon dioxide removed from it, a process called hematosis. Other secondary circuits are the hepatic portal system and the hypophyseal portal system.

**BLOOD DISTRIBUTION DURING CIRCULATION**

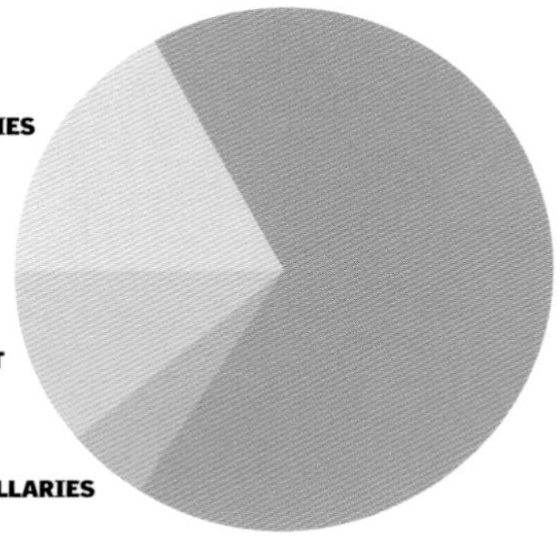


**60,000 miles (100,000 km)**

**THE TOTAL LENGTH OF THE BLOOD VESSELS. NINETY-EIGHT PERCENT OF THEM ARE CAPILLARIES.**

**1 inch (2.5 cm)**

**THE EXTERNAL DIAMETER OF THE AORTA (THE LARGEST ARTERY) AND THE VENA CAVA (THE LARGEST VEIN).**

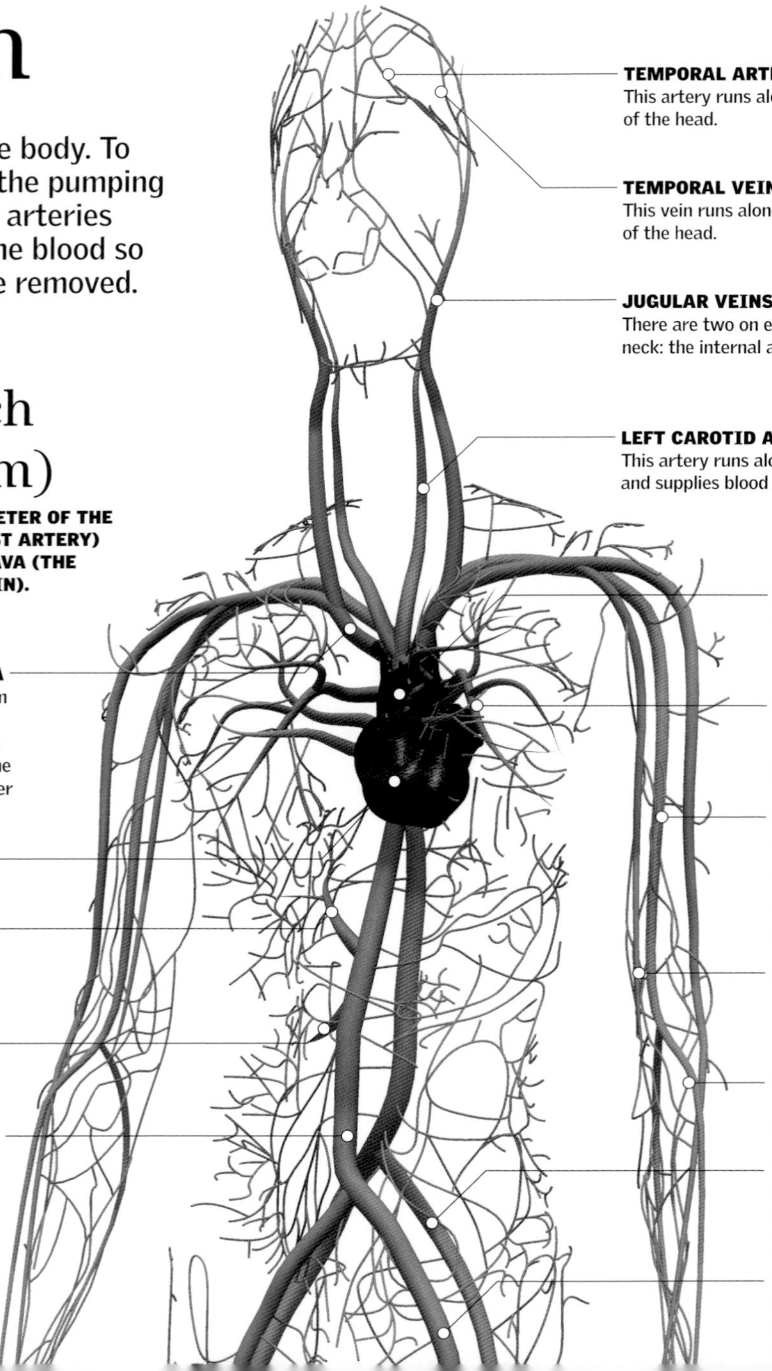

SUPERIOR VENA CAVA
This brings the blood from the upper part of the body for purification. The superior vena cava and the inferior vena cava together form the largest vein.
HEART
The great engine.
TRUNCUS OF THE PORTAL VEIN
It terminates in the sinusoids of the liver.
RENAL VEIN
Blood exits the kidneys through this vein.
INFERIOR VENA CAVA
This takes blood arriving from the area below the diaphragm and brings it up to the heart.
TEMPORAL ARTERY
This artery runs along the side of the head.
TEMPORAL VEIN
This vein runs along the side of the head.
JUGULAR VEINS
There are two on each side of the neck: the internal and the external.
LEFT CAROTID ARTERY
This artery runs along the neck and supplies blood to the head.
AORTIC ARTERY (AORTA)
The body's principal artery.
PULMONARY ARTERY
This artery carries blood to the lungs.
HUMERAL ARTERY
(Axillary) The right one arises from the brachiocephalic trunk and the left from the aortic arch.
SUBCLAVIAN VEIN
This vein connects the axillary with the superior vena cava.
RADIAL ARTERY
Runs along the radial side of the forearm.
LEFT PRIMITIVE ILIAC ARTERY
Provides blood to the pelvis and the legs.
LEFT PRIMITIVE ILIAC VEIN
This is the primary vein of the hip area.

## Arteries

Muscular elastic blood vessels. Their function is to bring oxygenated blood from the heart (from the primary artery, the aorta) to all the cells of the body. Arteries have thick walls, allowing them to withstand the high pressure of the blood.

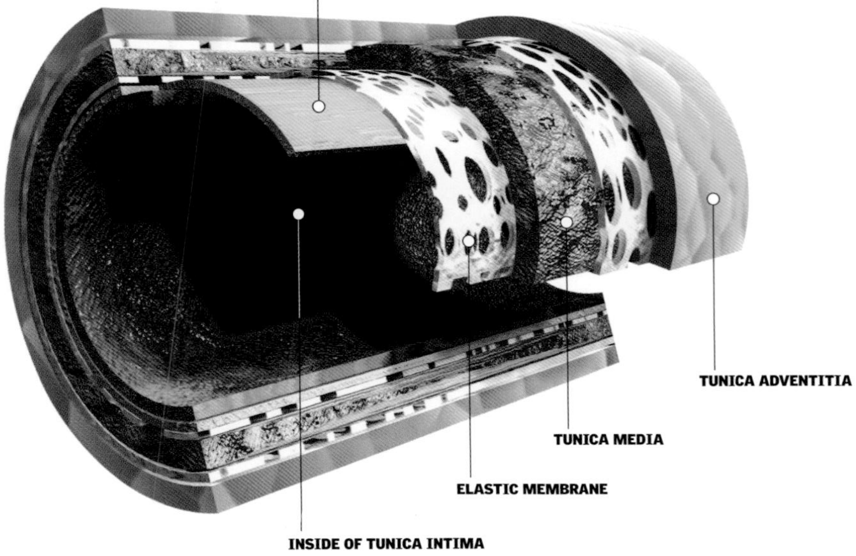


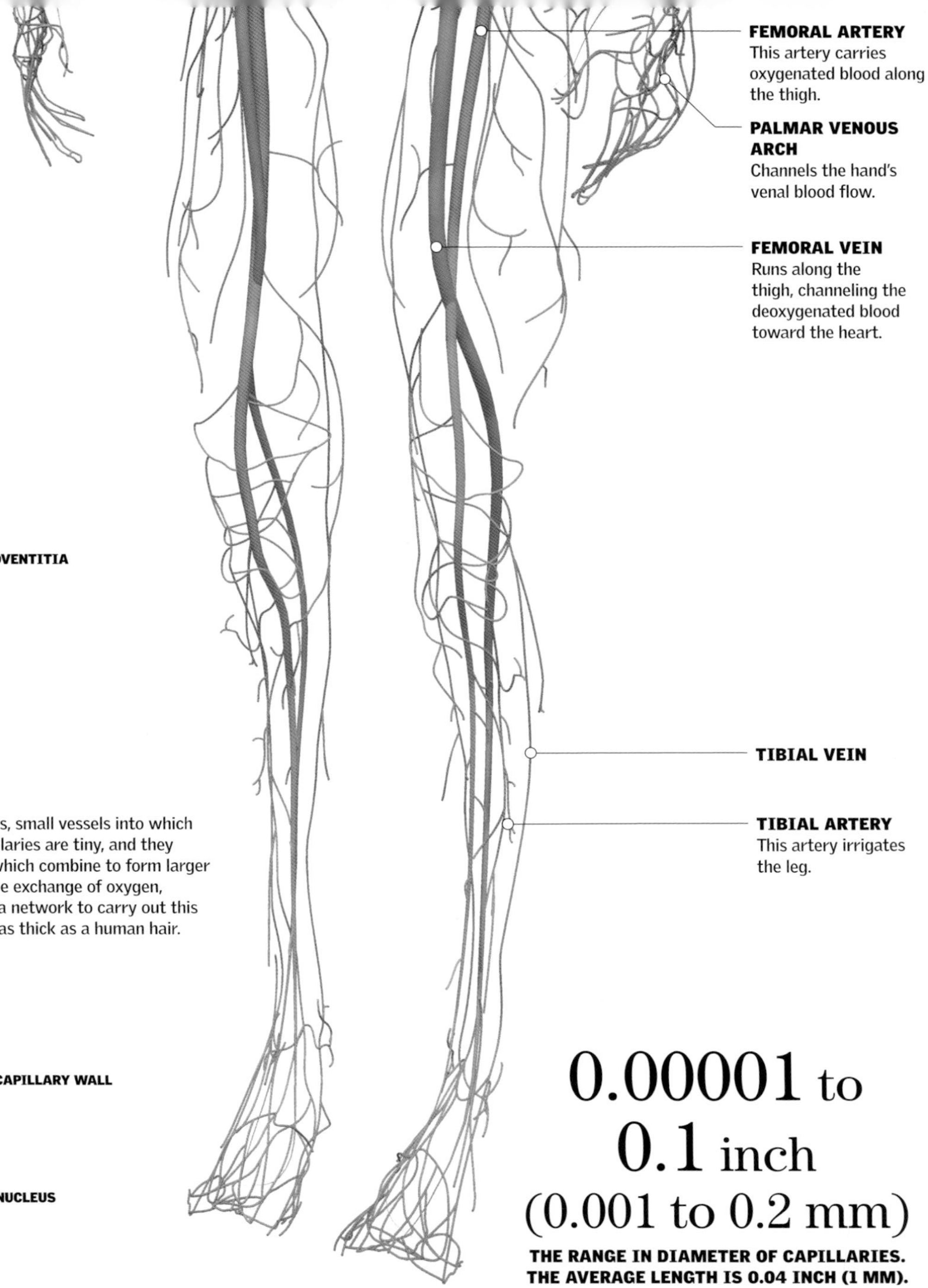


## Veins

The veins are the conduits that transport deoxygenated blood back toward the heart after it has traveled to different parts of the body. The veins have thin walls with less muscular fiber and less elasticity than the arteries. The principal veins have valves to prevent the reflux of blood, forcing it to travel in only one direction.

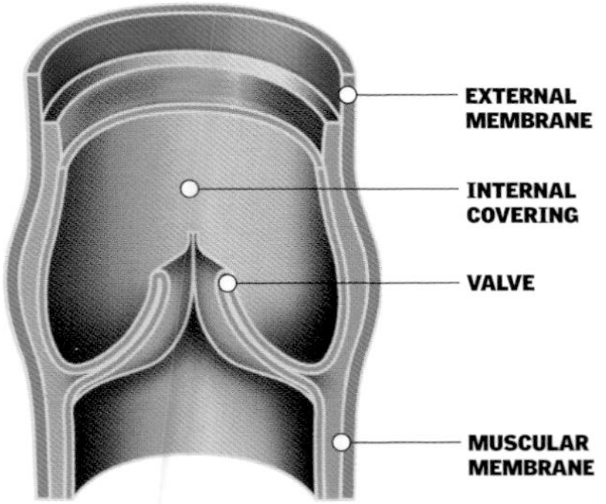


## Capillaries

These are branchings of the arterioles, small vessels into which the arteries are subdivided. The capillaries are tiny, and they come together to form small veins, which combine to form larger veins. The capillaries are crucial in the exchange of oxygen, nutrients, and waste, and they form a network to carry out this activity. Ten capillaries together are as thick as a human hair.

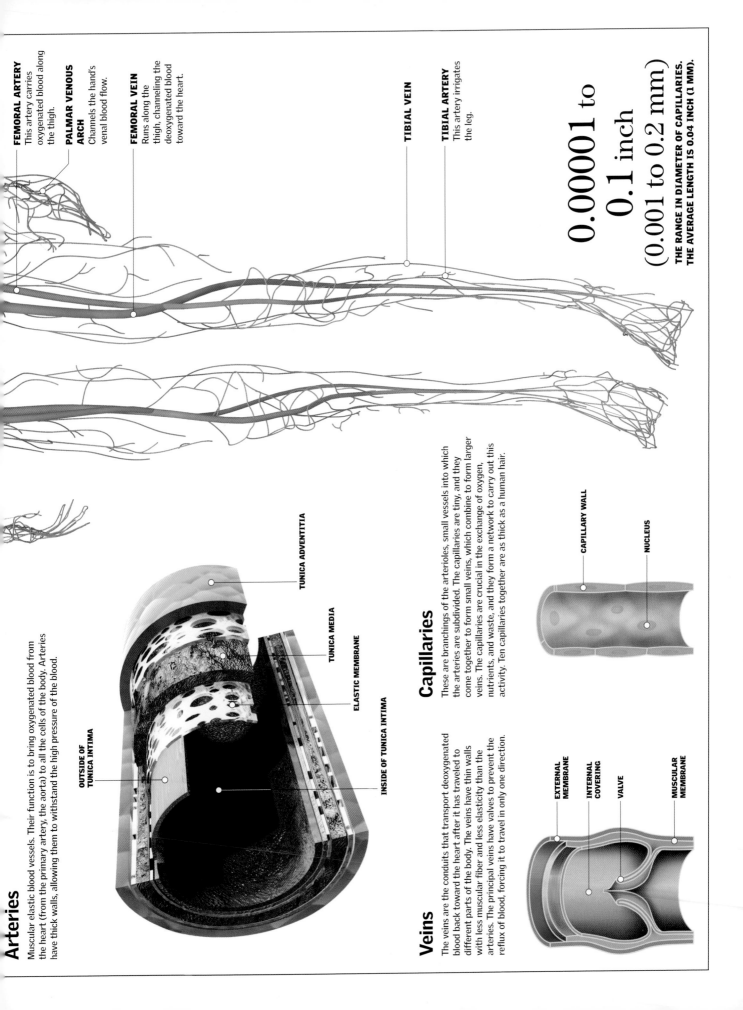

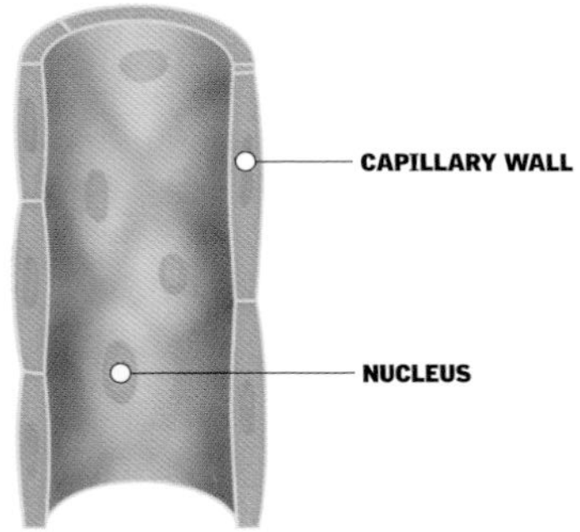


**0.00001 to 0.1 inch (0.001 to 0.2 mm)**

**THE RANGE IN DIAMETER OF CAPILLARIES. THE AVERAGE LENGTH IS 0.04 INCH (1 MM).**

# All About the Heart

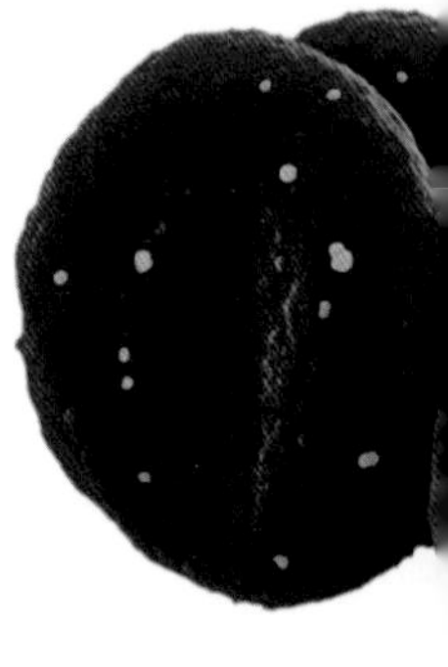

The heart is the engine of the circulatory apparatus: it supplies 10 pints (4.7 l) of blood per minute. Its rhythmic pumping ensures that blood arrives in every part of the body. The heart beats between 60 and 100 times per minute in a person at rest and up to 200 times per minute during activity. The heart is a hollow organ, the size of a fist; it is enclosed in the thoracic cavity in the center of the chest above the diaphragm. The name of the stomach's entrance, or cardias, comes from the Greek word for heart, *kardia*. Histologically, one can distinguish three layers of tissue in the heart, starting from the inside out: the endocardium, the myocardium, and the pericardium.

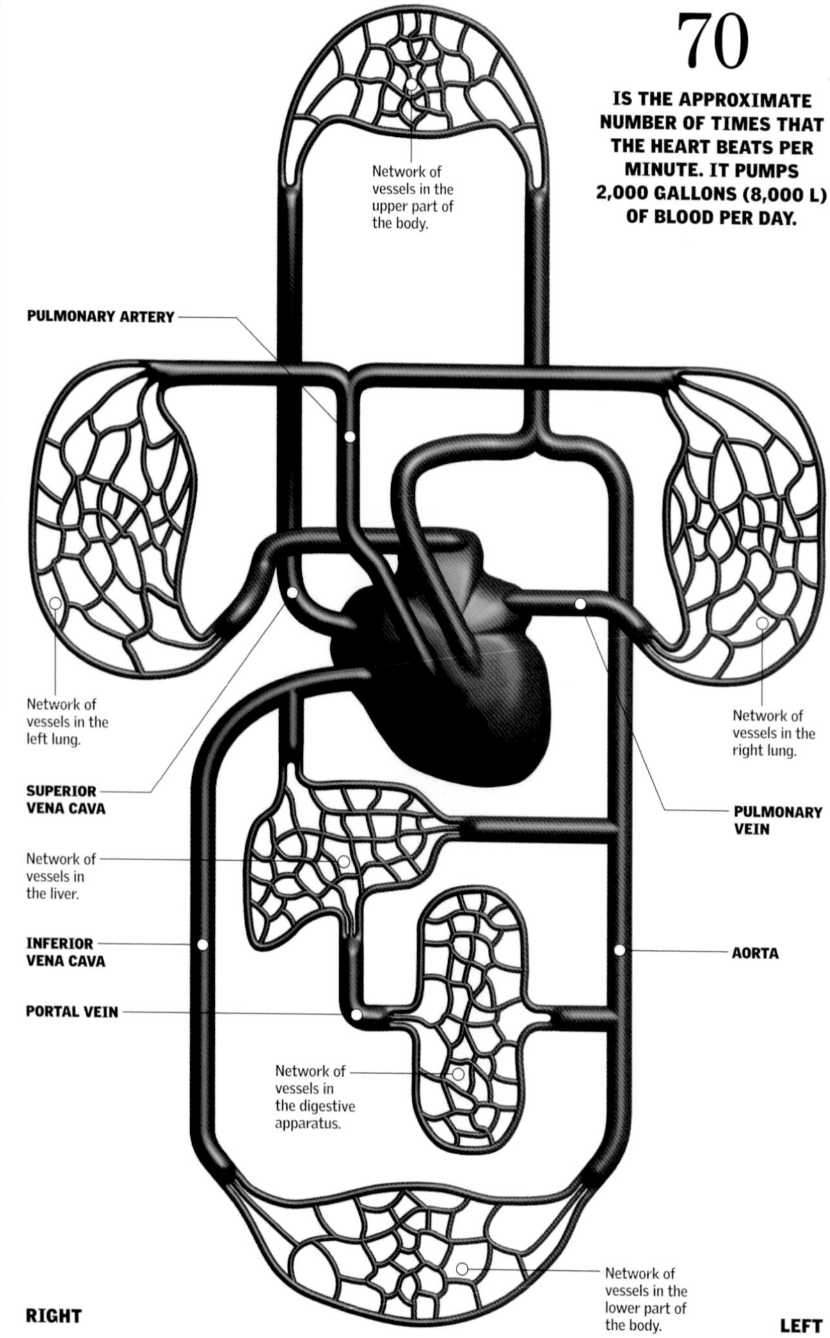

Network of vessels in the upper part of the body.
PULMONARY ARTERY
Network of vessels in the left lung.
SUPERIOR VENA CAVA
Network of vessels in the liver.
INFERIOR VENA CAVA
PORTAL VEIN
Network of vessels in the digestive apparatus.
Network of vessels in the right lung.
PULMONARY VEIN
AORTA
Network of vessels in the lower part of the body.
RIGHT
LEFT

## 70

**IS THE APPROXIMATE NUMBER OF TIMES THAT THE HEART BEATS PER MINUTE. IT PUMPS 2,000 GALLONS (8,000 L) OF BLOOD PER DAY.**

## The Sequence of the Heartbeat

**1 DIASTOLIC**
The atria and the ventricles are relaxed. The blood, supercharged with carbon dioxide, flows from all the corners of the body and enters the right atrium, while the blood that was oxygenated through the work of the lungs returns to the left part of the heart.

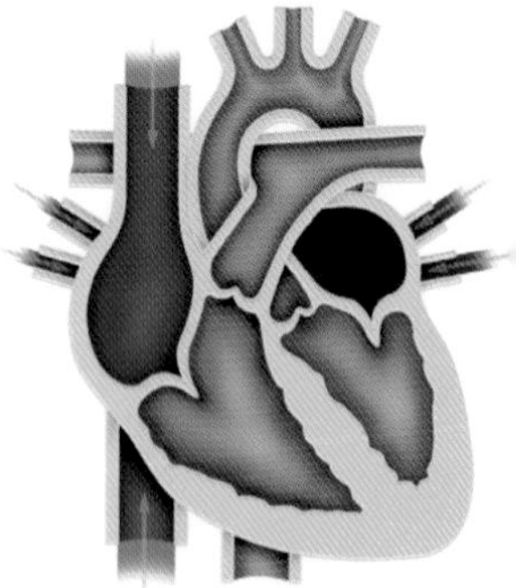

**2 ATRIAL SYSTOLE**
The atria contract to push the blood down toward the ventricles. The right ventricle receives the blood that will have to be sent to the lungs to be oxygenated. The left ventricle receives blood coming from the lungs, which is already oxygenated and must be pumped toward the aorta.

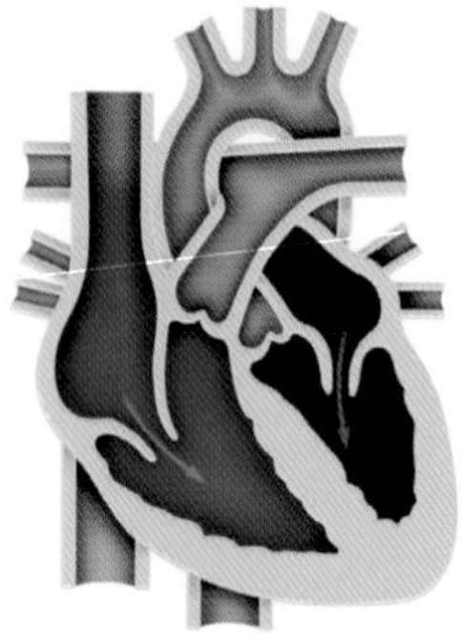

**3 VENTRICULAR SYSTOLE**
The ventricles contract after a brief pause. The systole, or contraction, of the right ventricle sends impure blood to the lungs. The contraction of the left ventricle pumps the already oxygenated blood toward the aorta; it is ready for distribution throughout the body.

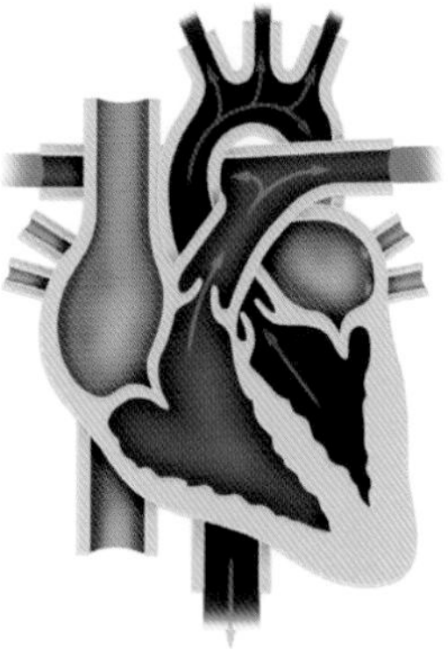

## 20 seconds

**A RED BLOOD CELL TRAVERSES THE BODY IN 20 SECONDS. THEREFORE, THE DISTANCE THAT IT TRAVELS AMOUNTS TO 7,500 MILES (12,000 KM).**

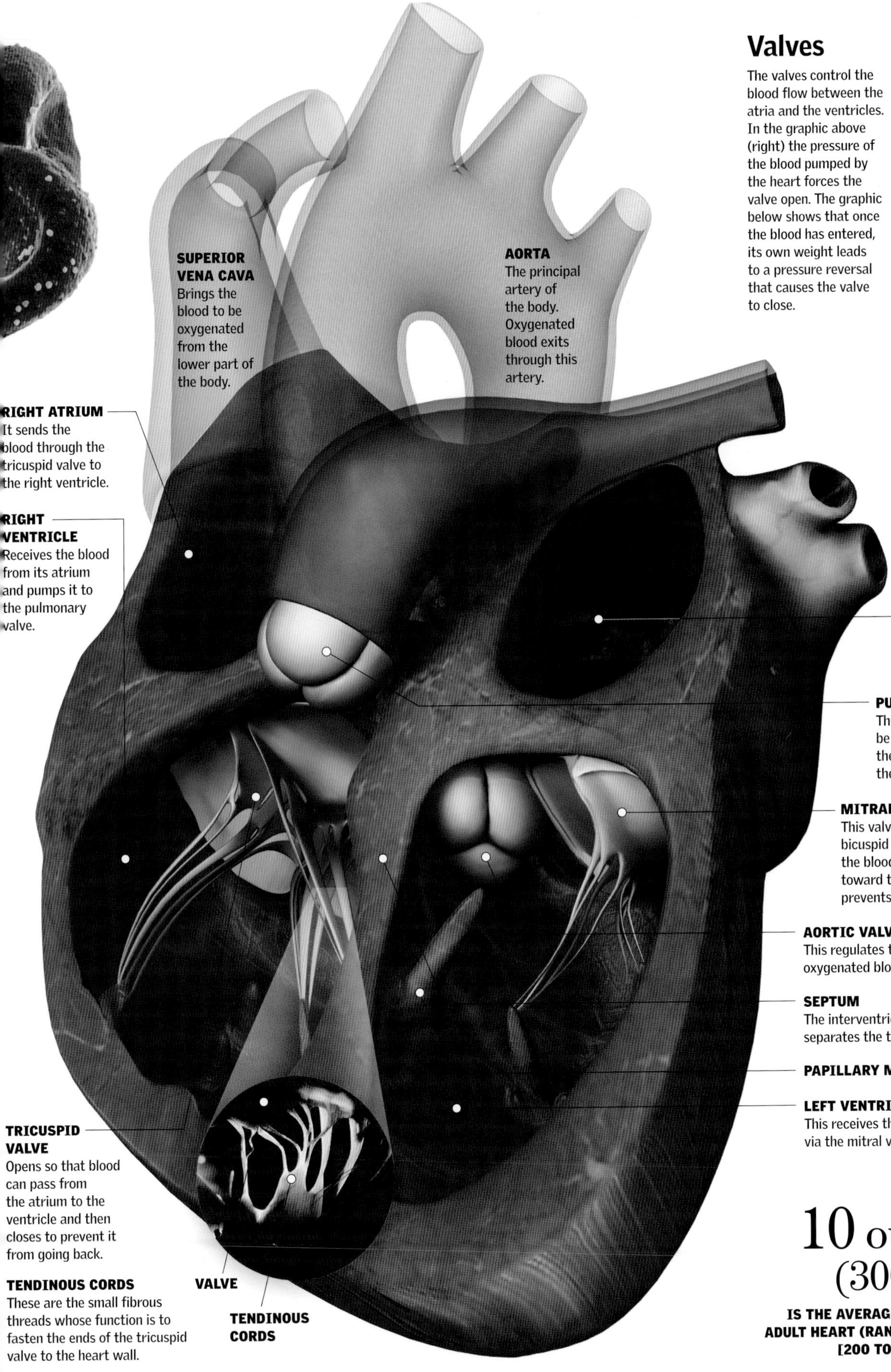

SUPERIOR VENA CAVA
Brings the blood to be oxygenated from the lower part of the body.
AORTA
The principal artery of the body. Oxygenated blood exits through this artery.
RIGHT ATRIUM
It sends the blood through the tricuspid valve to the right ventricle.
RIGHT VENTRICLE
Receives the blood from its atrium and pumps it to the pulmonary valve.
LEFT ATRIUM
This receives the oxygenated blood from the lungs.
PULMONARY VALVE
Through this valve blood to be oxygenated passes from the right ventricle toward the pulmonary artery.
MITRAL VALVE
This valve, also known as the bicuspid valve, opens the path for the blood from the left auricle toward the ventricle and then prevents it from returning.
AORTIC VALVE
This regulates the passage of the oxygenated blood toward the aorta.
SEPTUM
The interventricular wall that separates the two inferior cavities.
PAPILLARY MUSCLES
LEFT VENTRICLE
This receives the oxygenated blood via the mitral valve.
TRICUSPID VALVE
Opens so that blood can pass from the atrium to the ventricle and then closes to prevent it from going back.
TENDINOUS CORDS
These are the small fibrous threads whose function is to fasten the ends of the tricuspid valve to the heart wall.
VALVE
TENDINOUS CORDS

## Valves

The valves control the blood flow between the atria and the ventricles. In the graphic above (right) the pressure of the blood pumped by the heart forces the valve open. The graphic below shows that once the blood has entered, its own weight leads to a pressure reversal that causes the valve to close.

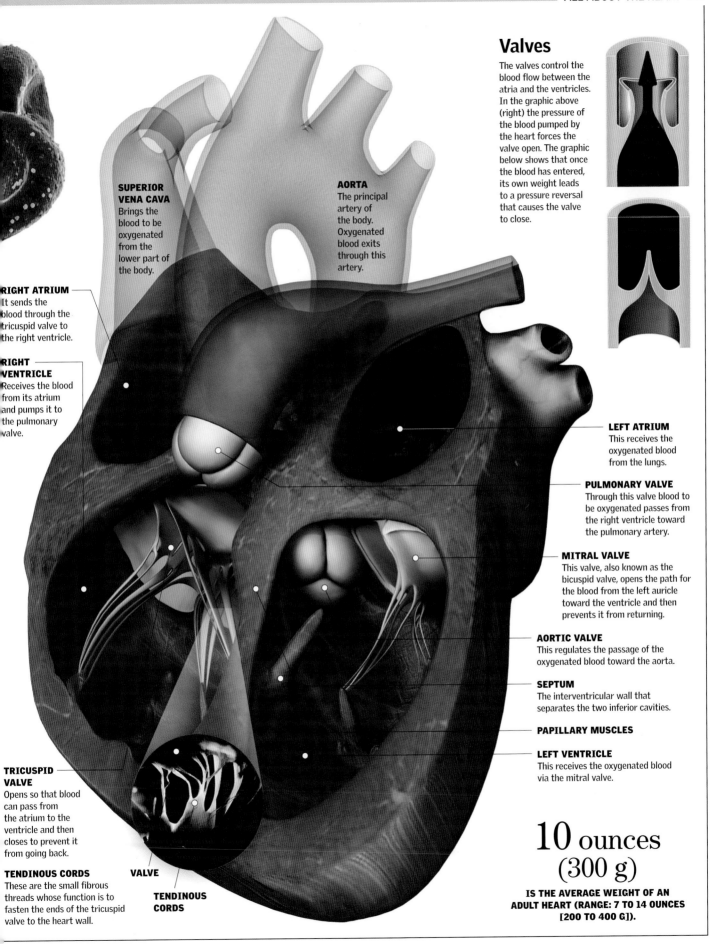

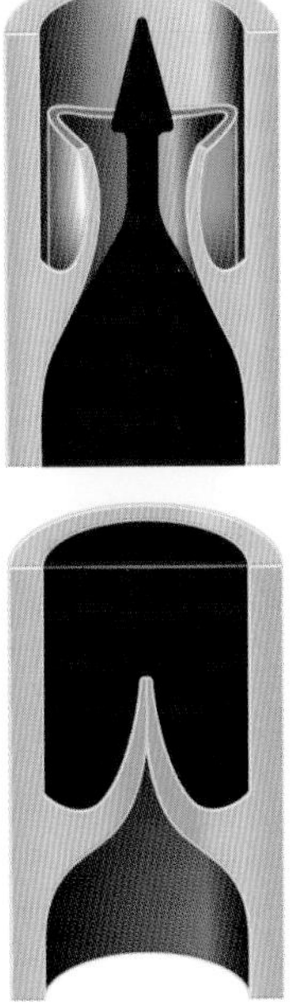

**10 ounces (300 g)**

**IS THE AVERAGE WEIGHT OF AN ADULT HEART (RANGE: 7 TO 14 OUNCES [200 TO 400 G]).**

# Components of the Blood

The blood is a liquid tissue composed of water, dissolved substances, and blood cells. The blood circulates inside the blood vessels thanks to the impulse it receives from the contraction of the heart. A principal function of the blood is to distribute nutrients to all the cells of the body. For example, the red blood cells (erythrocytes) carry oxygen, which associates with the hemoglobin, a substance in the cell responsible for the blood's red color. The blood also contains white blood cells and platelets that protect the body in various ways.

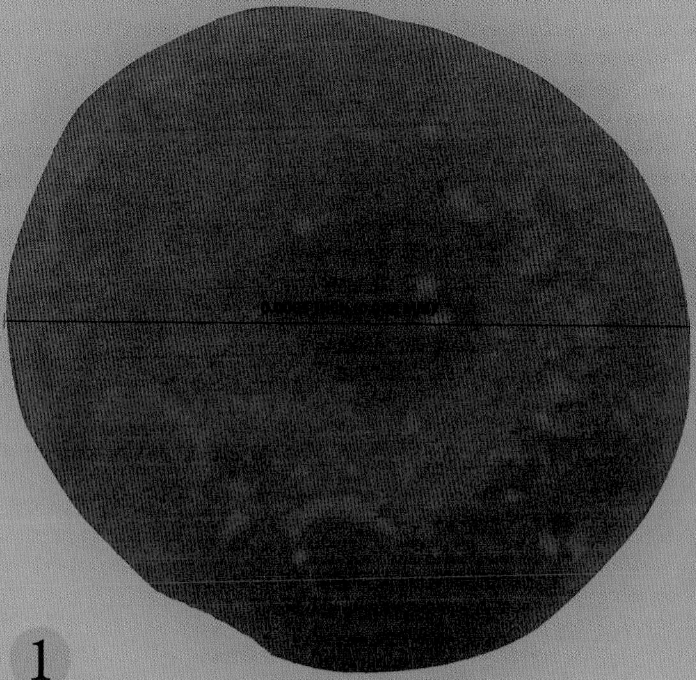

## 1 Red Blood Cells

These cells are phantom cells, because all they contain is a large amount of hemoglobin, a protein that has a great affinity for combining with oxygen. The red blood cells, which circulate in the blood, bring oxygen to the cells that need it, and they also remove a small part of the carbon dioxide that the cells are discarding as waste. Because they cannot reproduce themselves, they must be replaced by new red blood cells that are produced by the bone marrow.

5 quarts (4.7 l)

**THE APPROXIMATE VOLUME OF BLOOD PRESENT IN A HUMAN ADULT.**

**FLEXIBILITY**

Red blood cells are flexible and take on a bell shape in order to pass through the thinnest blood vessels.

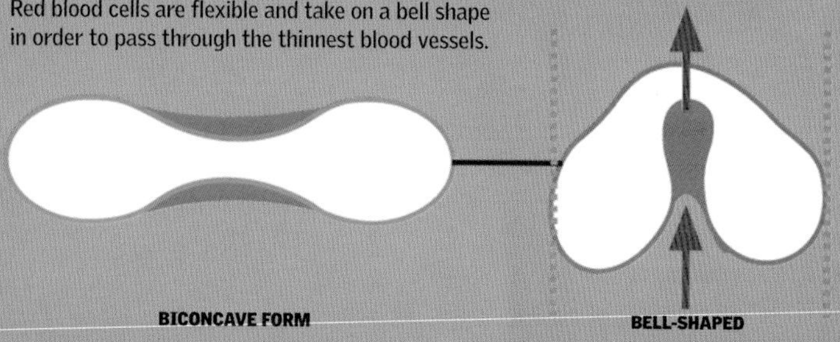


## The Blood Groups

Each person belongs to a blood group. Within the ABO system the groups are A, B, AB, and O. Each group is also identified with an antigen, or Rh factor, that is present in the red blood cells of 85 percent of the population. It is of vital importance to know what blood group a person belongs to so as to give only the right type during a blood transfusion. The immune system, via antibodies and antigens, will accept the body's own blood type but will reject the wrong type.

**GROUP A**

An individual with red blood cells with antigen A in its membranes belongs to blood group A, and that person's plasma has antibodies against type B. These antibodies recognize red blood cells with antigen B in their membranes as foreign.

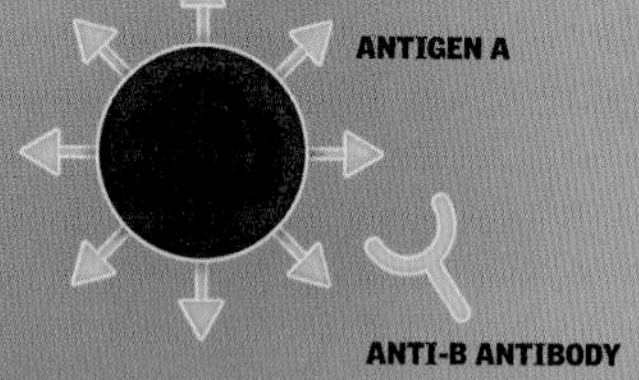

**GROUP B**

Members of this group have antigen B in the membrane of their red blood cells and anti-A antibodies in their blood plasma.

**GROUP AB**

Members of this group have antigen A and B in the membrane of their red blood cells and no antibodies in their blood plasma.

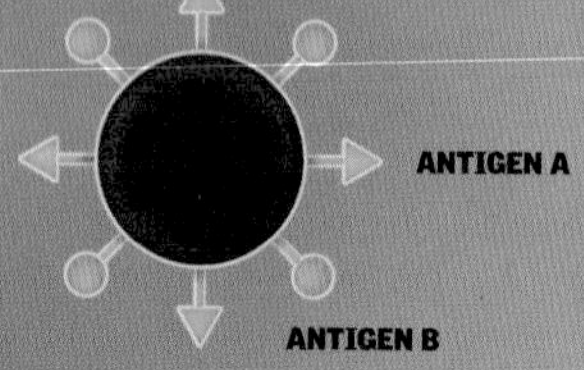


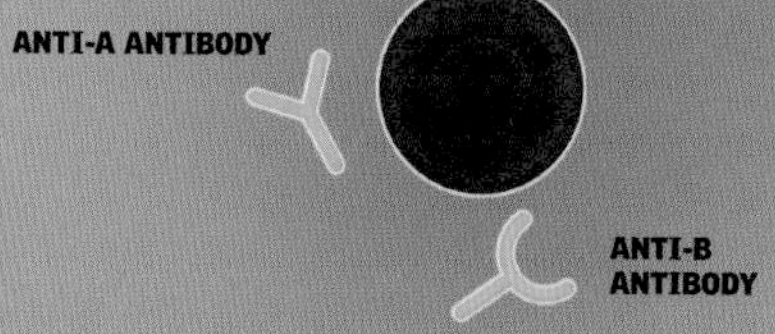


**GROUP O**

Members of this group have no antigens in the membranes of their erythrocytes and anti-A and anti-B antibodies in their blood plasma.

**COMPATIBILITY**

Donors of group O can give blood to any group, but group AB donors can give only to others with AB blood. The possibility of blood donation depends on the antibodies of the recipient.

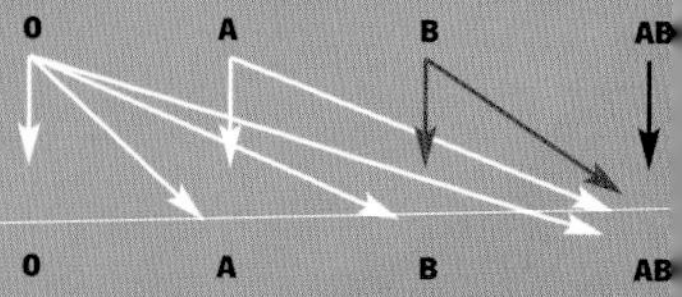

## 2

## White Blood Cells, or Leukocytes

This is what a leukocyte, or white blood cell, looks like swimming in blood plasma. They are called white because that is their color when viewed under a microscope.

| COMPOSITION | |
|---|---|
| **Granulocytes** | Neutrophils<br>Eosinophils<br>Basophils |
| **Agranulocytes** | Lymphocytes<br>Monocytes |

# 7%

**IS THE PORTION OF BODY WEIGHT REPRESENTED BY THE BLOOD.**

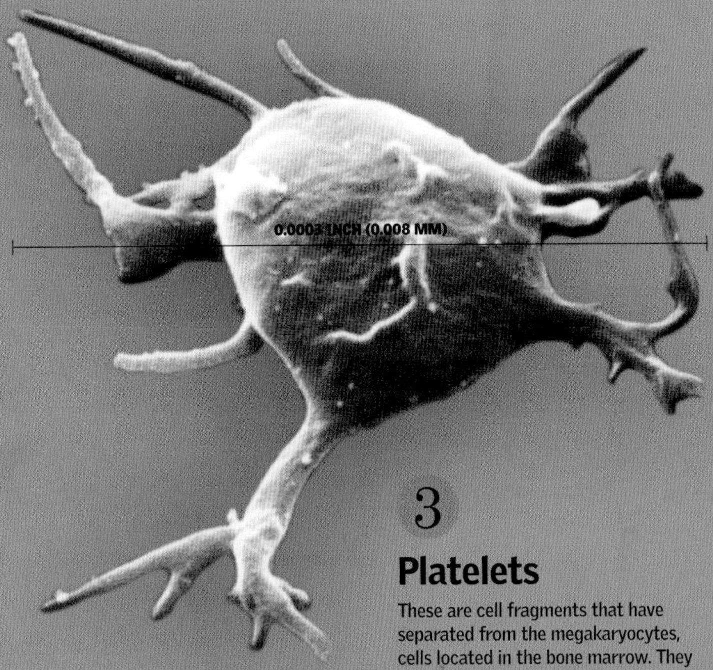


## Blood Components

The blood is a tissue, and as such it is characterized by the same type of cells and intercellular substance as tissue. It is distinguished from the rest of the tissues in the human body by an abundance of intercellular material, which consists primarily of water. The intercellular material, called plasma, is yellow, and it contains abundant nutrients and other substances, such as hormones and antibodies, that take part in various physiological processes.

| COMPONENTS OF THE BLOOD PER 0.00006 cubic inch (1 cu ml) | |
|---|---|
| **Red blood cells** | 4 to 6 million |
| **White blood cells** | 4,500 to 11,000 |
| **Platelets** | 150,000 to 400,000 |
| **Normal pH** | 7.40 |

**DAILY PRODUCTION IN MILLIONS**

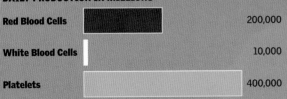

## 3

## Platelets

These are cell fragments that have separated from the megakaryocytes, cells located in the bone marrow. They have a role in blood coagulation. Next to the red blood cells, the platelets are the most abundant component of the blood.

## 4

## Plasma

Red and white blood cells and platelets (which contribute to coagulation) make up 45 percent of the blood. The remaining 55 percent is plasma, a fluid that is 90 percent water and the rest various nutrients.

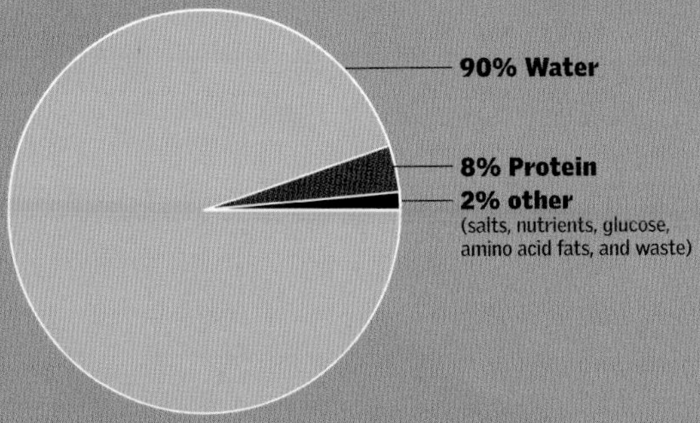


# 98.6° F (37° C)

**THE BLOOD MAINTAINS THE BODY AT THIS AVERAGE TEMPERATURE.**

# Lymphatic System

It accomplishes two basic functions: defense against foreign organisms (such as bacteria) and aid with transport of liquid and matter via the circulation of the lymph from the interstices of the tissue and from the digestive apparatus to the blood. About 3 to 4 quarts (2.8-3.7 l) of the liquid circulating in the system do not return. This liquid is known as lymph, and it is reabsorbed into the plasma only through the lymphatic vessels. The lymph contains cells called lymphocytes and macrophages, which are part of the immune system.

## Lymphatic Network

This network contains vessels that extend throughout the body and that filter the liquid that comes from the area surrounding the cells. The lymph circulates in only one direction and returns to the blood through the walls of small blood vessels. There are valves that prevent the lymph from flowing in the opposite direction. The lymph nodes filter harmful microorganisms from the lymph, which returns via blood vessels to maintain the equilibrium of the body's fluids. Together with the white blood cells, the lymph nodes are in charge of maintaining the immune system.

## Lymphatic Tissue

One part of the liquid that exits from blood flow and distributes itself in the body returns only through the action of the lymphatic tissue, which reabsorbs it via the lymphatic capillaries and returns it to the blood via the lymphatic vessels.

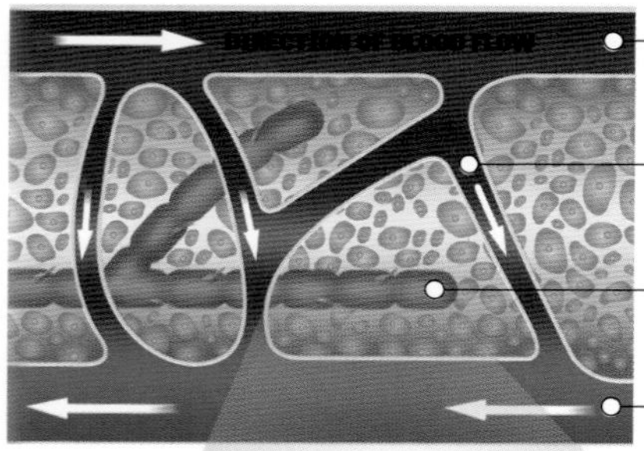
ARTERIOLE
BLOOD CAPILLARY
LYMPHATIC CELL
LYMPHATIC CAPILLARY
VENULE

# 6 gallons (24 l)

**THE AMOUNT OF LIQUID THAT LEAVES THE BLOOD AND PASSES THROUGH THE SYSTEM DAILY, MOVING THROUGH THE TISSUES AND RETURNING TO THE BLOODSTREAM**

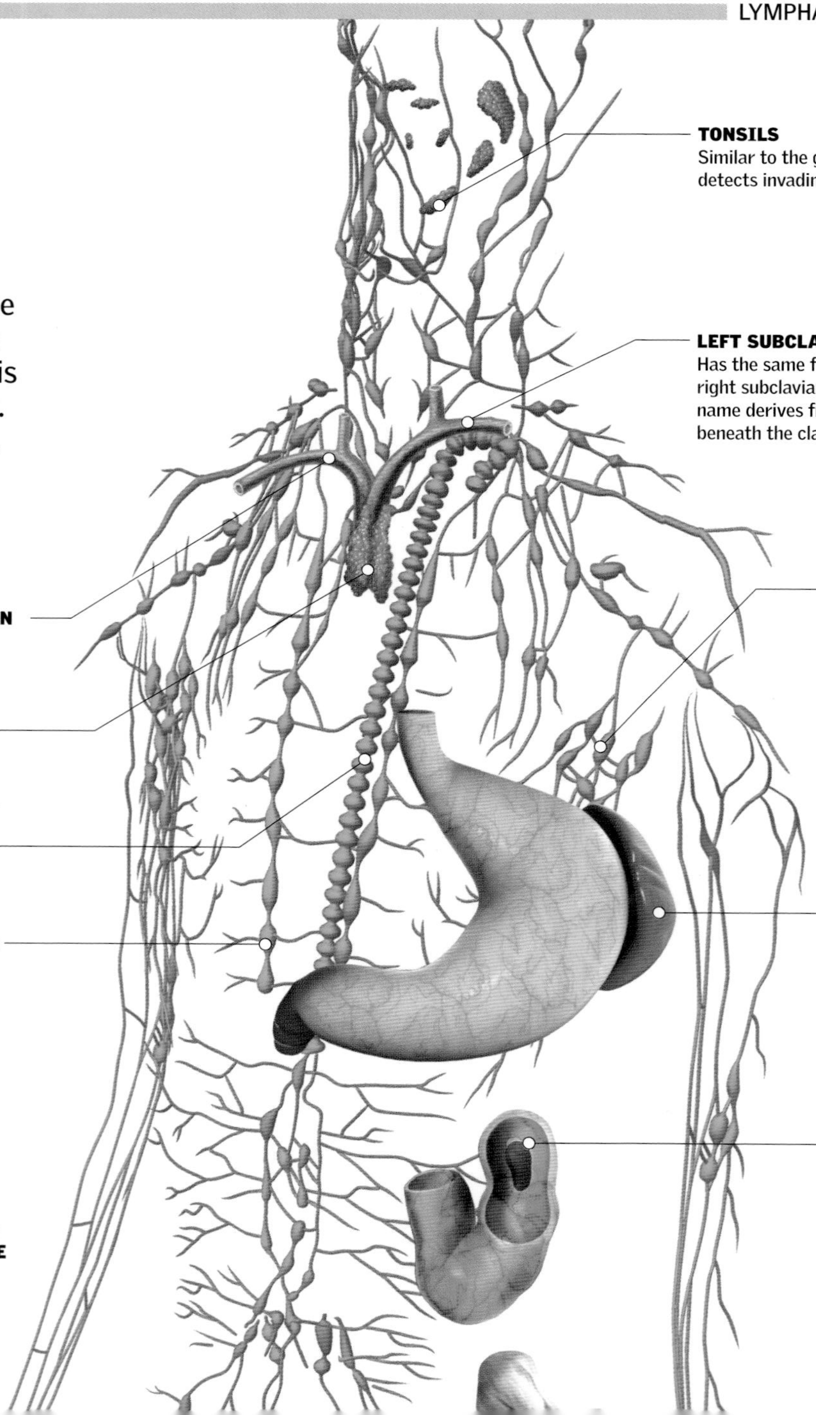
**TONSILS**
Similar to the ganglia, their tissue detects invading organisms.
**LEFT SUBCLAVIAN VEIN**
Has the same function as the right subclavian vein. The name derives from its location beneath the clavicle.
**RIGHT SUBCLAVIAN VEIN**
This vein brings the lymph from the upper part of the body to the lymphatic duct.
**THYMUS**
This transforms the white blood cells in the bone marrow into T lymphocytes.
**AXILLARY LYMPHATIC GANGLIA**
The lymph from the chest and the arms is filtered just above the armpits.
**THORACIC DUCT**
This sends the lymph to the left subclavian vein.
**LATERAL AORTIC NODES**
**SPLEEN**
The main lymph organ for the entire body.
**PEYER'S PATCH**
Lymphatic tissue located in the lower region of the small intestine.

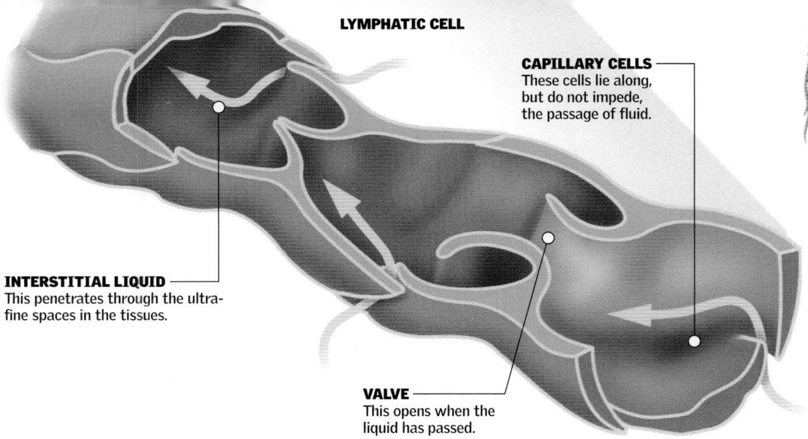

LYMPHATIC CELL
CAPILLARY CELLS
These cells lie along, but do not impede, the passage of fluid.
INTERSTITIAL LIQUID
This penetrates through the ultra-fine spaces in the tissues.
VALVE
This opens when the liquid has passed.

# Immune Response

1 The lymphatic system generates lymphocytes (also found in the blood and in other tissue) and macrophages. Together they constitute the immune system. Here invading bacteria are devoured by a macrophage, and the B lymphocytes take information from the surface of the bacteria that they need to "recognize" other similar bacteria.

2 The B lymphocytes are activated and upon recognizing a pathogen divide themselves into plasmatic cells and memory cells. The plasmatic cells secrete thousands of antibody molecules per second, which are carried by the blood to the site of the infection. The memory cells retain the antigen information, and, when faced with a new invasion, will once again divide rapidly in order to deal with it.

3 The antibodies, also called "immunoglobin," are protein molecules in the form of a "Y," with arms unique to each specific type of antibody. It is this feature that attaches them to a specific antigen. Their function is to "mark" invaders, which can then be destroyed by the macrophages.

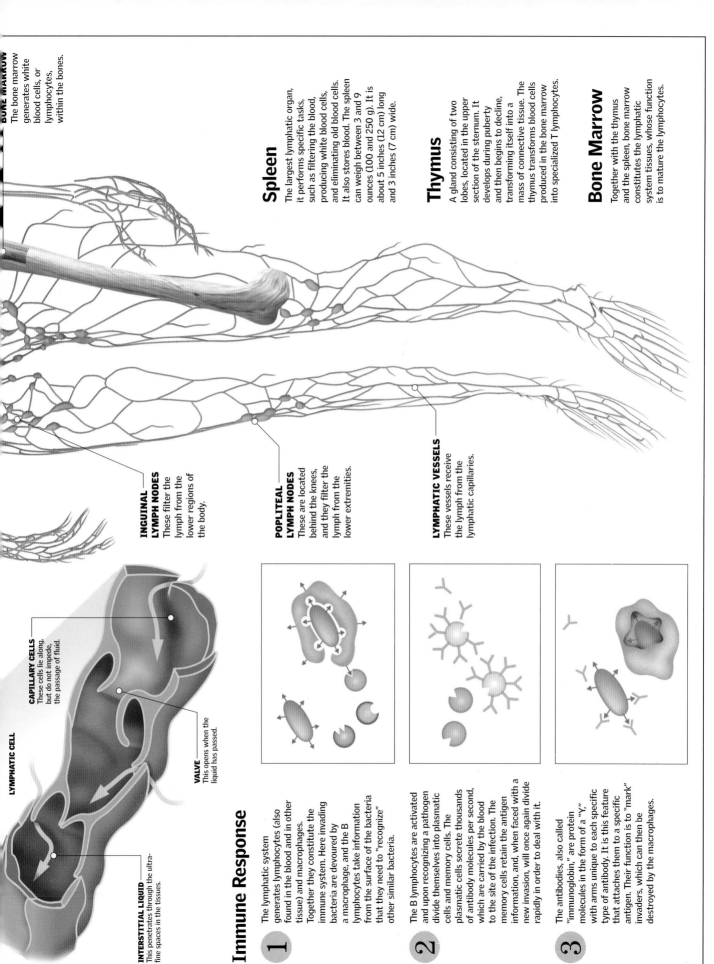

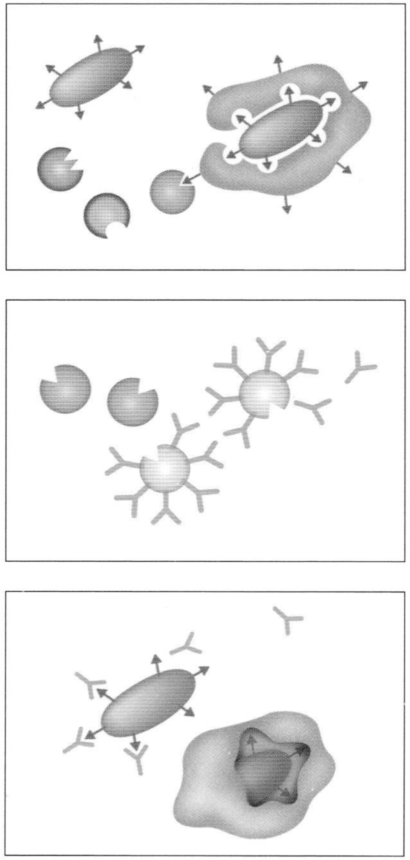

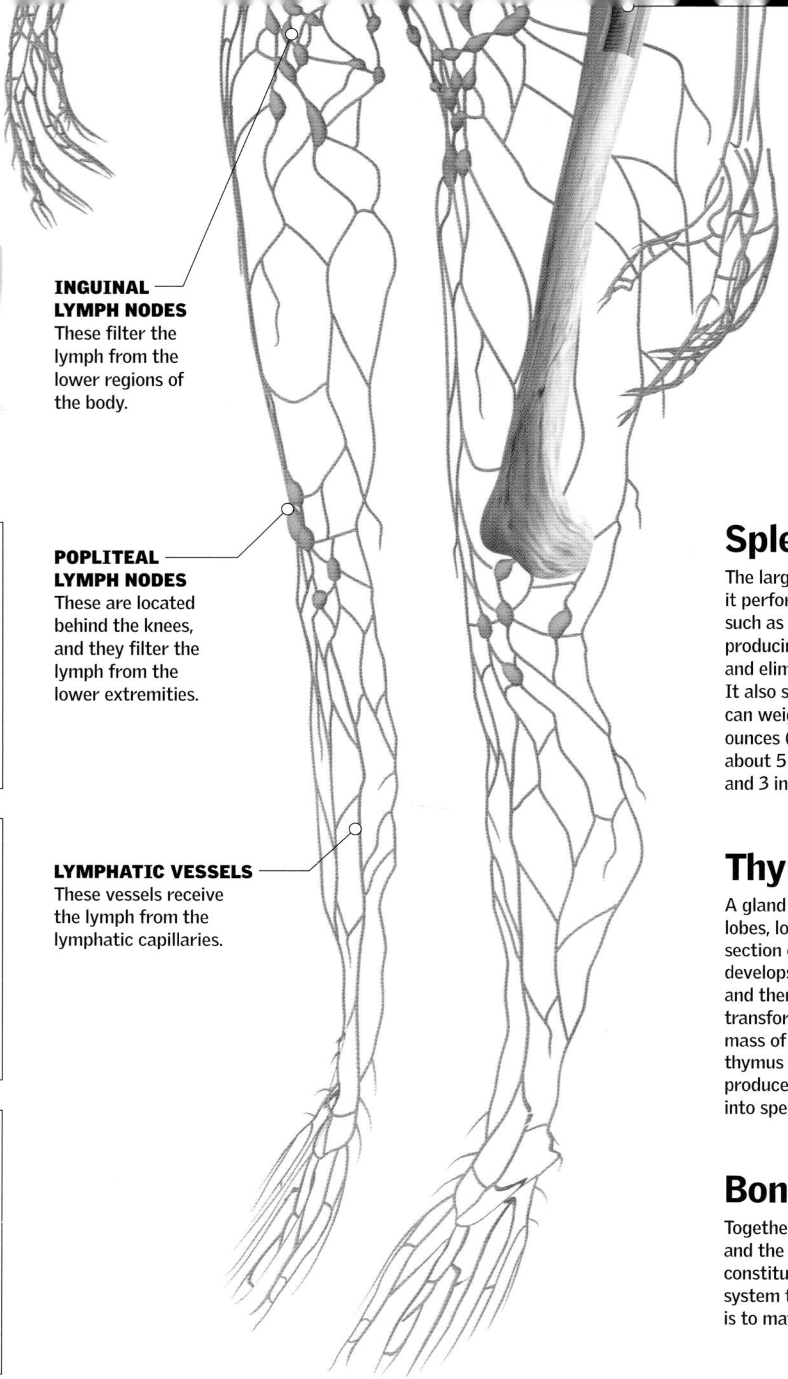


## Spleen

The largest lymphatic organ, it performs specific tasks, such as filtering the blood, producing white blood cells, and eliminating old blood cells. It also stores blood. The spleen can weigh between 3 and 9 ounces (100 and 250 g). It is about 5 inches (12 cm) long and 3 inches (7 cm) wide.

## Thymus

A gland consisting of two lobes, located in the upper section of the sternum. It develops during puberty and then begins to decline, transforming itself into a mass of connective tissue. The thymus transforms blood cells produced in the bone marrow into specialized T lymphocytes.

## Bone Marrow

Together with the thymus and the spleen, bone marrow constitutes the lymphatic system tissues, whose function is to mature the lymphocytes.

# Lymph Node

Also called a lymph gland, this node has a round shape and is about 0.4 inch (1 cm) in diameter. Lymph nodes are distributed throughout the body—in the neck, armpits, groin, and popliteal bone (behind the knees), as well as in the thorax and abdomen. The lymphatic vessels are the ducts for the lymph and the pathways for communication among the lymph nodes. The battle of the immune system against invading germs takes place within the nodes, which then enlarge because of inflammation.

## Natural Defenses

Besides the immune system, composed in part by the lymphatic system, the body has another group of resources called natural defenses, which people possess from birth. The body's first defensive barrier is the skin. If pathogenic agents succeed in passing through its filters, however, both the blood and the lymph possess specialized antimicrobial cells and chemical substances.

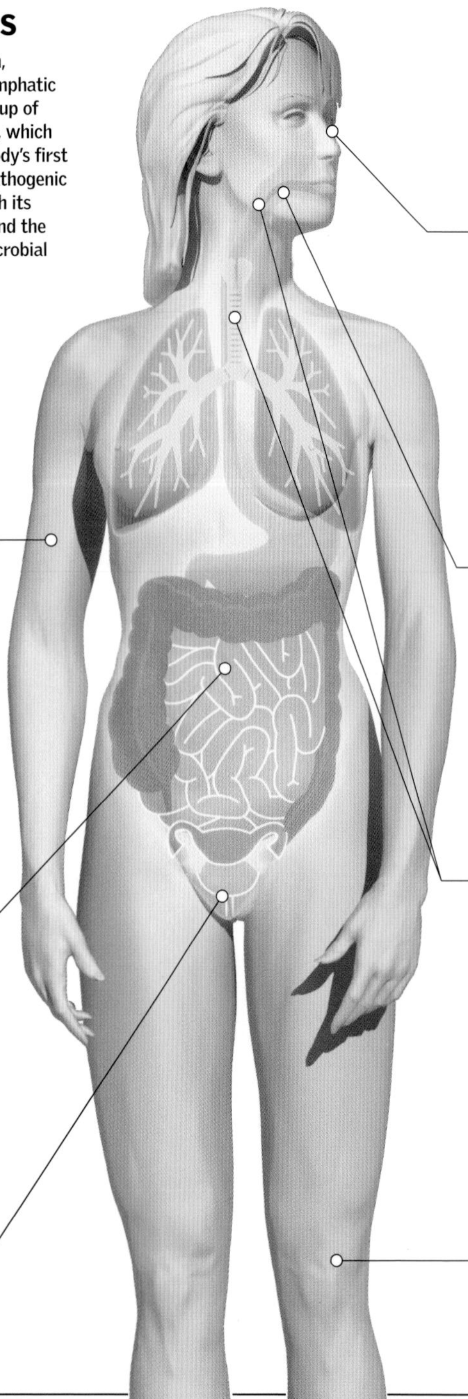

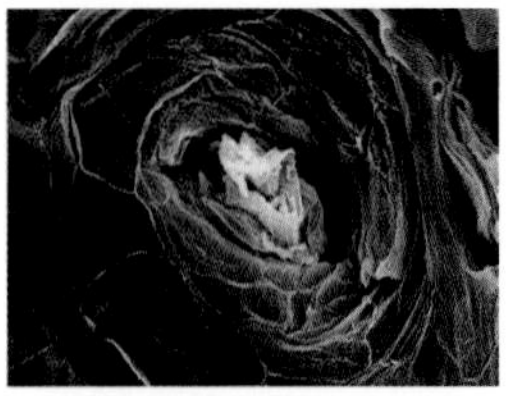

**SEBACEOUS GLAND**
Located on the surface of the skin, this gland secretes a fatty substance called sebo.

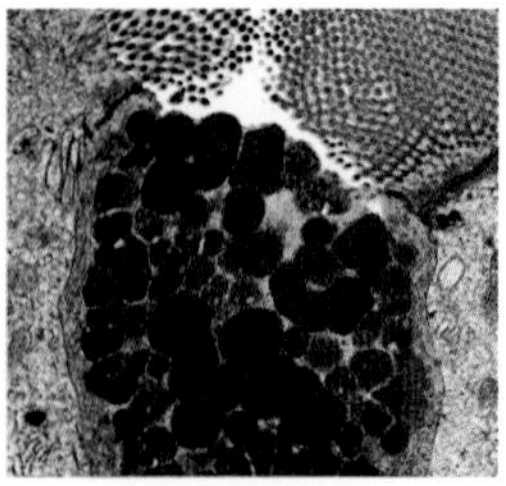

**INTESTINAL MUCOSA**
The goblet cells in this membrane produce a defensive mucus.

**VAGINAL BACTERIA**
Under normal conditions, these are inoffensive, and they occupy areas that could be invaded by pathogenic bacteria.

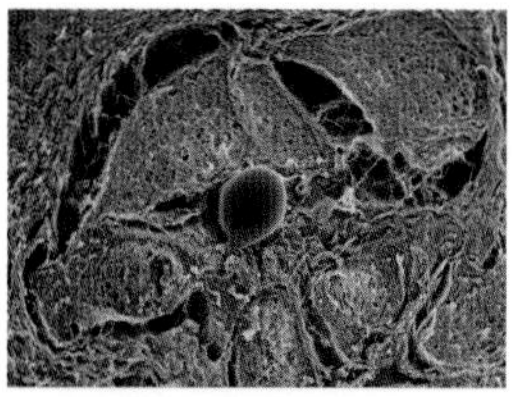

**LACHRYMAL GLAND**
Secretes tears that protect the eyes. Tears, like saliva and perspiration, kill bacteria.

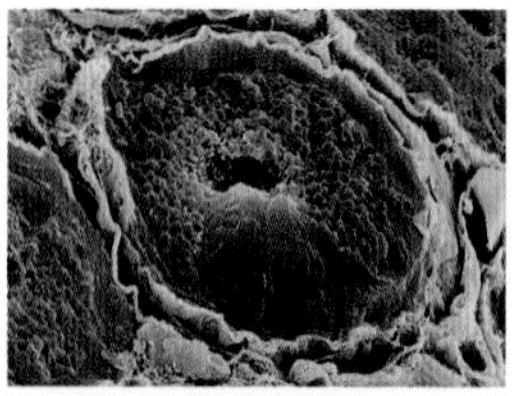

**SALIVARY GLAND**
This gland produces saliva, which contains bactericidal lysozymes.

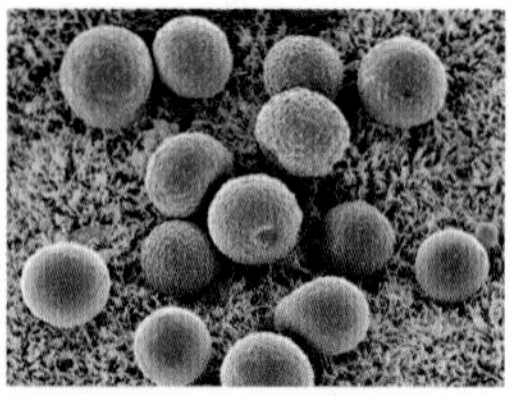

**MUCOUS SECRETIONS**
These secretions, called mucus, form in the upper and lower respiratory tracts, where they capture bacteria and carry them to the throat to be spit out.

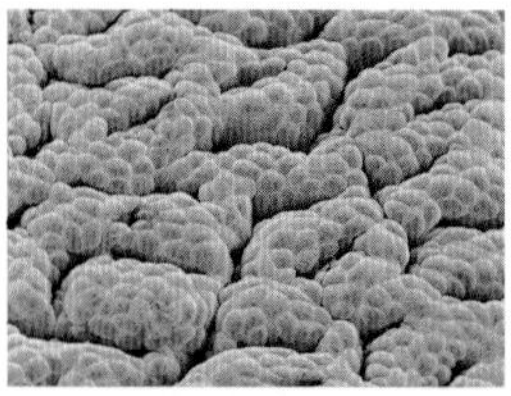

**SWEAT GLAND**
This secretes sweat, which helps to control body temperature, both to eliminate toxins, and to protect the skin immunologically.

## A Defensive Filter

The glands are covered with a sheath of connective tissue, which in turn forms an interior network that consists of clusters filled with lymphocytes. Their immunological functions are to filter the fluid that arrives via both the sanguine and lymphatic afferent veins, which then goes toward the heart to be returned to circulation via the efferent vessels and to produce immune cells for attacking and removing bacteria and carcinogenic cells.

**100 square inches (600 sq cm)**

**THE AREA OF THE SKIN COVERED BY SWEAT GLANDS, A PART OF THE NATURAL DEFENSES THAT COMPLEMENT THE WORK OF THE GANGLIA IN THE IMMUNE SYSTEM.**

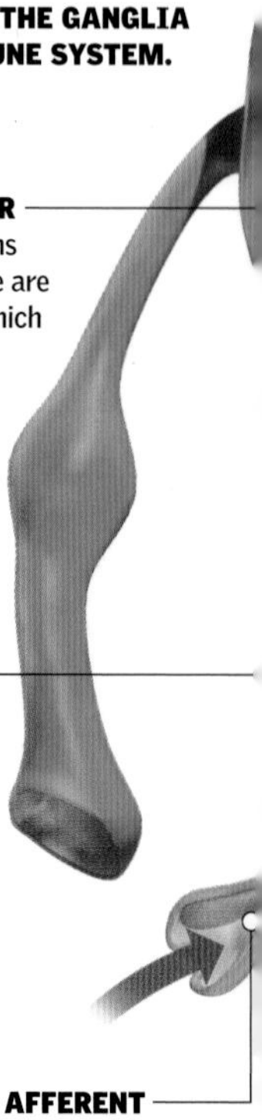

**GERMINAL CENTER**
The area that contains B lymphocytes. There are two types: B cells, which produce antibodies, and T cells.

**MACROPHAGES**
Together with the lymphocytes, they are the basis of the immune system. They devour the invading bodies that are detected.

**AFFERENT LYMPHATIC VESSEL**
The afferent vessels carry the lymphatic liquid from the blood to the ganglia, or lymphatic nodes.

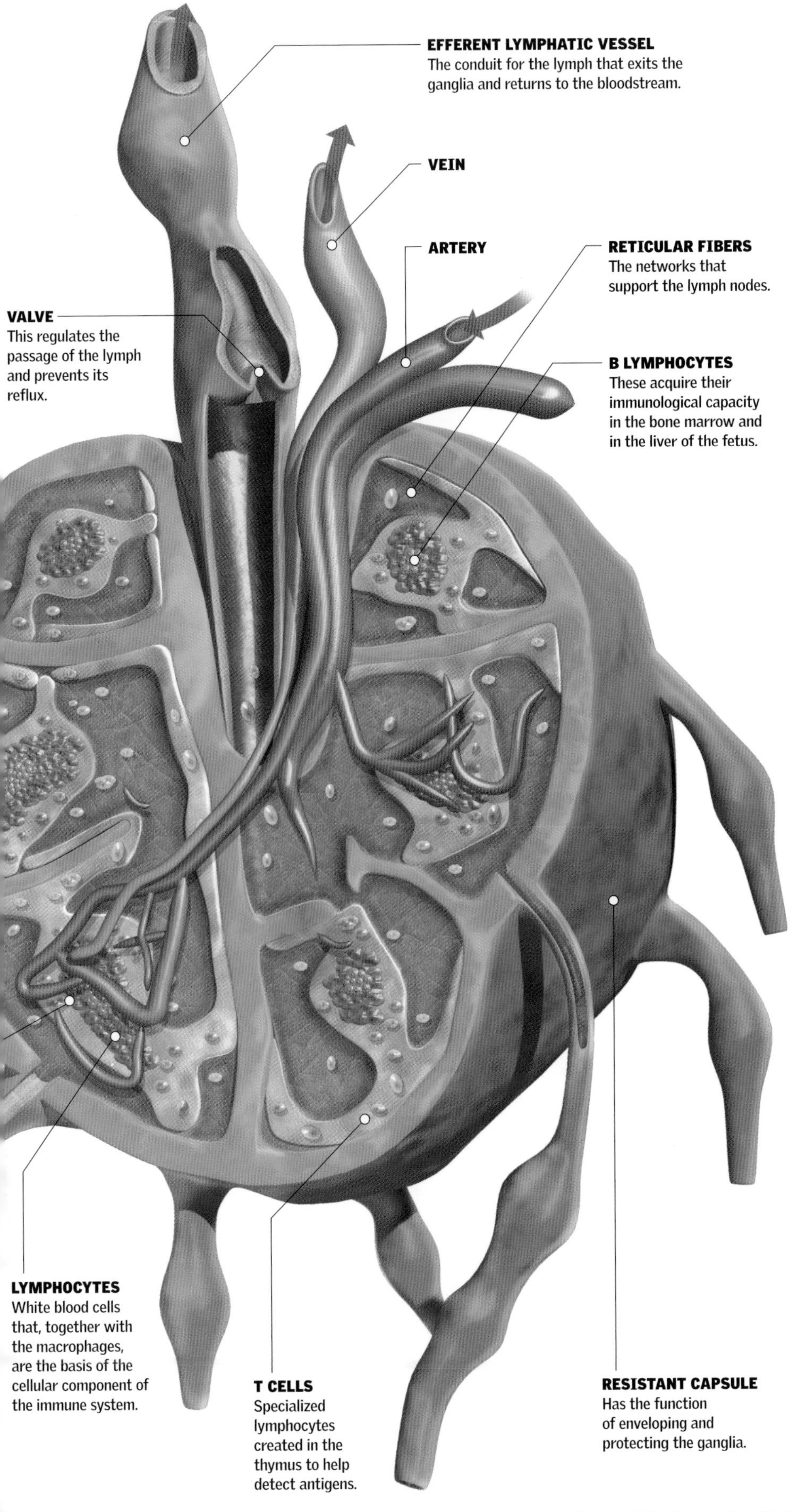

EFFERENT LYMPHATIC VESSEL
The conduit for the lymph that exits the ganglia and returns to the bloodstream.
VEIN
ARTERY
RETICULAR FIBERS
The networks that support the lymph nodes.
VALVE
This regulates the passage of the lymph and prevents its reflux.
B LYMPHOCYTES
These acquire their immunological capacity in the bone marrow and in the liver of the fetus.
LYMPHOCYTES
White blood cells that, together with the macrophages, are the basis of the cellular component of the immune system.
T CELLS
Specialized lymphocytes created in the thymus to help detect antigens.
RESISTANT CAPSULE
Has the function of enveloping and protecting the ganglia.

## Invaders

Disequilibrium can be caused in the homeostatic mechanisms of the human body, causing disease that may or may not be infectious. Noninfectious disease is usually produced by heredity, external factors, or lifestyle. Infections are brought on by parasitic organisms, such as bacteria, viruses, fungi, and protozoa (single-celled organisms belonging to the protist kingdom).

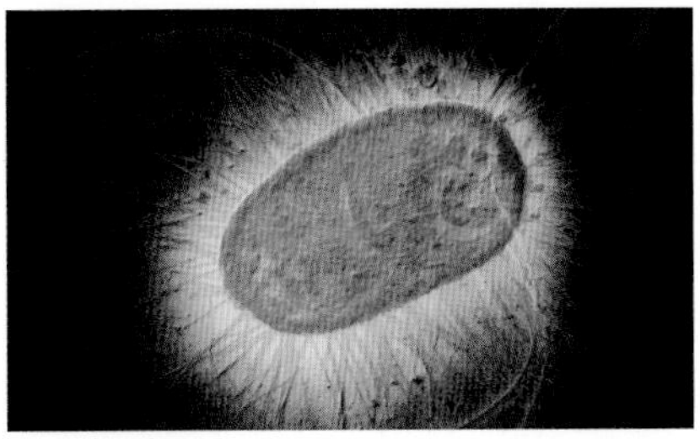

**A BACTERIA**
These life-forms are found by the billions in any medium. Not all of them are harmful. Bacteria known as germs are pathogenic and release poisonous substances called toxins.

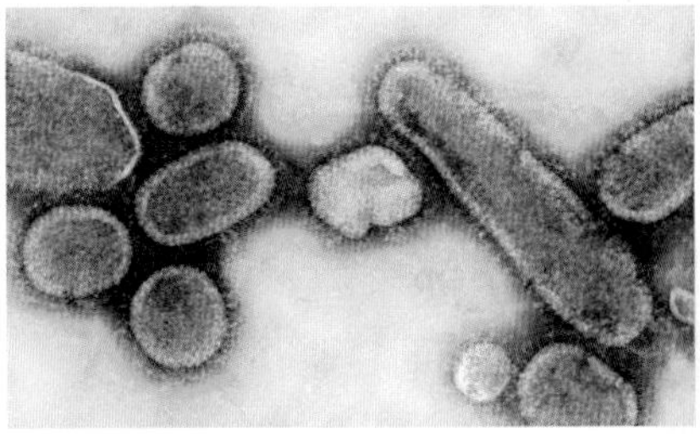

**B VIRUSES**
These are not really living beings but chemical packages. They consist of genetic material. When they enter the body, they invade a cell, where they reproduce and then spread.

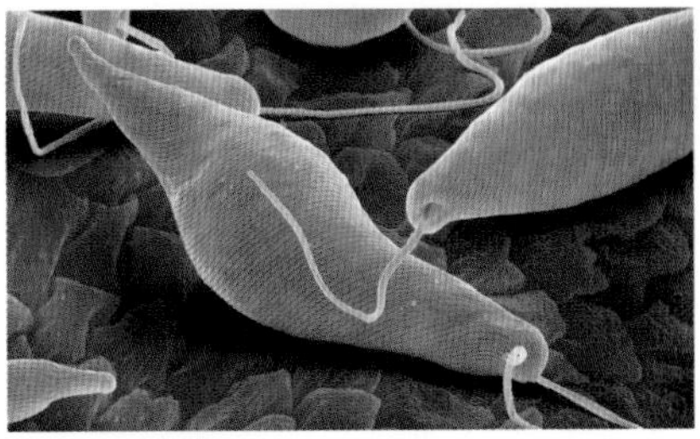

**C PROTOZOA**
These are organisms that typically live in water and in soil. There are about 30 pathogenic species, which can produce a range of diseases from sleeping sickness and severe diarrhea to malaria.

# Red

**THE COLOR OF INFLAMED SKIN WHEN BACTERIAL ACTION IN A WOUND CAUSES VASODILATION. THIS OCCURS BECAUSE THE BLOOD VESSELS EXPAND TO INCREASE BLOOD FLOW AS A MEANS OF DEFENSE.**

# Respiratory System

The respiratory system organizes and activates respiration, a process by which the human body takes in air from the atmosphere, extracts the oxygen that the circulation will bring to all the cells, and returns to the air products it does not need, such as carbon dioxide. The basic steps are inhalation, through which air enters the nose and mouth, and exhalation, through which air is expelled. Both actions are usually involuntary and automatic. Respiration involves the airway that begins in the nose and continues through the pharynx, larynx, trachea, bronchi, bronchioles, and alveoli; however, respiration occurs primarily in the two lungs, which are essentially bellows whose job it is to collect oxygen from the air. The oxygen is then distributed to the entire body via the blood.

## What Enters and What Exits

| Component | Percentage in Inhaled Air | Percentage of Exhaled Air |
|---|---|---|
| **Nitrogen** | 78.6 | 78.6 |
| **Oxygen** | 20.8 | 15.6 |
| **Carbon Dioxide** | 0.04 | 4 |
| **Water Vapor** | 0.56 | 1.8 |
| **Total** | 100 | 100 |

6 quarts (5.5 l)

**THE APPROXIMATE VOLUME OF AIR THAT ENTERS AND EXITS THE LUNGS DURING ONE MINUTE OF BREATHING.**

## Larynx

The resonance box that houses the vocal cords; it consists of various components of cartilaginous tissue. One of these components can be identified externally: it is the Adam's apple, or thyroid cartilage, located in the middle of the throat. The larynx is important for respiration because it links the pharynx with the trachea and ensures the free passage of air entering and leaving the lungs. It closes the epiglottis like a door when the organism is ingesting food in order to prevent food from entering the airway.

## Route

1. The air enters the nasal cavity, where it is heated, cleaned, and humidified (it also enters through the mouth).
2. The air passes through the pharynx, where the tonsils intercept and destroy harmful organisms.
3. The air passes through the larynx, whose upper part, the epiglottis, a cartilaginous section, prevents food from passing into the larynx when swallowing. From the larynx the air goes into the esophagus.
4. The air passes through the trachea, a tube lined with cilia and consisting of rings of cartilage that prevent its deformation. The trachea transports air to and from the lungs.
5. In the thoracic region the trachea branches into two bronchi, which are subdivided into smaller branches, the bronchioles, which in turn carry the air to the pulmonary alveoli, elastic structures shaped like sacs where gas exchange occurs.
6. From the alveoli the oxygen passes into the blood and then from the blood to the tissues of the body. The carbon dioxide exits the bloodstream and travels toward the alveoli to be subsequently exhaled. Exhaled air contains more carbon dioxide and less oxygen than inhaled air.

15

**WE NORMALLY BREATHE BETWEEN 15 AND 16 TIMES A MINUTE.**

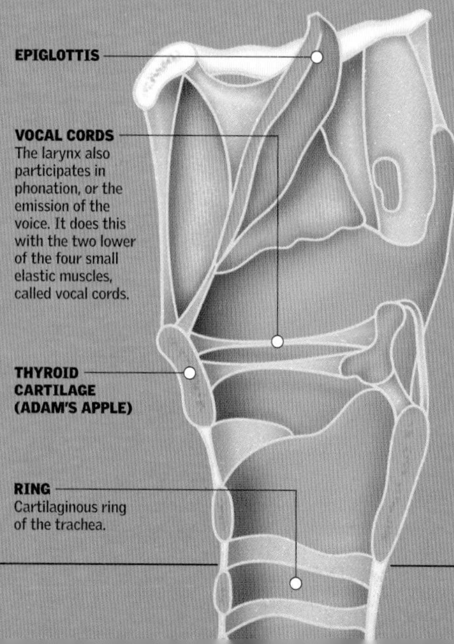

EPIGLOTTIS
VOCAL CORDS
The larynx also participates in phonation, or the emission of the voice. It does this with the two lower of the four small elastic muscles, called vocal cords.
THYROID CARTILAGE (ADAM'S APPLE)
RING
Cartilaginous ring of the trachea.

**HAIRS**
The interior of the trachea is covered with hairs (cilia), which, like the hairs in the nose, capture dust or impurities carried by the air.

## Respiratory Process

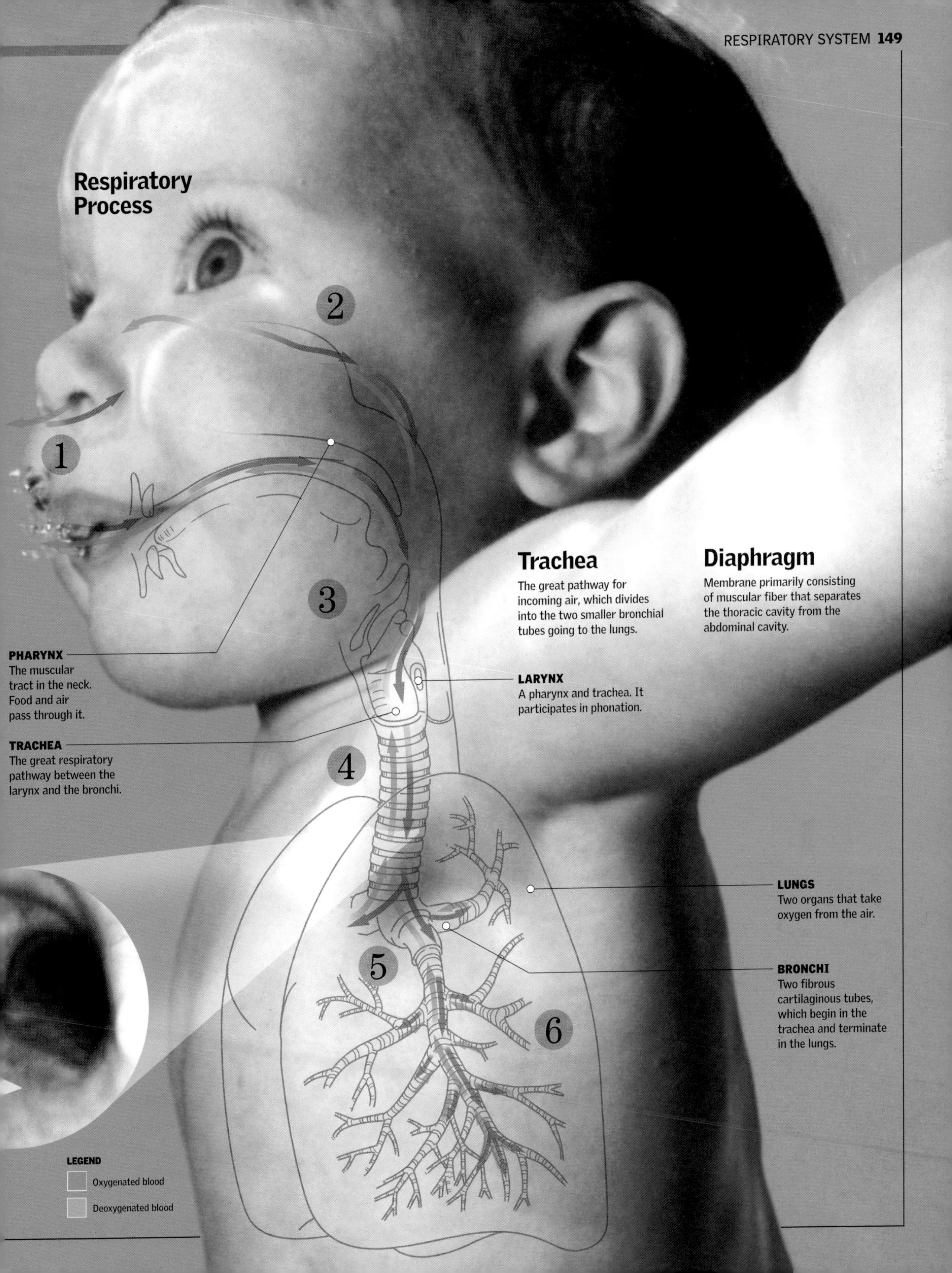


**PHARYNX**
The muscular tract in the neck. Food and air pass through it.

**TRACHEA**
The great respiratory pathway between the larynx and the bronchi.

## Trachea

The great pathway for incoming air, which divides into the two smaller bronchial tubes going to the lungs.

## Diaphragm

Membrane primarily consisting of muscular fiber that separates the thoracic cavity from the abdominal cavity.

**LARYNX**
A pharynx and trachea. It participates in phonation.

**LUNGS**
Two organs that take oxygen from the air.

**BRONCHI**
Two fibrous cartilaginous tubes, which begin in the trachea and terminate in the lungs.

# Lungs

Their principal function is to exchange gases between the blood and the atmosphere. Inside the lungs, oxygen is taken from the air, and carbon dioxide is returned to the air. There are two lungs. The left lung has two lobes and one lingula, and it weighs approximately 30 ounces (800 g); the right lung has three lobes and weighs 35 ounces (1,000 g). Both lungs process the same amount of air. In men each lung has a capacity of 3 quarts (3.2 l), and in women, 2 quarts (2.1 l). The lungs fill most of the space in the thoracic cage surrounding the heart. Their major motions are inhalation (taking in air) and exhalation (expulsion). The pleural membranes, intercostal muscles, and diaphragm all make this mobility possible.

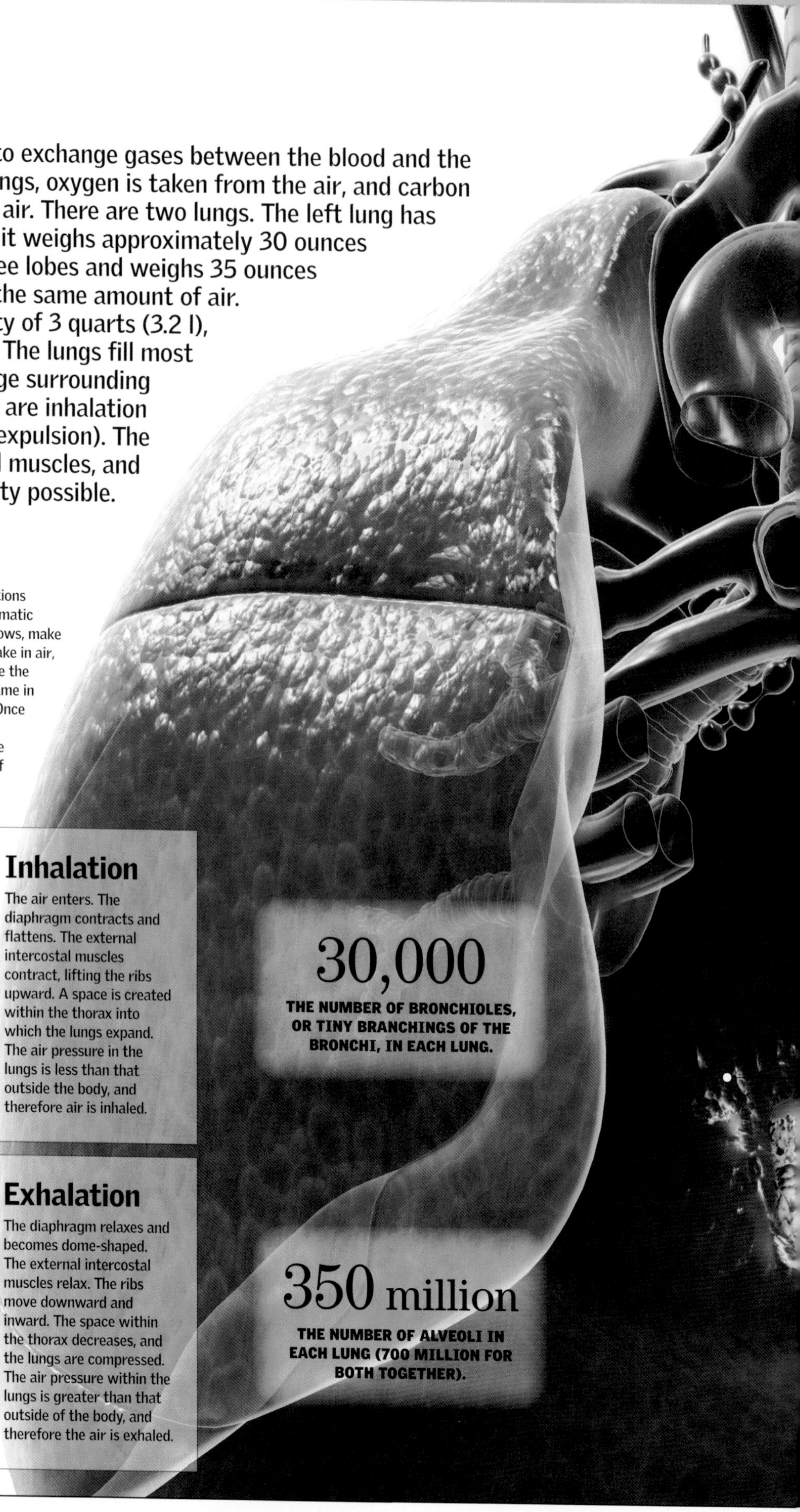

## A Marvelous Pump

The respiratory system accomplishes its functions by combining a series of involuntary and automatic movements. The lungs, opening and closing like bellows, make inhalation possible by increasing their capacity to take in air, which is then exhaled when the bellows close. Inside the lungs the first stage of processing the gases that came in through the nose and the trachea is accomplished. Once the exchange of oxygen to be absorbed and carbon dioxide to be expelled occurs, the next stages can be accomplished: transport of the gases and delivery of oxygen to the cells and tissues.

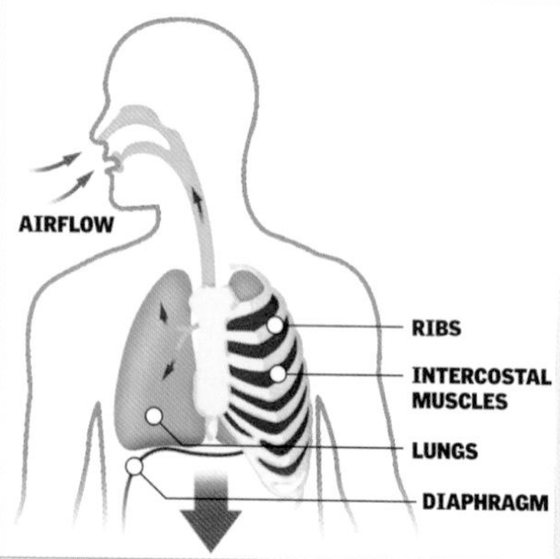


### Inhalation

The air enters. The diaphragm contracts and flattens. The external intercostal muscles contract, lifting the ribs upward. A space is created within the thorax into which the lungs expand. The air pressure in the lungs is less than that outside the body, and therefore air is inhaled.

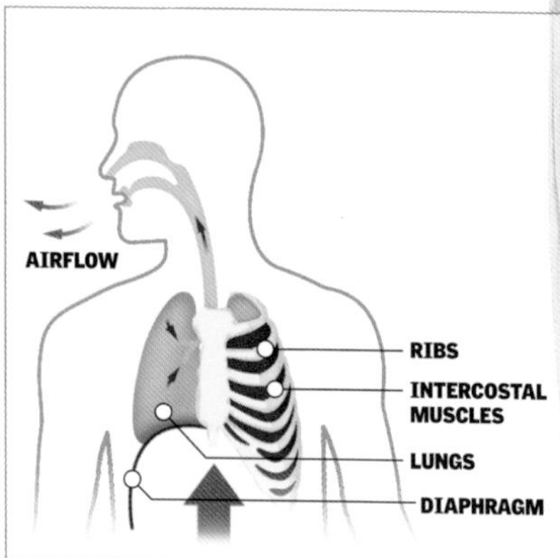


### Exhalation

The diaphragm relaxes and becomes dome-shaped. The external intercostal muscles relax. The ribs move downward and inward. The space within the thorax decreases, and the lungs are compressed. The air pressure within the lungs is greater than that outside of the body, and therefore the air is exhaled.

**30,000**

**THE NUMBER OF BRONCHIOLES, OR TINY BRANCHINGS OF THE BRONCHI, IN EACH LUNG.**

**350 million**

**THE NUMBER OF ALVEOLI IN EACH LUNG (700 MILLION FOR BOTH TOGETHER).**

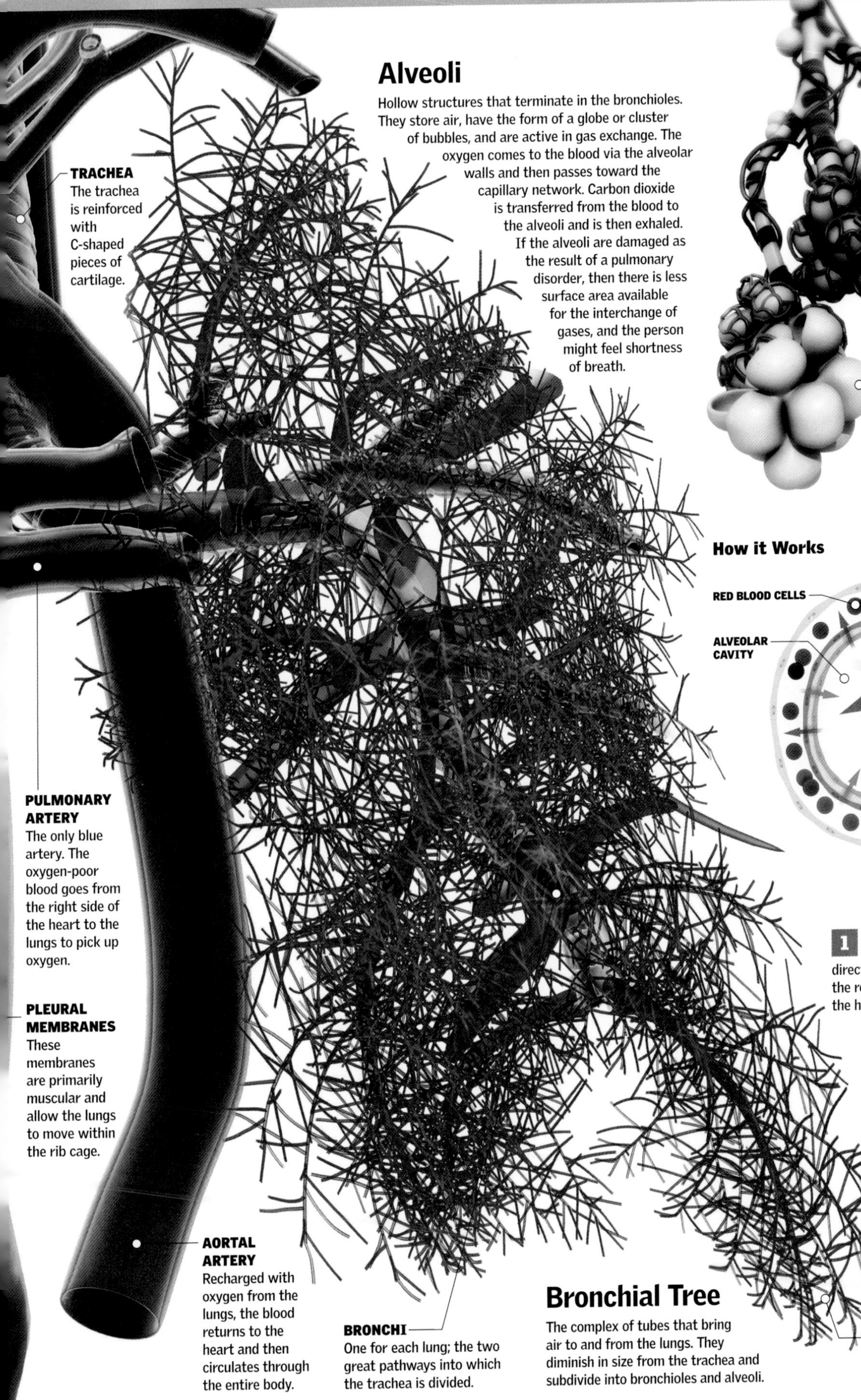

## Alveoli

Hollow structures that terminate in the bronchioles. They store air, have the form of a globe or cluster of bubbles, and are active in gas exchange. The oxygen comes to the blood via the alveolar walls and then passes toward the capillary network. Carbon dioxide is transferred from the blood to the alveoli and is then exhaled. If the alveoli are damaged as the result of a pulmonary disorder, then there is less surface area available for the interchange of gases, and the person might feel shortness of breath.

**ALVEOLI**
If dust or microorganisms enter, the macrophage cells defend against them.

**TRACHEA**
The trachea is reinforced with C-shaped pieces of cartilage.

**PULMONARY ARTERY**
The only blue artery. The oxygen-poor blood goes from the right side of the heart to the lungs to pick up oxygen.

**PLEURAL MEMBRANES**
These membranes are primarily muscular and allow the lungs to move within the rib cage.

**AORTAL ARTERY**
Recharged with oxygen from the lungs, the blood returns to the heart and then circulates through the entire body.

**BRONCHI**
One for each lung; the two great pathways into which the trachea is divided.

### How it Works

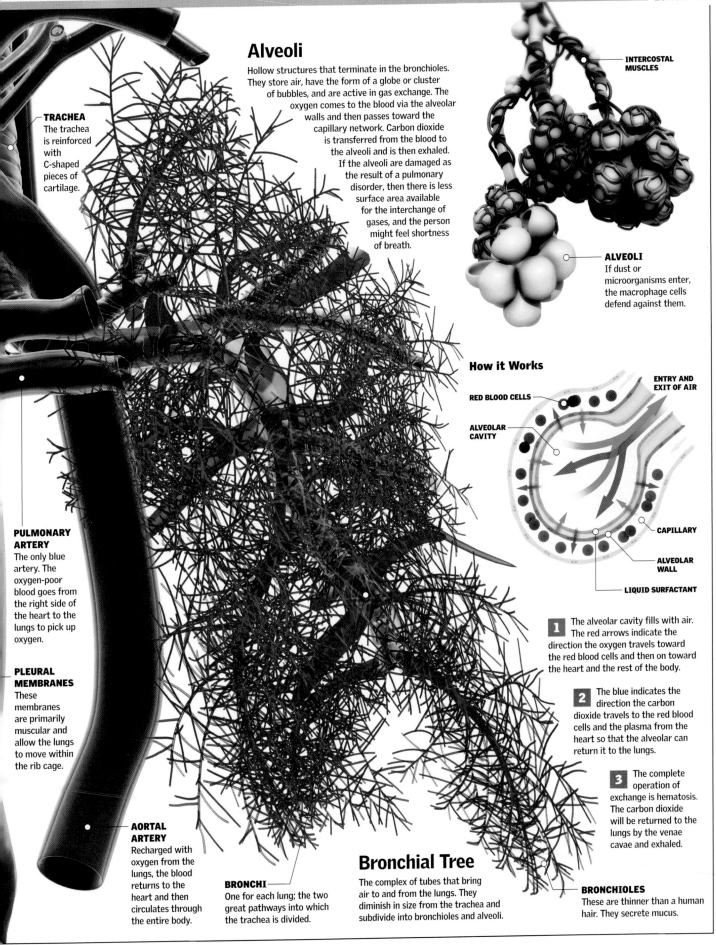

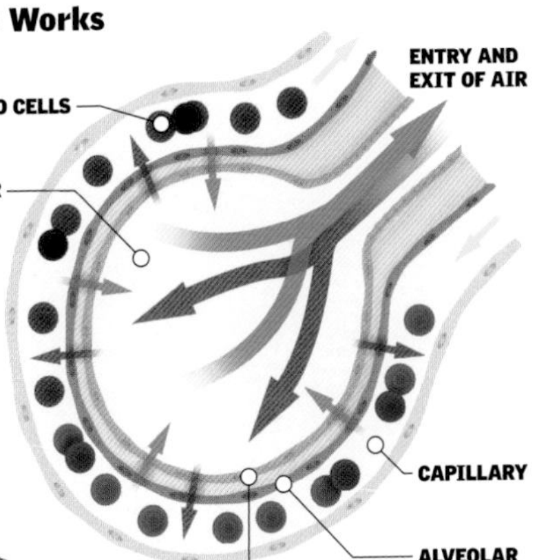


**1** The alveolar cavity fills with air. The red arrows indicate the direction the oxygen travels toward the red blood cells and then on toward the heart and the rest of the body.

**2** The blue indicates the direction the carbon dioxide travels to the red blood cells and the plasma from the heart so that the alveolar can return it to the lungs.

**3** The complete operation of exchange is hematosis. The carbon dioxide will be returned to the lungs by the venae cavae and exhaled.

## Bronchial Tree

The complex of tubes that bring air to and from the lungs. They diminish in size from the trachea and subdivide into bronchioles and alveoli.

**BRONCHIOLES**
These are thinner than a human hair. They secrete mucus.

# Digestive System

The digestive system is the protagonist of a phenomenal operation that transforms food into fuel for the entire body. The process begins with ingestion through the mouth and esophagus and continues with digestion in the stomach, the small intestine, and the large intestine, from which the feces are evacuated by the rectum and anus. By then the task will have involved important chemical components, such as bile, produced by the liver, and other enzymes, produced by the pancreas, by which the food is converted into nutrients. Separating the useful from the useless requires the filtering of the kidneys, which discard the waste in urine.

## The First Step: Ingestion

The digestive process begins with the mouth, the entry point to the large tract that changes in form and function and ends at the rectum and anus. The tongue and teeth are the first specialists in the task. The tongue is in charge of tasting and positioning the food, which is cut and ground by the teeth. This synchronized activity includes the maxillary bones, which are controlled by their corresponding muscles. The palate, in the upper part of the mouth, prevents food from passing into the nose. The natural route of the food is down the esophagus to the stomach.

### The Mouth

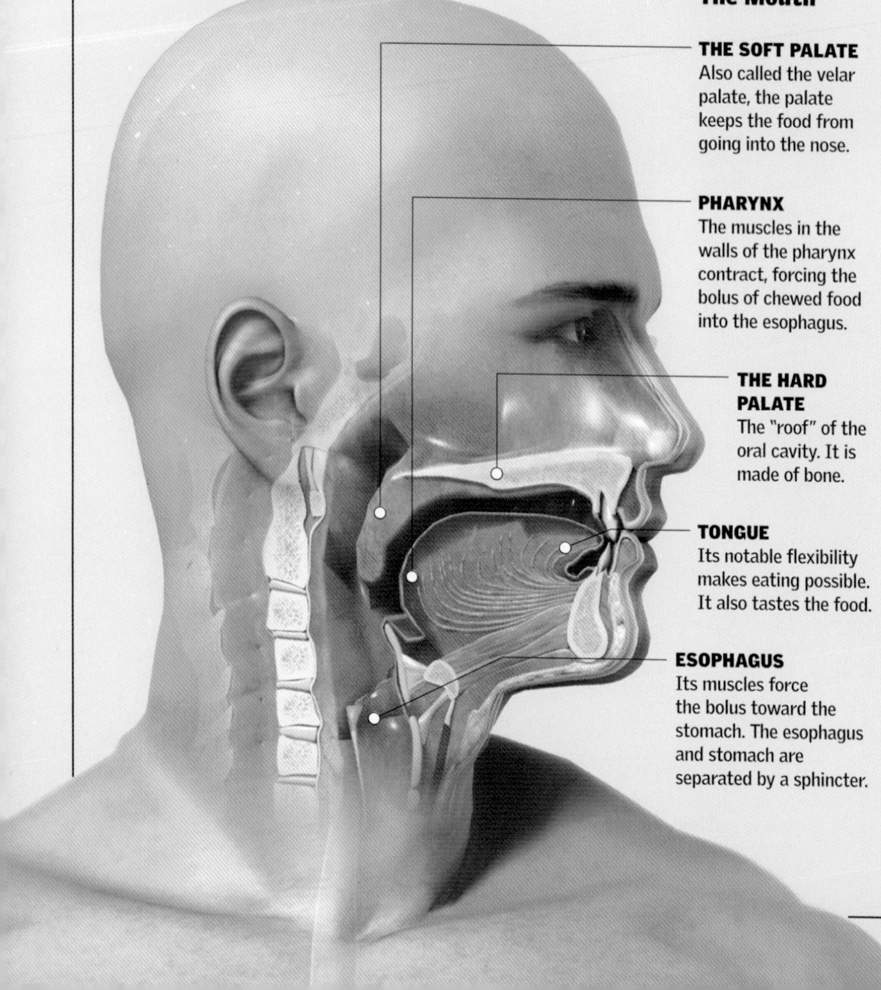

THE SOFT PALATE
Also called the velar palate, the palate keeps the food from going into the nose.
PHARYNX
The muscles in the walls of the pharynx contract, forcing the bolus of chewed food into the esophagus.
THE HARD PALATE
The "roof" of the oral cavity. It is made of bone.
TONGUE
Its notable flexibility makes eating possible. It also tastes the food.
ESOPHAGUS
Its muscles force the bolus toward the stomach. The esophagus and stomach are separated by a sphincter.

## Teeth

There are 32 teeth, and they are extremely hard, a condition necessary for chewing food. There are eight incisors, four canines, eight premolars, and 12 molars. Humans develop two sets of teeth, a provisional or temporary set (the baby teeth) and a permanent set (adult teeth). The first temporary teeth appear between six and 12 months of age. At 20 years of age the process of replacement that began at about age five or six is complete.

### A Set of Teeth

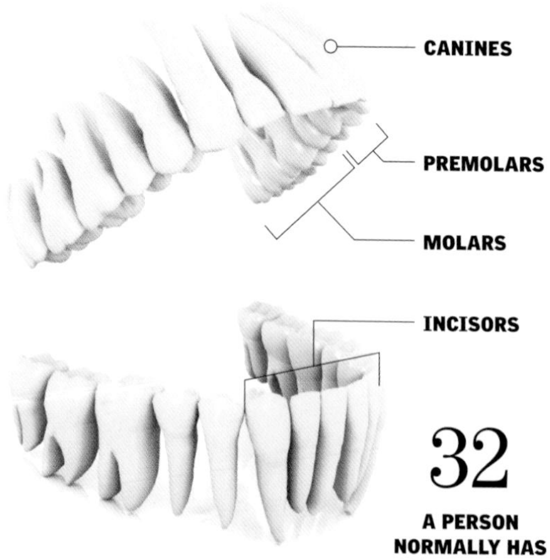


**32**

**A PERSON NORMALLY HAS THIS NUMBER OF TEETH.**

### The Inside of a Tooth

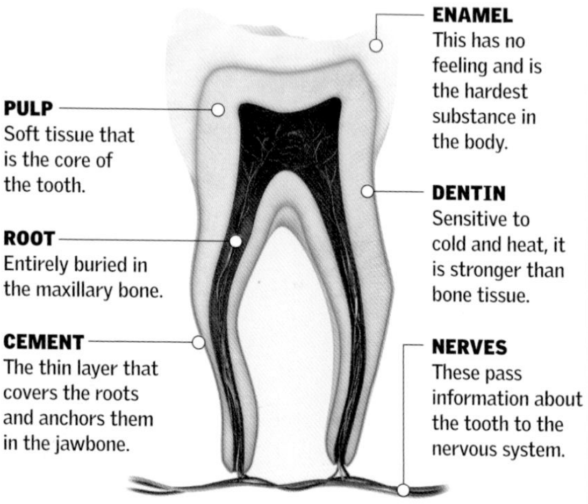

ENAMEL
This has no feeling and is the hardest substance in the body.
PULP
Soft tissue that is the core of the tooth.
DENTIN
Sensitive to cold and heat, it is stronger than bone tissue.
ROOT
Entirely buried in the maxillary bone.
CEMENT
The thin layer that covers the roots and anchors them in the jawbone.
NERVES
These pass information about the tooth to the nervous system.

## Enzymes and Hormones

The complex chemical processes that transform food are essentially accomplished by enzymes and hormones. Both types of substances are secreted by various glands of the digestive system, such as the salivary glands. Enzymes are substances that act as catalysts. Hormones are substances that regulate processes such as growth metabolism, reproduction, and organ function.

# Tract

**THE MUSCULAR MOVEMENT CALLED PERISTALSIS PUSHES THE FOOD ALONG. THAT IS WHY IT IS POSSIBLE TO EAT UPSIDE DOWN OR DURING WEIGHTLESSNESS, AS ASTRONAUTS DO.**

## Digestion Chronology

The process that converts food into nutrients begins a few seconds after the food is raised to the mouth and chewing begins. The average digestion time is about 32 hours, though digestion can range from 20 to 44 hours.

### 1 00:00:00

The process begins when the food reaches the mouth. The entire organism is involved in the decision, but it is the digestive system that plays the main role. The first steps are taken by the teeth and the tongue, aided by the salivary glands, which provide saliva to moisten the alimentary bolus. The morsels are chewed so that they can pass through the esophagus.

### 2 00:00:10

About 10 seconds after chewing has begun, the food is transformed into a moist alimentary bolus that makes its way through the pharynx to the esophagus and then to the stomach, where other changes will take place.

### 3 03:00:00

Three hours after its arrival, the food leaves the stomach, which has accomplished its function. The first phase of digestion is over. The bolus now has a liquid and creamy consistency.

### 4 06:00:00

Three hours later, the food that has been digested in the stomach arrives at the midpoint of the small intestine. At this point it is ready to be absorbed.

### 5 08:00:00

Two hours later, the non-digested, watery residue arrives at the junction of the small and large intestines. The useless material rejected by the body's chemical selectors continues its course, and it is now prepared to be expelled from the organism in the form of feces.

### 20:00:00

The alimentary residue remains in the large intestine between 12 and 28 hours. In this part of the process the residue is converted into semisolid feces.

### 6 24:00

Between 20 and 44 hours after having entered the mouth as food, the residue that was converted into semisolid feces in the previous stage arrives at the rectum. The waste will be evacuated through the anus as excrement.

**10 inches (25 cm)**

**IS THE LENGTH OF THE ESOPHAGUS.**

# Stomach

The part of the digestive tract that is a continuation of the esophagus. It is sometimes thought of as an expansion of the esophagus. It is the first section of the digestive system that is located in the abdomen. It has the shape of an empty bag that is curved somewhat like a bagpipe, the handle of an umbrella, or the letter "J." In the stomach, gastric juices and enzymes subject the swallowed food to intense chemical reactions while mixing it completely. The stomach connects with the duodenum through the pylorus. Peristalsis, the muscular contractions of the alimentary canal, moves the food from the stomach to the duodenum, the next station in the progress of the alimentary bolus.

## How We Swallow

Although swallowing is a simple act, it does require the coordination of multiple parts. The soft palate moves backward when the alimentary bolus passes through the esophagus. The epiglottis moves downward to close the trachea and prevent the food from entering the respiratory pathways. The alimentary bolus is advanced by the muscular motions of peristalsis.

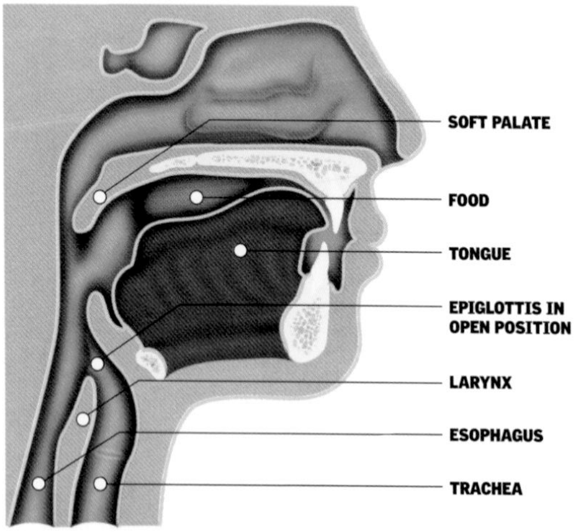


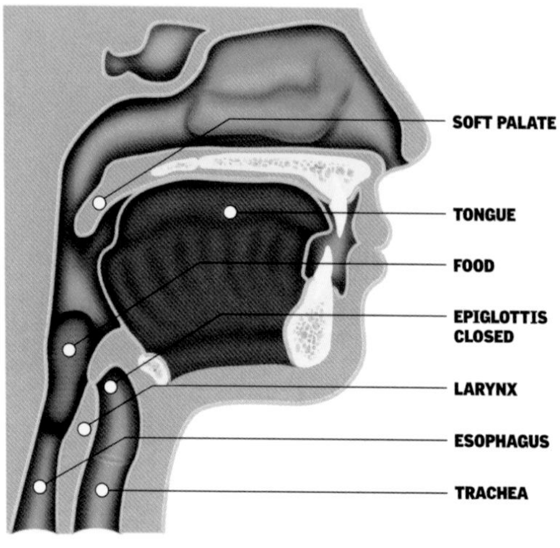


## X-ray of the Stomach

The stomach is the best known of the internal body organs, but it is also the most misunderstood. This J-shaped sac stretches to fill up with food, but it does not absorb any of the nutrients. Its work consists of starting the digestion process, storing semi-digested food, and releasing the food slowly and continuously. Internal gastric juices make it possible for the enzymes to decompose the proteins, while muscular contractions mix the food.

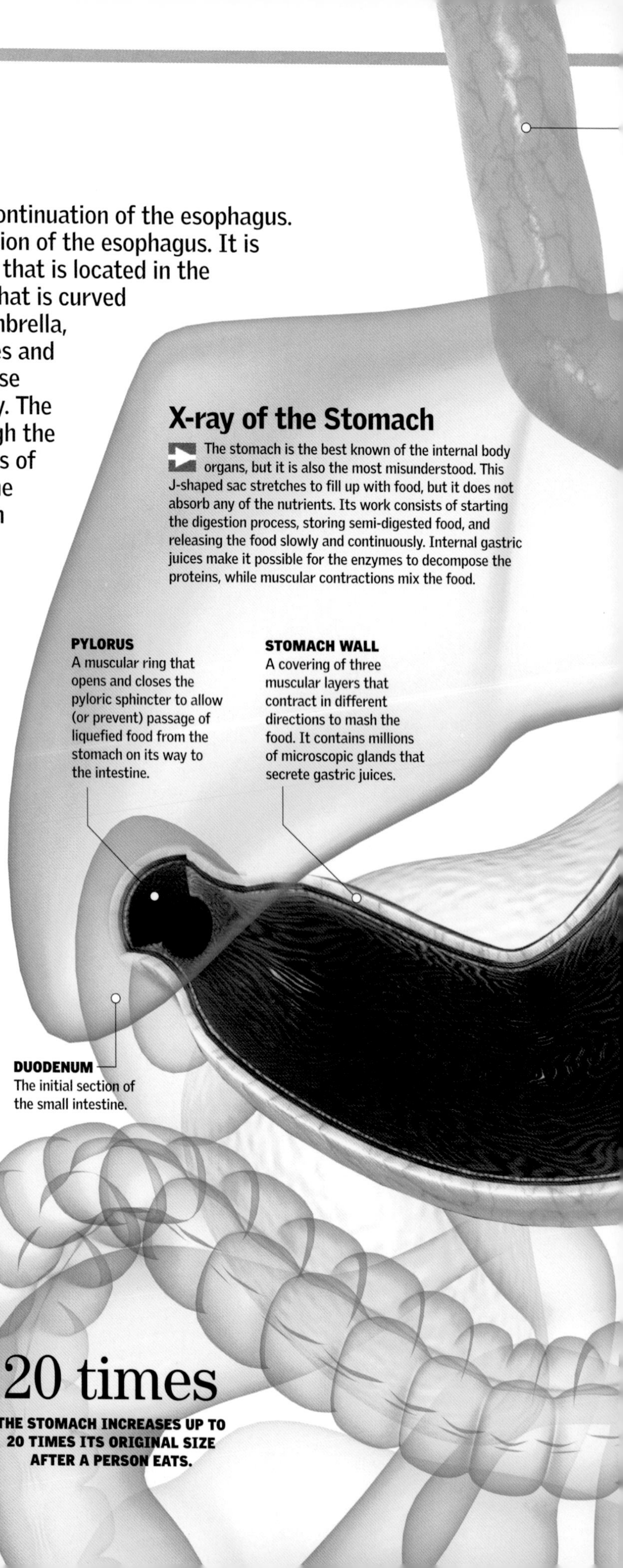

PYLORUS
A muscular ring that opens and closes the pyloric sphincter to allow (or prevent) passage of liquefied food from the stomach on its way to the intestine.
STOMACH WALL
A covering of three muscular layers that contract in different directions to mash the food. It contains millions of microscopic glands that secrete gastric juices.
DUODENUM
The initial section of the small intestine.

# 20 times

**THE STOMACH INCREASES UP TO 20 TIMES ITS ORIGINAL SIZE AFTER A PERSON EATS.**

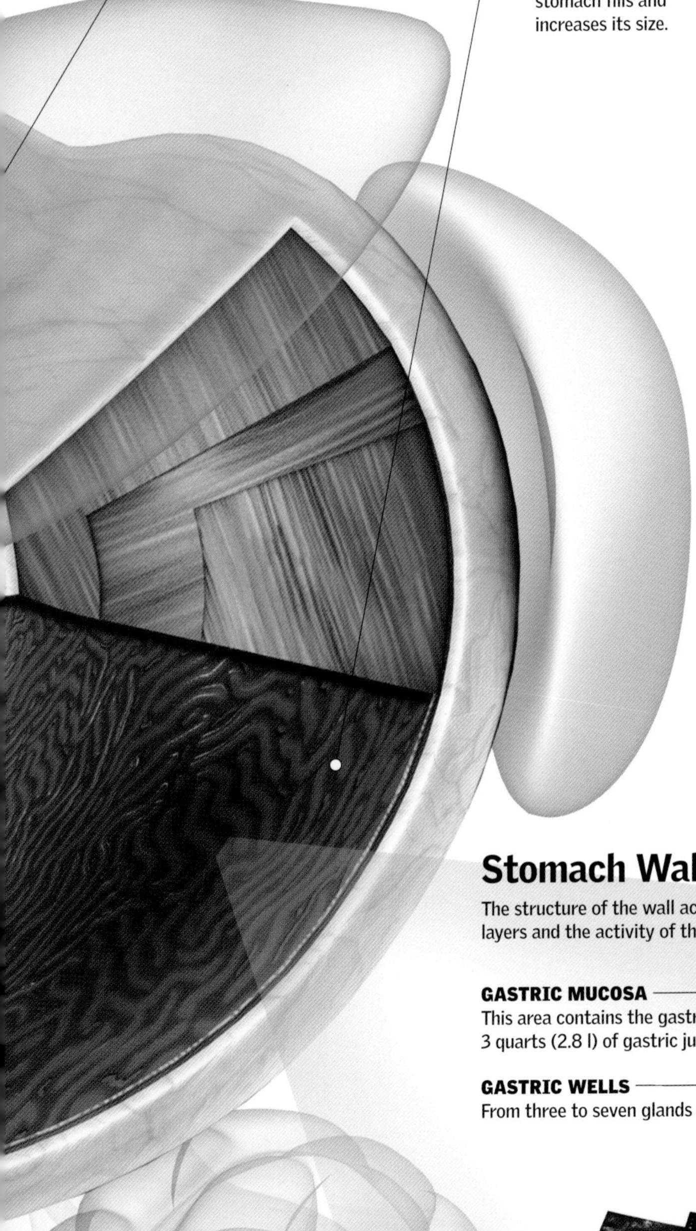

**ESOPHAGUS**
This carries chewed food to the stomach.

**INFERIOR ESOPHAGEAL SPHINCTER**
This closes the junction between the esophagus and the stomach to prevent reflux of the stomach contents.

**WRINKLES OR FOLDS**
These are formed when the stomach is empty, but they stretch out as the stomach fills and increases its size.

## Peristalsis: Muscles in Action

Peristalsis is the group of muscular actions that moves the food toward the stomach and, once the digestive stage has been completed, moves it on to the small intestine. The sphincters are stationary, ring-shaped muscular structures whose opening and closing regulates the passage of the bolus.

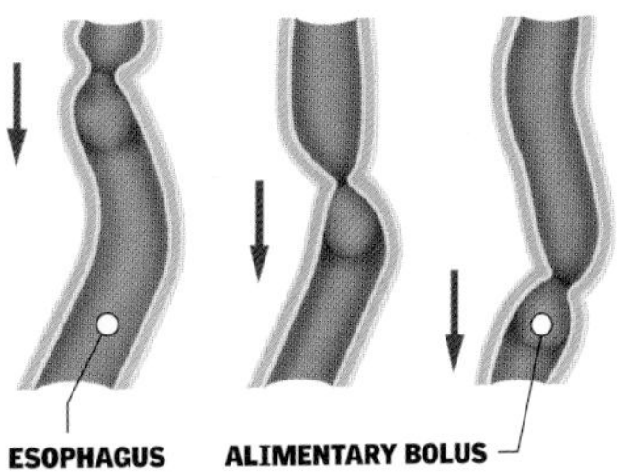


Food is sent toward the stomach, pumped by the muscular contractions of the esophageal walls. Gravity helps accomplish this downward journey.

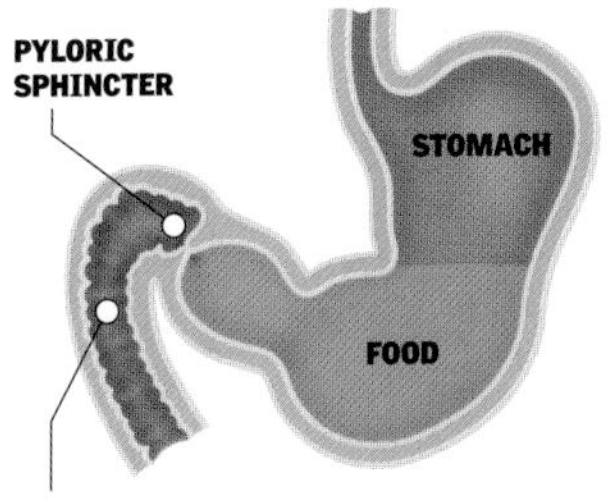


Full stomach. Food enters. The pyloric sphincter remains closed. The gastric juices kill bacteria and are mixed with the food through muscular motions.

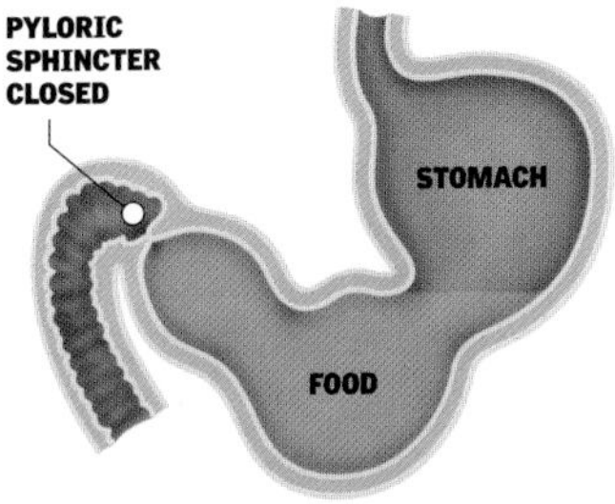


The stomach in full digestive action. The peristaltic muscles mix the food until it becomes a creamy, viscous liquid (chyme).

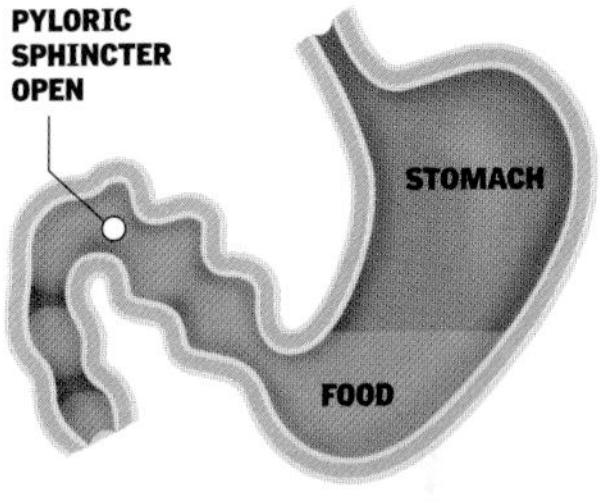


The stomach is being emptied. The pyloric sphincter relaxes, the muscles move the food, and small quantities of food exit toward the duodenum.

## Stomach Wall

The structure of the wall accounts for the two important functions of the stomach: the muscular layers and the activity of the gastric glands guarantee that digestion will run its course.

**GASTRIC MUCOSA**
This area contains the gastric glands, which produce 3 quarts (2.8 l) of gastric juice per day.

**GASTRIC WELLS**
From three to seven glands open to form a groove.

**GASTRIC GLANDS**

**MUSCULAR LAYERS OF THE MUCOSA**
Two fine layers of muscular fibers extend under the mucosa.

**SUBMUCOSA**
Tissue that connects the mucosa to the layers of muscle.

**THREE LAYERS OF MUSCLE**
They are the circular, the longitudinal, and the oblique.

**SUBSEROSA**
Layer that connects the serosa to the muscles.

**SEROSA**
Layer that covers the outer surface.

# Liver, Pancreas, Bile

The liver is the largest gland of the human body and the second largest organ (the skin is the largest). It has numerous functions, and a large part of the body's general equilibrium depends on it. The liver produces bile, a yellowish-green fluid that helps in the digestion of fats. The liver is the great regulator of the glucose level of the blood, which it stores in the form of glycogen. Glycogen can be released when the organism requires more sugar for activity. The liver regulates the metabolism of proteins. Proteins are the essential chemical compounds that make up the cells of animals and plants. The liver is also a large blood filter and a storage site for vitamins A, D, E, and K. The pancreas is a gland that assists in digestion, secreting pancreatic juice.

## Liver

Among its numerous functions, the liver rids the blood of potentially harmful chemical substances, such as drugs and germs. It filters out toxins, starting in the small intestine, and it is involved in maintaining the equilibrium of proteins, glucose, fats, cholesterol, hormones, and vitamins. The liver also participates in coagulation.

## Lobules

Among its other functions, the liver processes nutrients to maintain an adequate level of glucose in the blood. This task requires hundreds of chemical processes that are carried out by the hepatocytes, or liver cells. These are arranged in columns, forming structures called lobules. They produce bile and a sterol (a solid steroid alcohol) called cholesterol. They also eliminate toxins that might be present in food.

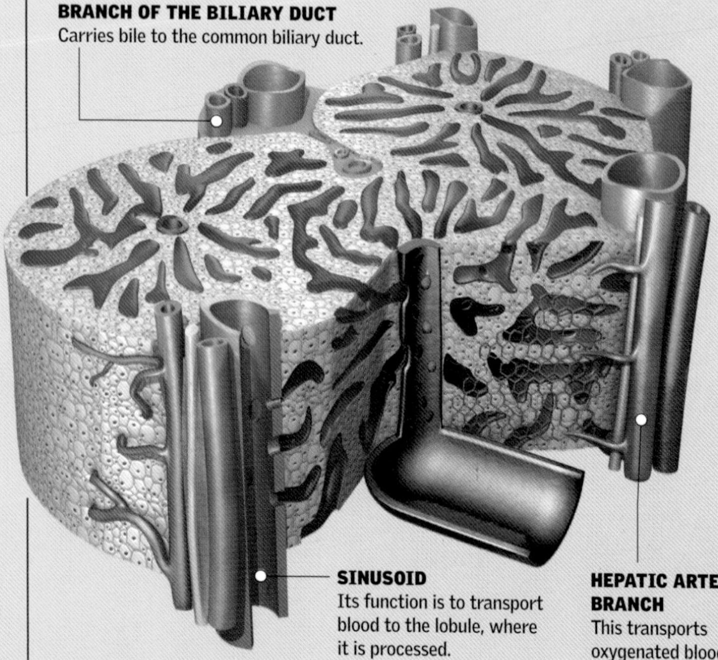

BRANCH OF THE BILIARY DUCT
Carries bile to the common biliary duct.
SINUSOID
Its function is to transport blood to the lobule, where it is processed.
HEPATIC ARTERIAL BRANCH
This transports oxygenated blood to the lobule.

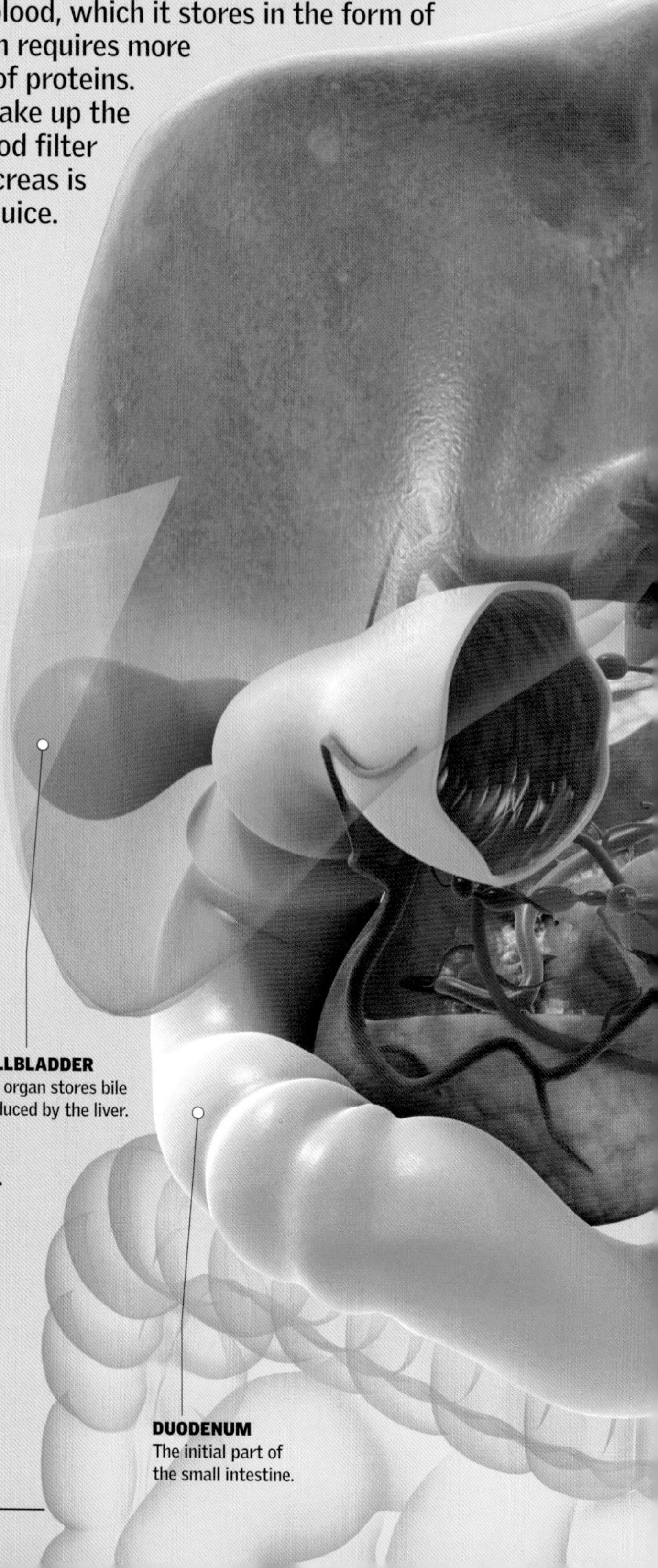

GALLBLADDER
This organ stores bile produced by the liver.
DUODENUM
The initial part of the small intestine.

## Vesicle and Bile

The biliary system stores bile that is produced by the hepatocytes in a specialized pouch called the gallbladder. The path the bile takes from the liver to the gallbladder leads through little canals, biliary ducts, and hepatic ducts, whose diameter increases as the bile moves along. When the body ingests fat, the bile is sent from the gallbladder to the small intestine to accomplish its main function: emulsifying fats to help promote their later absorption.

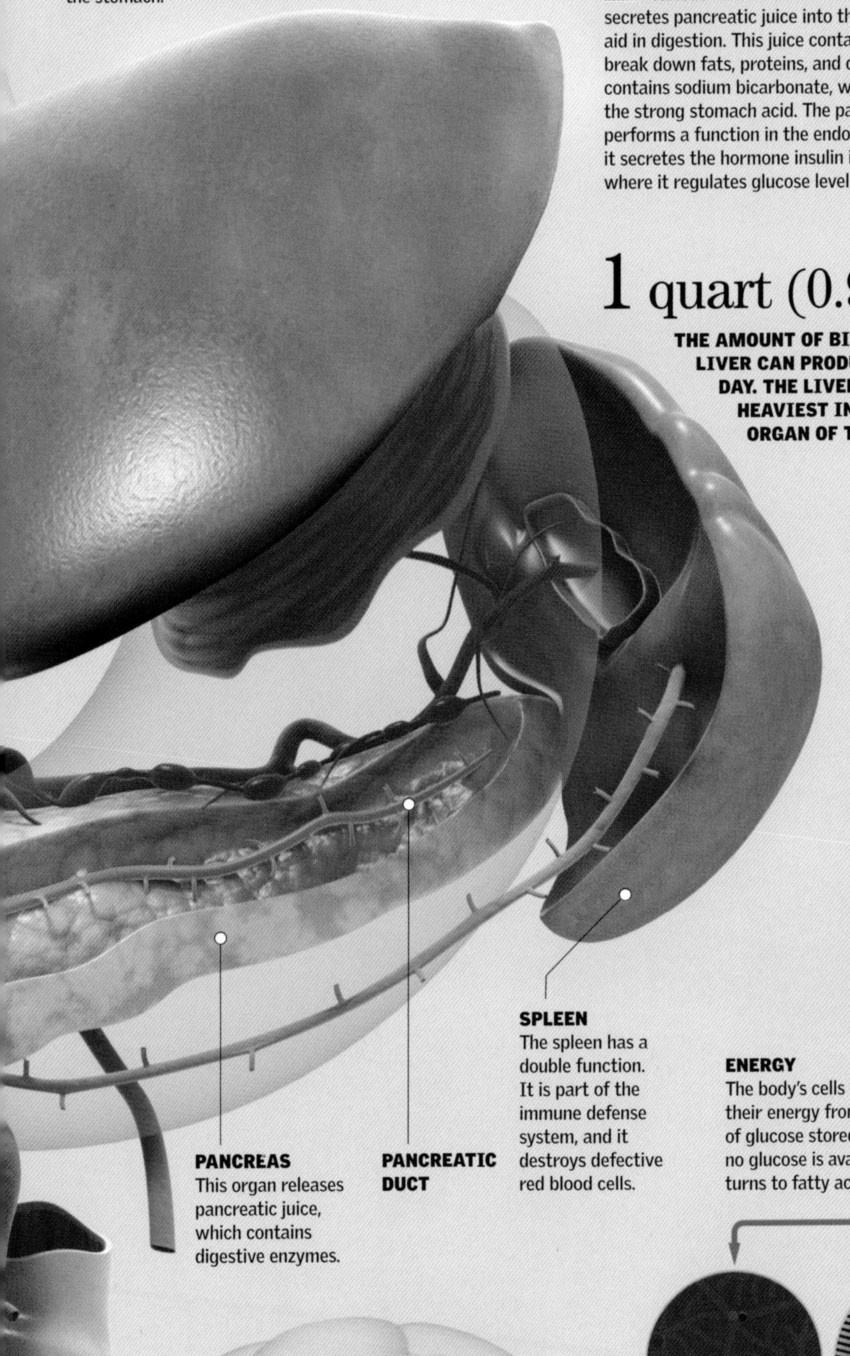
ESOPHAGUS
This conduit brings food to the stomach.
SPLEEN
The spleen has a double function. It is part of the immune defense system, and it destroys defective red blood cells.
PANCREAS
This organ releases pancreatic juice, which contains digestive enzymes.
PANCREATIC DUCT

## Pancreas

The pancreas is a gland that accomplishes various functions. Its exocrine component secretes pancreatic juice into the duodenum to aid in digestion. This juice contains enzymes that break down fats, proteins, and carbohydrates. It contains sodium bicarbonate, which neutralizes the strong stomach acid. The pancreas also performs a function in the endocrine system: it secretes the hormone insulin into the blood, where it regulates glucose levels.

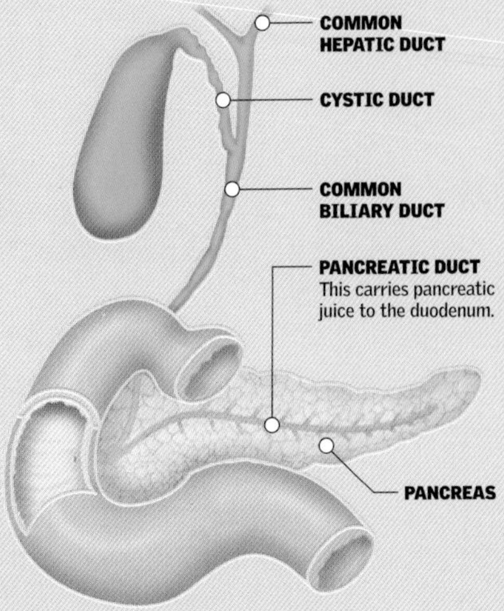
COMMON HEPATIC DUCT
CYSTIC DUCT
COMMON BILIARY DUCT
PANCREATIC DUCT
This carries pancreatic juice to the duodenum.
PANCREAS

# 1 quart (0.9 l)

**THE AMOUNT OF BILE THE LIVER CAN PRODUCE IN A DAY. THE LIVER IS THE HEAVIEST INTERNAL ORGAN OF THE BODY.**

## Metabolism

The complex of chemical reactions that occur in the cells of living beings, transforming simple substances into complex substances and vice versa. When the nutrients are absorbed into the bloodstream and passed to the liver, the liver breaks down proteins into amino acids, fats into fatty acids and glycerol, and carbohydrates into smaller components. A normal diet includes carbohydrates, proteins, fats, vitamins, and minerals.

**ENERGY**
The body's cells basically obtain their energy from the breakdown of glucose stored in the liver. When no glucose is available, the body turns to fatty acids for energy.

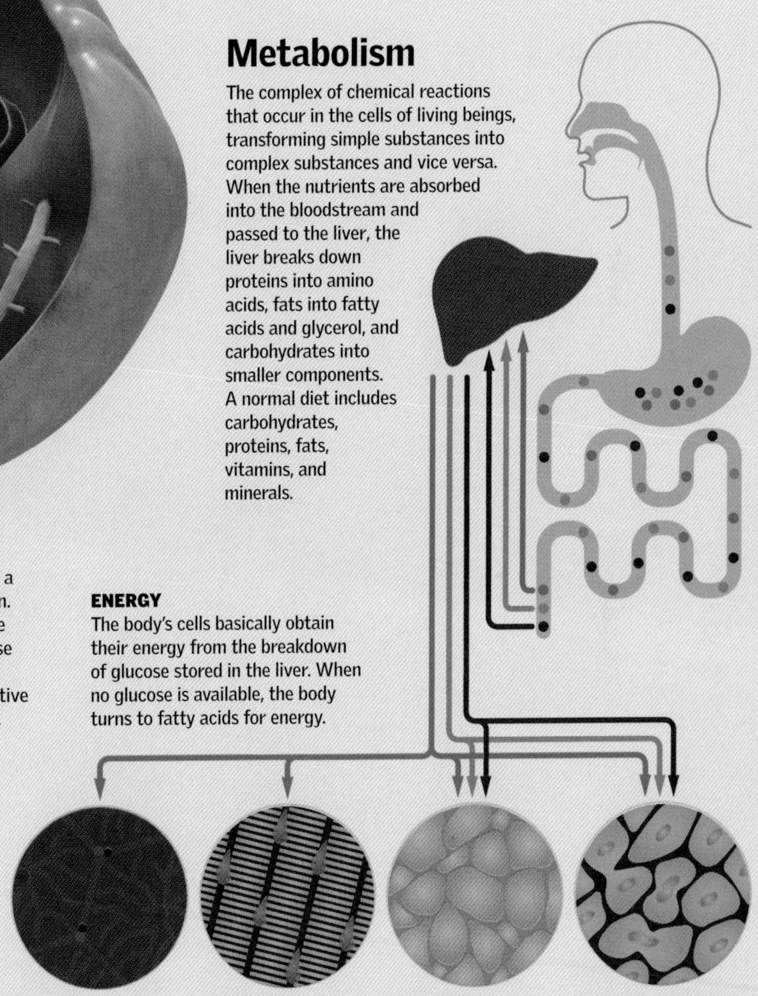

**HEPATIC TISSUE**
Excess glucose in the organism is stored as glycogen in the cells of the liver.

**MUSCULAR FIBER**
Muscle cells in the liver together with the hepatic cells store glycogen.

**ADIPOSE CELLS**
These are cells in which the organism stores excess fatty acids in the form of fat.

**CELLULAR GROWTH AND REPAIR**
Amino acids are converted into proteins by a process called anabolism. Proteins are fundamental for mitosis, cellular regeneration, and enzyme production.

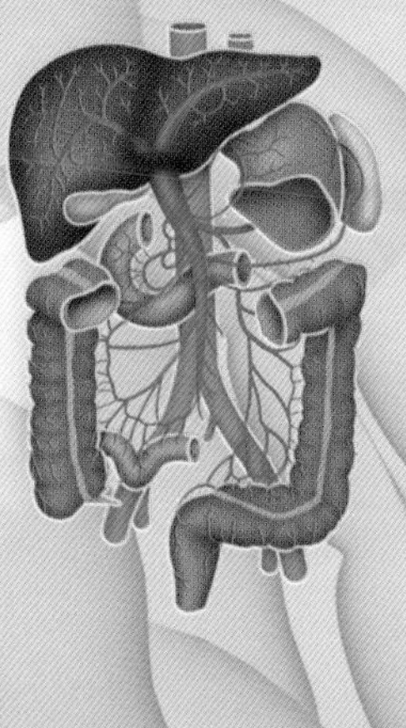

**THE CONNECTION**
The esophagus, stomach, gallbladder, spleen, and small intestine are linked functionally and by their position in the body. They constitute the great crossroads of digestion.

# Large and Small Intestine

The longest part of the digestive tract. It is about 26 to 30 feet (8 to 9 m) long and runs from the stomach to the anus. The small intestine receives the food from the stomach. Digestion continues through enzyme activity, which completes the chemical breakdown of the food. Then the definitive process of selection begins: the walls of the small intestine absorb the nutrients derived from the chemical transformation of the food. The nutrients then pass into the bloodstream. Waste substances, on the other hand, will go to the large intestine. There the final stage of the digestive process will occur: the formation of the feces to be excreted.

## The Union of Both

The small and large intestines join at the section called the ileum (which is the final section of the small intestine; the duodenum and jejunum come before the ileum). The iliac valve acts as a door between the small intestine and large intestine, or colon. The ileum terminates in the caecum (of the large intestine). The ileum measures approximately 13 feet (4 m) in length. Its primary function is the absorption of vitamin B12 and biliary salts. The primary function of the large intestine is the absorption of water and electrolytes that arrive from the ileum.

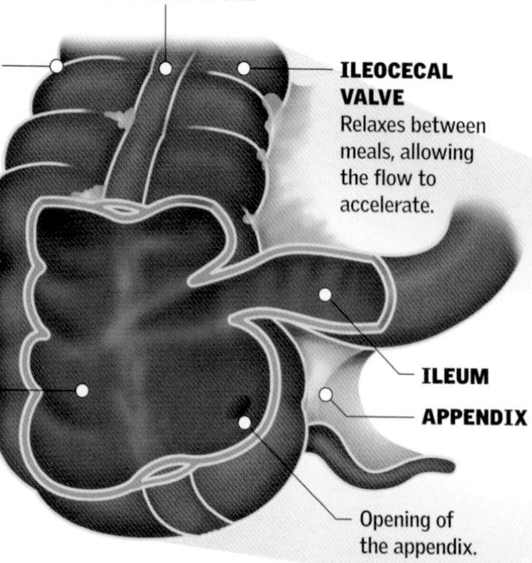
TAENIA MUSCLE
HAUSTRUM OF THE COLON
ILEOCECAL VALVE
Relaxes between meals, allowing the flow to accelerate.
CAECUM
Initial section of the large intestine.
ILEUM
APPENDIX
Opening of the appendix.

ASCENDING COLON
The water and mineral salts are absorbed along the length of the large intestine in a process that removes water from the digestive waste.
DUODENUM
The initial section of the small intestine, to which the secretions of the pancreas and the liver are directed.
CAECUM
Initial section of the large intestine.
ILEUM
Final section of the small intestine, linked with the large intestine.
RECTUM
The final point of the accumulation of the feces. Its storage capacity is small.
ANUS
Opening in the large intestine through which the feces exit.

**WATER THAT ENTERS THE ALIMENTARY CANAL**
In fluid ounces:

| | |
|---|---|
| **Saliva** | 34 (1 l) |
| **Water from Drinking** | 77 (2.3 l) |
| **Bile** | 34 (1 l) |
| **Pancreatic Juice** | 68 (2 l) |
| **Gastric Juice** | 68 (2 l) |
| **Intestinal Juice** | 34 (1 l) |
| **TOTAL** | 313 (9.3 l) |

**WATER REABSORBED BY THE ALIMENTARY CANAL**
In fluid ounces:

| | |
|---|---|
| **Small Intestine** | 280 (8.3 l) |
| **Large Intestine** | 30 (0.9 l) |
| **SUBTOTAL** | 310 (9.2 l) |
| **Water Lost in the Feces** | 3 (0.1 l) |
| **TOTAL** | 313 (9.3 l) |

**TRANSVERSE COLON**
The undigested remains begin to be transformed into feces.

**DESCENDING COLON**
The feces are solidified and accumulate before being expelled.

**JEJUNUM**
The intermediate part of the small intestine, which links the duodenum with the ileum.

**SIGMOID COLON**
This part of the intestine contains a structure that permits the gases to pass without pushing the feces.

## Differences and Similarities

The small intestine is longer than the large intestine. The length of the small intestine is between 20 and 23 feet (between 6 and 7 m), and the large intestine averages 5 feet (1.5 m). Their respective composition and functions are complementary.

**SMALL INTESTINE**

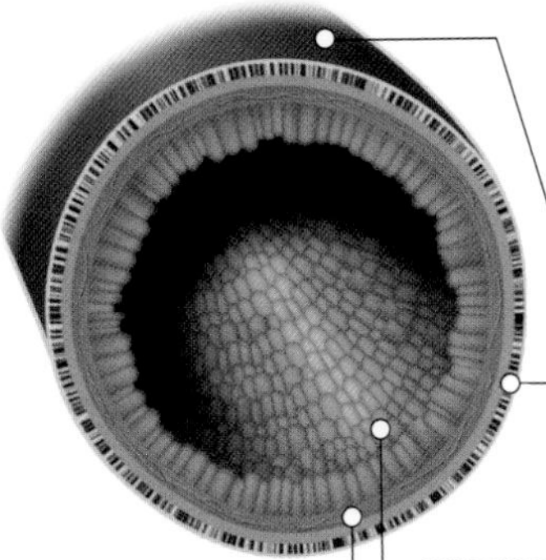

**LARGE INTESTINE**

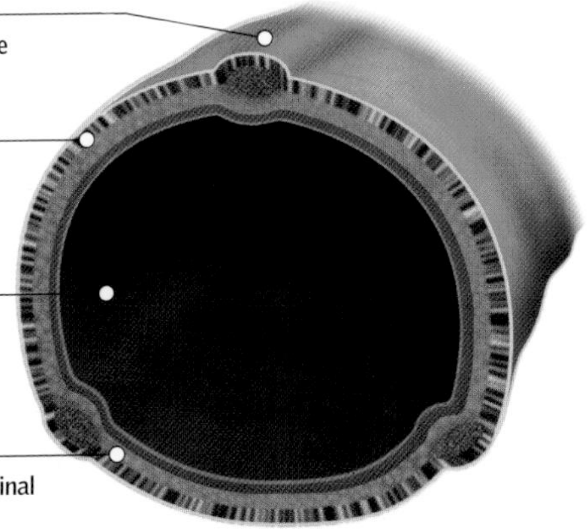

**SEROSA**
The external protective membrane in both.

**SUBMUCOSA**
In both, the loose covering with vessels and nerves.

**MUCOSA**
It is thin and absorbs nutrients via projections or hairs.
Absorbent fat that excretes mucus.

**MUSCULAR**
Thin muscle fibers that are longitudinal externally and circular internally.
The fibers are also covered with hairs, maximizing the area of the mucosa.
Fatty rigid layer that mixes and pushes the feces.

## Villi

The internal wall of the small intestine is covered with millions of hairlike structures called villi. Each one has a lymphatic vessel and a network of vessels that deliver nutrients to it. Each villus is covered by a cellular layer that absorbs nutrients. Together with epithelial cells, the villi function to increase the surface area of the intestine and optimize the absorption of nutrients.

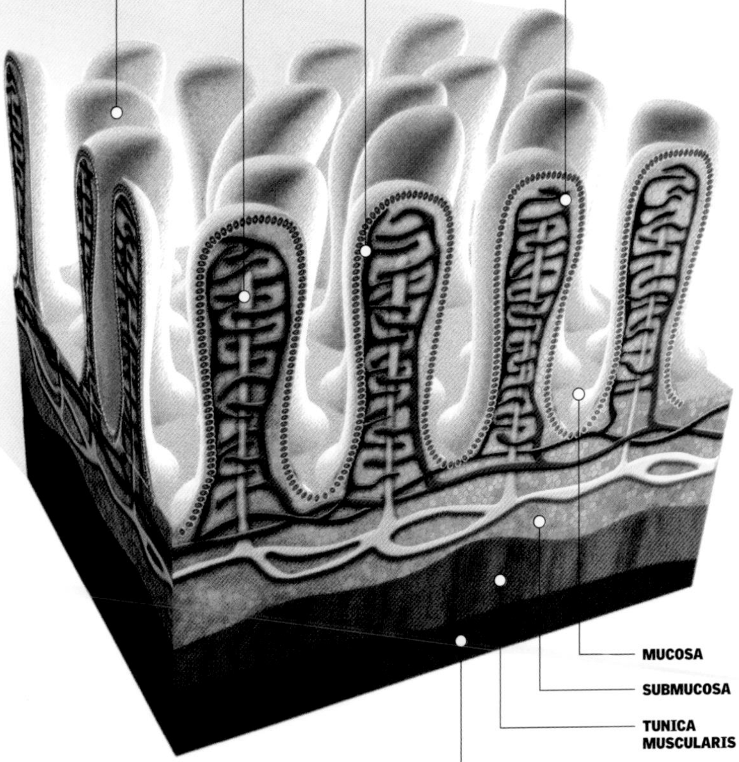

# Urinary System

Its basic organs are the kidneys (2), the ureters (2), the bladder, and the urethra. Its function is to regulate homeostasis, maintaining the equilibrium between the water and the chemicals in the body. The first phase of this objective is accomplished when the kidneys produce and secrete urine, a liquid that is eliminated from the body. Urine is essentially harmless, only containing about 2 percent urea, and is sterile: it is composed primarily of water and salts, and it normally does not contain bacteria, viruses, or fungi. The ureters are channels that carry the urine through the body. The bladder is a sac that stores the urine until it is passed to the urethra, a duct through which it will be expelled from the body.

## The Urinary Tract

The glomerulus is a grouping of vessels located in the cortex of the kidneys. Most of the filtering that takes place in the nephron is performed in the glomerulus. Wide arterioles carry blood to the glomerulus. Other, thinner arterioles exit from the glomerulus, carrying away blood. So much pressure is generated inside the kidney that the fluid exits from the blood via the porous capillary walls.

## The Bladder in Action

The bladder is continually filled with urine and then emptied periodically. When full, the bladder stretches to increase its capacity. When the muscle of the internal sphincter is relaxed, the muscles of the wall contract, and the urine exits through the urethra. In adults this occurs voluntarily in response to an order issued by the nervous system. In infants, on the other hand, this evacuation occurs spontaneously, as soon as the bladder is filled.

**FILLING**

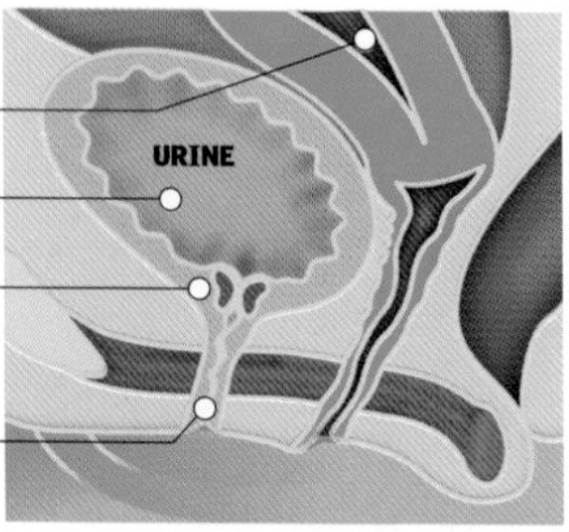


**EMPTYING**

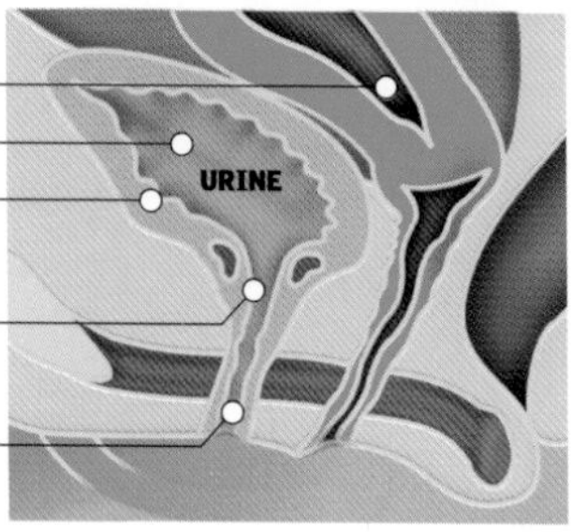


**Legend**

**1. BLOOD FILTERING**
The blood enters the kidney via the renal artery.

**2. TRANSFER**
The artery carries the blood into the kidney, where it is filtered by the kidney's functional units, the nephrons.

**3. STORAGE**
A certain amount of urine is obtained from the filtrate in the nephrons, and that urine is sent to the renal pelvis.

**4. ELIMINATION**
The urine passes from the renal pelvis to the ureter and then to the bladder, where it accumulates until it is eliminated through the tube-shaped urethra.

**15 minutes**

**IT TAKES 15 MINUTES FOR LIQUIDS TO CIRCULATE THROUGH THE NEPHRONS.**

**COMPONENTS OF URINE**

| | |
|---|---|
| **95%** | Water |
| **2%** | Urea, a toxic substance |
| **2%** | Chloride salts, sulfates, phosphates of potassium and magnesium |
| **1%** | Uric acid |

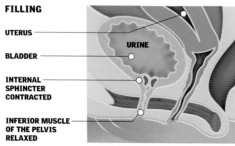

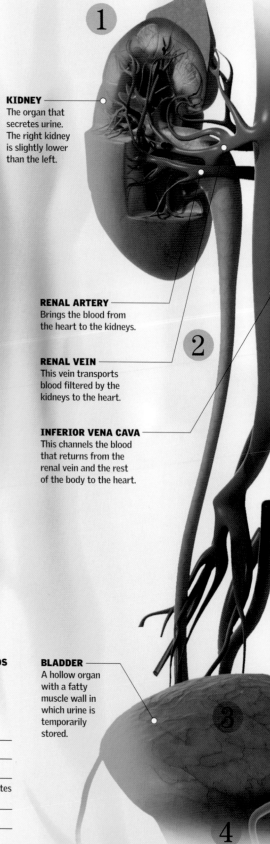

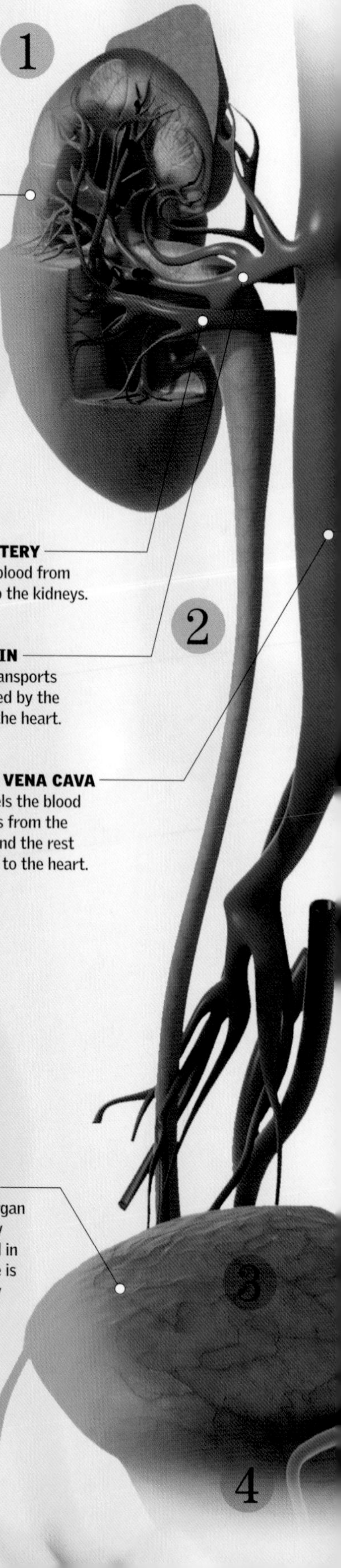

1
KIDNEY
The organ that secretes urine. The right kidney is slightly lower than the left.
RENAL ARTERY
Brings the blood from the heart to the kidneys.
2
RENAL VEIN
This vein transports blood filtered by the kidneys to the heart.
INFERIOR VENA CAVA
This channels the blood that returns from the renal vein and the rest of the body to the heart.
BLADDER
A hollow organ with a fatty muscle wall in which urine is temporarily stored.
3
4

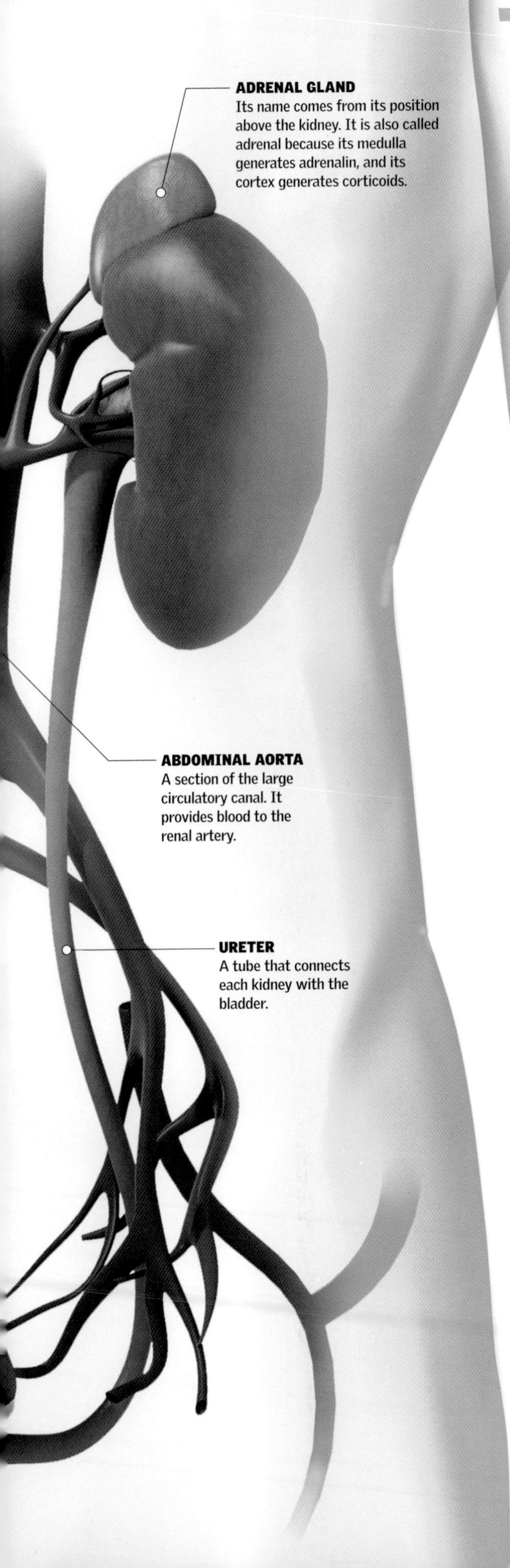

ADRENAL GLAND
Its name comes from its position above the kidney. It is also called adrenal because its medulla generates adrenalin, and its cortex generates corticoids.
ABDOMINAL AORTA
A section of the large circulatory canal. It provides blood to the renal artery.
URETER
A tube that connects each kidney with the bladder.

# Differences by Sex

The urinary system has a double relationship to the reproductive system. The two systems are linked by their close physical proximity, but they are also linked functionally. For example, the ureter is a vehicle for secretions produced by the glands of both systems. The urinary systems in men and women are different. A man's bladder is larger, and the male ureter is also larger than a woman's, because in a man the ureter extends to the end of the penis, for a total length of about 6 inches (20 cm); in a woman, on the other hand, the bladder is located at the front of the uterus, and the length of the ureter is approximately 1.5 inches (4 cm).

**In a Man** / **In a Woman**

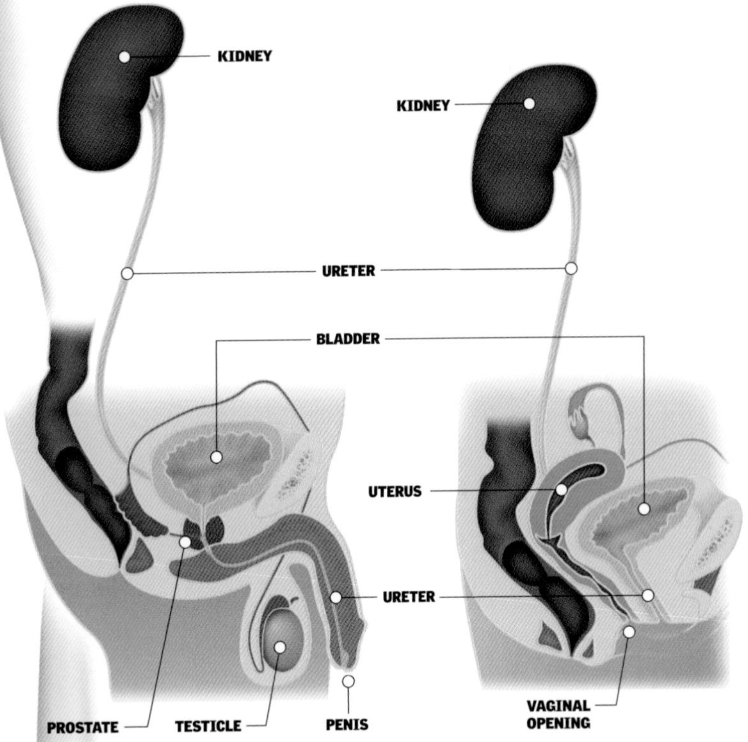


# Fluid Exchange

The volume of urine that a person expels every day is related to the person's consumption of liquids. Three quarts (2.5 l) a day would be excessive, but a significant decrease in the production of urine can indicate a problem. The table details the relationship between the consumption of liquid and its expulsion by the different glands of the human body.

| CONSUMPTION OF WATER | |
|---|---|
| **Drinking** | 60%<br>50 fluid ounces<br>(1,500 ml) |
| **Food** | 30%<br>25 fluid ounces<br>(750 ml) |
| **Metabolic water** | 10%<br>16 fluid ounces<br>(250 ml) |
| **TOTAL** | 3 quarts<br>(2,500 ml) |

| EXPULSION OF WATER | |
|---|---|
| **Urine** | 60%<br>50 fluid ounces<br>(1,500 ml) |
| **Losses through the lungs and the skin** | 28%<br>25 fluid ounces<br>(700 ml) |
| **Sweat** | 8%<br>16 fluid ounces<br>(200 ml) |
| **Feces** | 4%<br>3 fluid ounces<br>(100 ml) |
| **TOTAL** | 3 quarts<br>(2,500 ml) |

# Kidneys

Located on either side of the spinal column, the kidneys are the fundamental organs of the urinary system. They regulate the amount of water and minerals in the blood by producing urine that carries away the waste the kidneys discard. They keep the composition of the bodily fluids constant, regulate the pressure of the arteries, and produce important substances such as the precursor of vitamin D and erythropoietin. Every day they process 500 gallons (1,750 l) of blood and produce 2 quarts (1.5 l) of urine. The kidneys measure approximately 5 inches (12 cm) long and 3 inches (6 cm) wide. Their weight is only 1 percent of the total body weight, but they consume 25 percent of its energy. If one kidney ceases to function, the body is able to survive with the activity of the other.

## The Renal Circuit

Urine is produced in the nephrons in each kidney; there are thought to be a million nephrons in each kidney. From the nephrons the urine flows into the proximal convoluted tubule, where all the nutrients, such as glucose, amino acids, and most of the water and salts, are reabsorbed into the blood. After passing through the nephron the urine is filtered, and it arrives at the common collecting duct where only the residues and excess water are retained.

**1. ENTRY OF BLOOD**
The blood enters the kidney via the renal artery.

**2. FILTRATION**
The blood is filtered in the nephrons, the functional units of the kidneys.

**3. URINE IS OBTAINED**
A certain amount of urine is obtained from the filtrate in the nephrons, and it is sent to the renal pelvis. The filtered blood, free from waste, is sent to the renal vein and reenters the bloodstream.

**4. URINE**
The urine passes through the renal pelvis to the ureter and from there to the bladder, where it accumulates until it is eliminated through the tube-shaped urethra.

**5. CLEAN BLOOD**
The clean blood exits the kidney via the renal vein, which is connected to the vena cava. The blood then returns to the heart.

**1 million**

**ONE KIDNEY HAS ABOUT ONE MILLION NEPHRONS.**

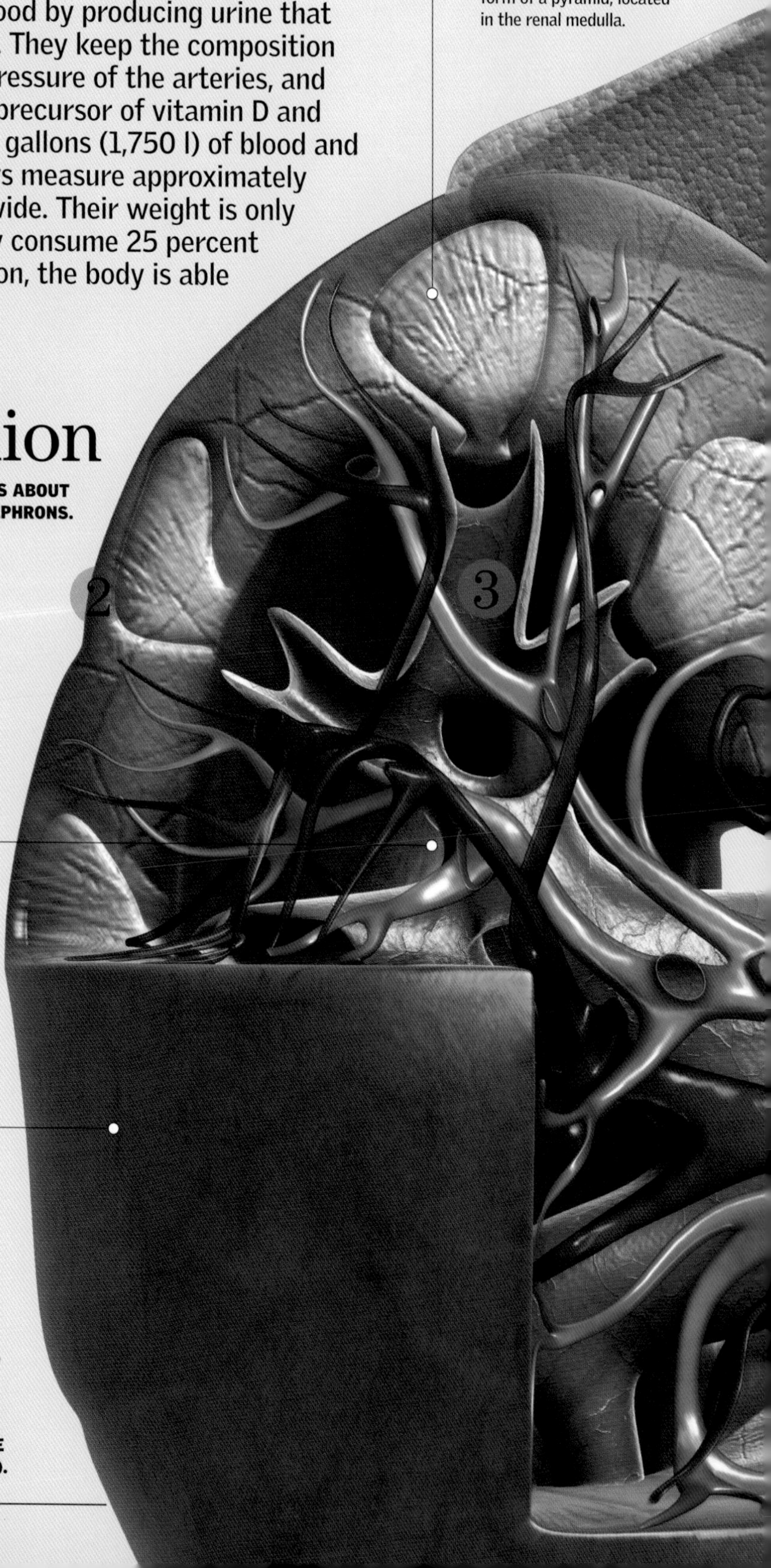

RENAL PYRAMID
A fluted structure in the form of a pyramid, located in the renal medulla.
2
3
RENAL PELVIS
This transports the urine to the ureter.
RENAL CAPSULE
Protective layer that covers each kidney. It consists of white fibrous tissue.

**45 minutes**

**THE FRENCH PHYSIOLOGIST CLAUDE BERNARD (1813-78) WAS THE FIRST TO NOTE THE IMPORTANCE OF THE KIDNEYS. AT THAT TIME IT WAS NOT KNOWN THAT THE KIDNEYS FILTER ALL THE WATER CONTENT OF THE BLOOD IN THE BODY EVERY 45 MINUTES AND THAT, EVEN SO, IT IS POSSIBLE TO SURVIVE WITH ONLY ONE KIDNEY (OR NONE, IN THE CASE OF DIALYSIS).**

# 41 to 51 fluid ounces (1,200 to 1,500 cc)

**IS THE AMOUNT OF URINE ELIMINATED EACH DAY BY AN ADULT.**

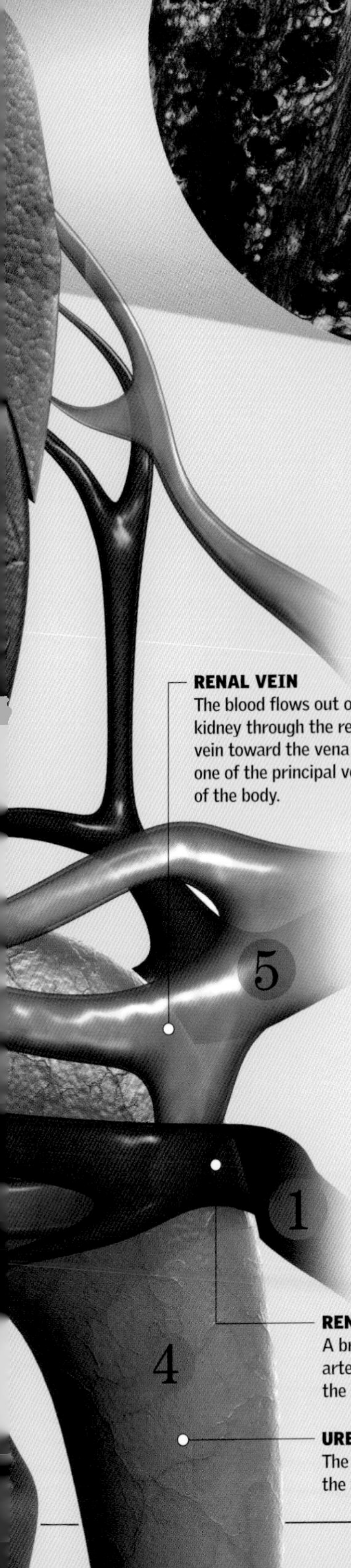

## Glomerulus

A grouping of vessels and capillaries in the kidney's cortex, or sheath. Most of the filtering done by the nephrons takes place there. The wide, afferent arterioles bring blood to the glomerulus. Other, narrower, efferent arterioles lead out of the glomerulus, transporting blood. Inside the glomerulus so much pressure is generated that substances in the blood pass out through the porous capillary walls.

## Nephrons

The functional units of the kidney that filter the blood and produce urine. The basic structure of the nephron consists of two parts: (1) the renal or Malpighian corpuscle, where filtration occurs, including the glomerulus and Bowman's capsule that envelops it; and (2) the renal tubule, a tube that collects the filtered liquid (urine) that is to be eliminated from the body.

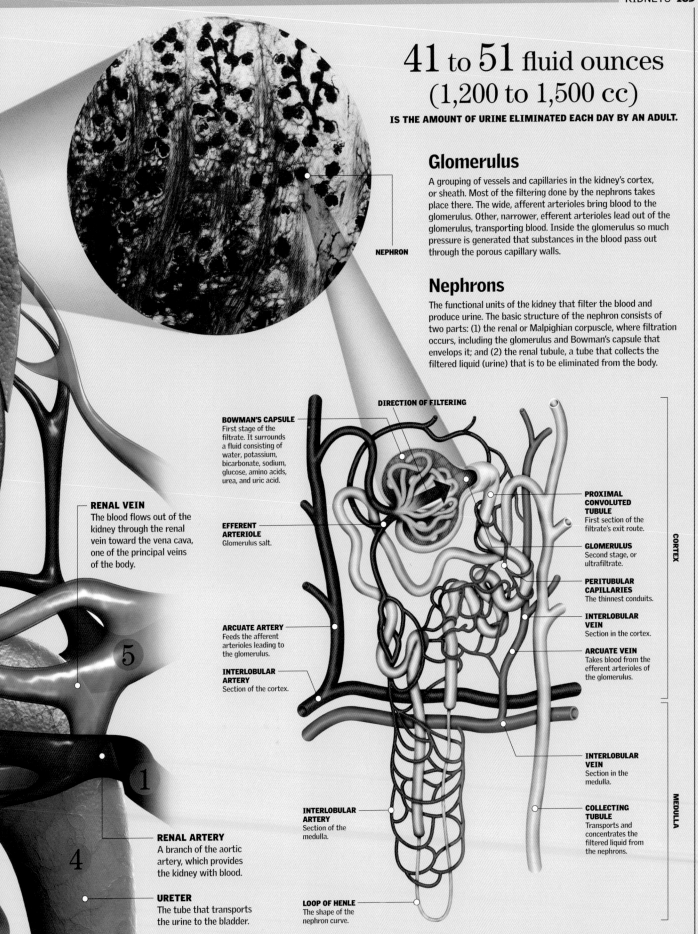

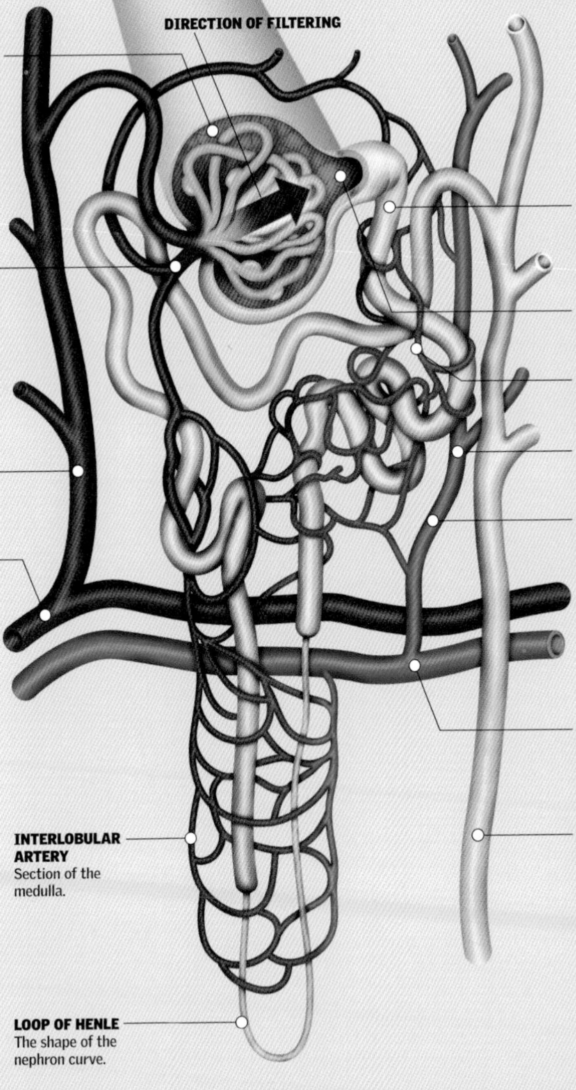


**BOWMAN'S CAPSULE**
First stage of the filtrate. It surrounds a fluid consisting of water, potassium, bicarbonate, sodium, glucose, amino acids, urea, and uric acid.

**EFFERENT ARTERIOLE**
Glomerulus salt.

**ARCUATE ARTERY**
Feeds the afferent arterioles leading to the glomerulus.

**INTERLOBULAR ARTERY**
Section of the cortex.

**INTERLOBULAR ARTERY**
Section of the medulla.

**LOOP OF HENLE**
The shape of the nephron curve.

**PROXIMAL CONVOLUTED TUBULE**
First section of the filtrate's exit route.

**GLOMERULUS**
Second stage, or ultrafiltrate.

**PERITUBULAR CAPILLARIES**
The thinnest conduits.

**INTERLOBULAR VEIN**
Section in the cortex.

**ARCUATE VEIN**
Takes blood from the efferent arterioles of the glomerulus.

**INTERLOBULAR VEIN**
Section in the medulla.

**COLLECTING TUBULE**
Transports and concentrates the filtered liquid from the nephrons.

**RENAL VEIN**
The blood flows out of the kidney through the renal vein toward the vena cava, one of the principal veins of the body.

**RENAL ARTERY**
A branch of the aortic artery, which provides the kidney with blood.

**URETER**
The tube that transports the urine to the bladder.

# Endocrine System

Consists of the glands inside the body that secrete approximately 50 specific substances called hormones into the blood. The hormones activate or stimulate various organs and control reproduction, development, and metabolism. These chemicals control many of the body's processes and even meddle in our love lives.

## A Kiss

**KISSING IS CONSIDERED TO BE HEALTHY BECAUSE, AMONG OTHER THINGS, IT STIMULATES THE PRODUCTION OF NUMEROUS HORMONES AND CHEMICAL SUBSTANCES.**

## The Hormonal Message

The endocrine system is made up of the so-called endocrine glands. This complex, controlled by the pituitary (hypophysis), or master, gland, includes the thyroid, parathyroid, pancreas, ovaries, testicles, adrenals, pineal, and hypothalamus. The role of these glands is to secrete the many hormones needed for body functions. The word "hormone" comes from the Greek *hormon*, which means to excite or incite. The term was suggested in 1905 by the British physiologist Ernest Starling, who in 1902 assisted in the isolation of the first hormone, secretin, which stimulates intestinal activity. Hormones control such functions as reproduction, metabolism (digestion and elimination of food), and the body's growth and development. However, by controlling an organism's energy and nutritional levels, they also affect its responses to the environment.

## The Master Gland

The pituitary gland, or hypophysis, is also called the master gland because it controls the rest of the endocrine glands. It is divided into two parts, the anterior lobe and the posterior lobe. The pituitary hormones stimulate the other glands to generate specific hormones needed by the organism.

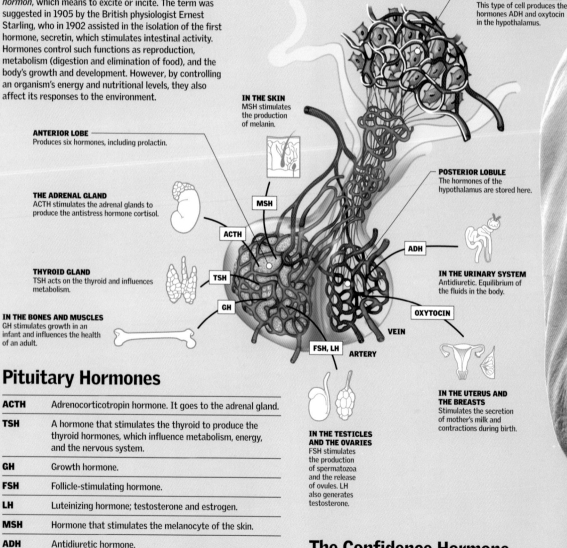

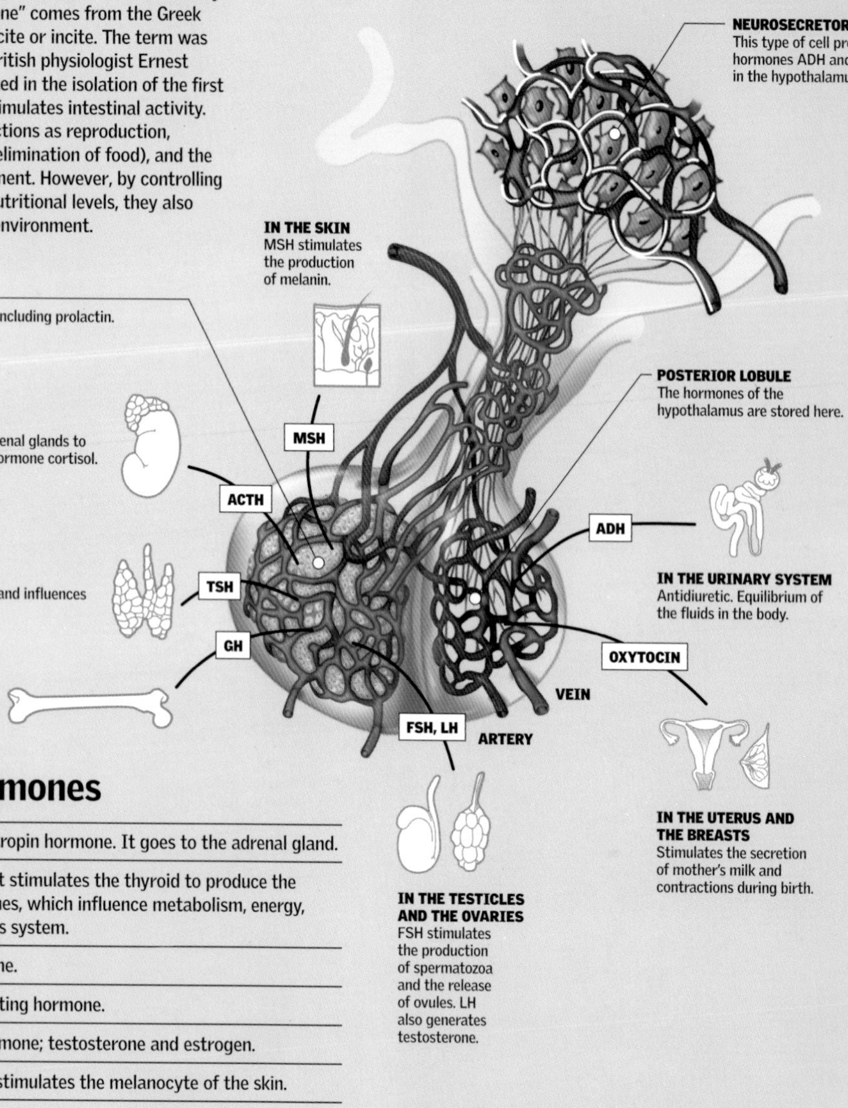

**NEUROSECRETORY CELLS**
This type of cell produces the hormones ADH and oxytocin in the hypothalamus.
**IN THE SKIN**
MSH stimulates the production of melanin.
**ANTERIOR LOBE**
Produces six hormones, including prolactin.
**POSTERIOR LOBULE**
The hormones of the hypothalamus are stored here.
**THE ADRENAL GLAND**
ACTH stimulates the adrenal glands to produce the antistress hormone cortisol.
MSH
ACTH
ADH
**IN THE URINARY SYSTEM**
Antidiuretic. Equilibrium of the fluids in the body.
**THYROID GLAND**
TSH acts on the thyroid and influences metabolism.
TSH
GH
OXYTOCIN
**IN THE BONES AND MUSCLES**
GH stimulates growth in an infant and influences the health of an adult.
VEIN
FSH, LH
ARTERY
**IN THE UTERUS AND THE BREASTS**
Stimulates the secretion of mother's milk and contractions during birth.
**IN THE TESTICLES AND THE OVARIES**
FSH stimulates the production of spermatozoa and the release of ovules. LH also generates testosterone.

## Pituitary Hormones

| | |
|---|---|
| **ACTH** | Adrenocorticotropin hormone. It goes to the adrenal gland. |
| **TSH** | A hormone that stimulates the thyroid to produce the thyroid hormones, which influence metabolism, energy, and the nervous system. |
| **GH** | Growth hormone. |
| **FSH** | Follicle-stimulating hormone. |
| **LH** | Luteinizing hormone; testosterone and estrogen. |
| **MSH** | Hormone that stimulates the melanocyte of the skin. |
| **ADH** | Antidiuretic hormone. |
| **PRL** | Prolactin; stimulates milk production by the mother. |
| **Oxytocin** | Stimulates the release of milk by the mother, as well as the contractions needed during labor. |

## The Confidence Hormone

Oxytocin, the hormone that influences basic functions, such as being in love, orgasm, birth, and breast-feeding, is associated with affection and tenderness. It is a hormone that stimulates the formation of bonds of affection.

**PHEROMONES**
are chemical substances released by the glands distributed in the skin that are related to sexual attraction. They act like hormones (whether or not they are actually hormones is a matter of dispute). They transmit sensations of attraction, excitation, and rejection.

**PITUITARY GLAND OR HYPOPHYSIS**
The pituitary gland is located at the base of the brain, and it is the most important control center of the endocrine system. It releases oxytocin in anticipation of a kiss; it is the hormone that stimulates orgasm, birth, and breast-feeding; and it is also associated with psychological behaviors such as affection and tenderness.

**MAMMARY GLANDS**
The LH hormone excites the production of estrogen hormones, which regulate female sexuality; the activity of the mammary glands; and the menstrual cycle. Puberty is marked by an increase of estrogen production.

**ADRENAL GLANDS**
The hormone adrenaline "awakens" the body before a risk—or before a kiss. It increases the cardiac rhythm, the arterial pressure, the level of glucose in the blood, and the flow of blood to muscles.

**PANCREAS**
Before a kiss, it increases the glucose level in the blood. The pancreas produces the two hormones that control the blood sugar level: insulin and glycogen.

**SEXUAL GLANDS**
The reproductive system responds to the same pituitary hormones in men and women: luteinizing hormone (LH) and follicle-stimulating hormone (FSH). (Both are released and activated in anticipation of a kiss.)

# Male Reproductive System

The male reproductive system is the complex of organs that leads to a man's production of one of two types of cells necessary for the creation of a new being. The principal organs are the two testicles, or male gonads, and the penis. The testicles serve as a factory for the production of millions of cells called spermatozoa, which are minute messengers of conception bearing the genetic information for the fertilization of the ovum. The penis is linked to the urinary apparatus, but for reproduction it is the organ that functions as a vehicle for semen, a liquid through which the spermatozoa can reach their destination. The word "semen" comes from Greek and means "seed."

## Testicles and Spermatozoa

The seminiferous tubes in the testicles are covered with spermatogenic cells. By a process of successive cellular divisions called meiosis, the spermatogenic cells are transformed into spermatozoa, the term for the gametes, or male sexual cells, the bearers of half of the genetic information of a new individual. The spermatozoa fertilize the ovum, or female gamete, which contains the other half of the genetic information. The number of chromosomes is kept constant because the spermatozoa and the ovum are both haploid cells (cells that possess half of the genetic information of other cells). When the two haploid cells unite, the fertilized egg, or zygote, is a diploid cell (which contains a total of 46 chromosomes).

### The Testicles

The sexual organs that produce sperm.

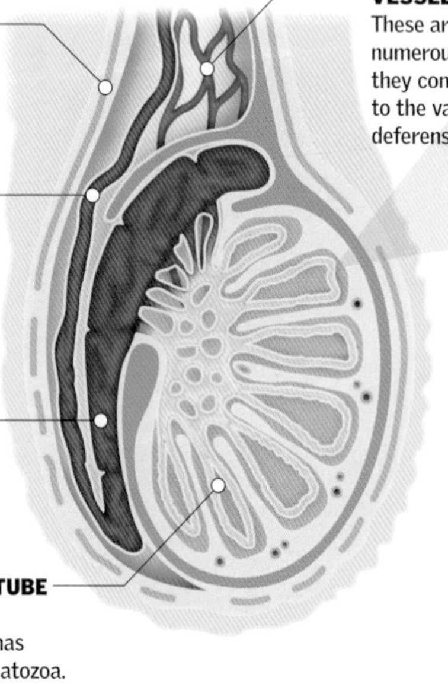

**SPERMATIC CORD**
Connects the testicles to the body.

**BLOOD VESSELS**
These are numerous, and they connect to the vas deferens.

**DEFERENT DUCT (DUCTUS DEFERENS)**
Connects the epididymis with the seminal vesicle.

**EPIDIDYMIS**
The tube where the semen matures and enters the deferent duct.

**SEMINIFEROUS TUBE**
Semen is produced here. Each testicle has thousands of spermatozoa.

### Seminiferous Tubule

Where spermatozoa are produced.

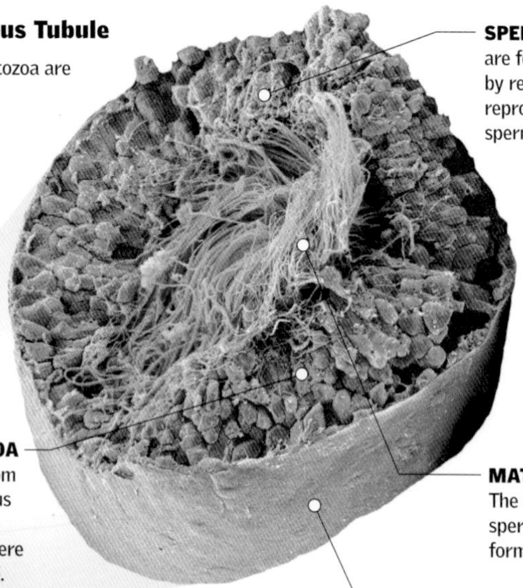

**SPERMATOCYTES**
are formed by repeated reproduction of the spermatogonia.

**SPERMATOZOA**
These move from the seminiferous tubules to the epididymis, where they are stored.

**MATURE SPERM**
The division of the spermatocytes forms spermatozoa.

**SEMINIFEROUS TUBULE**

### Spermatozoon

Male reproductive cell.

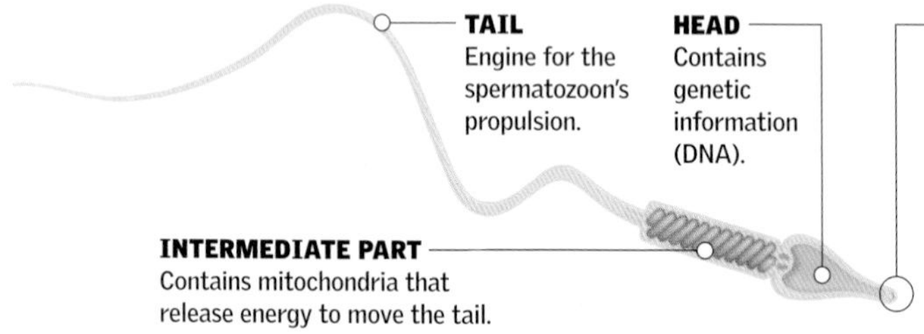

**TAIL**
Engine for the spermatozoon's propulsion.

**HEAD**
Contains genetic information (DNA).

**POINT**
(or acrosome). Contains enzymes that help the spermatozoon penetrate the external membrane of the ovum.

**INTERMEDIATE PART**
Contains mitochondria that release energy to move the tail.

## Internal Structure of the Penis

The most characteristic organ of a man's body, the penis has a cylindrical form with a double function for the urinary system and the reproductive system. In its normal, or relaxed, state the penis carries urine from the body via the urethra during urination. In its erect state its rigidity permits it to be introduced into the female vagina and to release sperm through ejaculation. The penis consists of spongy tissue supplied with blood vessels. The circulatory system supplies abundant blood to these vessels during sexual arousal so that the spongy tissue becomes swollen because of the filled blood vessels. This produces an erection, which makes copulation possible. The body of the penis surrounds the urethra and is connected to the pubic bone. The prepuce covers the head (glans) of the penis, which is located above the scrotum.

### The Penis

Transfers the sperm to the woman.

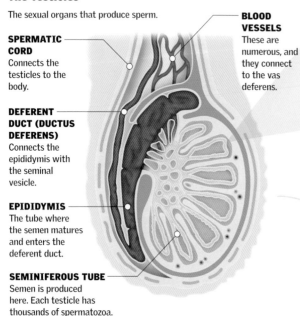

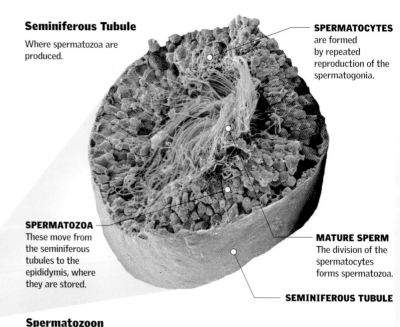

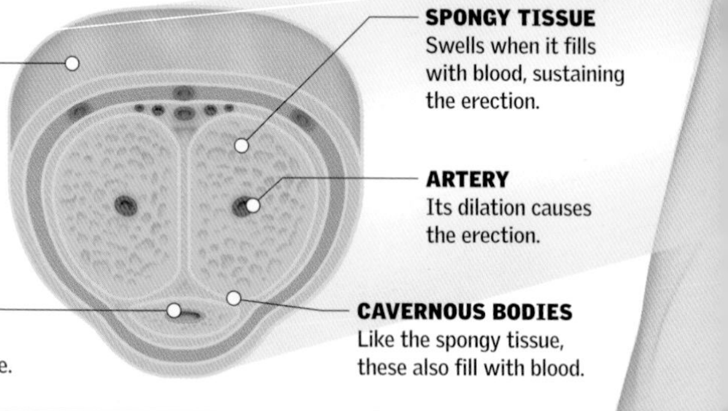

**EXTERNAL SKIN**
Covers the whole organ.

**SPONGY TISSUE**
Swells when it fills with blood, sustaining the erection.

**ARTERY**
Its dilation causes the erection.

**URETHRA**
Extends through the spongy tissue.

**CAVERNOUS BODIES**
Like the spongy tissue, these also fill with blood.

# 93° F
# (34° C)

**IS THE IDEAL APPROXIMATE TEMPERATURE REQUIRED BY THE TESTICLES TO PRODUCE SEMEN. IT IS LOWER THAN THE NORMAL BODY TEMPERATURE OF 98.6° F (37° C) BECAUSE THAT TEMPERATURE WOULD BE TOO WARM FOR THIS FUNCTION. THIS EXPLAINS WHY THE TESTICLES ARE OUTSIDE OF THE BODY. DEPENDING ON THE AMBIENT TEMPERATURE, THEY EXTEND OR RETRACT.**

## Prostate and Epididymis

The prostate is a gland located in front of the rectum and below the bladder. It is the size of a walnut, and it surrounds the urethra, a tube that carries urine from the bladder. The prostate produces the liquid for the semen, which carries the spermatozoa. During orgasm, muscular contractions occur that send the liquid from the prostate out through the urethra. The epididymis is a duct that, when stretched out to its full length, is approximately 20 feet (5 m) long. In the male body it is extremely coiled and lies on the back surface of the testicles, where it is connected with the corresponding vas deferens. The vas deferens stores spermatozoa and provides them with an exit route. The seminal vesicles are two membranous receptacles that connect to both sides of the vas deferens and form the ejaculatory duct.

### The Glands

**PITUITARY GLAND**

**FSH** → Sustentacular Cells.

**LH** → Interstitial Cells.

**TESTICLES**

Interstitial Cells. → **TESTOSTERONE** → Sustentacular Cells.

Sustentacular Cells. → Production of Spermatozoa.

**TESTOSTERONE** → Maintains male reproductive structures and secondary sexual characteristics.

# 150 million

**THE NUMBER OF SPERMATOZOA THAT EACH 0.06 CUBIC INCH (1 ML) OF SEMEN CAN CONTAIN.**

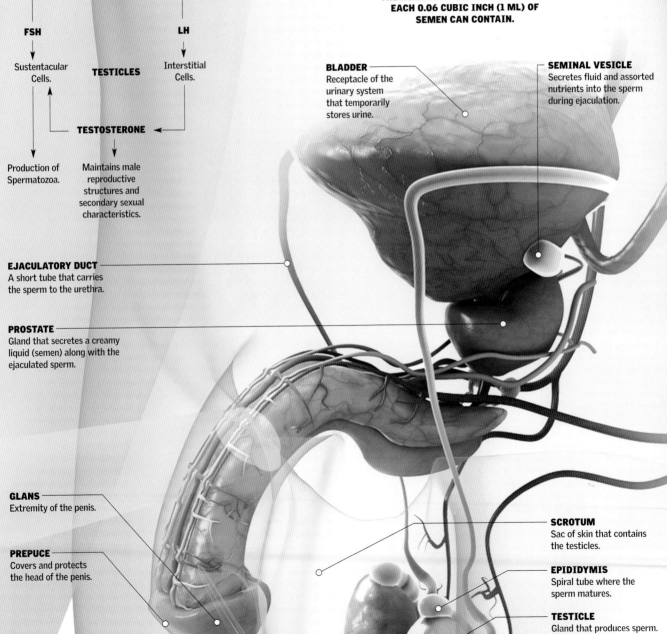

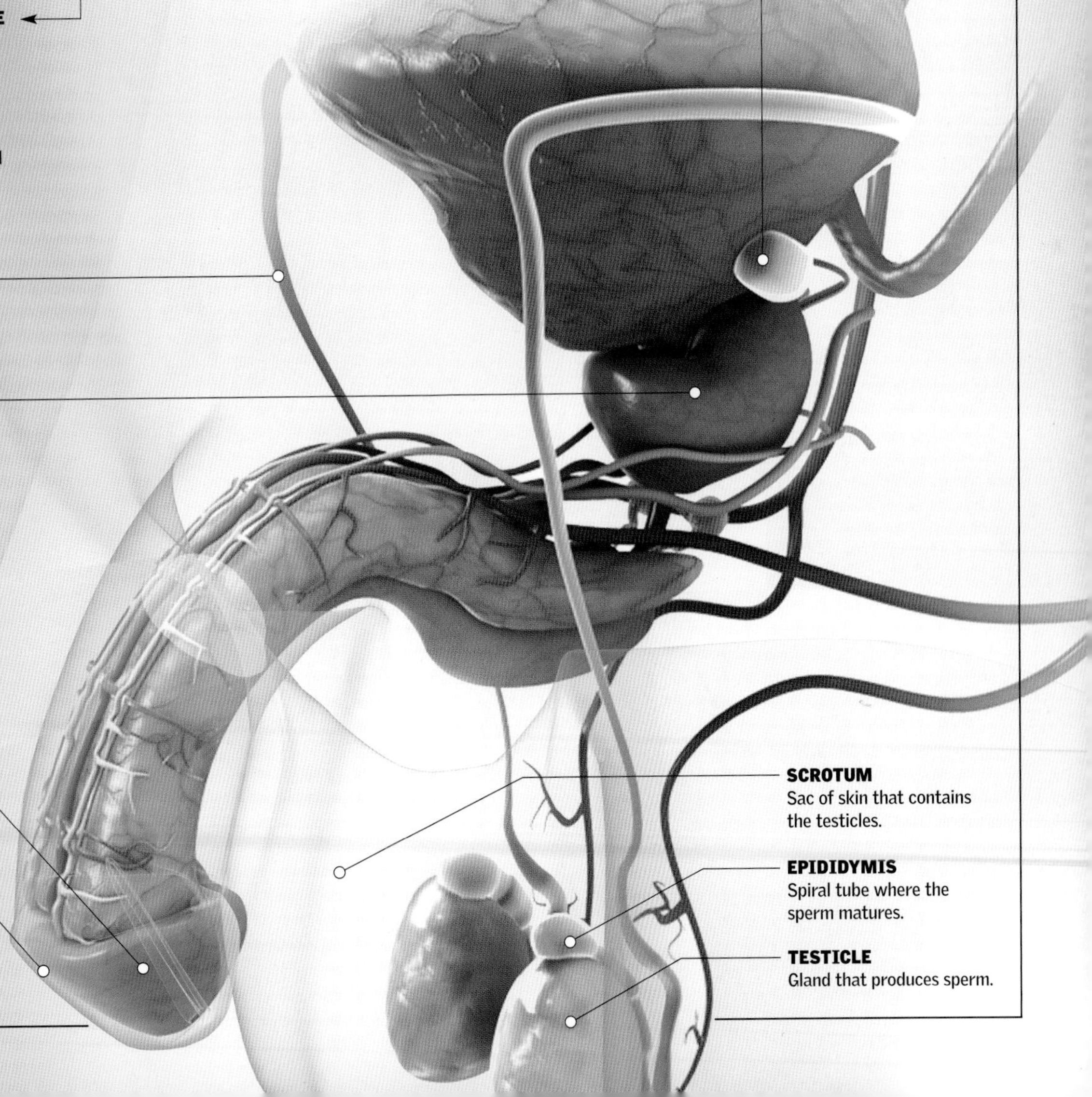

BLADDER
Receptacle of the urinary system that temporarily stores urine.
SEMINAL VESICLE
Secretes fluid and assorted nutrients into the sperm during ejaculation.
EJACULATORY DUCT
A short tube that carries the sperm to the urethra.
PROSTATE
Gland that secretes a creamy liquid (semen) along with the ejaculated sperm.
GLANS
Extremity of the penis.
PREPUCE
Covers and protects the head of the penis.
SCROTUM
Sac of skin that contains the testicles.
EPIDIDYMIS
Spiral tube where the sperm matures.
TESTICLE
Gland that produces sperm.

# Female Reproductive System

Its primary function is the production of ova, and its organs are arranged so as to allow the fertilization of the ovum by a spermatozoon of the male reproductive system and from that moment to facilitate a series of processes known collectively as pregnancy for the creation of a new being. The internal organs of the female reproductive system are the vagina, the uterus, the ovaries, and the fallopian tubes. The external genitalia, generally referred to as the vulva, are relatively hidden and include the labia majora and minora, the clitoris, the urinary meatus, Bartholin's glands, and the vaginal orifice that leads to the vagina. The menstrual cycle governs the system's function.

## The 28 Days of the Menstrual Cycle

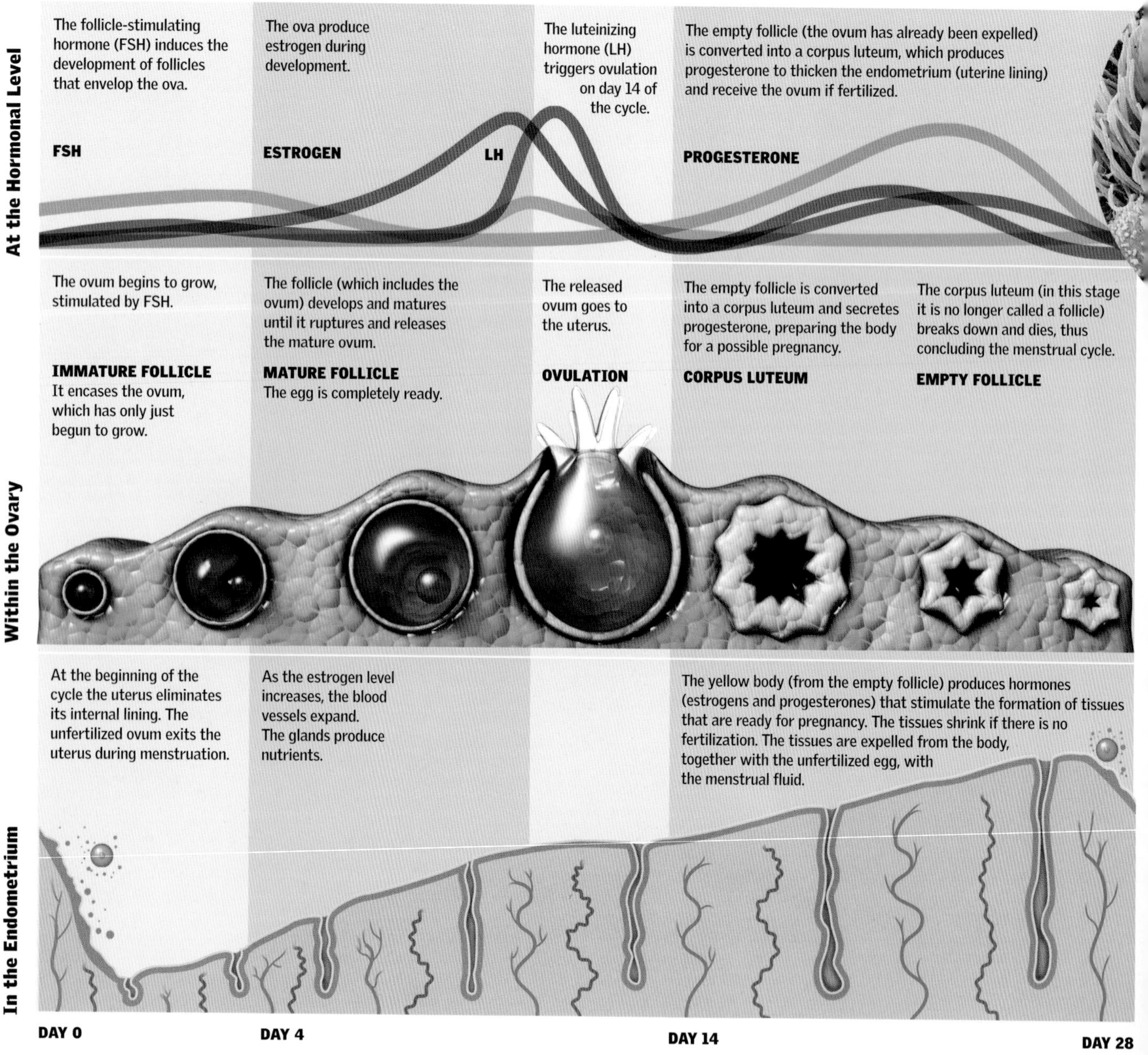

At the Hormonal Level
The follicle-stimulating hormone (FSH) induces the development of follicles that envelop the ova.
The ova produce estrogen during development.
The luteinizing hormone (LH) triggers ovulation on day 14 of the cycle.
The empty follicle (the ovum has already been expelled) is converted into a corpus luteum, which produces progesterone to thicken the endometrium (uterine lining) and receive the ovum if fertilized.
FSH
ESTROGEN
LH
PROGESTERONE
Within the Ovary
The ovum begins to grow, stimulated by FSH.
The follicle (which includes the ovum) develops and matures until it ruptures and releases the mature ovum.
The released ovum goes to the uterus.
The empty follicle is converted into a corpus luteum and secretes progesterone, preparing the body for a possible pregnancy.
The corpus luteum (in this stage it is no longer called a follicle) breaks down and dies, thus concluding the menstrual cycle.
IMMATURE FOLLICLE
It encases the ovum, which has only just begun to grow.
MATURE FOLLICLE
The egg is completely ready.
OVULATION
CORPUS LUTEUM
EMPTY FOLLICLE
In the Endometrium
At the beginning of the cycle the uterus eliminates its internal lining. The unfertilized ovum exits the uterus during menstruation.
As the estrogen level increases, the blood vessels expand. The glands produce nutrients.
The yellow body (from the empty follicle) produces hormones (estrogens and progesterones) that stimulate the formation of tissues that are ready for pregnancy. The tissues shrink if there is no fertilization. The tissues are expelled from the body, together with the unfertilized egg, with the menstrual fluid.
DAY 0
DAY 4
DAY 14
DAY 28

# 2 million

**IS THE APPROXIMATE NUMBER OF OVA THAT AN INFANT GIRL HAS IN HER BODY AT BIRTH. BETWEEN THE AGES OF 10 AND 14, ABOUT 300,000 TO 400,000 OVA REMAIN, OF WHICH ONLY 400 WILL MATURE COMPLETELY OVER HER LIFETIME.**

Cilia, tiny hairlike structures, move the ova very smoothly.

## Menstruation: The Key to Female Reproduction

The female reproductive system is more protected than that of the male because the bony structure of the pelvis houses and shields it. Its development begins around the age of 10, when the female hormones begin a three- to four-year process during which the genital organs, the breasts, the pubic hair, and the general shape of the body change. Toward the age of 13, sometimes earlier or later, the first menstruation, called the menarche, occurs, signaling the beginning of a woman's fertility. She will normally remain fertile for several decades. During menopause, when fertilization is no longer possible, a woman's sexual life is usually not affected and can continue normally.

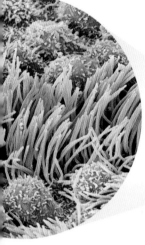

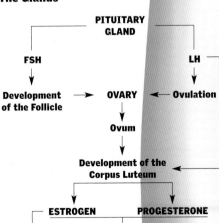

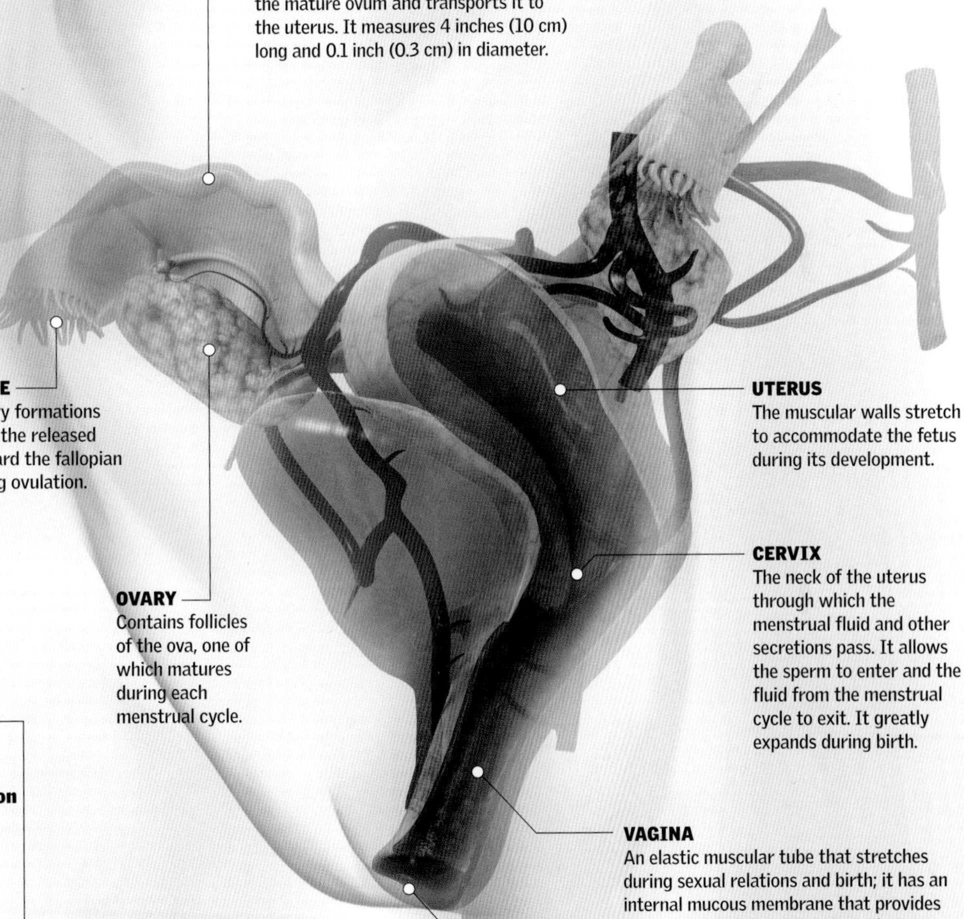
**FALLOPIAN TUBE**
A tube close to each ovary that receives the mature ovum and transports it to the uterus. It measures 4 inches (10 cm) long and 0.1 inch (0.3 cm) in diameter.
**FIMBRIAE**
Filamentary formations that guide the released ovum toward the fallopian tube during ovulation.
**OVARY**
Contains follicles of the ova, one of which matures during each menstrual cycle.
**UTERUS**
The muscular walls stretch to accommodate the fetus during its development.
**CERVIX**
The neck of the uterus through which the menstrual fluid and other secretions pass. It allows the sperm to enter and the fluid from the menstrual cycle to exit. It greatly expands during birth.
**VAGINA**
An elastic muscular tube that stretches during sexual relations and birth; it has an internal mucous membrane that provides lubrication and an acid medium that acts as a defense against infection. It serves as the pathway from the uterus to the exterior.
**CLITORIS**
A sensitive protuberance of tissue that responds to sexual stimulation.

### The Glands

**PITUITARY GLAND**

**FSH** → **Development of the Follicle** → **OVARY**

**LH** → **Ovulation** → **OVARY**

**OVARY** → **Ovum** → **Development of the Corpus Luteum** (← **LH**)

**Development of the Corpus Luteum** → **ESTROGEN** / **PROGESTERONE**

**ESTROGEN**: Maintains the reproductive structures and secondary sexual characteristics.

**PROGESTERONE**: Prepares for the implementation and control of the menstrual cycle.

# The Senses and Speech

Everything we know about the world comes to us through the senses. Traditionally it was thought that we had only five: vision, hearing, touch, smell, and taste. However, for some time now we have known that we have many additional classes of sensations—such as pain, pressure, temperature, muscular sensation, and a sense of motion—that are generally included in the sense of touch. The areas of the brain involved are called somatosensory areas. Although we often take our senses for granted, each one of them is delicate and irreplaceable. Without them it is nearly impossible to understand our surroundings. They are a bridge between us and everything alive on the Earth.



**HEALTHY SKIN AND EYES** (opposite)
The health of the skin and eyes depends upon a diet that provides them with a sufficient amount of proteins and minerals.

# Smell and Taste

These two senses of the body function as powerful allies of the digestive system. Taste involves the perception of dissolved chemical substances arriving, for example, in the form of food. Taste sensation is principally seated on the upper surface of the tongue, and saliva is a fundamental ingredient for dissolving and tasting. Smell involves the perception of these chemicals when they take the form of dispersed aromas. The sense of smell operates at a greater distance than that of taste and can capture substances floating in the environment. It is thought that smell is some 10,000 times more sensitive than any of our other senses.

## Olfactory Cells

These are located deep in the nasal cavity, extended over the so-called olfactory epithelium. It is calculated that some 25 million cells are located there. Their useful life is, on average, 30 days, after which they are replaced by new cells. They have a dual function. One end of each olfactory receptor is connected to the olfactory bulb and transmits the sensations it records, so that the bulb is able to send the nerve impulses to the brain with the necessary information. The other end terminates in a group of cilia, or microscopic hairs, which serve a protective function within the mucosa.

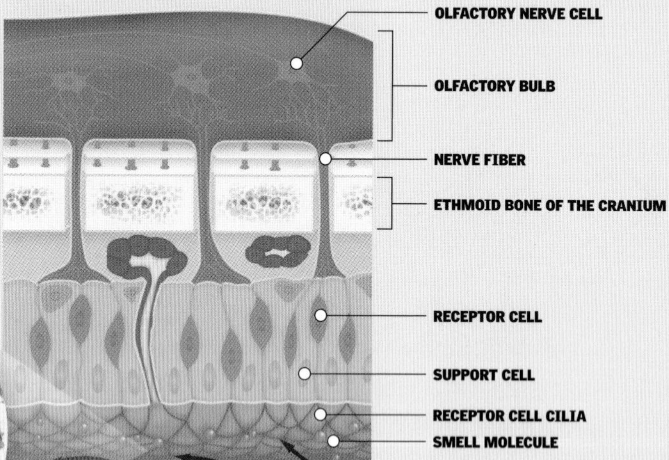


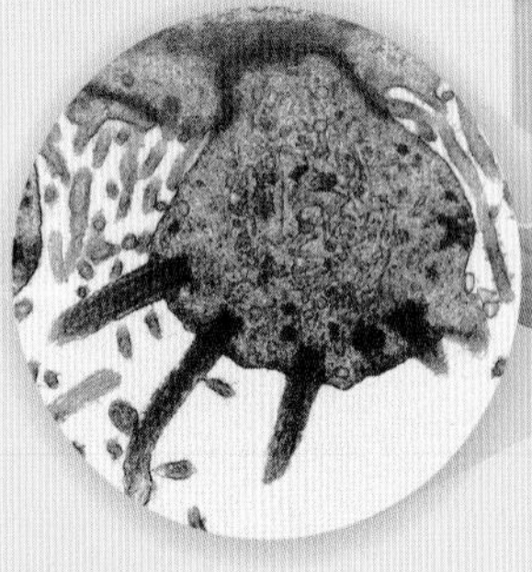

**10,000**

**THE NUMBER OF ODORS THE SENSE OF SMELL CAN DISTINGUISH.**

**4 Flavors**

**THE SURFACE OF THE TONGUE CAN DISTINGUISH: SWEET, SALTY, SOUR, AND BITTER.**

## Gustatory Papillae

The tongue is the principal seat of the sense of taste. It has great mobility at the bottom of the mouth and contains between 5,000 and 12,000 gustatory papillae. Each of these papillae has approximately 50 sensory cells, which have an average life span of 10 days. The salivary glands are activated by the ingestion of food or just before ingestion. They generate an alkaline liquid called saliva, a chemical solvent that, together with the tongue, breaks down the substances of which food is composed and makes it possible to differentiate between them by taste. The tongue takes charge of perceiving these tastes via the fungiform papillae, which give the tongue its rough appearance.

### Gustatory Papilla

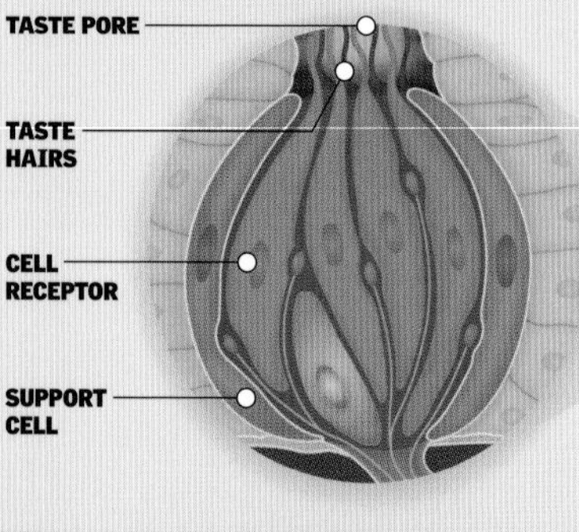


### Surface of the Tongue

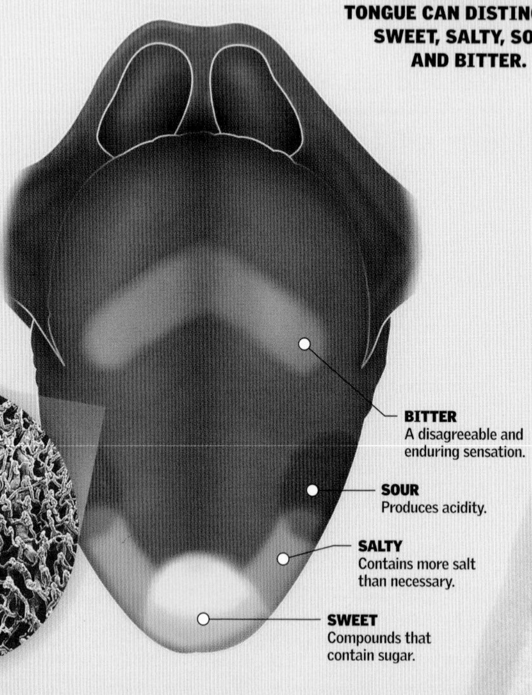

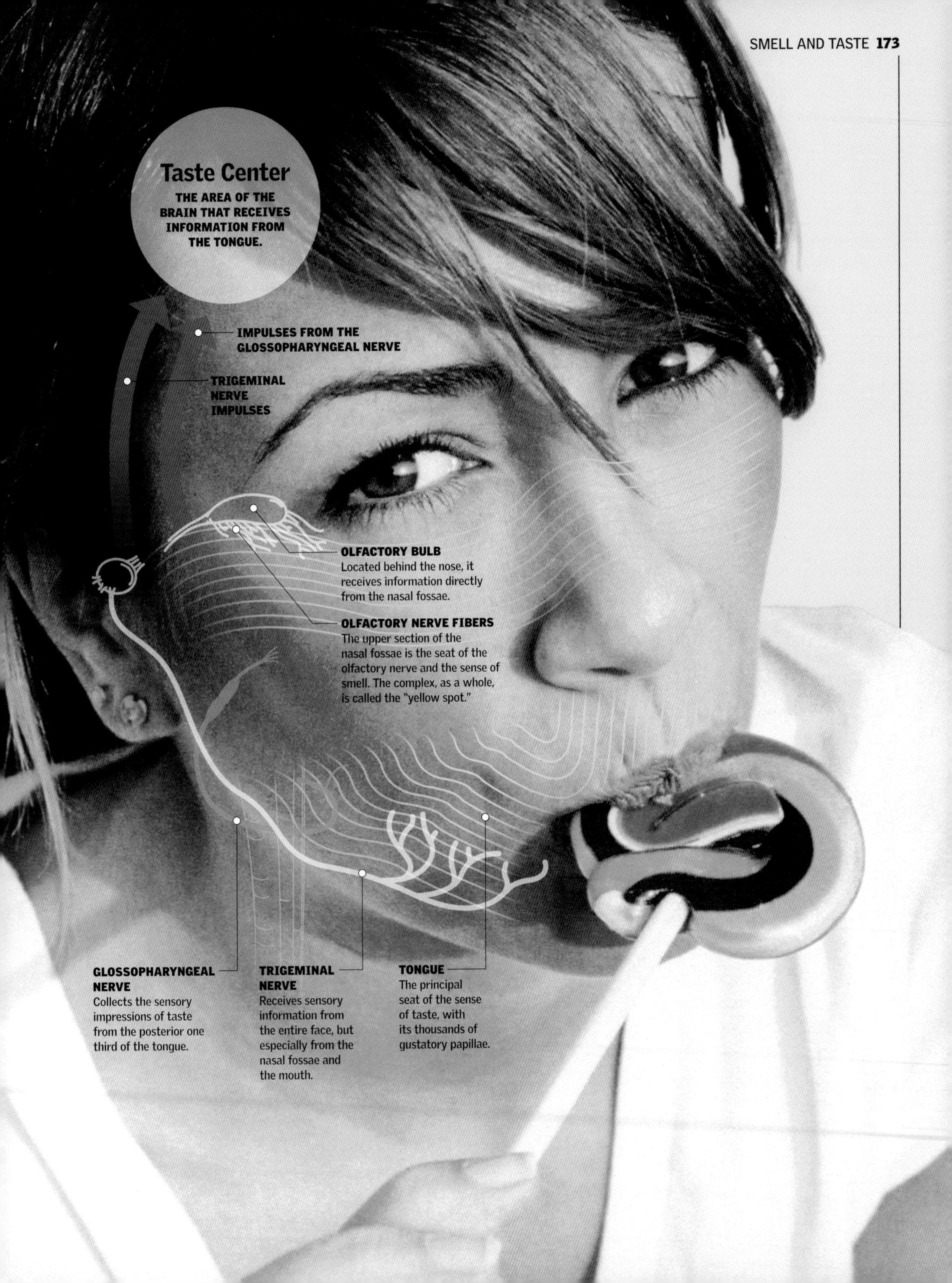
Taste Center
THE AREA OF THE BRAIN THAT RECEIVES INFORMATION FROM THE TONGUE.
IMPULSES FROM THE GLOSSOPHARYNGEAL NERVE
TRIGEMINAL NERVE IMPULSES
OLFACTORY BULB
Located behind the nose, it receives information directly from the nasal fossae.
OLFACTORY NERVE FIBERS
The upper section of the nasal fossae is the seat of the olfactory nerve and the sense of smell. The complex, as a whole, is called the "yellow spot."
GLOSSOPHARYNGEAL NERVE
Collects the sensory impressions of taste from the posterior one third of the tongue.
TRIGEMINAL NERVE
Receives sensory information from the entire face, but especially from the nasal fossae and the mouth.
TONGUE
The principal seat of the sense of taste, with its thousands of gustatory papillae.

# Touch and the Skin

Touch is one of the five senses. Its function is to perceive sensations of contact, pressure, and temperature and to send them to the brain. It is located in the skin (the integument), the organ that covers the entire outside of the body for protection. The cellular renewal of the skin is continuous, and when recording external changes (of temperature, for example), it activates reflexive mechanisms to open or close the pores and, thus, to maintain the required body temperature. Secretions, such as those of the sweat glands, also contribute to this process by reducing heat. Like the sebaceous glands, they are important for hydration and hygiene in the areas where they are located.

## The Thinnest and the Thickest

The thinnest skin on the body is that of the eyelids. The thickest is that of the sole of the foot. Both provide, like all the skin of the body, a protective function for muscles, bones, nerves, blood vessels, and interior organs. It is thought that hair and fingernails are modified types of skin. Hair grows over the whole body, except for the palms of the hands, the soles of the feet, the eyelids, and the lips.

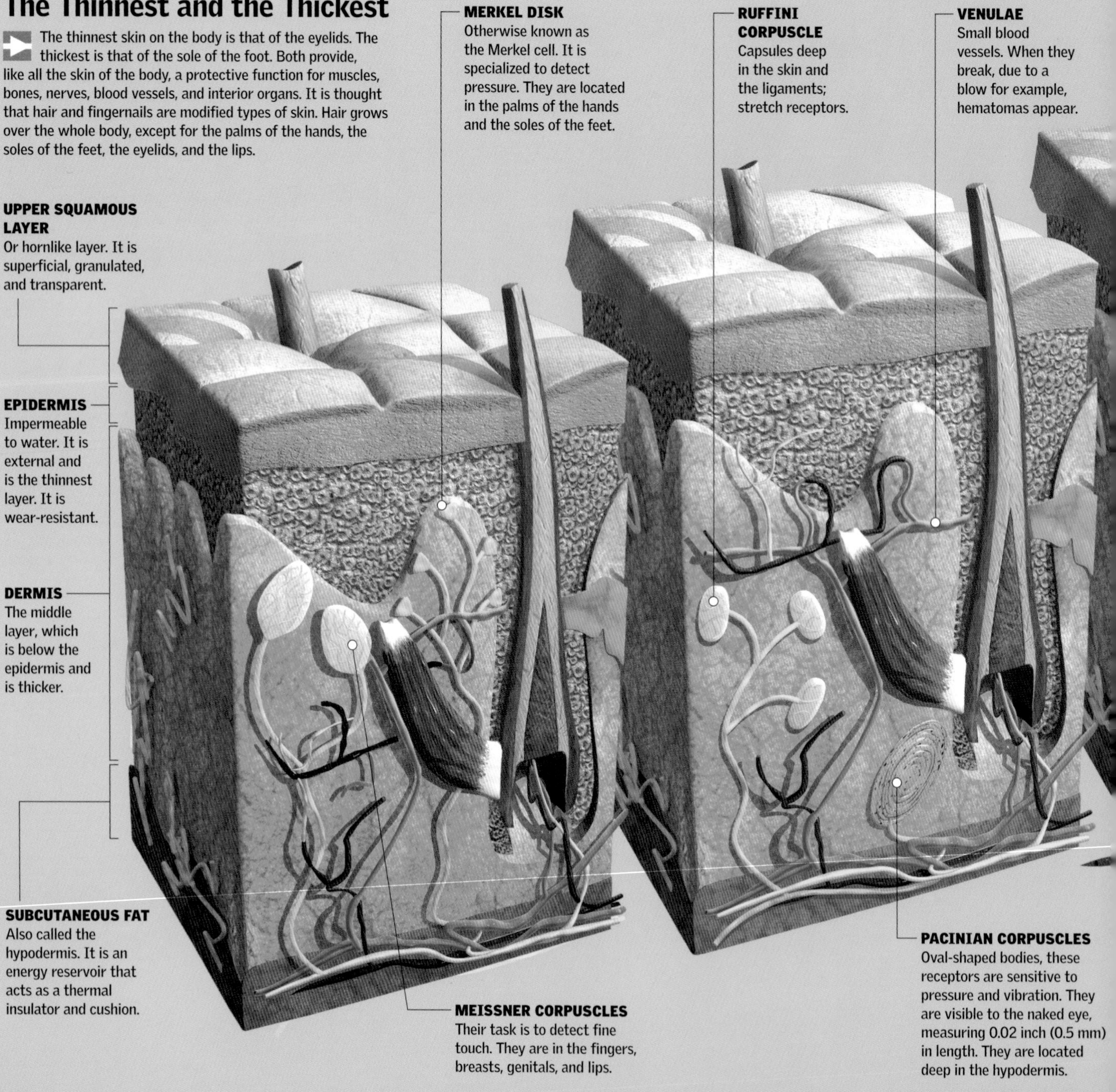
**MERKEL DISK**
Otherwise known as the Merkel cell. It is specialized to detect pressure. They are located in the palms of the hands and the soles of the feet.
**RUFFINI CORPUSCLE**
Capsules deep in the skin and the ligaments; stretch receptors.
**VENULAE**
Small blood vessels. When they break, due to a blow for example, hematomas appear.
**UPPER SQUAMOUS LAYER**
Or hornlike layer. It is superficial, granulated, and transparent.
**EPIDERMIS**
Impermeable to water. It is external and is the thinnest layer. It is wear-resistant.
**DERMIS**
The middle layer, which is below the epidermis and is thicker.
**SUBCUTANEOUS FAT**
Also called the hypodermis. It is an energy reservoir that acts as a thermal insulator and cushion.
**MEISSNER CORPUSCLES**
Their task is to detect fine touch. They are in the fingers, breasts, genitals, and lips.
**PACINIAN CORPUSCLES**
Oval-shaped bodies, these receptors are sensitive to pressure and vibration. They are visible to the naked eye, measuring 0.02 inch (0.5 mm) in length. They are located deep in the hypodermis.

# Skin

**A MAN'S SKIN PRODUCES A GREATER QUANTITY OF SEBUM, OR OILY SECRETION, THAN THAT OF A WOMAN. THEREFORE, A MAN'S SKIN IS TOUGHER AND GREASIER THAN A WOMAN'S.**

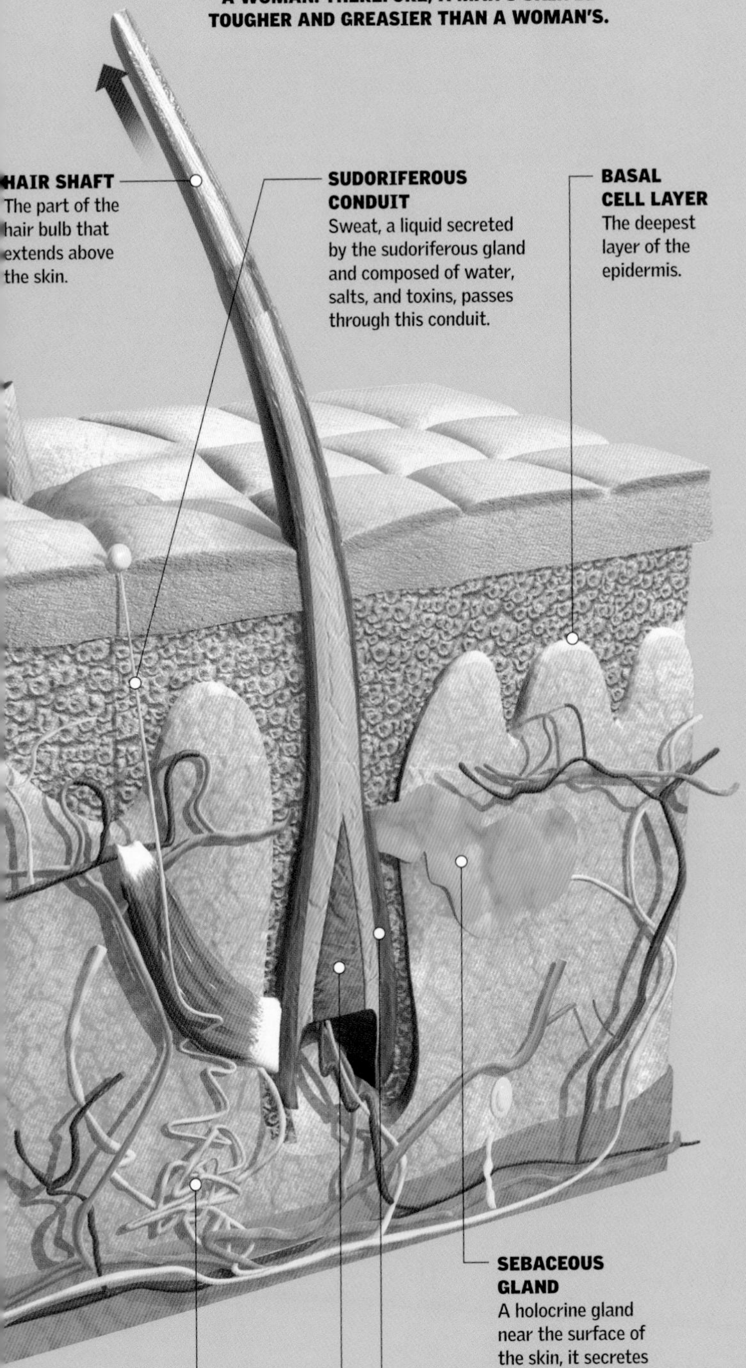

HAIR SHAFT
The part of the hair bulb that extends above the skin.
SUDORIFEROUS CONDUIT
Sweat, a liquid secreted by the sudoriferous gland and composed of water, salts, and toxins, passes through this conduit.
BASAL CELL LAYER
The deepest layer of the epidermis.
SEBACEOUS GLAND
A holocrine gland near the surface of the skin, it secretes an oily substance that coats the skin and keeps it soft and flexible.
SUDORIFEROUS GLANDS
These regulate the temperature of the body. The eccrine glands are tubular and cover the entire surface of the body. The apocrine glands are specialized; they are located only in the armpits and the genital area. They are large and do not empty directly onto the skin but into the pilous follicle.
HAIR FOLLICLE
The sheath that covers a hair.
BULBUS PILI (HAIR BULB)
The lower extremity of the hair. It is thick and surrounds the nerve papilla.

## Responding to Temperature

When the skin perceives the sensation of cold, the blood vessels and the muscles contract. The purpose of this is to prevent the escape of heat; as a consequence, the hairs stand on end, resulting in what is commonly called goose bumps. The opposite happens in response to heat: the blood vessels dilate because the skin has received instructions from the brain to dissipate heat, and the vessels emit heat as if they were radiators. The sudoriferous glands exude sweat onto the surface of the skin. The evaporation of sweat removes heat from the skin.

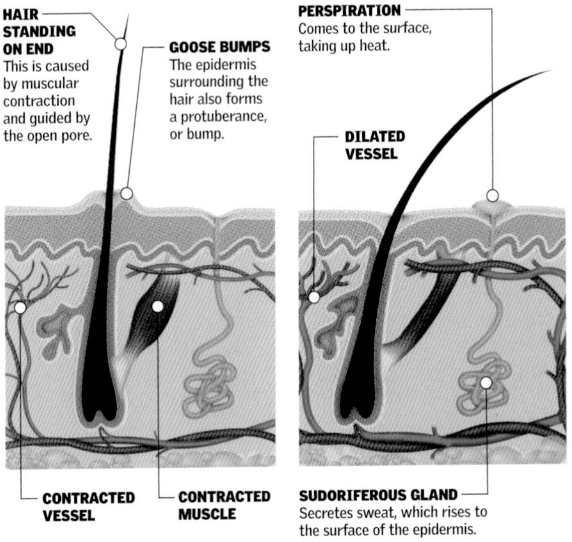

HAIR STANDING ON END
This is caused by muscular contraction and guided by the open pore.
GOOSE BUMPS
The epidermis surrounding the hair also forms a protuberance, or bump.
PERSPIRATION
Comes to the surface, taking up heat.
DILATED VESSEL
CONTRACTED VESSEL
CONTRACTED MUSCLE
SUDORIFEROUS GLAND
Secretes sweat, which rises to the surface of the epidermis.

**A COLD**
As with fear, cold puts a person's hair on end—literally! The contraction of both the blood vessels and the muscles causes the hair on the skin to stand on end.

**B HEAT**
This causes the secretion of sweat, which increases as the temperature rises. Cooling is caused by the evaporation of the sweat, which carries heat away from the body.

## Nails

They are hard and hornlike. Their principal component is keratin, a protein that is also present in the skin and the hair. Their function is to cover and protect the ends of the fingers and toes. Their cells arise from the proliferative matrix and advance longitudinally. Once outside the body, they die. That is why there is no pain when you cut them.

### A Shield for the Fingers and Toes

The fingernail can be seen with the naked eye, but the protective structure of the fingers and toes also includes their hidden matrix and bone structure.

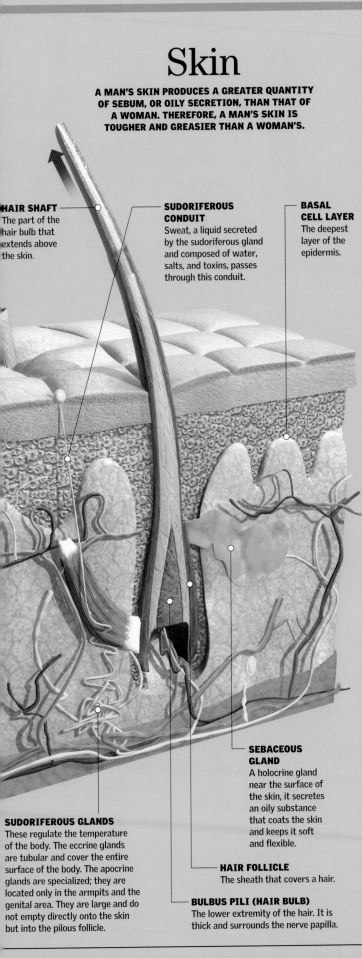

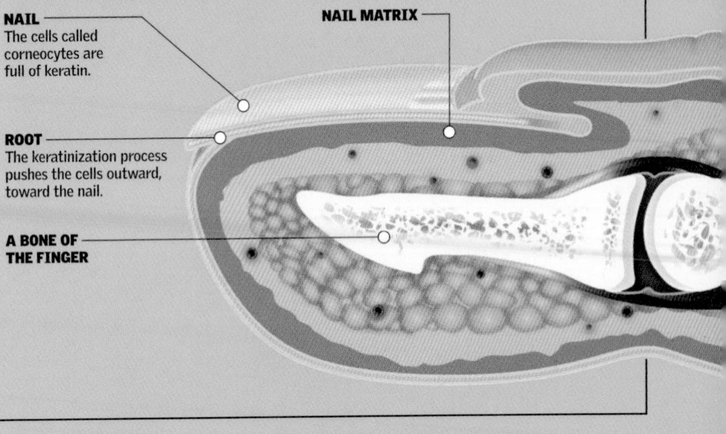

# Anatomy of the Eye

Almost all the information that comes from the world into the brain depends on vision. The eye, one of the most complex organs of the body, allows us to judge the size and texture of an object even before we touch it or to know how far away it is. More than 100 million cells are activated instantaneously in the presence of light, converting the image perceived into nerve impulses that are transmitted to the brain. For this reason 70 percent of all the body's sensory receptors are concentrated in the eyes. It is vital that the brain receive information in a correct form: otherwise, things would appear to be distorted.

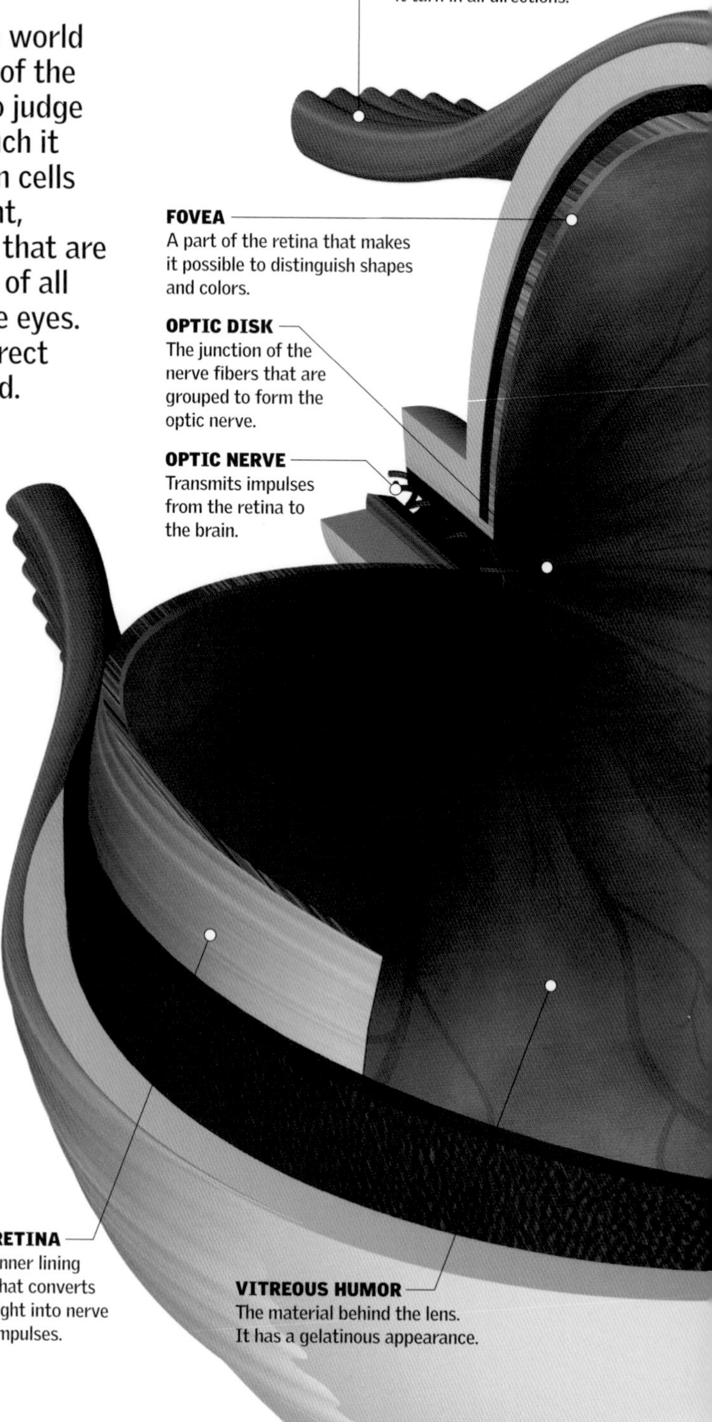

EYE MUSCLE
One of the six muscles that envelops the eye and makes it turn in all directions.
FOVEA
A part of the retina that makes it possible to distinguish shapes and colors.
OPTIC DISK
The junction of the nerve fibers that are grouped to form the optic nerve.
OPTIC NERVE
Transmits impulses from the retina to the brain.
RETINA
Inner lining that converts light into nerve impulses.
VITREOUS HUMOR
The material behind the lens. It has a gelatinous appearance.

## How Does the Eye See?

An object reflects light in all directions. The light is partially focused by the cornea, which refracts the entering rays. The lens focuses the rays of light, changing its shape to give the light the focus it needs. The rays cross the inside of the eye. The light arrives at the retina, and the rays perceived produce an inverted image of the object. The retina sends this information to the brain, which processes it and constructs a correct image of the object. Thanks to the fovea the eye can perceive details such as the shape and color of objects.

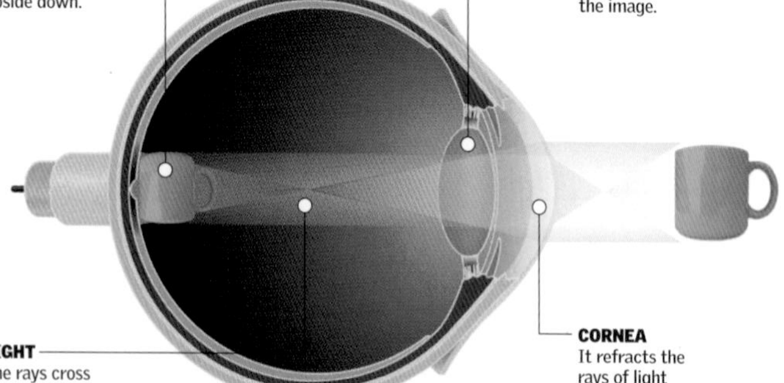

IMAGE
The object is perceived upside down.
LENS
Its function is to focus and construct the image.
LIGHT
The rays cross inside the eye.
CORNEA
It refracts the rays of light passing through it.

## Seeing in Three Dimensions

When the eyes look ahead, the field of vision is binocular because both eyes see at the same time, each one from a different perspective. The images are superimposed at an angle of approximately 120°. This allows stereoscopic vision (two images of the same object from different angles, without deformation). The brain perceives the image in three dimensions.

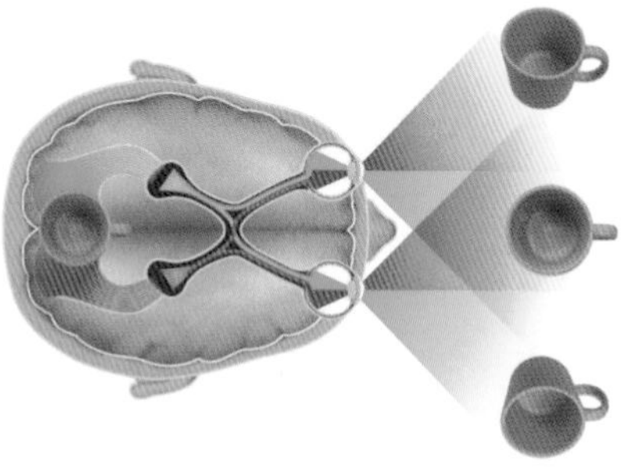


## Iris

A colored membranous disk, with a pupil in the center. It has two types of muscular fibers: circular and radial. In response to bright light the circular fibers contract and the radial fibers relax: the pupil diminishes in size to reduce the amount of light that enters. When there is less light, the circular muscles relax, and the radial ones contract. The pupil then dilates so that more light will enter the eye to facilitate vision.

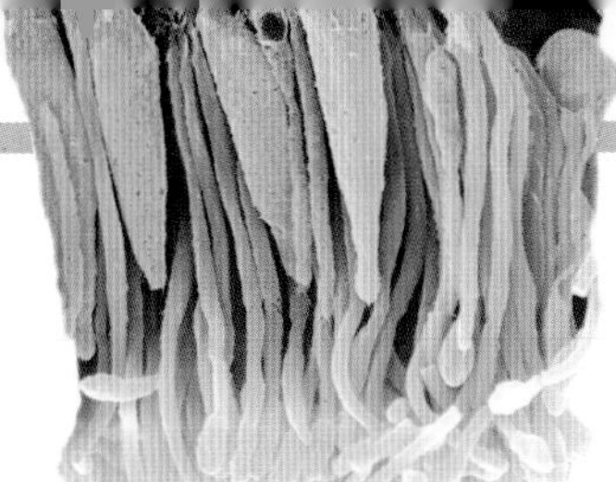

## Rods and Cones

There are two types of photosensitive cells that transform light into electrical impulses. The rods function to "see" only in black and white. The cones are found in the fovea (the part of the retina where the light is focused with the most precision) and allow us to see colors in detail. The impulses of both types of cells pass through the nerve-cell connectors and arrive at the optic nerve.

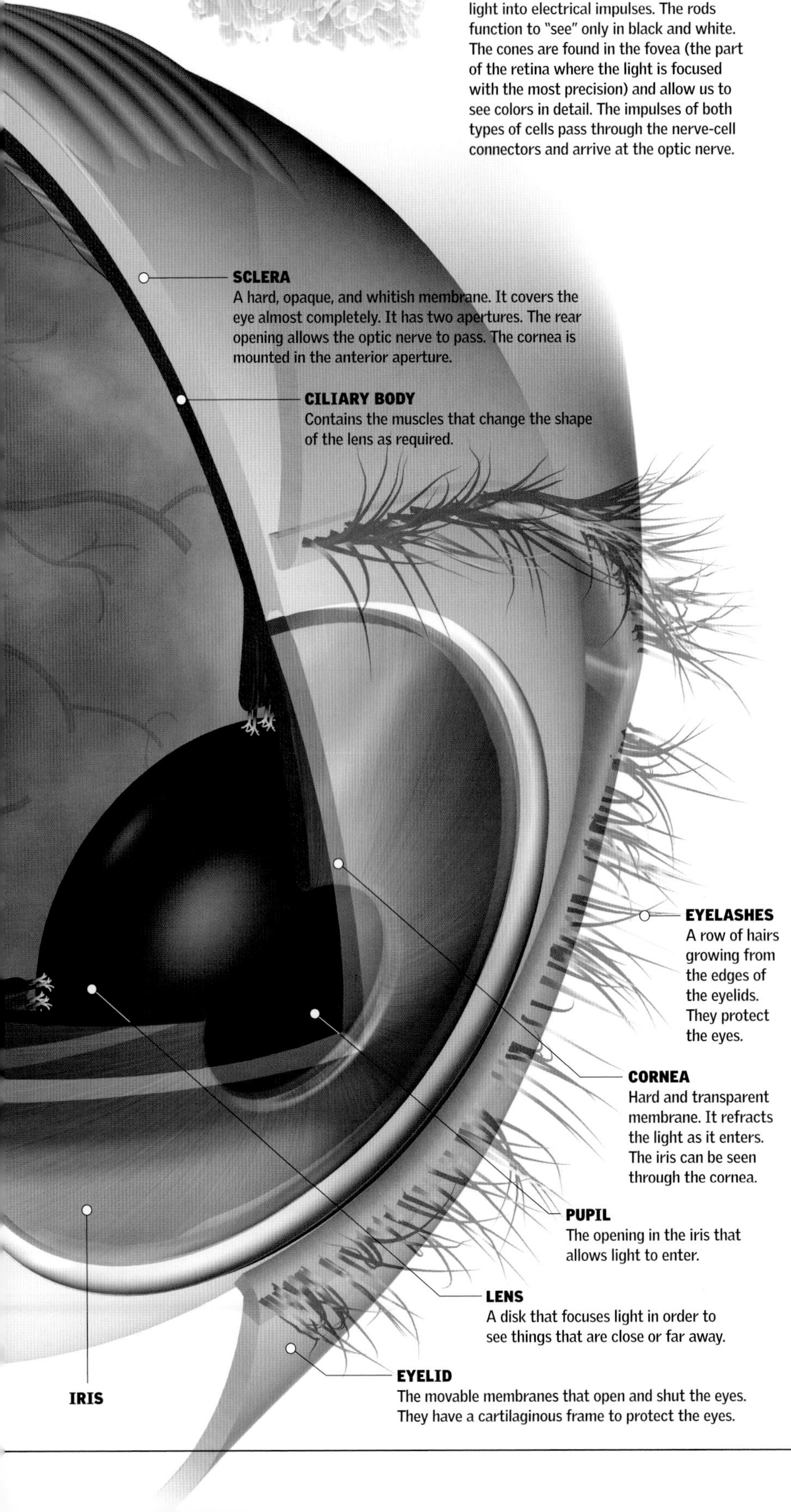

SCLERA
A hard, opaque, and whitish membrane. It covers the eye almost completely. It has two apertures. The rear opening allows the optic nerve to pass. The cornea is mounted in the anterior aperture.
CILIARY BODY
Contains the muscles that change the shape of the lens as required.
EYELASHES
A row of hairs growing from the edges of the eyelids. They protect the eyes.
CORNEA
Hard and transparent membrane. It refracts the light as it enters. The iris can be seen through the cornea.
PUPIL
The opening in the iris that allows light to enter.
LENS
A disk that focuses light in order to see things that are close or far away.
EYELID
The movable membranes that open and shut the eyes. They have a cartilaginous frame to protect the eyes.
IRIS

## Vision Problems

The most common problems involve seeing things out of focus. These are hypermetropia and myopia. Both can be corrected by the use of lenses. A hereditary condition called color blindness, or Daltonism, is less frequent.

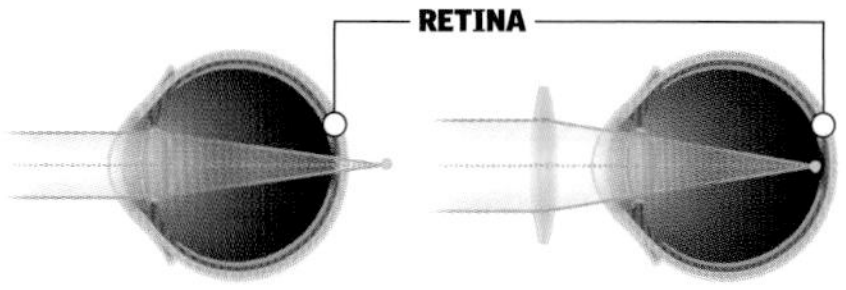


### A Hyperopia (Farsightedness)

This condition makes it difficult to see objects that are close to us. It happens when the image is focused behind the retina. It can be corrected by convex (converging) lenses, which make the rays of light strike the retina properly.

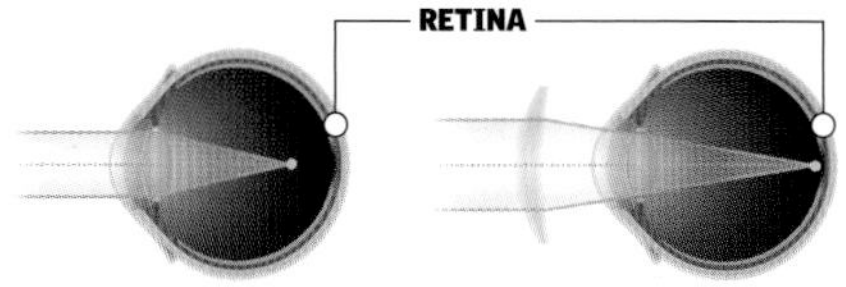


### B Myopia (Nearsightedness)

Here the image is formed in front of the retina. This usually occurs when the ocular sphere is longer than normal. The myopic person has difficulty seeing distant objects. Myopia is corrected with concave (diverging) lenses or by an operation using a laser.

### C Color Blindness

Persons who are color blind have problems distinguishing between certain colors. It is a hereditary illness caused by the absence of the types of cone cells that are sensitive to yellow, green, or blue.

# Protection

**THE EYELIDS PROTECT THE EYES FROM BRIGHT LIGHT AND DUST. THE EYELASHES REDUCE EXCESS LIGHT. THE EYEBROWS KEEP SWEAT OUT OF THE EYES. THE NASOLACHRYMAL DUCT TAKES THE TEARS FROM THE NASAL CAVITY TO THE LACHRYMAL DUCTS—THE OPENINGS AT THE EXTREMITIES OF THE EYES—WHERE THEY ARE SECRETED.**

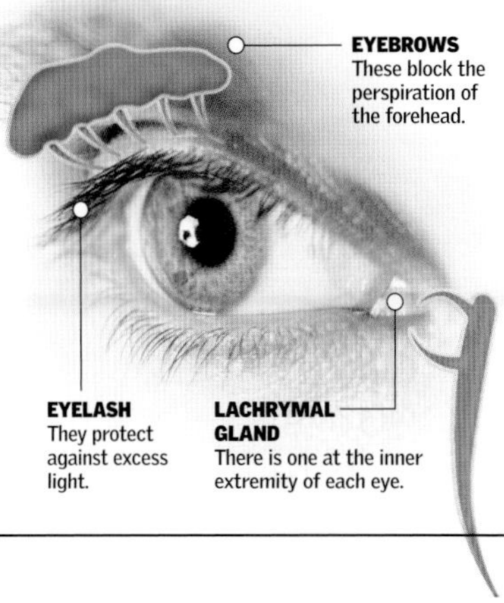

EYEBROWS
These block the perspiration of the forehead.
EYELASH
They protect against excess light.
LACHRYMAL GLAND
There is one at the inner extremity of each eye.

# The Mechanics of Hearing

The ears are the organs responsible for hearing and maintaining equilibrium. When the ears perceive sounds, they register their characteristics—volume, tone, and timbre—as well as the direction from which they come. A group of nerve terminals receives information about the body's motion and transmits this to the brain in order to maintain dynamic and static equilibrium. The ears are important for communication by means of speech or other means, such as music. They are capable of distinguishing a great range of volumes, from the buzzing of a mosquito to the roar of an airplane. The ears contain the smallest bones of the body.

## Frequencies

The frequency of a sound is the speed at which the sound makes the air vibrate. It is measured in units called hertz (Hz): one hertz corresponds to one vibration per second. High frequencies correspond to high sounds, and low frequencies to low sounds. The human ear can hear sounds between 20 and 20,000 vibrations per second.

**FREQUENCIES AUDIBLE TO HUMANS AND ANIMALS**

| Subject | Minimum | Maximum |
|---|---|---|
| **Person 10 years old** | 20 Hz | 20,000 Hz |
| **Person 60 years old** | 20 Hz | 12,000 Hz |
| **Dog** | 60 Hz | 45,000 Hz |
| **Frog** | 100 Hz | 3,000 Hz |
| **Bat** | 1,000 Hz | 120,000 Hz |
| **Cat** | 60 Hz | 65,000 Hz |

## Organ of Corti

Contains ciliary cells that collect vibrations and transform mechanical energy into energy of the nervous system. Next the impulses arrive at the brain via the cochlear nerve. The nerve cells do not have a regenerative capacity, so if they are lost hearing will be lost along with them.

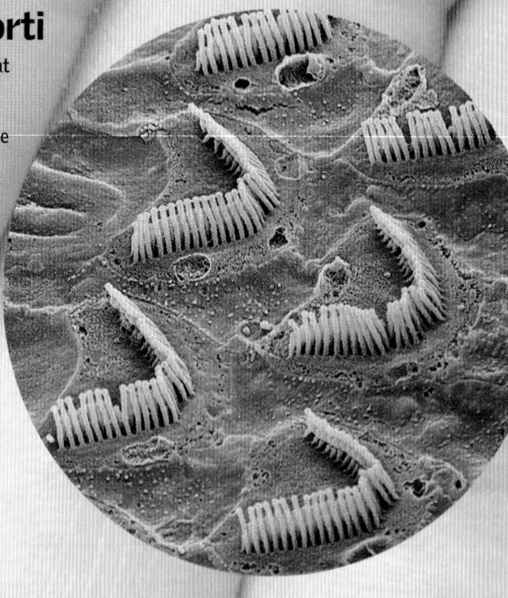

### The Processing of Sound

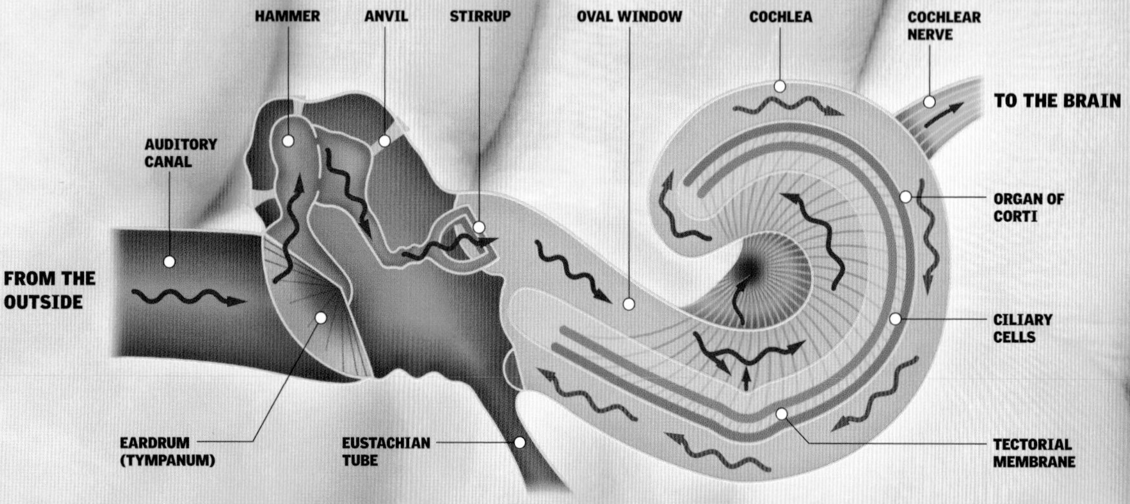


**1 ENTRANCE**
The sound wave is captured by the ear and enters via the auditory canal.

**2 VIBRATION**
The tympanum registers the intensity of the wave.

**3 TRANSMISSION**
The vibration of the eardrum is transmitted to the hammer, from the hammer to the anvil, from the anvil to the stirrup, from the stirrup to the oval window, from there to the cochlea, and from there to the cochlear nerve, whose electrical impulses are transmitted to the brain.

## Equilibrium

Dynamic and static equilibrium are maintained by the inner ear. Above the cochlea there are three semicircular canals, which are spiral-shaped conduits. Inside the canals are a gelatinous membrane and thousands of cilia, or hairlike structures, traversed by a cranial nerve that connects them to the brain. When the head moves, this gelatinous membrane is displaced, and the tiny cilia send the brain information about the velocity and the direction of this displacement. On that basis the body can move as required to maintain equilibrium. Excessive motion produces seasickness, because the cilia continue to move even when the motion stops.

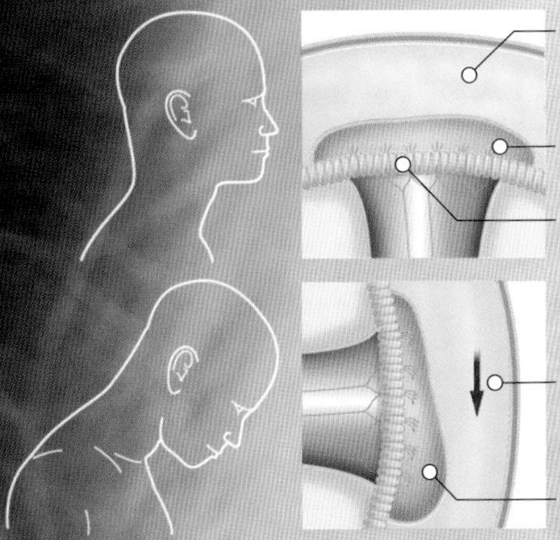


**LINEAR MOTION**
The displacement of the gelatinous membrane, caused by a difference in height, changes the structure of the auditory cilia.

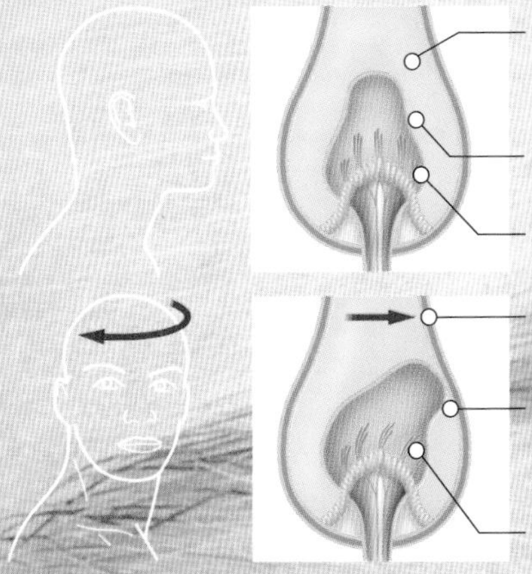


**ROTATIONAL MOTION**
The gelatinous membrane takes on the shape of a dome so that lateral motion will also disturb its equilibrium.

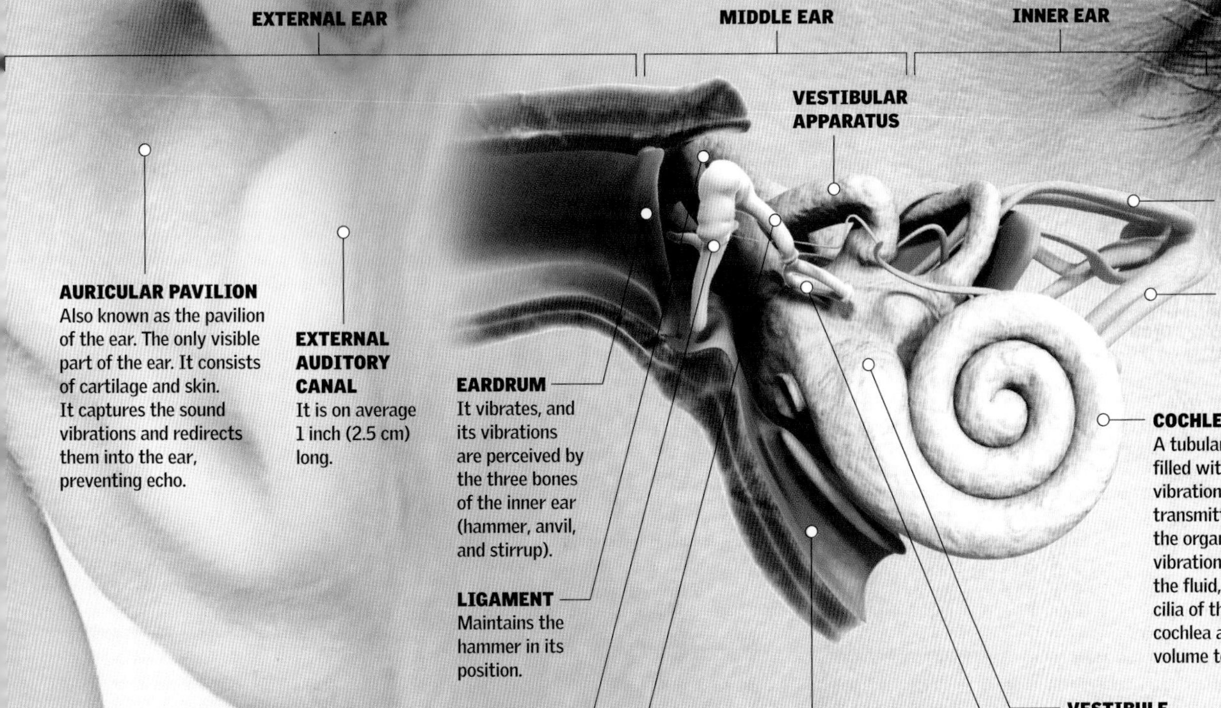


**AURICULAR PAVILION**
Also known as the pavilion of the ear. The only visible part of the ear. It consists of cartilage and skin. It captures the sound vibrations and redirects them into the ear, preventing echo.

**EXTERNAL AUDITORY CANAL**
It is on average 1 inch (2.5 cm) long.

**EARDRUM**
It vibrates, and its vibrations are perceived by the three bones of the inner ear (hammer, anvil, and stirrup).

**LIGAMENT**
Maintains the hammer in its position.

**HAMMER**
Transmits the eardrum's vibrations. It is 0.3 inch (8 mm) long.

**ANVIL**
Receives the hammer's vibrations.

**EUSTACHIAN TUBE**
Connects the middle ear with the back of the nose and the pharynx. It controls the air pressure in the ear, at times through yawning.

**VESTIBULAR NERVE**

**COCHLEAR NERVE**
Brings the nerve impulses of the inner ear to the brain.

**COCHLEA**
A tubular, spiral structure filled with fluid that receives vibrations, which are then transmitted to the brain by the organ of Corti. These vibrations produce waves in the fluid, which stimulate the cilia of the organ of Corti. The cochlea allows differences in volume to be identified.

**VESTIBULE**
Oval window or labyrinth. Encased in the temporal bone, one conduit goes to the cochlea (for hearing), and two go to the semicircular canals (for equilibrium).

**STIRRUP**
Transmits vibrations to the oval window. It is 0.15 inch (4 mm) long.

# Speech and Nonverbal Language

Speaking is the verbal expression of a language and includes articulation, which is the manner in which words are formed. However, one can make oneself understood by means other than the spoken word, such as with signs, facial expressions, or gestures. These are examples of what is called nonverbal communication, whereby even silence can be expressive.

## Language and Speech

Linguists explain that the organs of speech necessary to express language in sounds, which constitute the fundamental elements of speech, are just as independent of language as a telegraph apparatus is of the Morse code it transmits. Linguists also compare language (the verbal system of communication that is almost always written) with a symphony whose score exists independently of the musicians who play it. The vocal cords behave like instruments. They are folds of muscle that open and close to produce sounds. When they are not producing vocal sounds, normal breathing occurs. Under the control of the brain, the vocal cords produce sounds that are modified by the lips and the tongue to create speech.

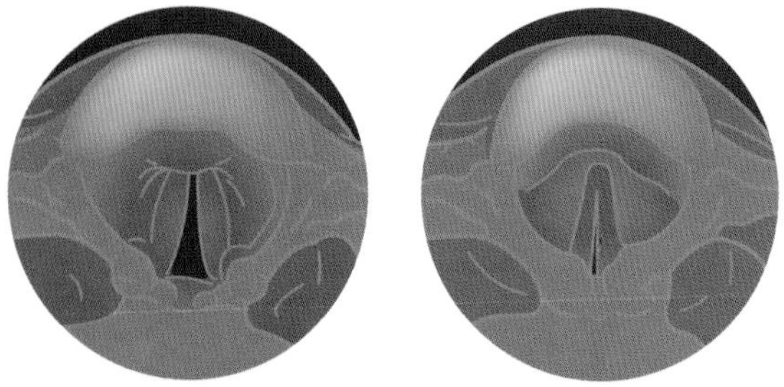

**A PASSAGE OF AIR**
The vocal cords relax and open to allow air to pass to and from the lungs. No sound is produced because the vocal cords do not vibrate, which is the basis for sound.

**B SOUND IS PRODUCED**
The vocal cords stretch horizontally above the larynx. They tighten when air flows past them. Sound is the vibration of the vocal cords.

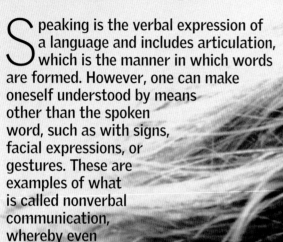

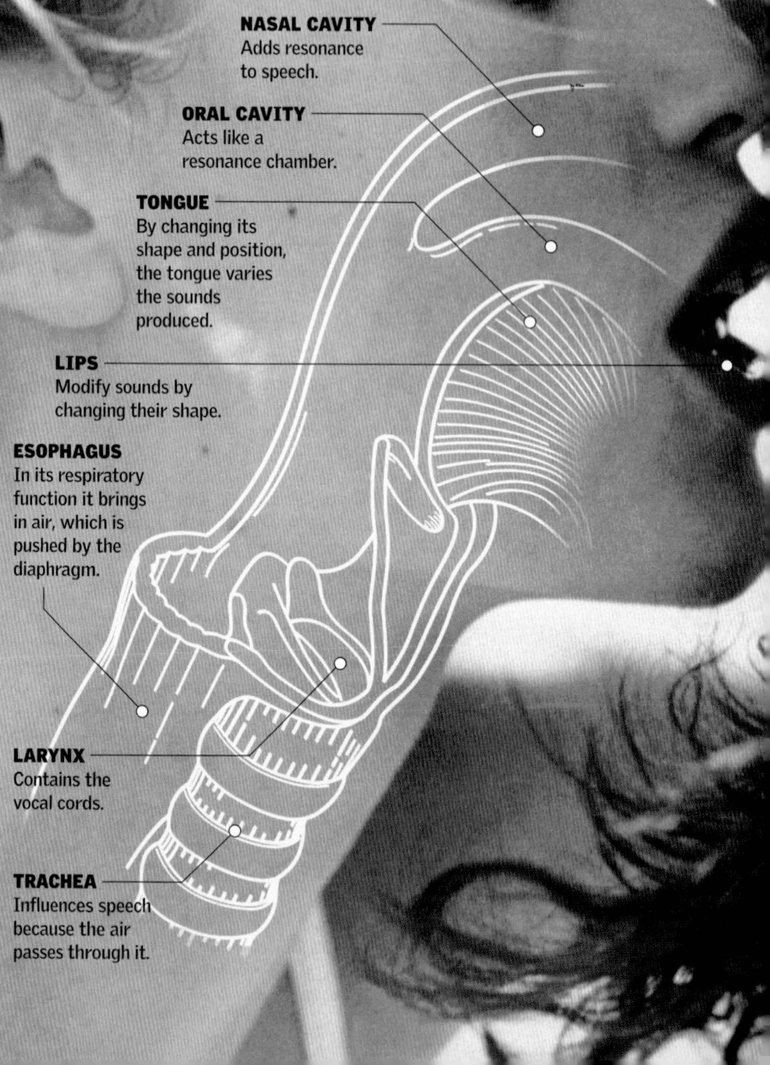

NASAL CAVITY
Adds resonance to speech.
ORAL CAVITY
Acts like a resonance chamber.
TONGUE
By changing its shape and position, the tongue varies the sounds produced.
LIPS
Modify sounds by changing their shape.
ESOPHAGUS
In its respiratory function it brings in air, which is pushed by the diaphragm.
LARYNX
Contains the vocal cords.
TRACHEA
Influences speech because the air passes through it.

# Language of Gesture

The expressivity of the human face is the result of more than 30 muscles that tense small areas of the skin when they contract. Most of them operate in pairs. Their use is reflexive in most cases, as in the gestures, facial expressions, and grimaces that often accompany the spoken word and are silent expressions in certain situations. In other cases, however, such as the art of acting, their use and mastery can be studied and practiced. The usual example of this is the art of mimes, who can stage complete dramas that are transmitted very effectively with no recourse to the spoken word or use of the voice.

**Broca**
Controls the articulation of speech.
**Visual**
Receives and analyzes the nerve impulses from the eye.
**Wernicke**
Controls the comprehension of language.

**FACIAL EXPRESSIONS**
The muscles of the face also serve to communicate feelings.

**FROWNING**
Action of the corrugator muscles on the eyebrows.

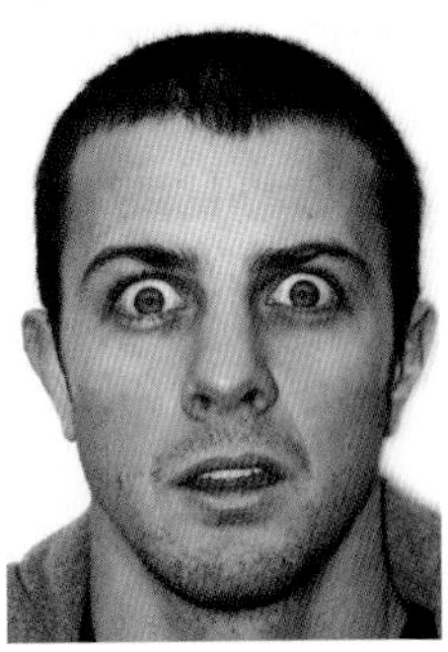

**SURPRISE**
The muscles of the forehead are contracted.

**SMILE**
Action of the smile muscles and the zygomaticus major.

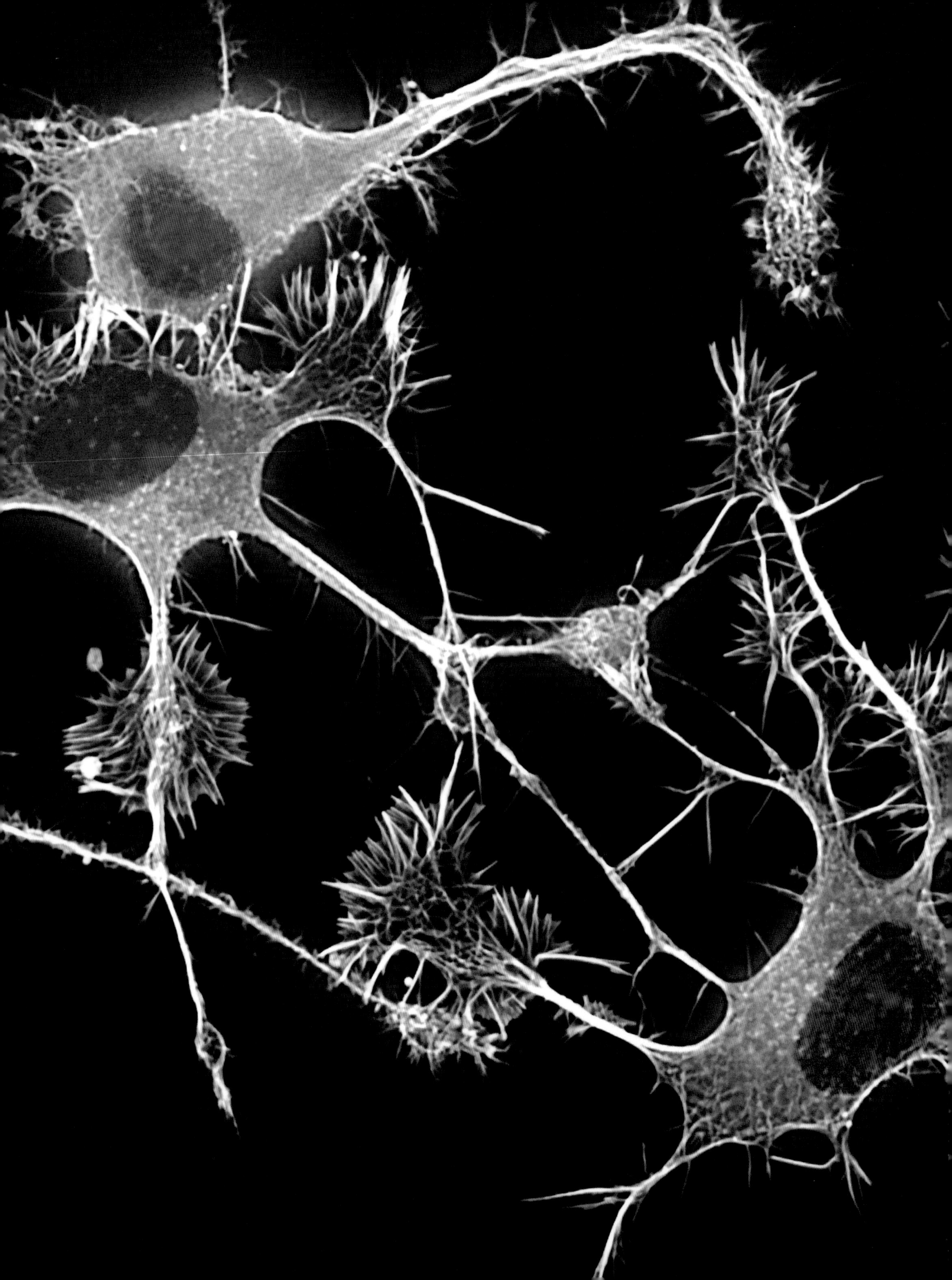

# Control Centers

Brain tissue consists of thousands of millions of neurons that continually send each other signals through connections called synapses. Thanks to this network the brain can remember, calculate, decide, think, dream, create, and express emotions. We invite you to understand the secrets about how these activities of the brain are accomplished. What determines the formation of synapses and neuronal networks? Where are intelligence and memory located? Is it possible to stimulate brain cells? What happens during a dream? What are nerves, and how are they formed? What functions are carried out by each region of the brain? You will find all this and much more in this chapter, including incredible images.



**NERVE CELLS** (opposite)
Microscope photograph of a group of neurons.

# Nervous System

The body's most complex system, many of whose characteristics and potentialities are still unknown. Together with the endocrine system, the brain has the job of controlling the organism. Its specific functions are rapid and intellectual activities, such as memory, emotions, and will. The brain is divided into three portions: the central (the brain and the spinal cord), the peripheral (nerves of the spinal cord and cranium), and the vegetative (or autonomic function).

## The Great Coordinator

The nervous system acts as the great coordinator of the functions of all the parts and organs of the body. In simpler organisms, such as unicellular organisms, the same cell receives sensations and responds to them without requiring intermediation or specialized coordination. However, in more complex organisms such as the human body, the cells of the different parts of the body are differentiated, as are the functions of the organs that these cells make up. Thus there are receptor cells, which receive stimuli (such as the cells of the organs linked to the eye or the senses). There are also effector cells (such as those of the muscles or the glands), which are involved in the organism's responses. The nervous system links these functions together through its three principal parts: the brain, the spinal cord, and the nerves in general. The nerves consist of numerous axons and dendrites, enveloped by a sheath of conjunctive tissue. These groups of neurons are called ganglia when they are outside the brain and the spinal cord, and they are called nuclei when they are inside.

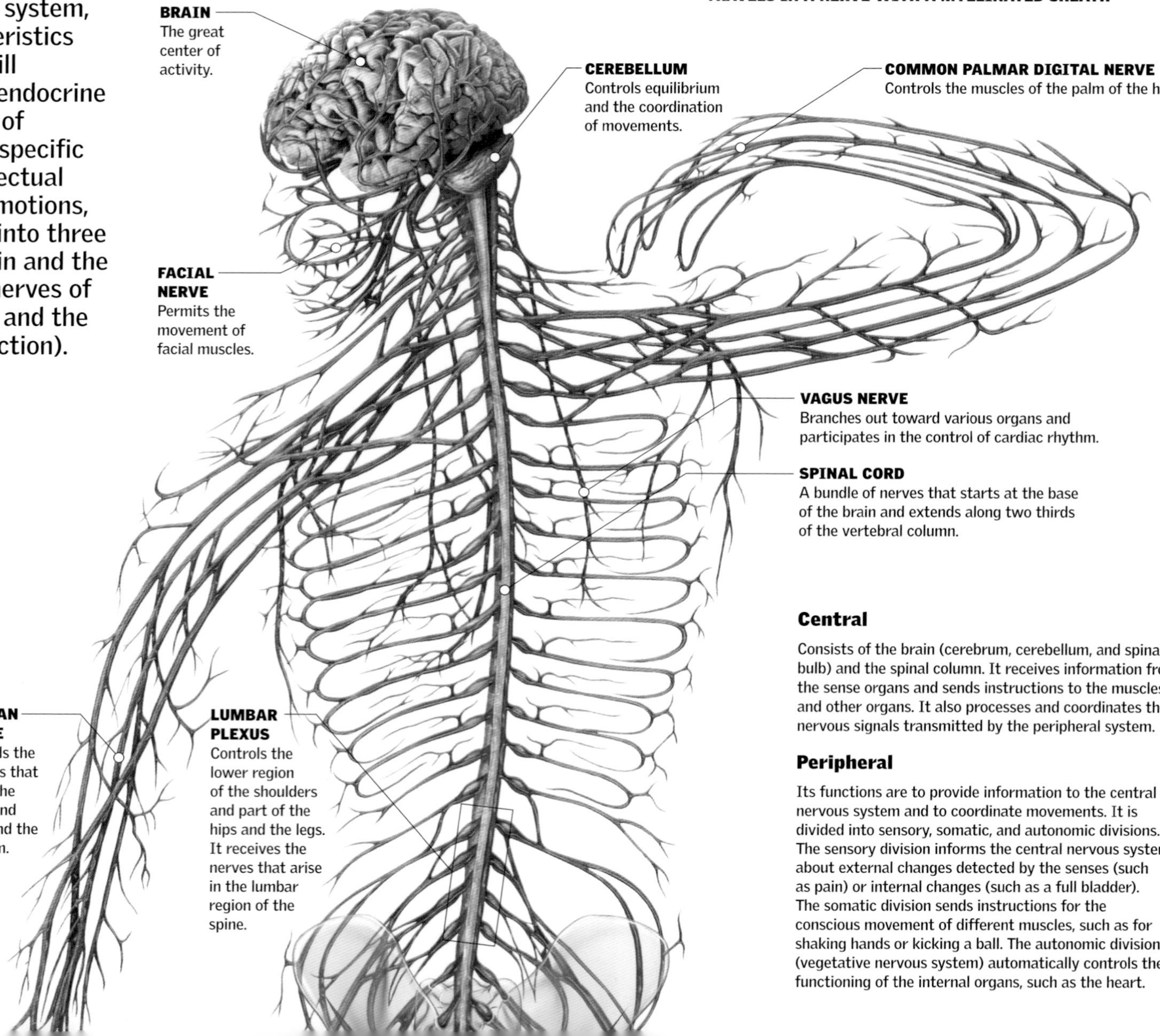

**BRAIN** The great center of activity.

**FACIAL NERVE** Permits the movement of facial muscles.

**MEDIAN NERVE** Controls the muscles that cover the wrist and surround the forearm.

**LUMBAR PLEXUS** Controls the lower region of the shoulders and part of the hips and the legs. It receives the nerves that arise in the lumbar region of the spine.

**CEREBELLUM** Controls equilibrium and the coordination of movements.

**COMMON PALMAR DIGITAL NERVE** Controls the muscles of the palm of the hand.

**VAGUS NERVE** Branches out toward various organs and participates in the control of cardiac rhythm.

**SPINAL CORD** A bundle of nerves that starts at the base of the brain and extends along two thirds of the vertebral column.

## 300 feet (90 m) per second

**THE SPEED AT WHICH IT IS CALCULATED THAT A NERVE IMPULSE TRAVELS IN A NERVE WITH A MYELINATED SHEATH**

### Central

Consists of the brain (cerebrum, cerebellum, and spinal bulb) and the spinal column. It receives information from the sense organs and sends instructions to the muscles and other organs. It also processes and coordinates the nervous signals transmitted by the peripheral system.

### Peripheral

Its functions are to provide information to the central nervous system and to coordinate movements. It is divided into sensory, somatic, and autonomic divisions. The sensory division informs the central nervous system about external changes detected by the senses (such as pain) or internal changes (such as a full bladder). The somatic division sends instructions for the conscious movement of different muscles, such as for shaking hands or kicking a ball. The autonomic division (vegetative nervous system) automatically controls the functioning of the internal organs, such as the heart.

## Structure of a Nerve

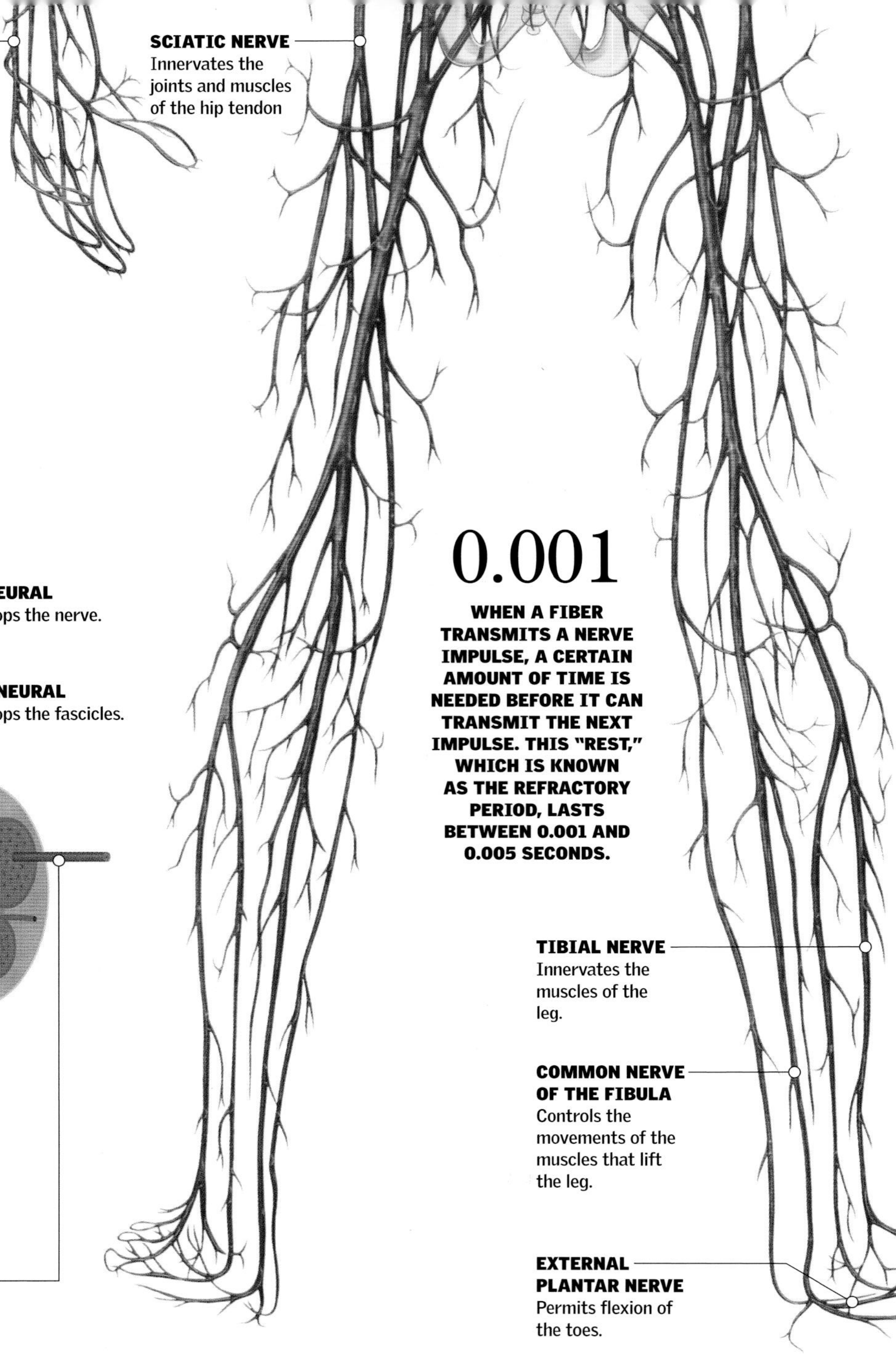


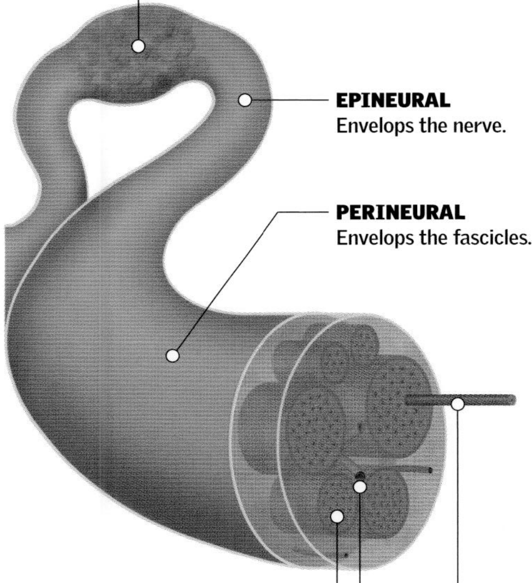


**0.001**

**WHEN A FIBER TRANSMITS A NERVE IMPULSE, A CERTAIN AMOUNT OF TIME IS NEEDED BEFORE IT CAN TRANSMIT THE NEXT IMPULSE. THIS "REST," WHICH IS KNOWN AS THE REFRACTORY PERIOD, LASTS BETWEEN 0.001 AND 0.005 SECONDS.**

### Pain and the Nervous Response

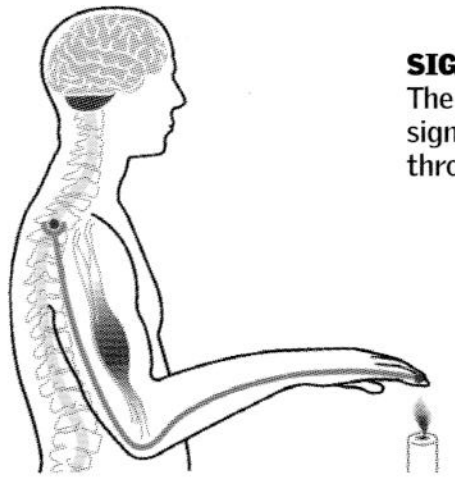

SIGNALS
The nerve endings receive pain signals that arrive at the brain through the spinal column.

**1** The reflex action of withdrawing the hand or another part of the body from an object that may cause pain (for example, by being pricked or subjected to heat) is an automatic response. Thus the pain receptors in the skin of the fingers detect the heat stimulus from a flame and send nerve impulses via the sensory nerves to the spinal cord. The impulses move at high speed through the medulla along pathways of associated neurons.

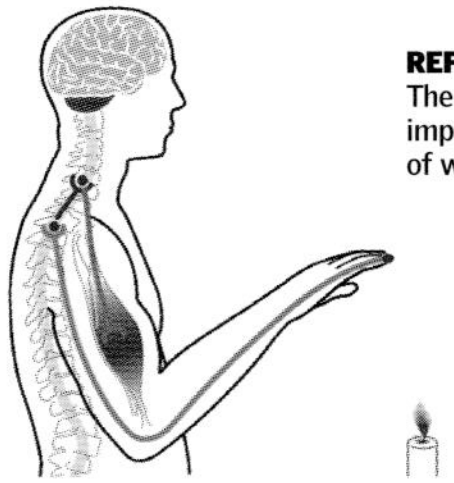

REFLEX
The retransmission of nerve impulses provokes the reflex of withdrawing the hand.

**2** Within thousandths of a second after detecting the pain stimulus, the nerve impulses reach the motor neurons. These neurons transmit the impulses to the flexor muscles in the upper part of the arm. Once the impulses have been received, the muscles contract, the arm bends, and the fingers move away from the flame before any pain is consciously felt.

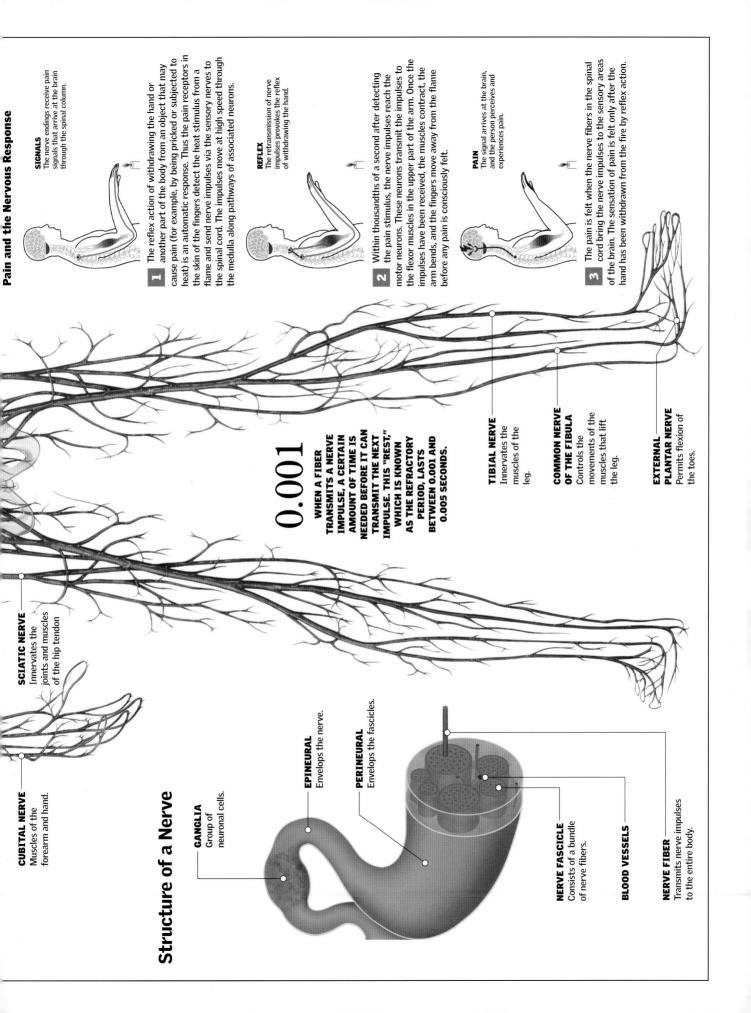

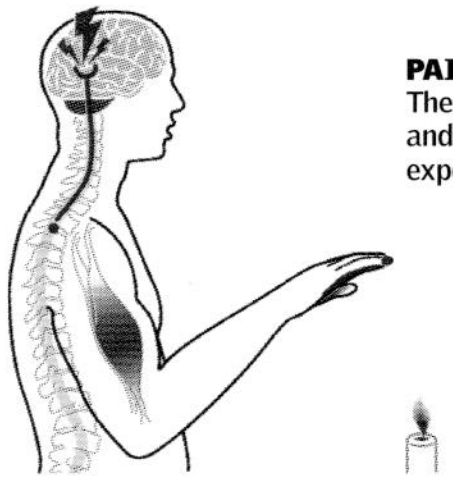

PAIN
The signal arrives at the brain, and the person perceives and experiences pain.

**3** The pain is felt when the nerve fibers in the spinal cord bring the nerve impulses to the sensory areas of the brain. The sensation of pain is felt only after the hand has been withdrawn from the fire by reflex action.

# Neurons

Neurons are cells that make up the nervous system. Their function is to transmit impulses in the form of electrical signals carrying information to the brain and from there to the periphery. The neurons provide the basis for the system's activities and form a highly complex communication network. They are surrounded and protected by other nerve cells that are not excitable, called glial cells, which constitute more than half of all an organism's nerve cells.

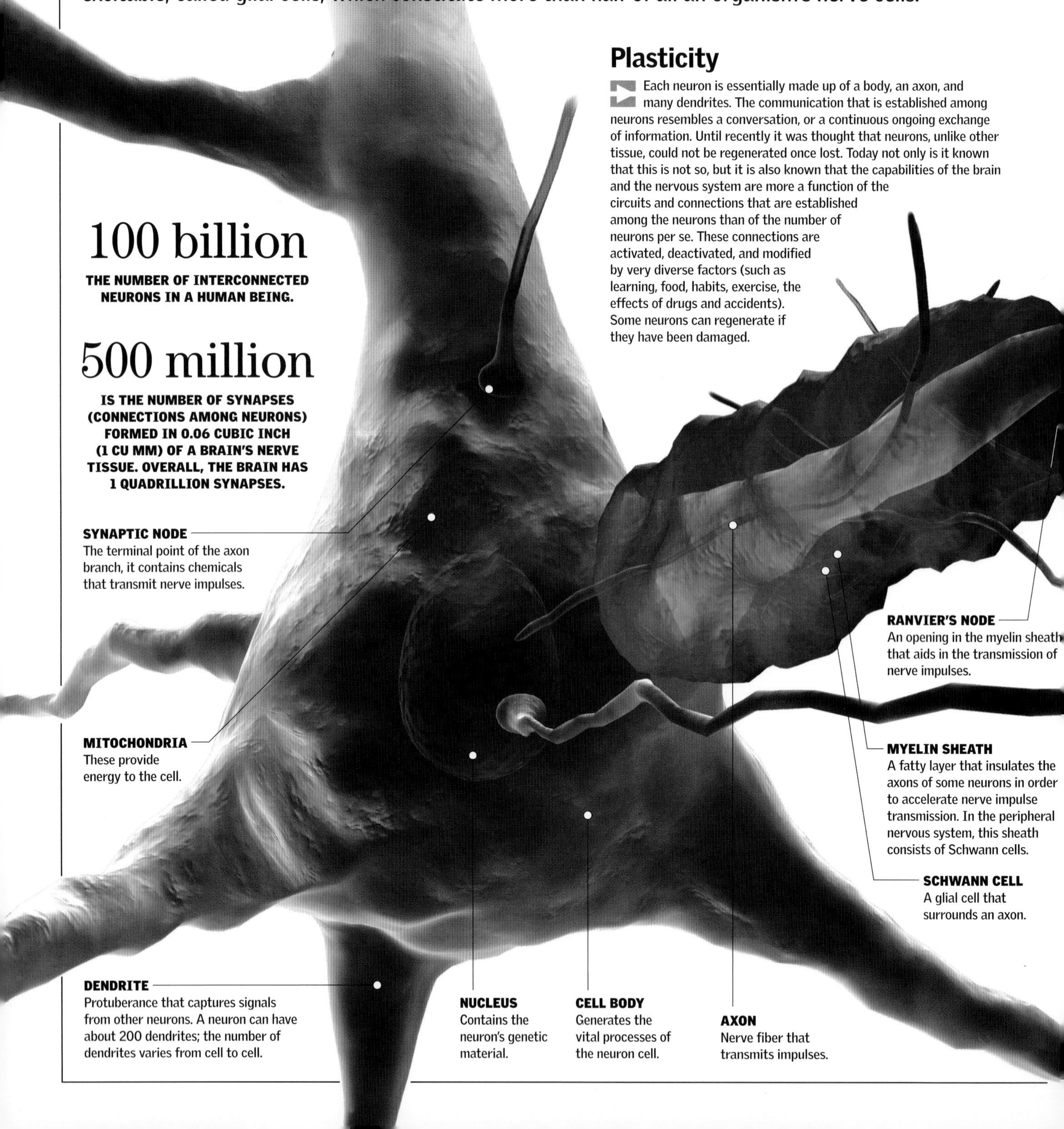

## Plasticity

Each neuron is essentially made up of a body, an axon, and many dendrites. The communication that is established among neurons resembles a conversation, or a continuous ongoing exchange of information. Until recently it was thought that neurons, unlike other tissue, could not be regenerated once lost. Today not only is it known that this is not so, but it is also known that the capabilities of the brain and the nervous system are more a function of the circuits and connections that are established among the neurons than of the number of neurons per se. These connections are activated, deactivated, and modified by very diverse factors (such as learning, food, habits, exercise, the effects of drugs and accidents). Some neurons can regenerate if they have been damaged.

**100 billion**

**THE NUMBER OF INTERCONNECTED NEURONS IN A HUMAN BEING.**

**500 million**

**IS THE NUMBER OF SYNAPSES (CONNECTIONS AMONG NEURONS) FORMED IN 0.06 CUBIC INCH (1 CU MM) OF A BRAIN'S NERVE TISSUE. OVERALL, THE BRAIN HAS 1 QUADRILLION SYNAPSES.**

**SYNAPTIC NODE**
The terminal point of the axon branch, it contains chemicals that transmit nerve impulses.

**MITOCHONDRIA**
These provide energy to the cell.

**DENDRITE**
Protuberance that captures signals from other neurons. A neuron can have about 200 dendrites; the number of dendrites varies from cell to cell.

**NUCLEUS**
Contains the neuron's genetic material.

**CELL BODY**
Generates the vital processes of the neuron cell.

**AXON**
Nerve fiber that transmits impulses.

**RANVIER'S NODE**
An opening in the myelin sheath that aids in the transmission of nerve impulses.

**MYELIN SHEATH**
A fatty layer that insulates the axons of some neurons in order to accelerate nerve impulse transmission. In the peripheral nervous system, this sheath consists of Schwann cells.

**SCHWANN CELL**
A glial cell that surrounds an axon.

## Transmission and Synapses

The synapse is the point of communication between neurons. It comprises a synaptic cleft, a synaptic knob, and a target to which the nerve signal is directed. In order for a neuron to be activated, there must be a stimulus that converts the electrical charge inside the membrane of the cell from negative to positive. The nerve impulse travels via the axon toward the synaptic knob and brings about the release of chemical substances called neurotransmitters. These in turn can elicit a response from the target to which the stimulus is directed.

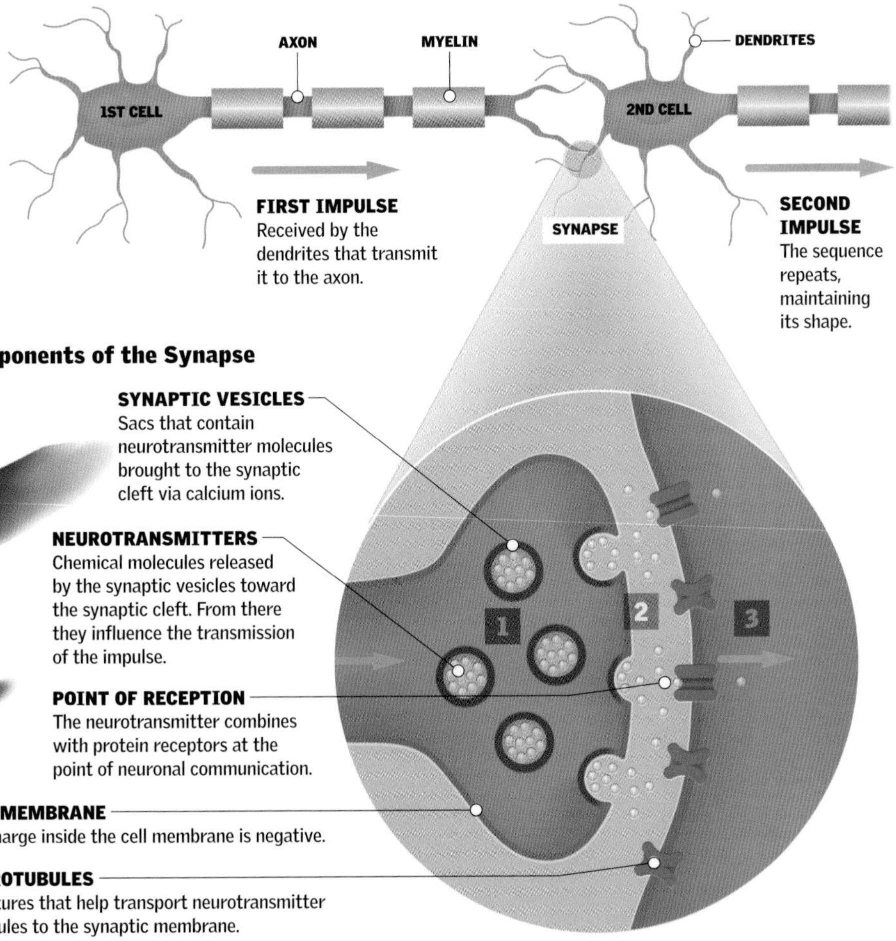


**FIRST IMPULSE** Received by the dendrites that transmit it to the axon.

**SECOND IMPULSE** The sequence repeats, maintaining its shape.

### Components of the Synapse

**SYNAPTIC VESICLES** Sacs that contain neurotransmitter molecules brought to the synaptic cleft via calcium ions.

**NEUROTRANSMITTERS** Chemical molecules released by the synaptic vesicles toward the synaptic cleft. From there they influence the transmission of the impulse.

**POINT OF RECEPTION** The neurotransmitter combines with protein receptors at the point of neuronal communication.

**CELL MEMBRANE** The charge inside the cell membrane is negative.

**MICROTUBULES** Structures that help transport neurotransmitter molecules to the synaptic membrane.

### Transmission of Nerve Impulses

**1 WITHOUT INFORMATION** When the neuron is at rest, the sodium ions inside it are uniformly distributed so that the electrical charge inside the cell membrane is permanently negative.

**2 THE IMPULSE ARRIVES** The arrival of the neurotransmissions at the dendrites causes a reversal of the charge, which becomes positive in this area, giving it a tendency to move in the direction of the negatively charged part of the cell.

**3 TRANSMISSION OF INFORMATION** The positive charge travels toward the negatively charged axon until it reaches the synapse and thus the other cell. The areas it has left return to their stable (negative) state.

### Types of Neurons According to Their Complexity

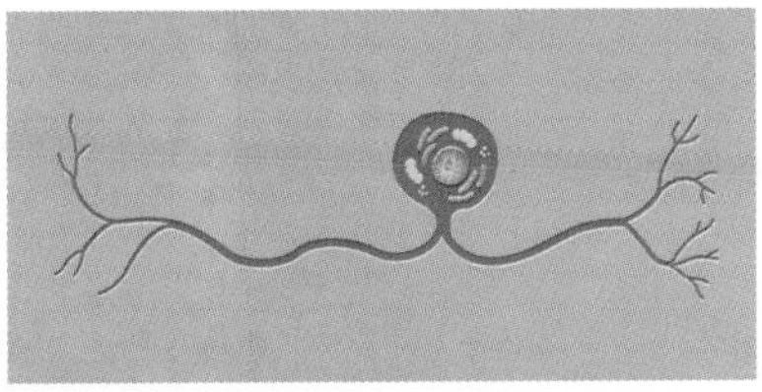

**UNIPOLAR**. Two branches of the same axon extend from one cell body.

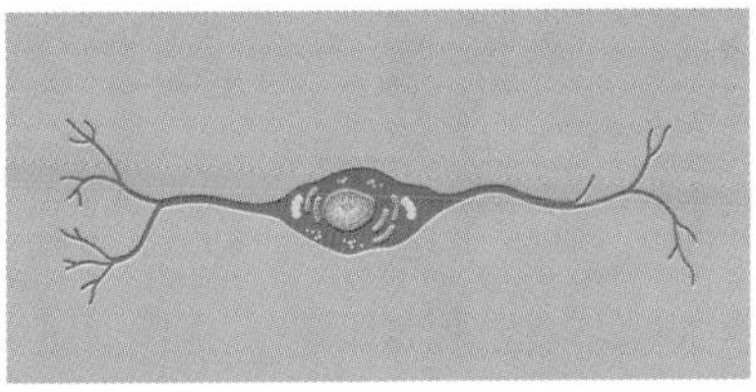

**BIPOLAR**. Two separate axons extend from each end of a cell body.

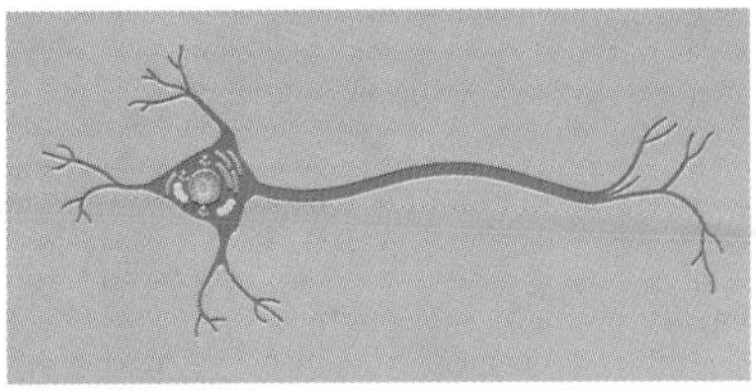

**MULTIPOLAR**. One axon and a number of dendrites extend from a cell body.

## Neuromuscular Union

This is a special kind of synapse between the neurons and the skeletal muscle fibers that causes voluntary contraction of the muscles.

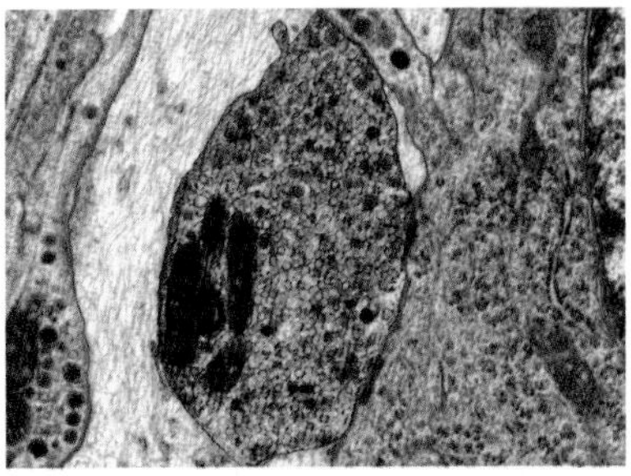

The axon of a neuron links itself with a muscle fiber. At the point of contact a chemical synapse is produced between the neuron and an effector, a muscle with electrically excitable tissue, and movement results.

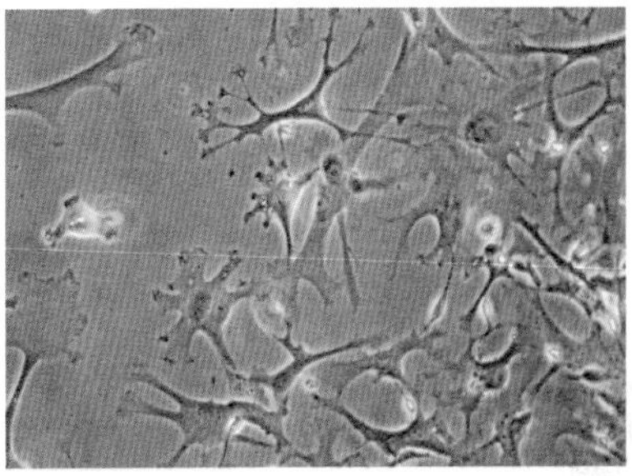

**ASTROCYTES** These are cells that are located in cerebral tissue, where they exceed neurons in number. Astrocytes have some delicate protuberances that are linked to the blood vessels and that regulate the flow of nutrients and waste between neurons and blood.

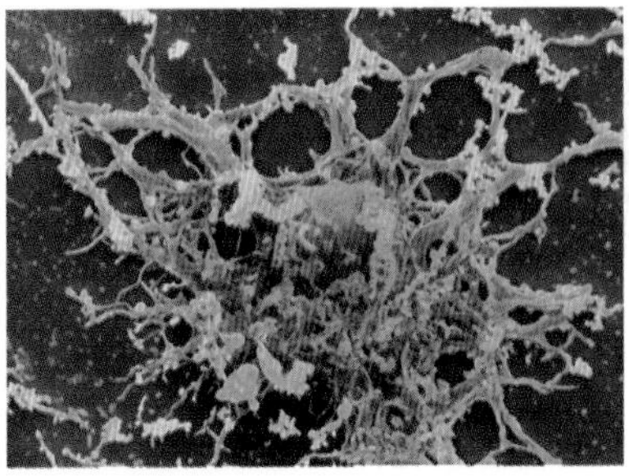

**OLIGODENDROCYTES** These are the cells that form the myelin sheath around the nerve fibers of the brain and the spinal column. Their function is similar to that of Schwann cells in the peripheral nervous system.

# The Brain

The brain is the body's control center. Underneath its folds more than 100 billion neurons organize and examine incoming information and act as a guide for the organism. In spite of amounting to only 2 percent of the total weight of a human body, the brain alone uses one fifth of the oxygen inhaled. It is one of the most fragile parts of the body and, therefore, one of the most protected. Along with the spinal cord, the brain forms the central nervous system, which gives instructions to the peripheral nervous system.

## 3 pounds (1.4 kg)

**AVERAGE WEIGHT OF AN ADULT BRAIN. AT BIRTH THE BRAIN WEIGHS BETWEEN 12 AND 14 OUNCES (350 AND 400 G).**

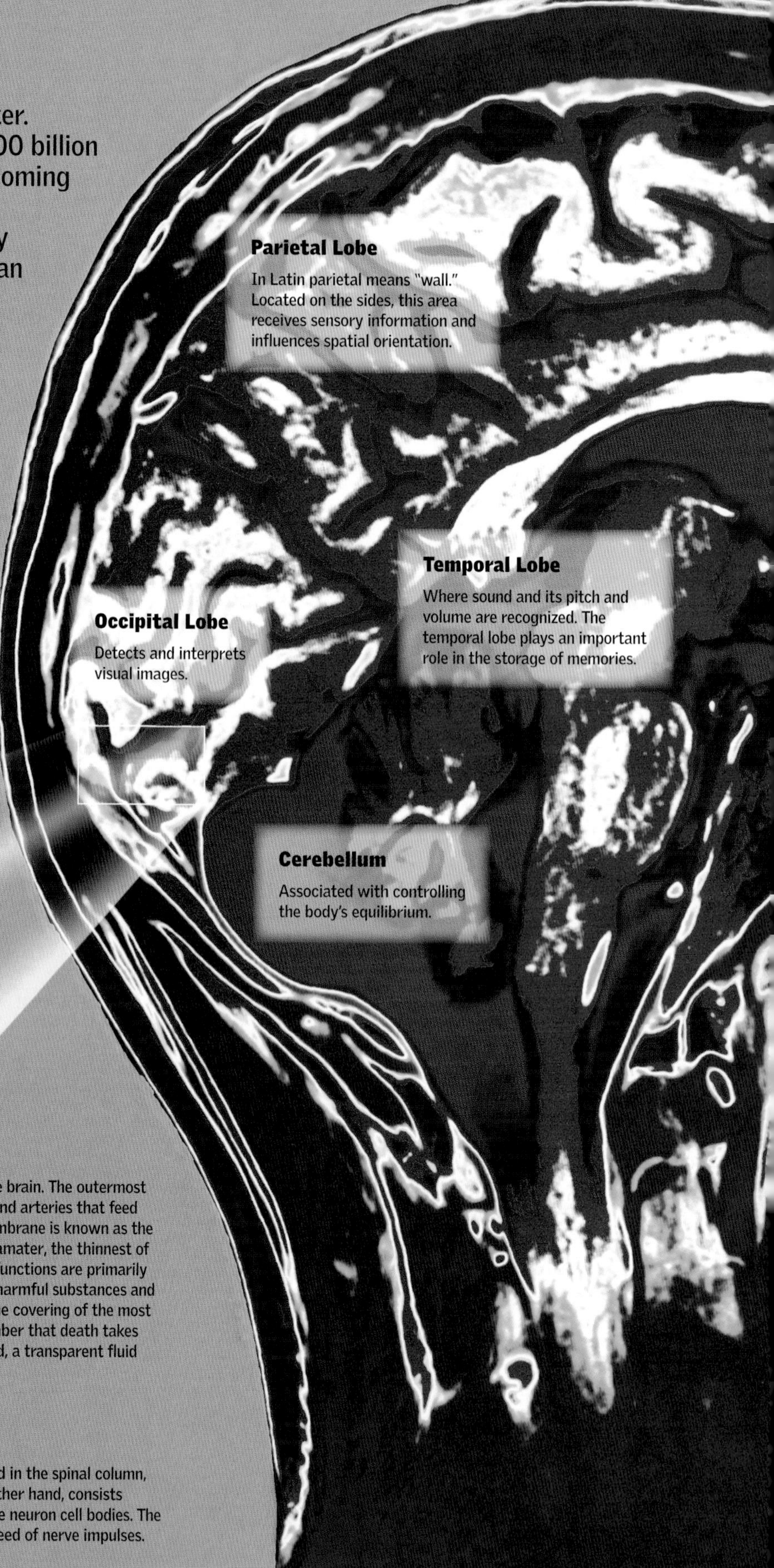

**Parietal Lobe**

In Latin parietal means "wall." Located on the sides, this area receives sensory information and influences spatial orientation.

**Temporal Lobe**

Where sound and its pitch and volume are recognized. The temporal lobe plays an important role in the storage of memories.

**Occipital Lobe**

Detects and interprets visual images.

**Cerebellum**

Associated with controlling the body's equilibrium.

**Meninges**

Protective membranes covering the brain.

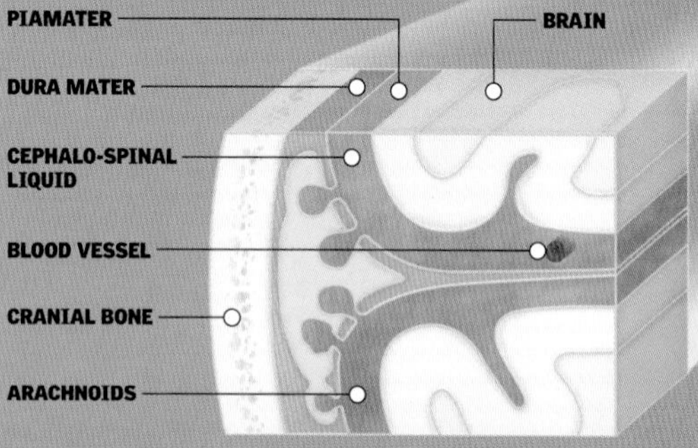


## Meninges

There are three membranes, called meninges, that cover the brain. The outermost one covers the inside of the cranium, and it contains veins and arteries that feed blood to the cranial bones. It is called dura mater. The middle membrane is known as the arachnoid and consists of netlike elastic connective tissue. The piamater, the thinnest of the three, is the closest to the surface of the cerebral cortex. Its functions are primarily protective. On one hand it acts as a filter to prevent the entry of harmful substances and microorganisms into the nervous system. On the other hand, as the covering of the most important organ of the body, it acts like an elastic helmet (remember that death takes place when the brain ceases to function). The cephalo-spinal liquid, a transparent fluid that acts like a shock absorber, circulates within the meninges.

## Gray and White Matter

The so-called gray matter, located in the cerebral cortex and in the spinal column, consists of groups of neuronal cells. White matter, on the other hand, consists primarily of myelin-sheathed axons or nerves that extend from the neuron cell bodies. The fatty layers of myelin allow for an increase in the transmission speed of nerve impulses.

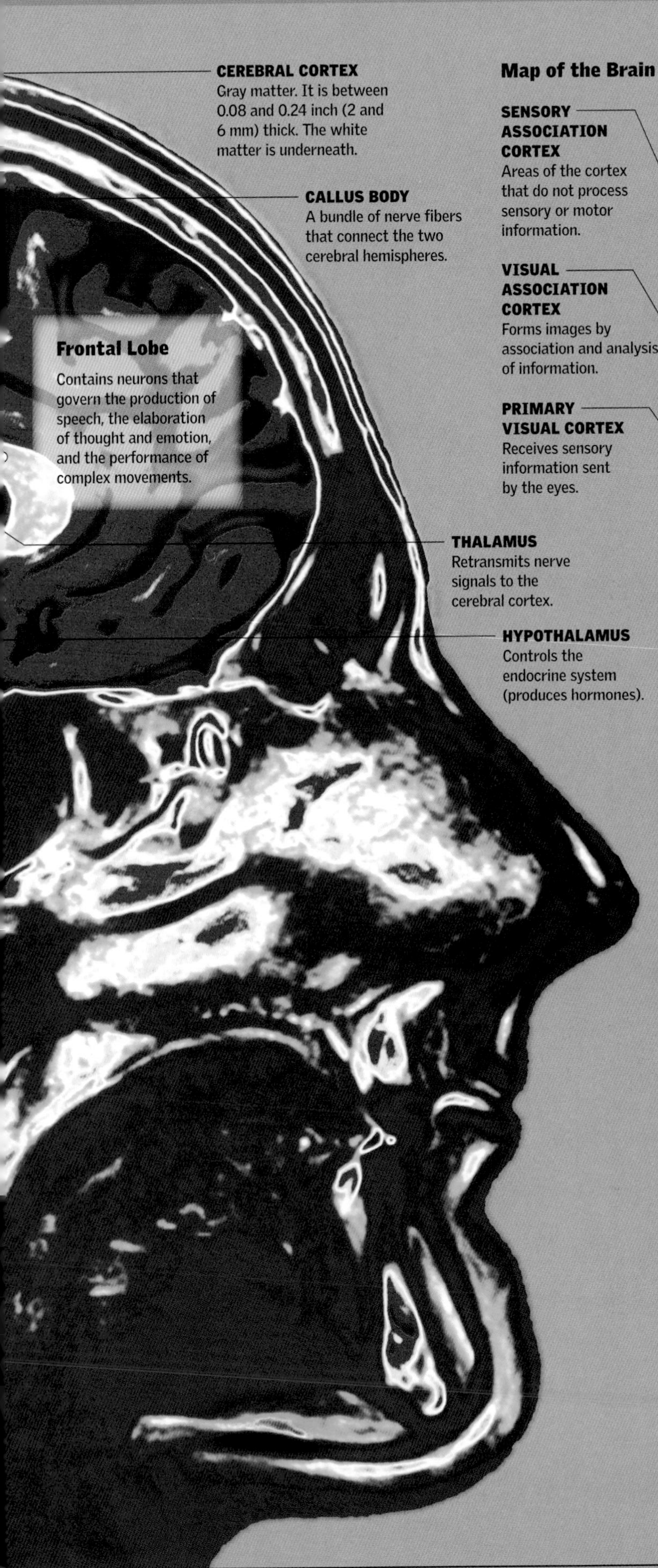

**CEREBRAL CORTEX**
Gray matter. It is between 0.08 and 0.24 inch (2 and 6 mm) thick. The white matter is underneath.

**CALLUS BODY**
A bundle of nerve fibers that connect the two cerebral hemispheres.

### Frontal Lobe

Contains neurons that govern the production of speech, the elaboration of thought and emotion, and the performance of complex movements.

**THALAMUS**
Retransmits nerve signals to the cerebral cortex.

**HYPOTHALAMUS**
Controls the endocrine system (produces hormones).

## Map of the Brain

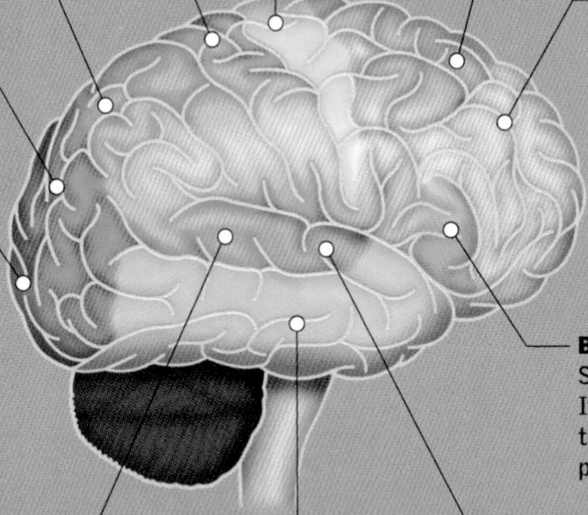

**SENSORY ASSOCIATION CORTEX**
Areas of the cortex that do not process sensory or motor information.

**PRIMARY SENSORY CORTEX**
Receives signals from the sensory receptors in the skin.

**MOTOR CORTEX**
Sends instructions to the muscles telling them to contract.

**PRE-MOTOR CORTEX**
Coordinates complex movements of the muscle motor area.

**VISUAL ASSOCIATION CORTEX**
Forms images by association and analysis of information.

**PRE-FRONTAL CORTEX**
Promotes the development of reasoning and planning (area of association and analysis of information).

**PRIMARY VISUAL CORTEX**
Receives sensory information sent by the eyes.

**BROCA'S AREA**
Speech production. It is a motor area that commands the phonation muscles.

**WERNICKE'S AREA**
Linguistic area for auditory decoding.

**ASSOCIATIVE AUDITORY CORTEX**
Area for association and analysis of sounds.

**PRIMARY AUDITORY CORTEX**
A sensory area. It receives information from the sensory receptors of the eyes.

# Spinal Medulla

The spinal medulla is the spinal cord, which goes from the cephalic trunk to the lumbar region. Together with the brain it forms the central nervous system. It can reach a length of 18 inches (45 cm). It is composed of gray and white matter. The gray matter is located in its core, in tissue consisting essentially of neurons. Surrounding the gray matter is white matter that contains the nerve fibers that transmit signals to and from the brain. The spinal nerves extend outward from the medulla to the body and its extremities. Paralysis in one or more parts of the body can result if the spinal cord is damaged.

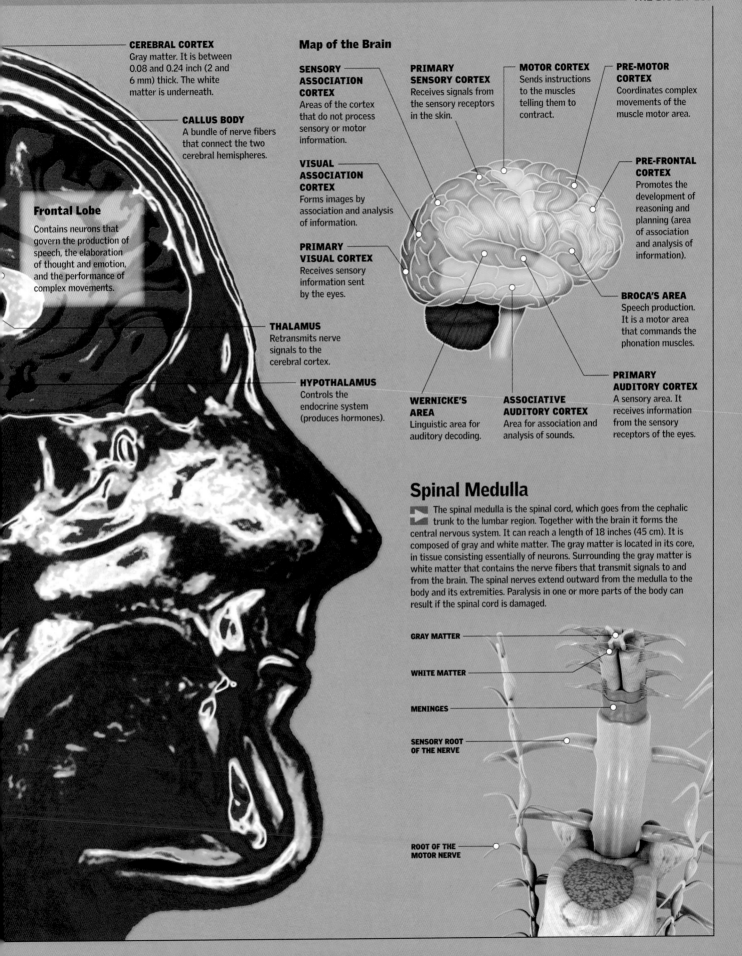

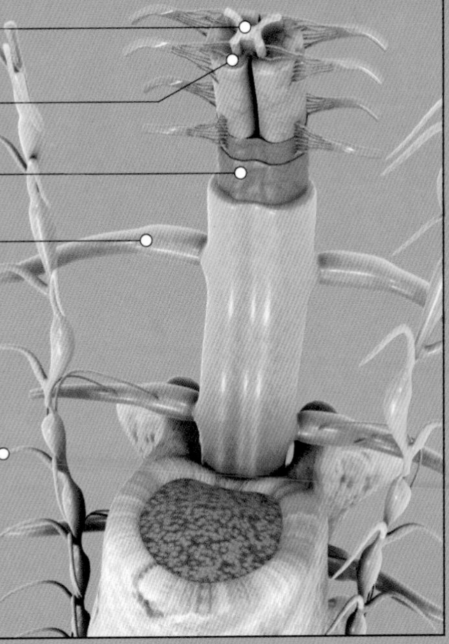

# The Peripheral Nerves

The peripheral nerves have the task of bringing information to and from the brain and spinal column. Depending on their location, they may be cranial or spinal nerves. The sensory fibers in the peripheral nerves receive information from the outside world, the skin, and the internal organs and transmit it to the central nervous system; the motor fibers begin to contract the skeletal muscles and transmit signals in the opposite direction from the sensors. The nerves are located deep in the body, with some exceptions, such as the cubital nerve in the elbow.

## Cranial Nerves

The 12 pairs of cranial nerves extend from the lower part of the brain, as can be seen in the main illustration. Except for the vagus nerve, the cranial nerves control the muscles of the head in the neck region or bring nerve impulses from sense organs, such as the eyes, to the brain. In the case of nerve impulses that come from the eyes, it is the pair of optical nerves that record the sensations from the retina of the eye. The olfactory nerve works the same way for the nose.

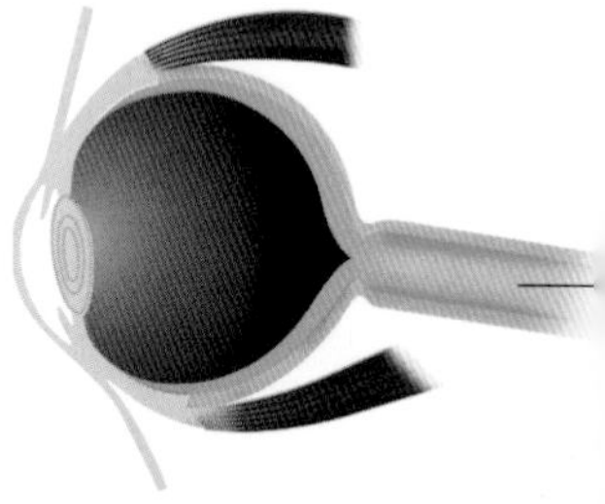

**PAIR II**
Optic nerve. Supplies the retina. Transmits signals, from the photo receptors, perceived as vision.

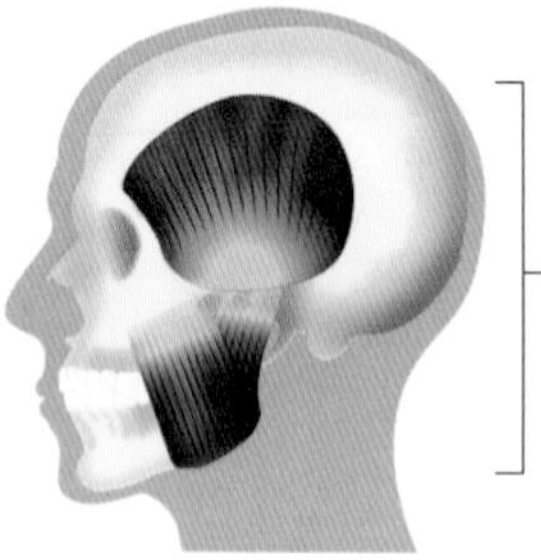

**PAIR V**
Trigeminal nerve. Controls the muscles involved in chewing and transmits sensory information from the eyes, the teeth, and the side of the face.

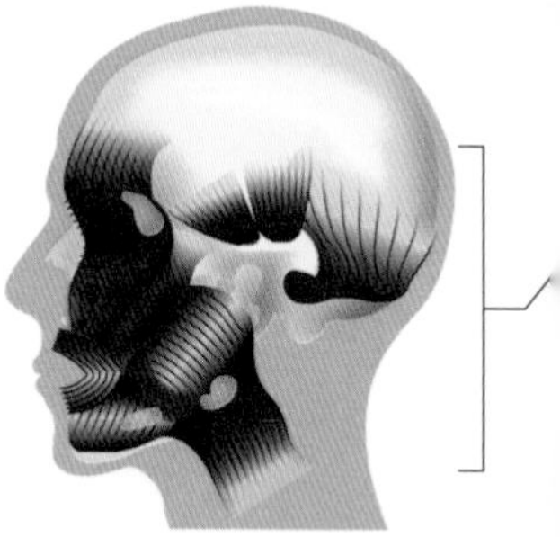

**PAIR VII**
Facial nerve. Controls the muscles of facial expressions and the salivary and tear glands. Transmits sensory information from the taste buds.

## Spinal Nerves

There are 31 pairs of spinal nerves that begin at the spinal cord and extend through the spaces between the vertebrae. Each nerve is divided into numerous branches. These nerves control most of the body's skeletal muscles, as well as the smooth muscles and the glands. The cervical nerves serve the muscles of the chest and shoulders. The lumbar nerves serve the abdomen and part of the legs, and the sacral nerves control the rest of the legs and the feet.

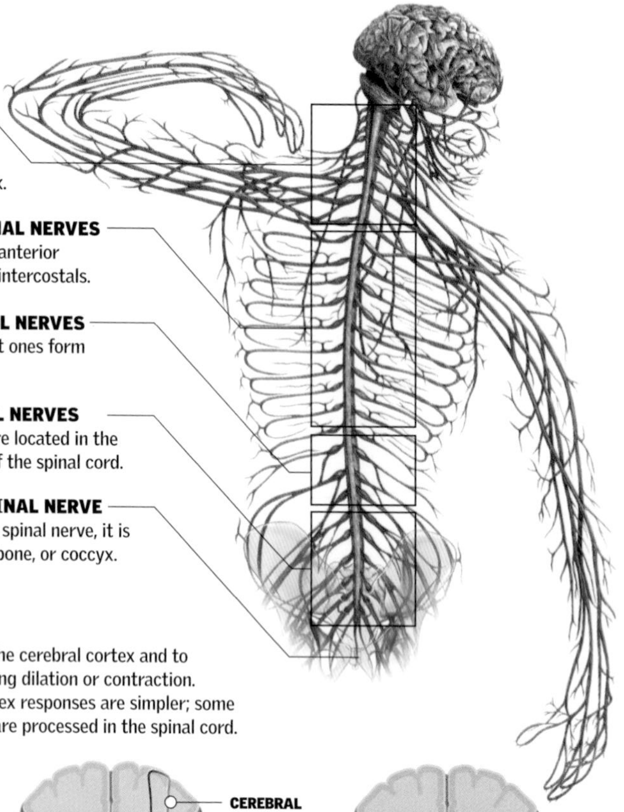

**SPINAL CERVICAL NERVES**
Eight pairs. They innervate the neck.

**THORACIC SPINAL NERVES**
Twelve pairs. The anterior branch forms the intercostals.

**LUMBAR SPINAL NERVES**
Five pairs. The last ones form the "horse's tail."

**SACRAL SPINAL NERVES**
Five pairs. They are located in the lowest segment of the spinal cord.

**COCCYGEAL SPINAL NERVE**
The only unpaired spinal nerve, it is located in the tailbone, or coccyx.

### The Three Responses

The nerve receptors gather information that goes to the cerebral cortex and to the spinal cord. The response can be automatic, ordering dilation or contraction. Voluntary response implies a complex nerve path. Reflex responses are simpler; some of them are processed in the brain, but most of them are processed in the spinal cord.

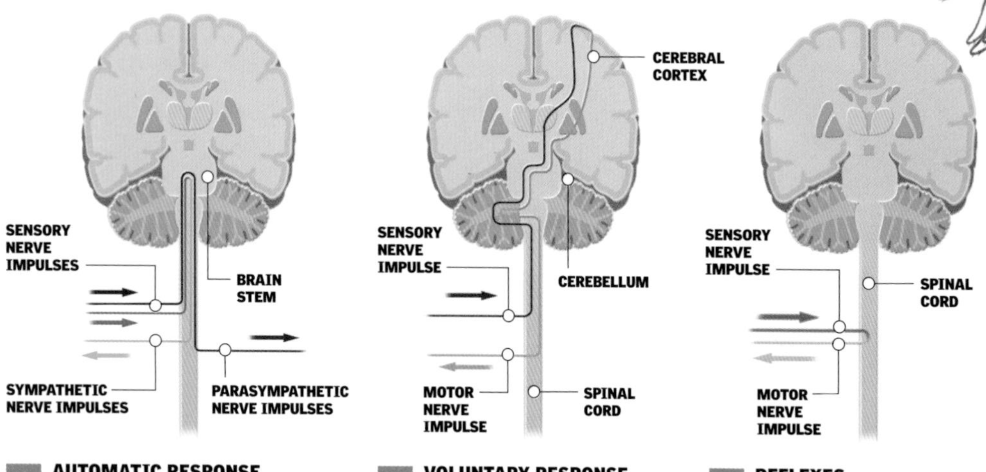


**A AUTOMATIC RESPONSE**
The impulses, or sympathetic (dilation) or parasympathetic (contraction) response signals, travel over separate pathways.

**B VOLUNTARY RESPONSE**
The sensory impulses that activate voluntary responses occur in various areas of the brain. The nerve path is complex.

**C REFLEXES**
Some are processed in the brain, but most of them are processed in the spinal cord, where the impulse is processed and the reply is sent.

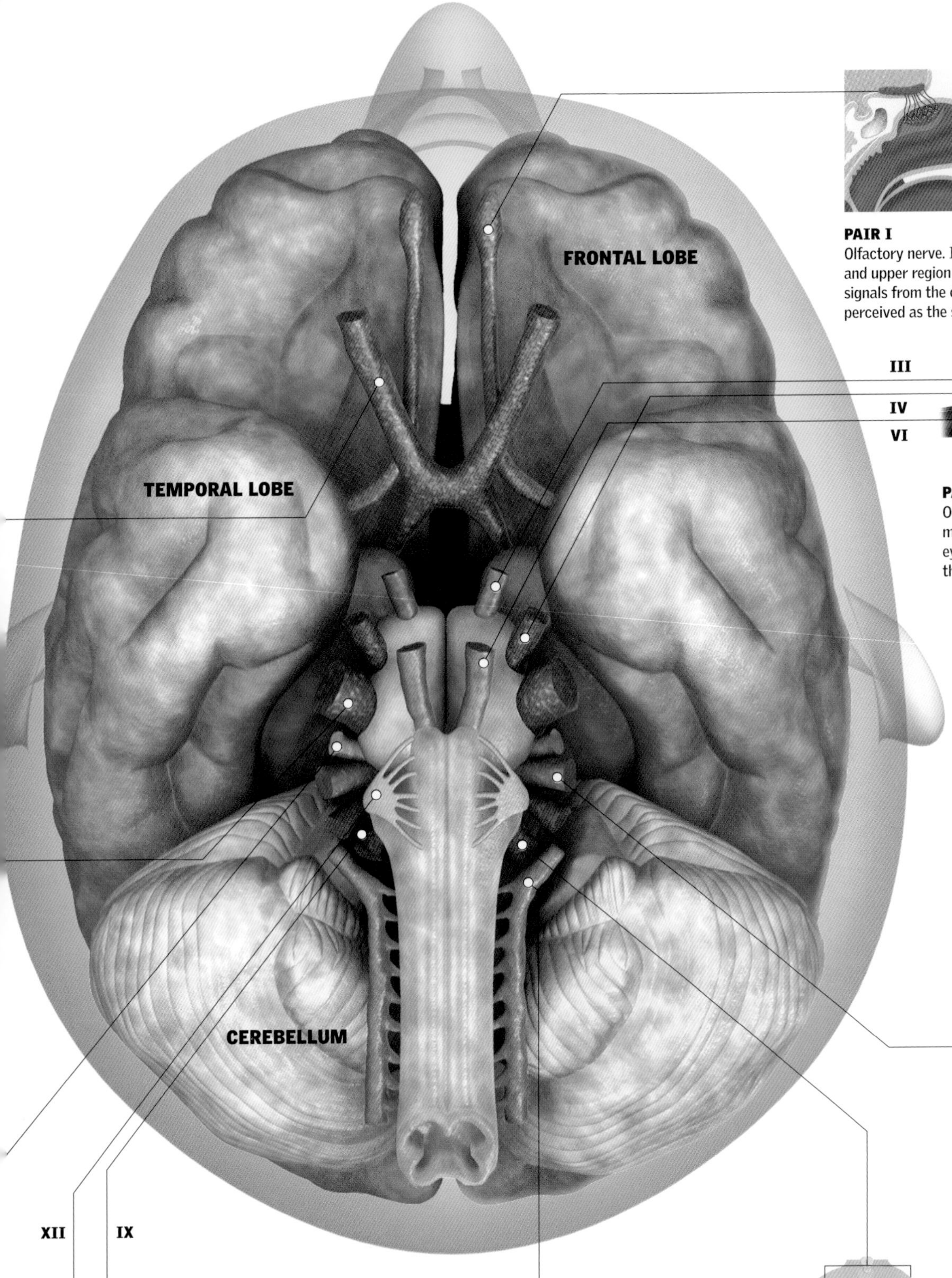


**PAIR I**
Olfactory nerve. Innervates the internal and upper region of the nose and transmits signals from the olfactory cells that are perceived as the sense of smell.

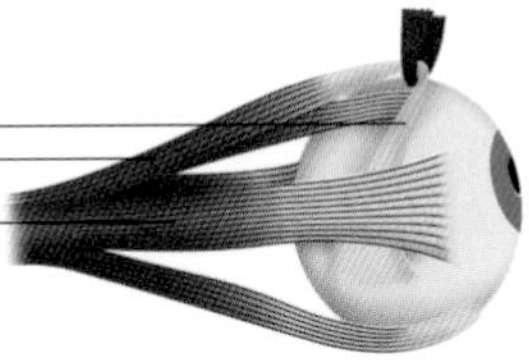


**PAIR III**
Oculomotor nerve. Controls the movements of the eye and the eyelid. It changes the shape of the pupil and the lens.

**PAIR IV**
Trochlear nerve. Controls the oblique muscle above the eye.

**PAIR VI**
Abducens nerve. The nerve that moves the external lateral rectus muscle of the eye.

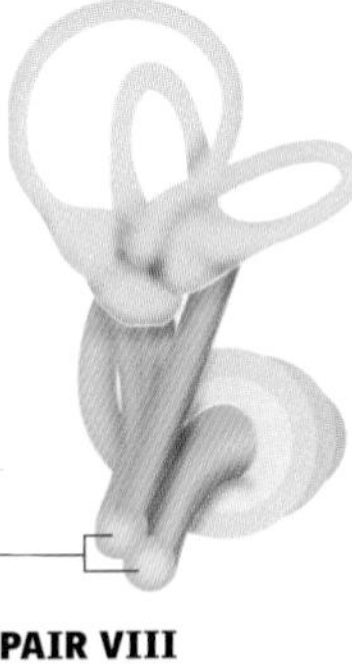

**PAIR VIII**
The cochlear vestibular nerve. Transmits sensory signals from the inner ear, which are perceived as sound; enables equilibrium.

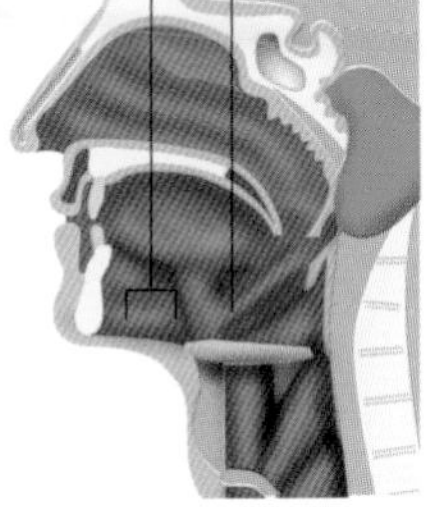

**PAIR IX**
Glossopharyngeal nerve. Controls the salivary glands and transmits sensory signals from the tongue and the pharynx.

**PAIR XII**
Hypoglossal nerve. Controls the movements of the tongue.

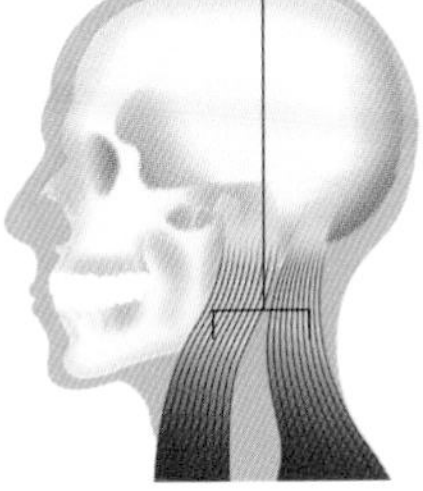

**PAIR XI**
Accessory nerve. Its function is to control the muscles involved in swallowing and moving the head.

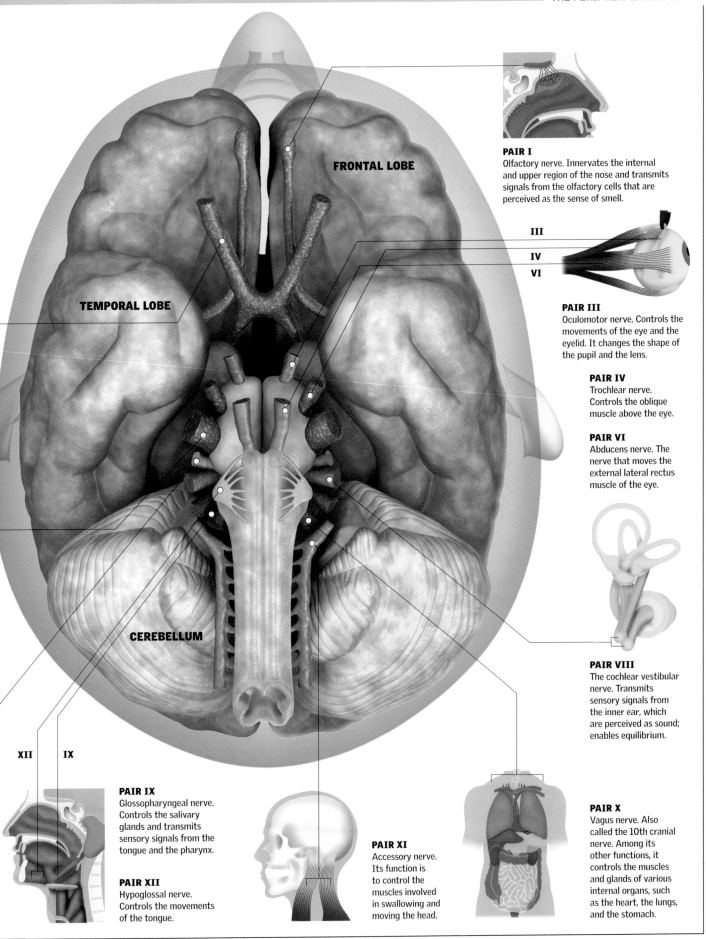

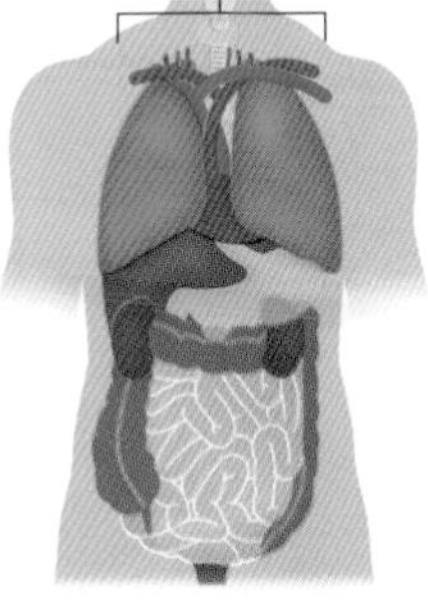

**PAIR X**
Vagus nerve. Also called the 10th cranial nerve. Among its other functions, it controls the muscles and glands of various internal organs, such as the heart, the lungs, and the stomach.

# Dreams and Memory

To be able to process the information gathered during the day, the brain takes advantage of periodic dream states. During a dream the brain reduces its activities, and its patterns of thought are disconnected from the external world. The passage from consciousness to dreaming (and from dreaming to consciousness) is the task of neurotransmitters, chemical substances that are manufactured and released from the reticular activator system, a regulator in the cephalic talus, which lies in the brain stem.

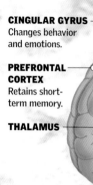

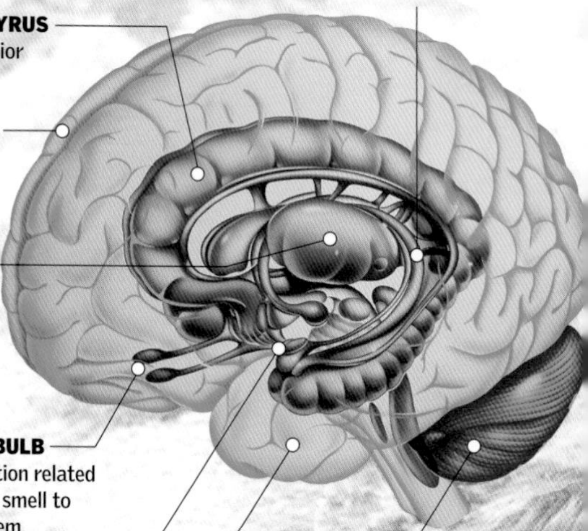
HIPPOCAMPUS
Stores short-term memory and converts it into long-term memory.
CINGULAR GYRUS
Changes behavior and emotions.
PREFRONTAL CORTEX
Retains short-term memory.
THALAMUS
OLFACTORY BULB
Sends information related to the sense of smell to the limbic system.
AMYGDALA
Stores fears and phobias.
TEMPORAL LOBE
Stores semantic memory.
CEREBELLUM
Controls movement and equilibrium.

## Formation of Memory

Memory is a set of processes in which unconscious associations are capable of retaining and recording highly varied information. This information can be perceived consciously or unconsciously and ranges from ideas and concepts to sensations that were previously experienced or processed. Memory has many forms, but the two basic ones are the long-term and short-term memory.

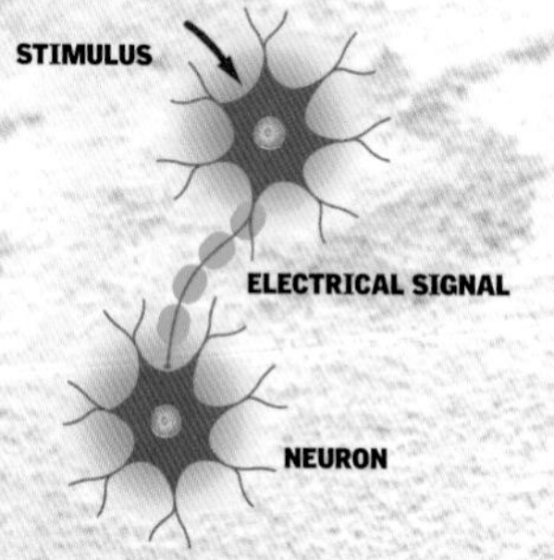
STIMULUS
ELECTRICAL SIGNAL
NEURON

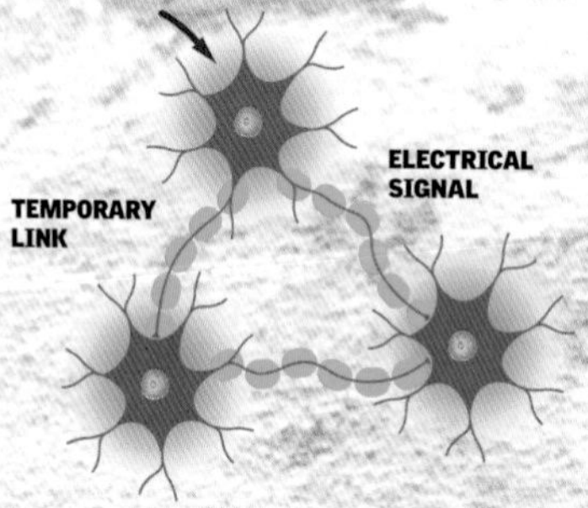
TEMPORARY LINK
ELECTRICAL SIGNAL

**1 CONNECTION.** An experience triggers a pattern (or model to be repeated), exciting two neutrons. To form long-term memory the template that was generated earlier by the short-term memory must be replicated. When a stimulus is received, the neuron reacts, sending an impulse to a neighboring neuron.

**2 LINK FORMATION.** The nerve impulses sent to the neighboring neurons generate a greater capacity for response from the cells that sent the impulses. A temporary union is formed among the cells. In the future, they will be more likely to trigger a nerve impulse together. A neuronal template is beginning to be created.

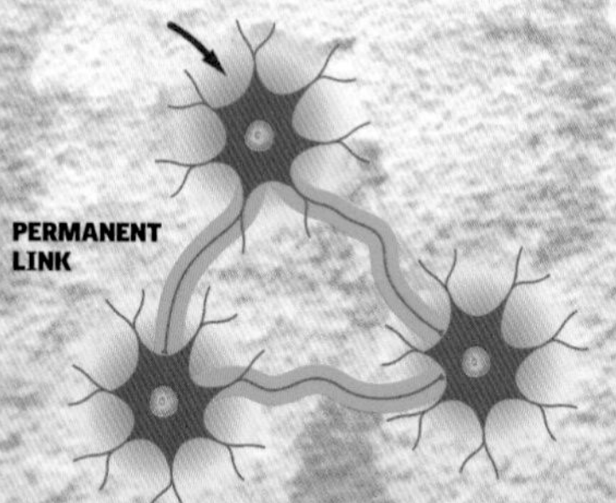
PERMANENT LINK

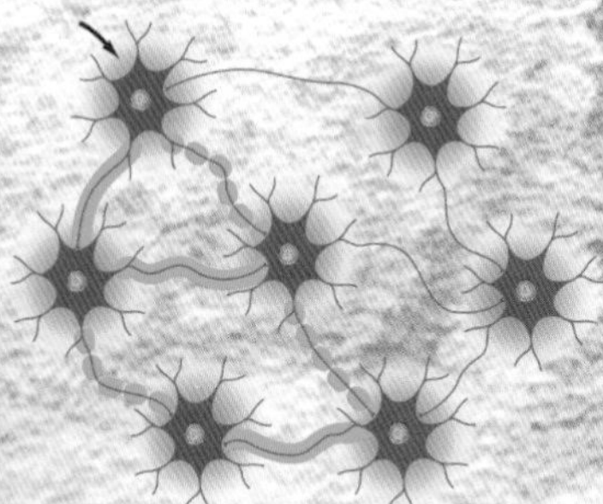

**3 DEEPER LINKS.** Every time an event is remembered, a nerve impulse is triggered. As a recollection is repeated, the neurons become more solidly connected. Then the neurons begin to send joint impulses, no matter which was excited first. The development of connections is strengthened with repetition or notable or stressful events.

**4 EXPANDING NETWORK.** With successive repetition, different groups of neurons begin to form a neuronal network that represents the long-term memory. The more complex the network, the more accessible and durable the memory will be. Each group of neuronal cells represents a different aspect through which one accesses the complete memory.

## Limbic System

Consists of a complex of structures that wrap around the upper part of the brain stem. These structures control emotions such as annoyance and happiness. They protect us from danger and play an important role in memory formation. For example, the amygdala produces fear when processing danger. The hippocampus permits us to store and remember short-term memories that are brought to the cortex. When the hippocampus is damaged, new memories cannot be incorporated.

## 20 seconds

**THE TIME AFTER WHICH SHORT-TERM MEMORY LOSES INFORMATION (SUCH AS A TELEPHONE NUMBER) THAT HAS NOT BEEN USED.**

## REM

**THE ACRONYM FOR RAPID EYE MOVEMENT. THE EYES MOVE, THOUGH THE BODY IS STATIONARY.**

## Dream Patterns

A pattern is a model that serves as a template, or mold, to obtain the same format. During sleep the two great patterns are REM and NREM, with their four phases. REM sleep is the most enigmatic; it is thought that dreams are produced during REM. During that time the human being lives out an inner experience, generally involuntary, where the mind provides representations of sensations, images, situations, dialogs, sounds, etc.

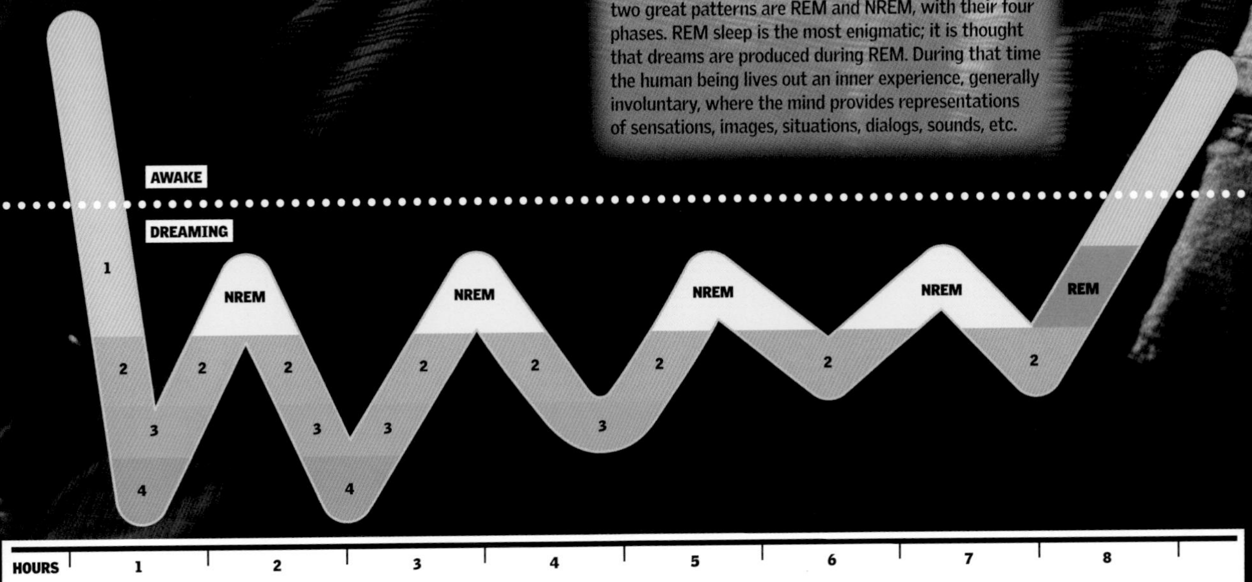


**PHASE 1**
Transition between waking and sleeping. The electroencephalograph (EEG), a device that measures cerebral activity, registers alpha waves. The body is relaxed, but if someone disturbs the sleeping person then he or she will wake up.

**PHASE 2**
Second-phase NREM. The EEG pattern is more irregular. Waking up the person is more difficult.

**PHASE 3**
Delta waves appear. The vital signs decrease: respiration and the heartbeat slow down, and the body temperature falls.

**PHASE 4**
Now the dream phase or phase of deep sleep occurs. The delta waves are dominant, and the vital signs drop to minimal levels.

**PHASE REM**
Rapid Eye Movement. The vital signs increase. The skeletal muscles become inhibited. Dreams enter the scene.

# 3 FIGHTING DISEASE AND DISORDERS

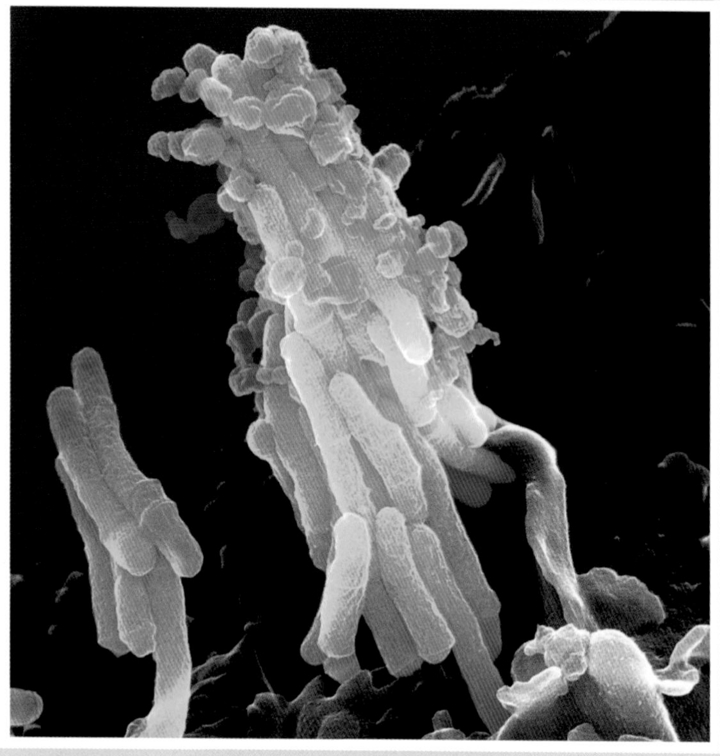

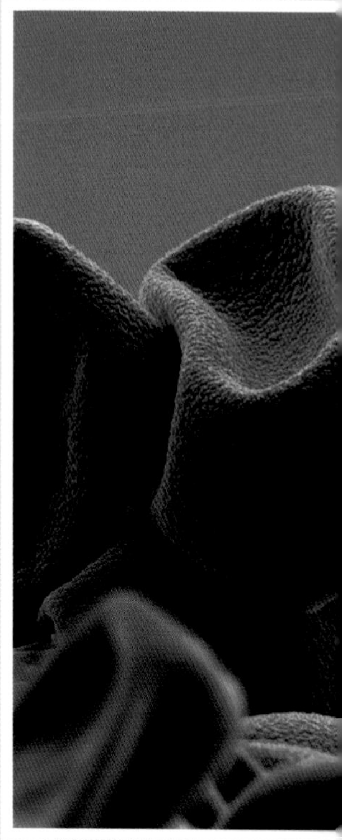

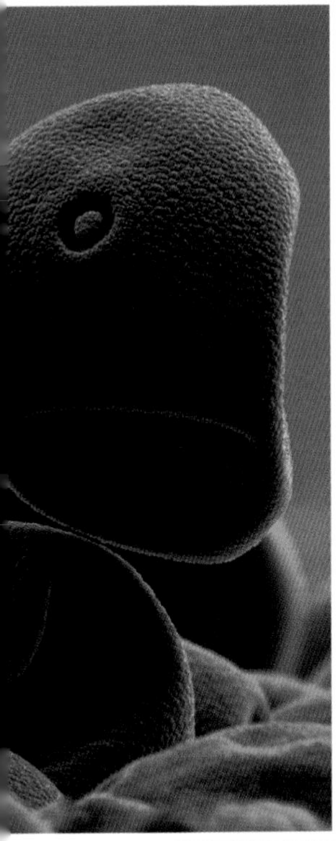

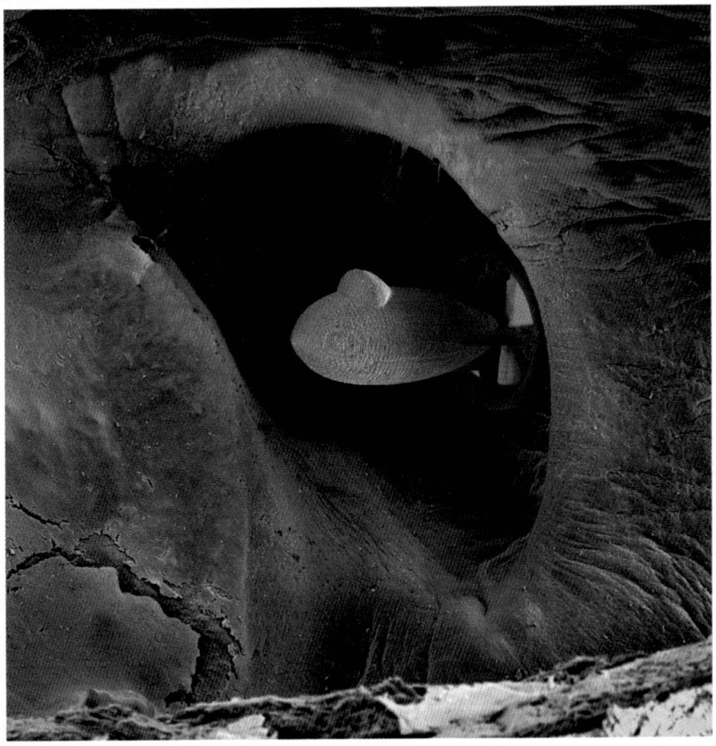

210

230

# FIGHTING DISEASE AND DISORDERS

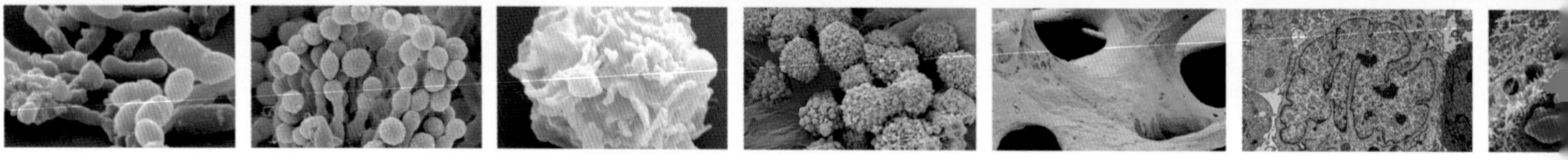

**This section of the book is divided into three chapters. The first is dedicated to a study of bacteria and other microlife, the second deals with everything that happens to the body when it gets sick, and the third reviews the latest advances in medical technology. The information, reviewed by professionals and accompanied by incredible illustrations, will captivate you from the first page.**

The surprising advances in molecular biology and genetics allow us to have new therapeutic and diagnostic tools that make it possible to think that, in the future, humans will live eternally. The themes covered and shown here have a scientific basis. We tell you, for example, which mechanisms regulate the operation of the genes and how these mechanisms can correct certain errors in the DNA. In a not too distant future, nanomachines (many times smaller than a cell) could be guided inside the body to eliminate obstructions in blood vessels or to kill cancer cells. Sick cells could be eliminated without damaging the healthy ones. Although the tests are still in development, it is expected that within the next few years this type of revolutionary treatment, which combines genetics with pharmacology, could be applied. Currently there are patients on whom certain drugs have toxic effects or do not have the desired effect; in the future, drugs could be made according to the genetic makeup of each patient.

Another aspect to highlight in the use of new technologies is related to medical informatics, in which all kinds of patient information is managed. Such a system already allows all hospitals in some cities, such as Vienna, Austria, to communicate digitally. That way a doctor can quickly access the medical record of the person through a remote communication network such as the Internet. There is real evidence that in the next decades the way that medicine will treat diseases requiring the replacement of organs or tissues will also change. The tissue created in the laboratory will be genetically

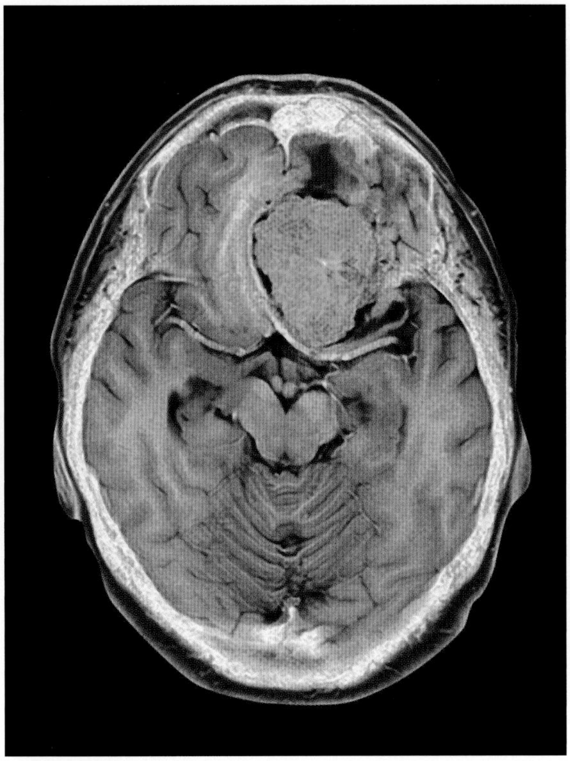

**MRI SCAN OF A BRAIN**
Magnetic resonance imaging (MRI) is a valuable technique for examining the organs of the body. For example, it allows doctors to study both the surface and the inside of the brain.

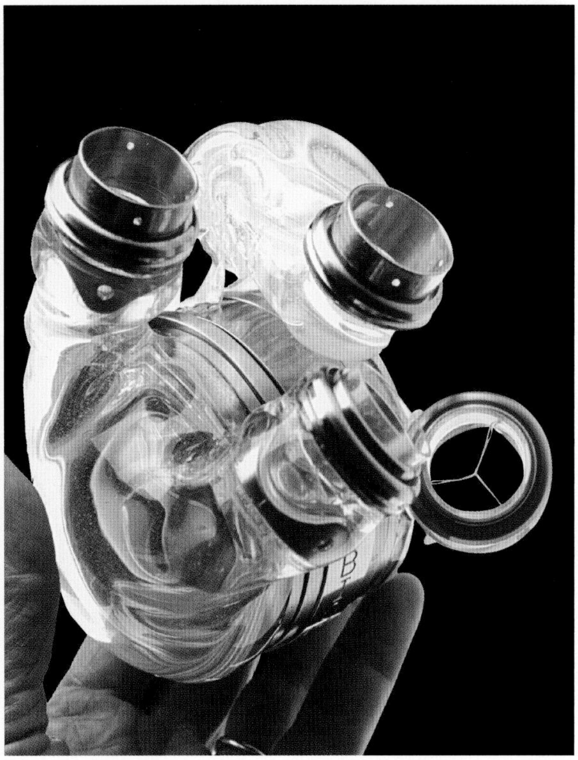

**ARTIFICIAL HEART**
Efforts continue to design artificial organs that could replace organs that have been damaged or affected by severe diseases. This prototype replacement heart would last at least five years.

identical to the patient's, so there will be no rejection. Once this is achieved, transplants and artificial organ implants will be a thing of the past.

In addition, the health of many people will be determined during fetal development. As a result of the advances in prenatal diagnostic methods, the possibility has appeared in recent years of carrying out surgical procedures to correct certain congenital problems inside the womb. It is not difficult to foresee the development of medical units in which the fetus is a patient. Even though many of these advances are still in the research stage, it is not unrealistic to believe that they will end up being useful. Several decades ago, nobody would have thought about replacing sick organs with healthy ones or the possibility of choosing a baby's sex.

In the decades to come, the application of genetic therapy will allow, among other things, the cure of genetic disorders caused by defective genes. In addition, the alternative of knowing beforehand what diseases a person could develop will be extremely valuable in the field of health, because we will be able to choose examinations and treatments according to individual needs. Do not wait any longer. Turn the page and begin to enjoy this section of the book, which may be a point of departure in your own adventure in learning.

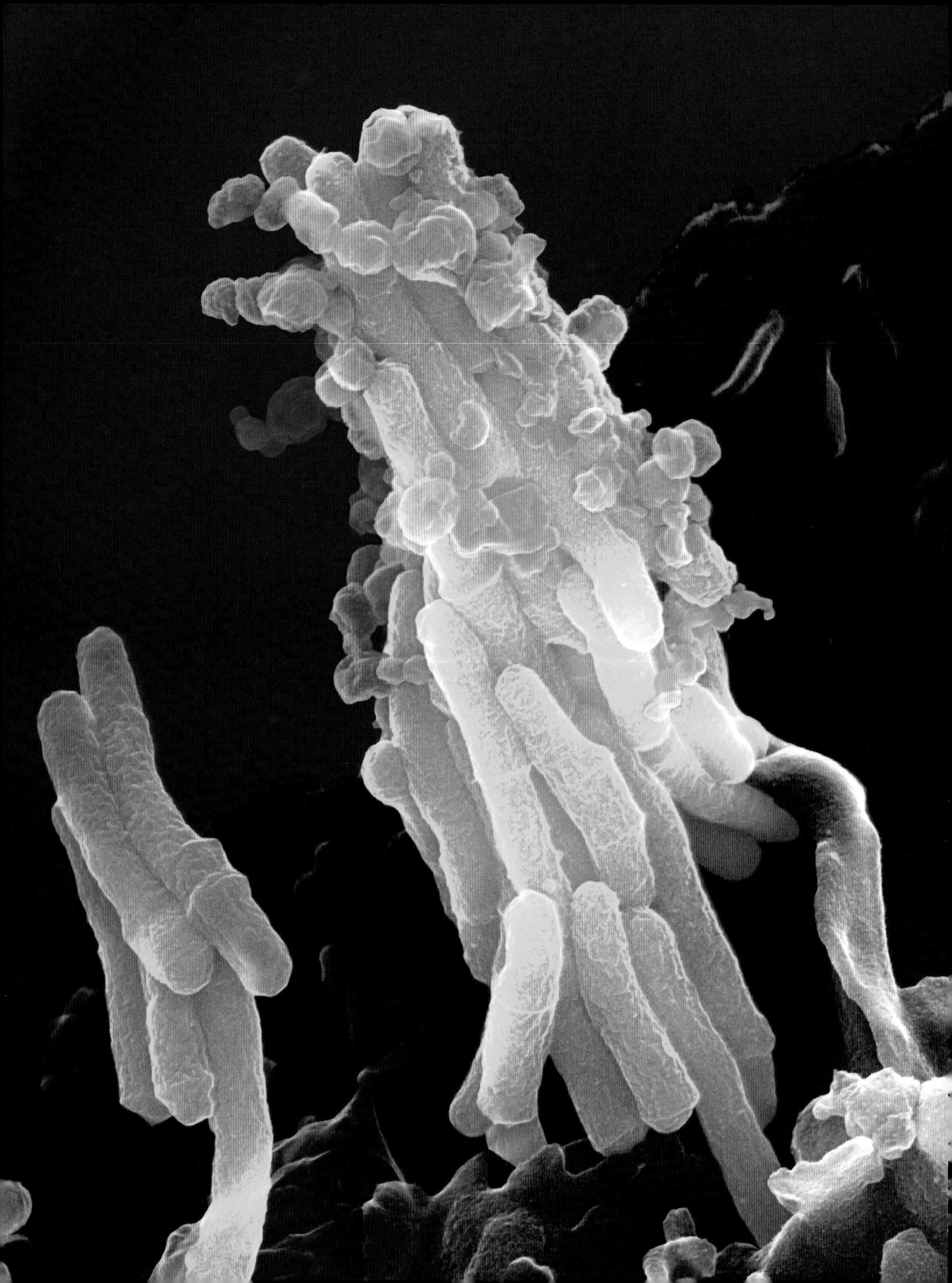

# Microlife

What is a bacterium? What is a virus? How do antibiotics act on them? What function do the red and white blood cells perform when they are in action? Did you know that white blood cells are bigger than the red ones and that, by changing shape, they can pass through capillary walls to reach different tissues and hunt down foreign organisms that are in the way, such as bacteria or cancer cells? In this chapter, we will also show you how platelets, another defense system of the body, prevent hemorrhages, or bleeding. While learning about our internal functions, you will be surprised and captivated by illustrations and so much more.



**TUBERCULOSIS BACTERIA** (opposite)
Image of the bacteria *Mycobacterium tuberculosis* (in yellow) infecting a blood cell.

# Bacteria

## 40 million

**BACTERIAL CELLS EXIST IN ONLY 0.06 CUBIC INCH (1 ML) OF SALIVA.**

Bacteria are the smallest, most abundant, and hardiest life-forms on Earth. They are so microscopic that 0.06 cubic inch (1 ml) of saliva may contain up to 40 million bacterial cells. They exist and live everywhere, from our skin to the smallest cracks in rocks. Most are benign and even vital to the survival of other living beings, but some are pathogenic and can cause diseases, some of them deadly. Almost all nourish themselves by absorbing substances from their surroundings, but some make use of the energy of the sun, and others use the chemical energy in volcanic emissions. All are made up of one cell and usually reproduce by dividing in two.

## What Are Bacteria?

Bacteria have the capacity to survive in extremely hostile environments, even at temperatures of 480° F (250° C). For this reason, they are the most ancient living organisms on the planet. In a common habitat, such as the human mouth, there can be as many as 25 different species of bacilli among the 40 million bacterial cells in just 0.06 cubic inch (1 ml) of saliva. And, if there are so many in just a small amount of saliva, imagine how many there might be in the entire world—millions and millions of species. However, only 1 percent of bacteria produce diseases. Likewise, 70 percent of antibiotics are produced through bacterial fermentation.

### Classification of Bacteria

Some 10,000 bacteria species have been identified, and it is estimated that there are still many left to be discovered. They are classified both by their shape and through chemical tests to help identify specific species.

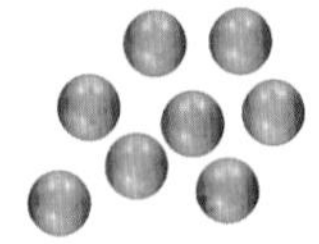

**A COCCUS**
Spherical cocci can live isolated, and others can group into pairs, chains, or branches.

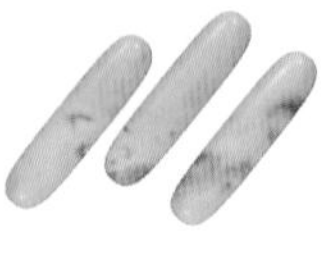

**B BACILLUS**
Many bacteria have this rod-shaped form.

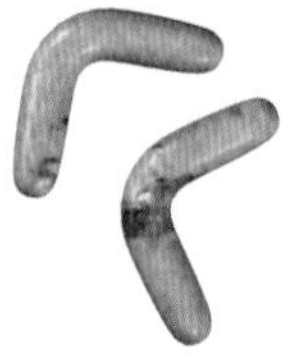

**C VIBRIO**
These bacteria have the shape of a comma or boomerang.

**D SPIRILLA**
This class of bacteria has a corkscrew shape.

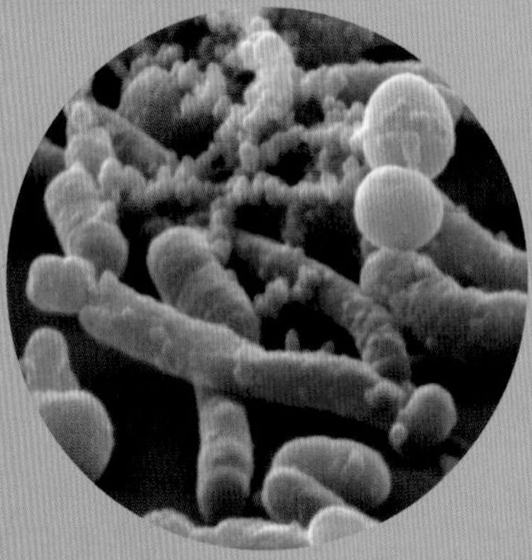

## Benign

Almost all bacteria are benign and even healthy for living beings. *Lactobacillus acidophilus*, for example, is a bacterium that transforms lactose into lactic acid to produce yogurt, and it is also present in the human body in the vagina and in the intestinal tract. The bacterium *Rhizobium*, on the other hand, allows roots of legume plants to absorb nitrogen from the soil.

## Harmful

Harmful bacteria are pathogenic and are present in all living beings and in agricultural products. They can transfer from food to people, from people to food, or among people or foodstuffs. In the 14th century, the *Yersinia pestis* bacterium, present in rats and fleas, caused many deaths in what was known as the plague.

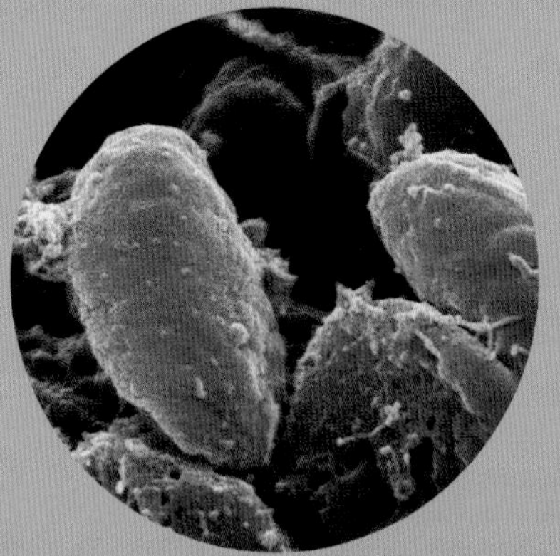

**CIRCULAR CHROMOSOME**
DNA molecule closed at its ends.

**CELL MEMBRANE**
This is involved in the transport of substances and contains elements that can be toxic when they come in contact with other beings.

**CELL WALL**
This keeps the cell from exploding if it absorbs too much water. The flagella are attached to it.

## Parts of a Bacterium

Bacteria are usually considered the most primitive type of cell there is, because their structure is simpler than most others. Many are immobile, but others have flagella (thin hairs that move like whips to propel the bacteria in liquid media). The cell wall is generally made up of carbohydrates, including murein, a peptidoglycan complex, lipids, and amino acids. No organelles or protoplasmic formations are found in their cytoplasm.

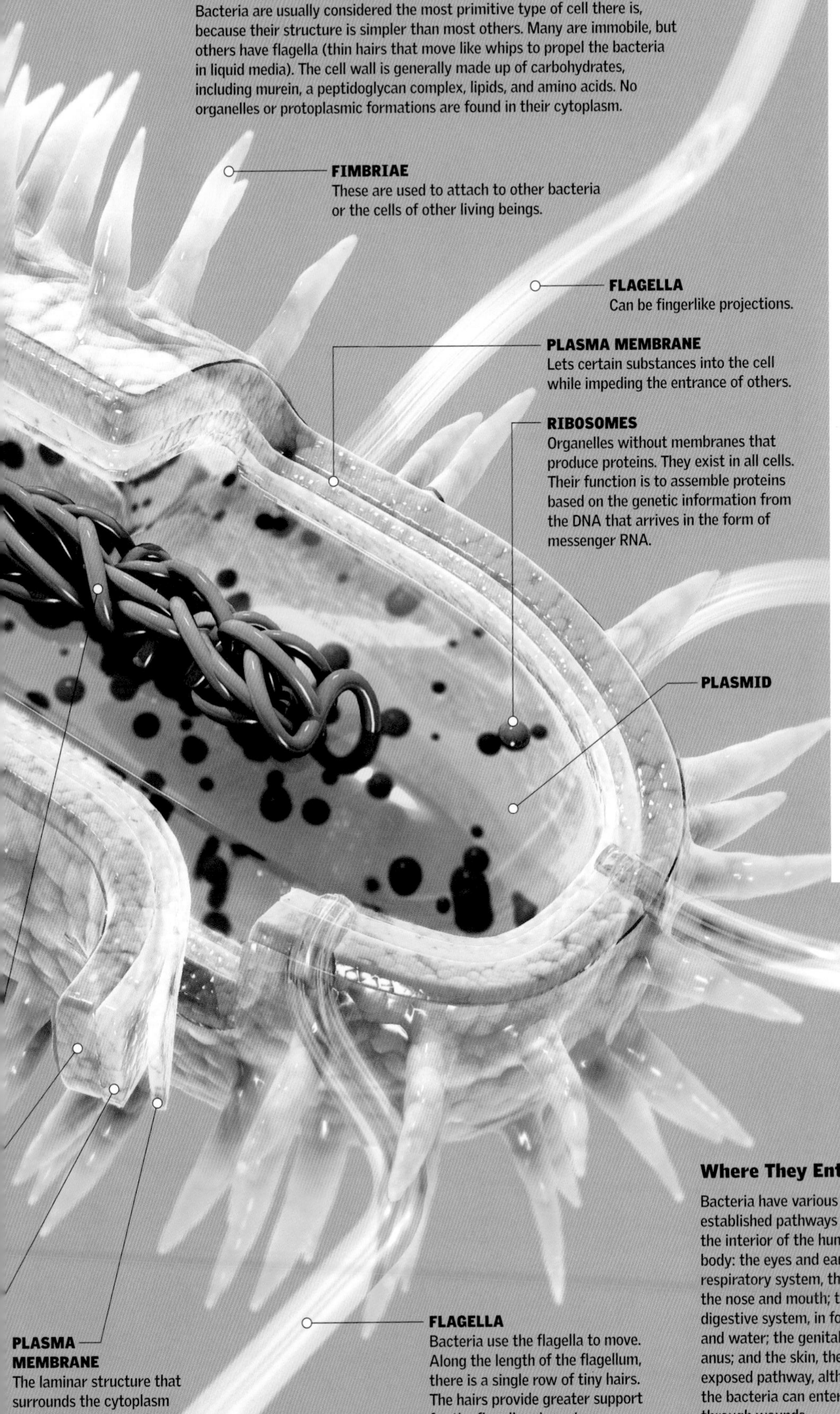

**FIMBRIAE**
These are used to attach to other bacteria or the cells of other living beings.

**FLAGELLA**
Can be fingerlike projections.

**PLASMA MEMBRANE**
Lets certain substances into the cell while impeding the entrance of others.

**RIBOSOMES**
Organelles without membranes that produce proteins. They exist in all cells. Their function is to assemble proteins based on the genetic information from the DNA that arrives in the form of messenger RNA.

**PLASMID**

**PLASMA MEMBRANE**
The laminar structure that surrounds the cytoplasm of all cells like bacteria.

**FLAGELLA**
Bacteria use the flagella to move. Along the length of the flagellum, there is a single row of tiny hairs. The hairs provide greater support for the flagellum in water.

### Antibiotic Action

Certain microorganisms—fungi or bacteria—produce chemical substances that are toxic for some specific bacteria; they cause their death or stop their growth or reproduction. Penicillin and streptomycin are examples. These substances are called antibiotics.

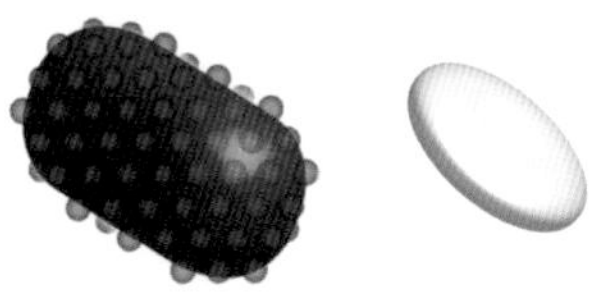

1. When a bacterium breaks through the body's barriers, the immune system recognizes it as an antigen and generates antibodies against it.

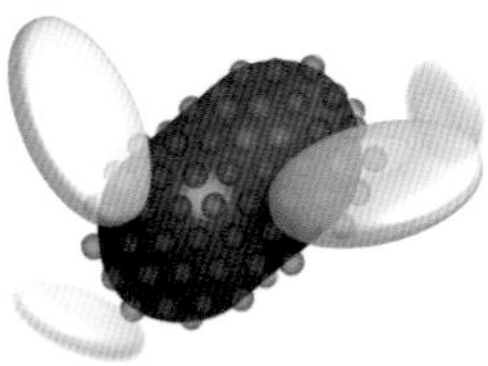

2. The leukocytes release cytokines, substances that attract more leukocytes, and by means of antibodies, they attach to the bacterium to destroy it.

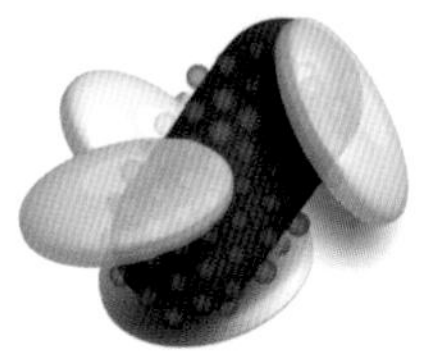

3. Once the leukocytes are attached to the bacterium, they eat it.

**70%**

**OF ANTIBIOTICS ARE PRODUCED FROM BACTERIAL FERMENTATION.**

### Where They Enter

Bacteria have various established pathways to the interior of the human body: the eyes and ears; the respiratory system, through the nose and mouth; the digestive system, in food and water; the genitals and anus; and the skin, the most exposed pathway, although the bacteria can enter only through wounds.

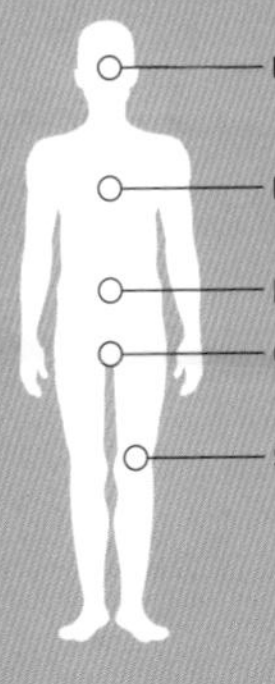

# Minuscule Life

Viruses are not, in a strict sense, life-forms. They cannot live independently and are at the limit of inert material. They lack systems to obtain and store energy and to synthesize protein. For this reason, they are symbiotes committed to the cells, both prokaryotes and eukaryotes, on which they depend for their reproduction. Their structure might be nothing more than a simple envelope of protein that surrounds a package of nucleic acid (DNA or just RNA). In the case of bacteriophages, they invade bacteria and inoculate their own DNA into them. New viruses are produced from the copy of the genetic material.

## "Filterable Viruses"

In 1898, while the origin of certain plant diseases was being studied, Dutch microbiologist Martinus Beijerinck discovered that some infections persisted even when filters for all known bacteria were used. He deduced that the responsible agents must be much smaller than bacteria. He called them "filterable virus," a word from the Latin related to "poison." They are so small that they cannot be seen with an optical microscope. Today we know that their structure does not even support the organelles of a cell: they are just chemical packages inserted in a protein coat.

### Ornate Shapes

The shape of a virus has a close relationship to the chemical composition of its envelope. The proteins that compose it are expressed in the form of crystals, which take on geometric shapes, primarily simple and complex polyhedrons.

## Anatomy of a Bacteriophage

This very small virus attacks bacteria exclusively. It has a capsid that contains the strand of DNA that is injected into the bacteria through a hollow tail body that has six fibers; these fibers allow it to attach to the cell wall.

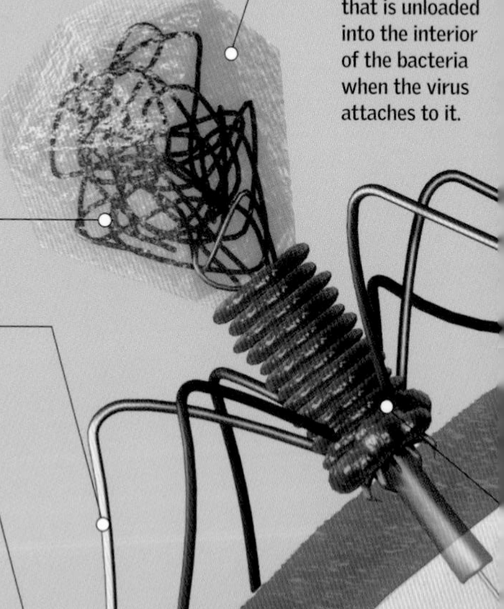

**CAPSID**
Contains a strand of DNA that is unloaded into the interior of the bacteria when the virus attaches to it.

**DNA**
This contains all the information necessary for the virus to replicate.

**FIBERS**
These help the virus attach to the surface of the cell that it attacks.

## Invaded Bacterium

When they reach the cell wall of a bacterium, bacteriophage viruses suddenly abandon their inert appearance: they attach to the surface of the live cell and inject their DNA, which allows the virus to make copies of itself. The life of the bacterium is altered by the takeover of the viral DNA, which gives instructions to manufacture different parts of new viruses. When the attacked cell dies, its remains are used by other nearby bacteria.

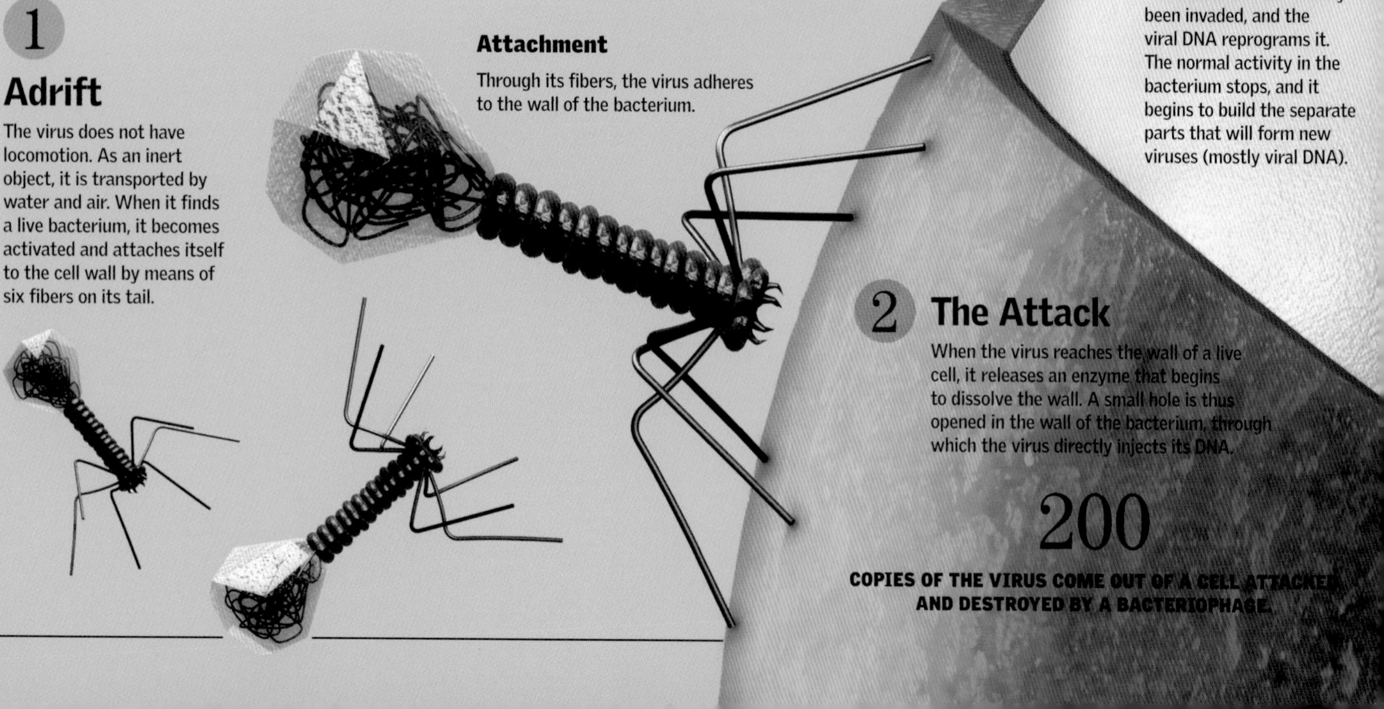

### 1 Adrift

The virus does not have locomotion. As an inert object, it is transported by water and air. When it finds a live bacterium, it becomes activated and attaches itself to the cell wall by means of six fibers on its tail.

**Attachment**

Through its fibers, the virus adheres to the wall of the bacterium.

### 2 The Attack

When the virus reaches the wall of a live cell, it releases an enzyme that begins to dissolve the wall. A small hole is thus opened in the wall of the bacterium, through which the virus directly injects its DNA.

### 3 DNA is Reproduced

The bacterium has already been invaded, and the viral DNA reprograms it. The normal activity in the bacterium stops, and it begins to build the separate parts that will form new viruses (mostly viral DNA).

**200**

**COPIES OF THE VIRUS COME OUT OF A CELL ATTACKED AND DESTROYED BY A BACTERIOPHAGE.**

## Notorious Families

### With RNA

These virus families do not have DNA in their genetic material.

### With DNA

Further divided into simple-strand and double-strand viruses.

**FILOVIRUSES**
One is the Ebola virus, which causes a type of hemorrhagic fever.

**RETROVIRUSES**
The best known is HIV, which produces AIDS. The HTLV retrovirus can cause leukemia.

**CORONAVIRUSES**
Cause diseases that range from the common cold to SARS and atypical pneumonia.

**FLAVIVIRUSES**
Very numerous, they cause hepatitis, West Nile fever, encephalitis, and dengue.

**HEPADNAVIRUSES**
Only the hepatitis B and D viruses belong to this family.

**HERPESVIRUSES**
The cause of chicken pox and herpes zoster, among others.

**POXVIRUSES**
In this group is the virus that causes smallpox.

**PAPILLOMAVIRUSES**
These viruses produce warts and are also associated with cervical cancer.

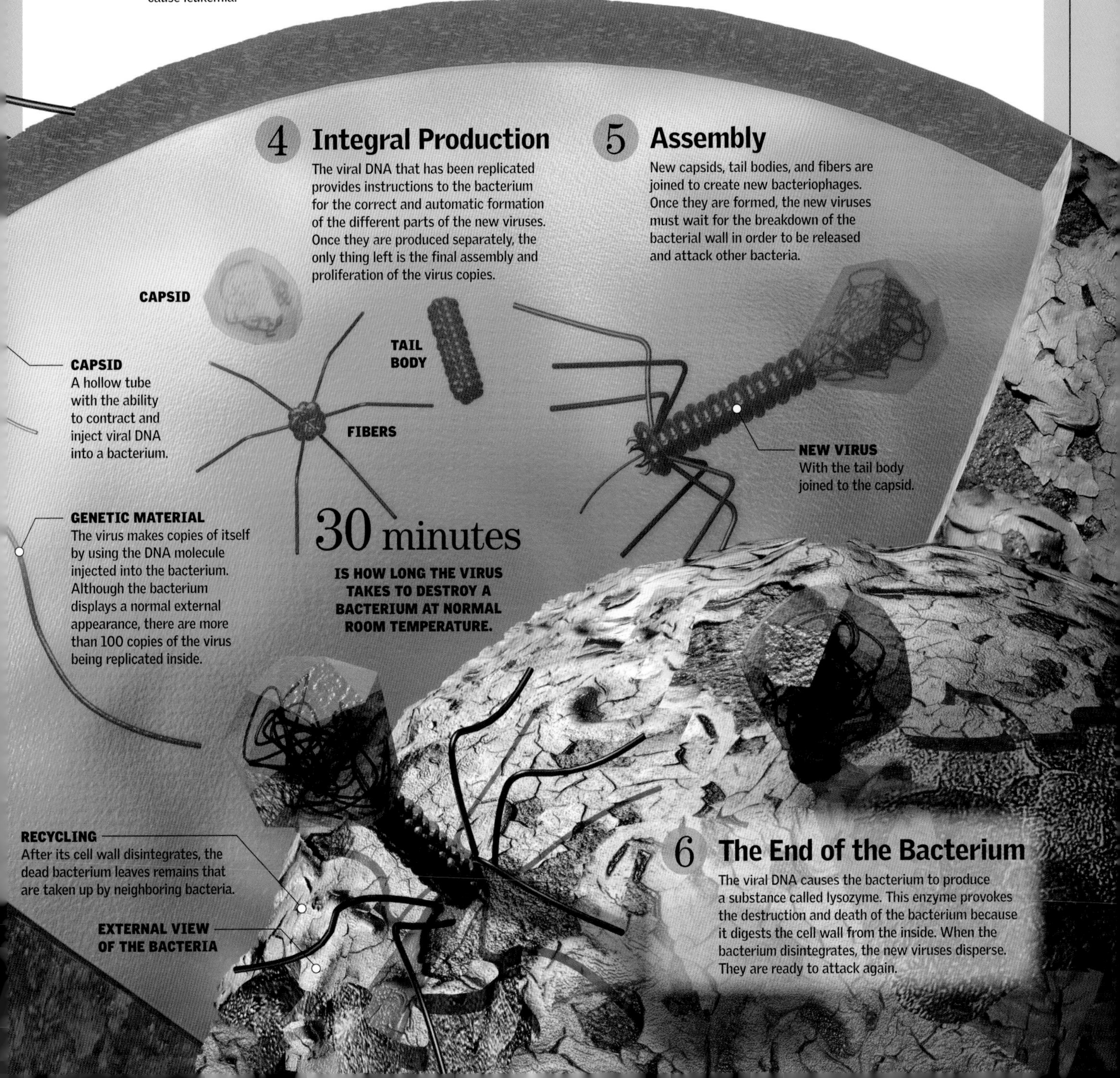

CAPSID
TAIL BODY
FIBERS
NEW VIRUS
With the tail body joined to the capsid.
CAPSID
A hollow tube with the ability to contract and inject viral DNA into a bacterium.
GENETIC MATERIAL
The virus makes copies of itself by using the DNA molecule injected into the bacterium. Although the bacterium displays a normal external appearance, there are more than 100 copies of the virus being replicated inside.
RECYCLING
After its cell wall disintegrates, the dead bacterium leaves remains that are taken up by neighboring bacteria.
EXTERNAL VIEW OF THE BACTERIA

## 4 Integral Production

The viral DNA that has been replicated provides instructions to the bacterium for the correct and automatic formation of the different parts of the new viruses. Once they are produced separately, the only thing left is the final assembly and proliferation of the virus copies.

## 5 Assembly

New capsids, tail bodies, and fibers are joined to create new bacteriophages. Once they are formed, the new viruses must wait for the breakdown of the bacterial wall in order to be released and attack other bacteria.

**30 minutes**

**IS HOW LONG THE VIRUS TAKES TO DESTROY A BACTERIUM AT NORMAL ROOM TEMPERATURE.**

## 6 The End of the Bacterium

The viral DNA causes the bacterium to produce a substance called lysozyme. This enzyme provokes the destruction and death of the bacterium because it digests the cell wall from the inside. When the bacterium disintegrates, the new viruses disperse. They are ready to attack again.

# Fungi

Fungi are living beings from the Fungi kingdom that are similar to plants, but they do not have the ability to synthesize their own food; this forces many of them to be parasites of other vegetables or animals and, of course, humans. Multicellular fungi tend to be formed by filaments and spores that reproduce very easily; others are unicellular. Infections by fungi (mycosis) tend to be superficial, such as ringworm or athlete's foot, caused by dermatophytes, but they can be systemic if, for example, they colonize the blood.

## 0.00020 inch (5 microns)

**SPORES LARGER THAN THIS SIZE TEND TO CAUSE SURFACE REACTIONS BECAUSE OF THEIR DIFFICULTY IN PENETRATING THE SKIN. THAT IS WHY THE SPECIES IN *ALTERNARIA*, *CLADOSPORIUM*, *ASPERGILLUS*, AND *PENICILLIUM* COMMONLY PRODUCE ALLERGIES.**

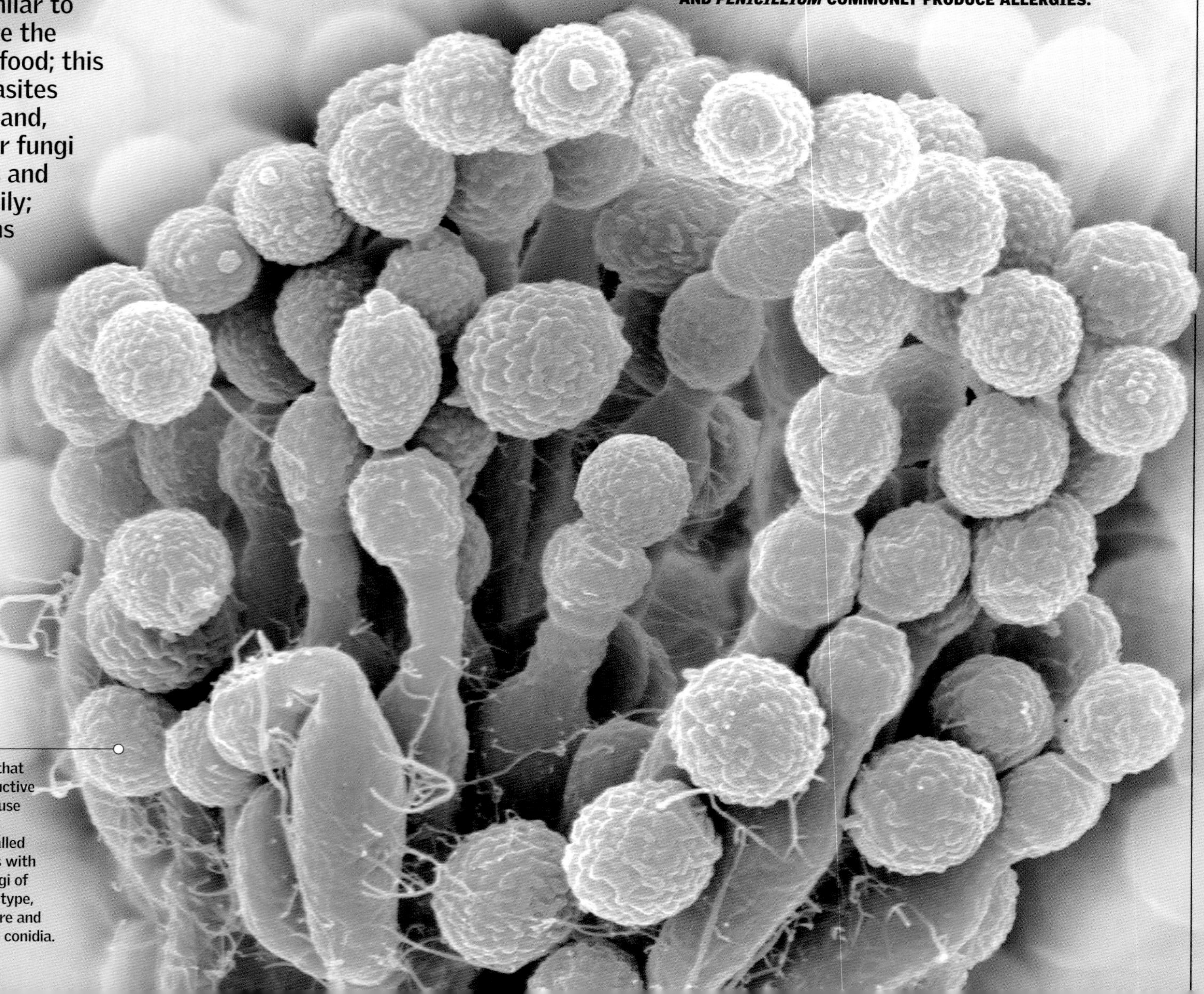

## Parasitic Cells

Not all fungi cause disease. Many, which are essentially saprophytes, have a beneficial purpose. They grow on organic matter that they decompose through exoenzymes, and then they absorb and recycle it. By not being able to carry out photosynthesis, their ability to obtain energy and biosynthesis depends on the organic material that they absorb.

## *Penicillium*

This microscopic fungus, very common in the domestic environment, is used in the production of blue cheeses and is the base for the first antibiotic created by man: penicillin. Its antibiotic properties were discovered by accident.

**SPORANGIA**
The spherical sacs that contain the reproductive cells (spores). Because these are small and asexual, they are called conidia. As happens with all multicellular fungi of the deuteromycota type, the sporangia mature and break, releasing the conidia.

# 0.7 ounce (20 g)

**THE AMOUNT OF PENICILLIN THAT CAN BE OBTAINED FOR EACH QUART (ABOUT 1 L) OF CULTURE OF THE *PENICILLIUM CHRYSOGENUM* FUNGUS WITH CURRENT BIOTECHNOLOGICAL METHODS. PENICILLIN ALTERS THE CELL WALLS OF BACTERIA AND DESTROYS THEM.**

**CONIDIOPHORES**
The branches of the stalk that have conidia on one of their ends and which together make up the reproductive organ of the fungus.

## Getting Rid of Fungi

Fungal infections respond to a variety of drug treatments. More superficial infections, such as oral candidiasis, respond to the local application of antimycotic substances. Deeper systemic infections, however, particularly in persons with some sort of immune system deficiency, can be more difficult to cure. Sometimes they require prolonged administration (as long as several months) of drugs that are taken orally and act systemically (throughout the entire body). These drugs frequently have a level of toxicity that must be taken into account when evaluating the advantages and disadvantages of each treatment.

**1 THE CELL**
Mycotic cells, which on their own are harder to treat than bacteria, look a lot like human cells. The drugs used must be sufficiently selective to attack only these cells and not human cells.

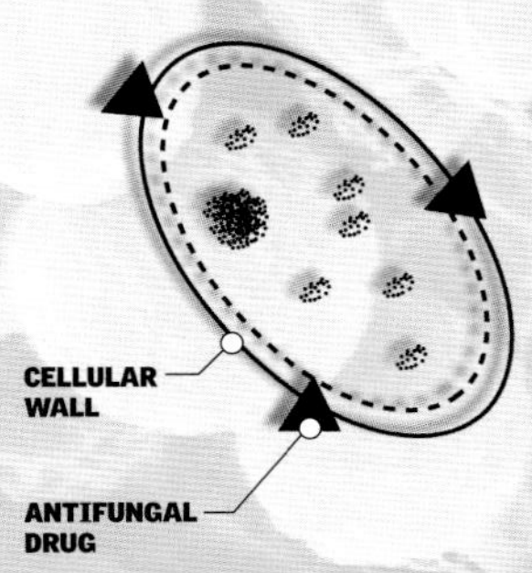


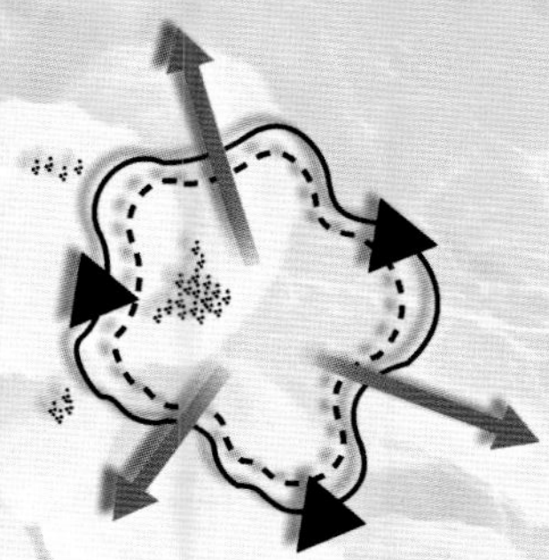

**2 THE DRUG**
The main action of antimycotic drugs is to damage the envelope of the mycotic cell, which makes up 90 percent of its mass. This way, the cytoplasm is left without support and dissolves in the bloodstream.

**ALMOST WITHOUT DIFFERENCE**
The cells that make up the different parts of a fungus are not very different from each other. Each has a polysaccharide wall that does not alter its permeability.

**HYPHA**
The hyphae are the filaments that make up the body of a multicellular fungus. Generally they form a networked structure (mycelium). The portion of the hypha that rises to branch off and form conidiophores is called the stalk. The fungus is the ensemble of all the hyphae and can have many stalks.

### Where They Commonly Invade

Fungi are very simple organisms. In human tissues, some species generate superficial wounds (in the toenails or fingernails, skin, or mucous membranes) or even fatal infections in some internal organs.

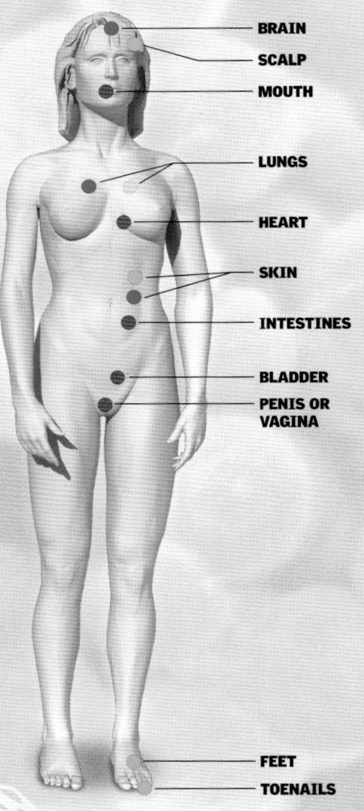


**CRYPTOCOCCOSIS**
This infection can cause certain forms of meningitis (inflammation of the meninges, the membranes that cover the brain) and pneumonias (lung infections). It can also affect the skin and the bones.

**ASPERGILLOSIS**
*Aspergillus fumigatus* is a fungus that tends to spread through air-conditioning systems. It attacks the lungs of persons with a suppressed immune system.

**DERMATOPHYTOSIS**
This fungal infection is the most common superficial mycotic infection and can affect toenails or fingernails (onychomycosis), the feet (athlete's foot), and the scalp (ringworm). Ringworm can cause hair loss.

**CANDIDIASIS**
*Candida* species prefer mucous membranes, so they attack such areas as the mouth or vagina. Alteration of the natural flora of the vagina can lead to this type of infection, and more than half of all women have suffered from such an infection at some time.

# Bad Company

Microorganisms can be habitual companions of the human body. There are bacteria that live in the digestive tract and interact in a positive way with humans because they exchange nutrients. However, there is a group of parasitic protists that obtain benefits from the relationship at the expense of the host's health. They are called endoparasites, and they can produce chronic diseases that, in some cases, can be deadly.

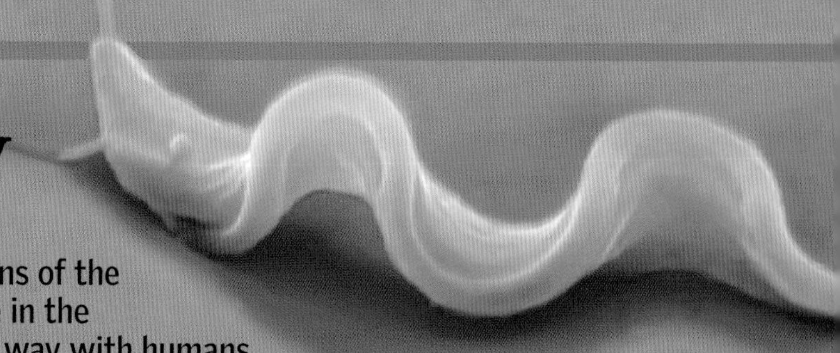


## Sleeping Sickness

This disease in humans is caused by two subspecies of protists of the genus *Trypanosoma*: *T. brucei gambiense* and *T. brucei rhodiense*. *T. brucei gambiense* causes a chronic disease that develops over several years and is found mostly in central and western Africa. The disorder caused by the *T. brucei rhodiense* has the same syndrome but develops in weeks in countries of southern and eastern Africa. The infection in humans is caused by the bite of an insect, the tsetse fly.

**MICROSCOPIC VIEW**
Trypanosomes are unicellular organisms. They are characterized by their elongated shape that ends in a prominent, free flagellum. Their cytoplasm contains a nucleus and mitochondria, among other organelles.

### *Trypanosoma brucei*

| | |
|---|---|
| **Location** | Africa |
| **Size** | 0.001 inch (30 microns) |
| **Disease** | Sleeping Sickness |

**DISTRIBUTION**
The tsetse fly, which transmits the trypanosome, is found in Africa between 15° N and 20° S. More than 60 million people are potential victims of sleeping sickness in this region.

## The Tsetse Fly

Tsetse flies are representative of the genus *Glossina*. These dipterous insects are grouped into 23 species of African flies that feed on human blood; in other words, they are hematophagous. The fly's bite and saliva deposited on the human skin cause victims to scratch themselves. This opens the way for the parasites present in the fly's saliva to enter the blood.

**ANATOMY**

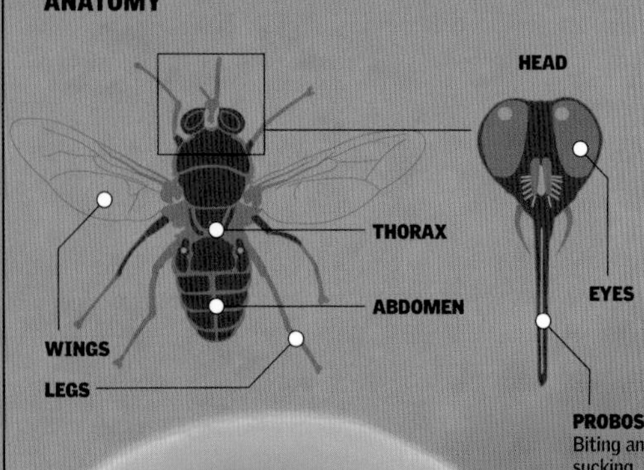


## Epidemic

Sleeping sickness is limited to the African continent. It is an epidemic that affects more than 36 countries. In 1999, the World Health Organization (WHO) confirmed 40,000 cases of the sickness but estimated between 300,000 and 500,000 persons were infected with the parasite. In 2005, following increased surveillance efforts, the number of actual cases was estimated between 50,000 and 70,000.

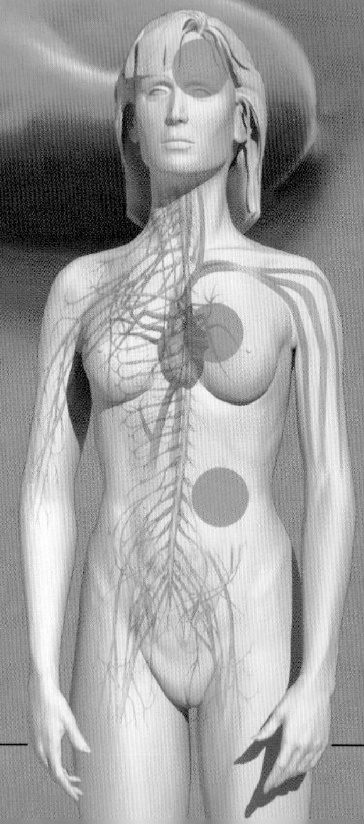

### The Disease, Step by Step

**1**
**FIRST SYMPTOMS**
The small wounds in the skin allow the parasite to enter into the blood.

**2**
**SLEEPINESS**
Through blood circulation, the trypanosome lodges in different organs of the human body.

**3**
**SERIOUS ILLNESS**
The endoparasite reproduces in bodily fluids, such as blood, lymph, and cerebrospinal liquid.

## Deadly Nightmare

*Trypanosoma brucei gambiense*, the tsetse fly, and the human body are the three players in this disease. The fly sucks human blood, which already contains the parasites. The parasites undergo a series of transformations inside the body of the fly and finally lodge themselves in its salivary glands. When the fly with parasites in its saliva searches for food and bites a person, it transfers the trypanosomes. The first phase of the sickness, similar to other diseases, includes itching, fever, headaches, and joint pain. Later the endoparasite crosses the hematoencephalic barrier and attacks the central nervous system. There it disrupts vital neurological processes—including the waking and sleeping cycle—which causes drowsiness and even death.

**BLOOD**
The first tissue to be invaded by the protozoan.

### Life Cycle

**FLY**
Bites and infects the mammal.

**1 METACYCLE IN HUMANS**
Upon feeding, the insect injects thousands of parasites in the metacyclic trypomastigote stage, which enter the human blood.

**BEGINNING**
The parasite enters the mammal.

DIVISION

**2 REPRODUCTION**
The trypomastigotes multiply through binary fission.

**3 CIRCULATION**
The new trypomastigotes circulate through the blood toward the different organs. The sickness can be diagnosed at this stage.

**4 INVASION OF THE NERVOUS SYSTEM**
The fluids present in the central nervous system are infected with trypomastigotes. The sickness already presents its characteristic syndrome.

MAMMALS

**ANOTHER FLY**
Bites and is infected by the infected mammal.

TSETSE FLY

DIVISION

**5 PROCYCLE**
The parasites transform themselves in the digestive tract of the fly and divide through binary fission.

**6 MIGRATION**
Procyclic trypomastigotes leave the digestive tract and migrate to the salivary glands of the fly. There they transform into epimastigotes.

**7 SALIVA**
The metacyclic trypomastigotes are part of the saliva. They can be injected into the blood.

**BINARY FISSION**
New reproduction. The metacyclic trypomastigotes form.

DIVISION

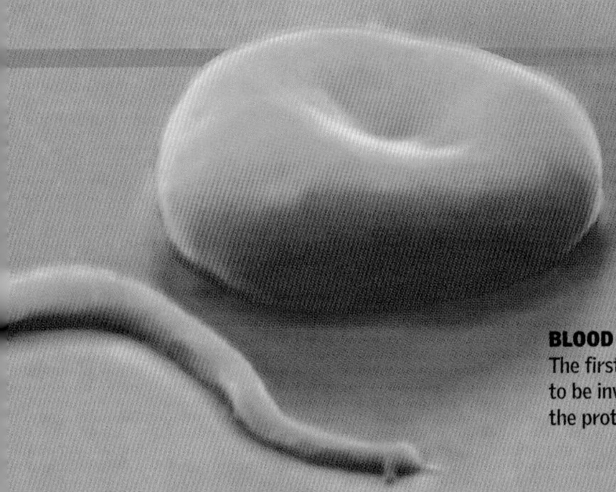

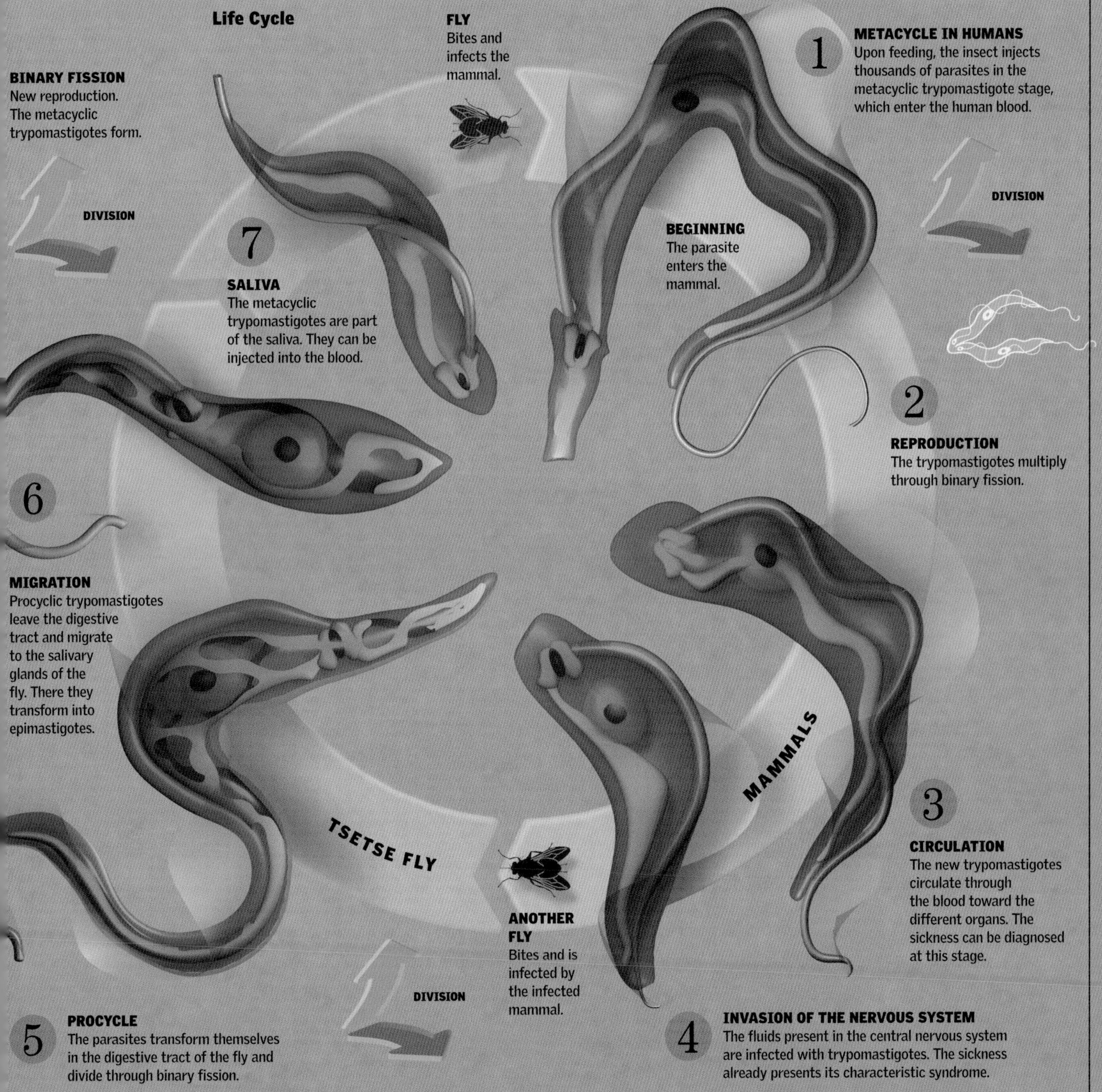

# Life and Protection

White and red blood cells are the main cellular components of blood, and they play important roles in the body. The red blood cells transport oxygen from the lungs to the tissues, and they carry carbon dioxide on their return. They live for about 120 days and then die in the spleen. The white blood cells have a smaller presence than the red ones, but they are in charge of protecting against infections, and they roam the body looking for viruses and bacteria.

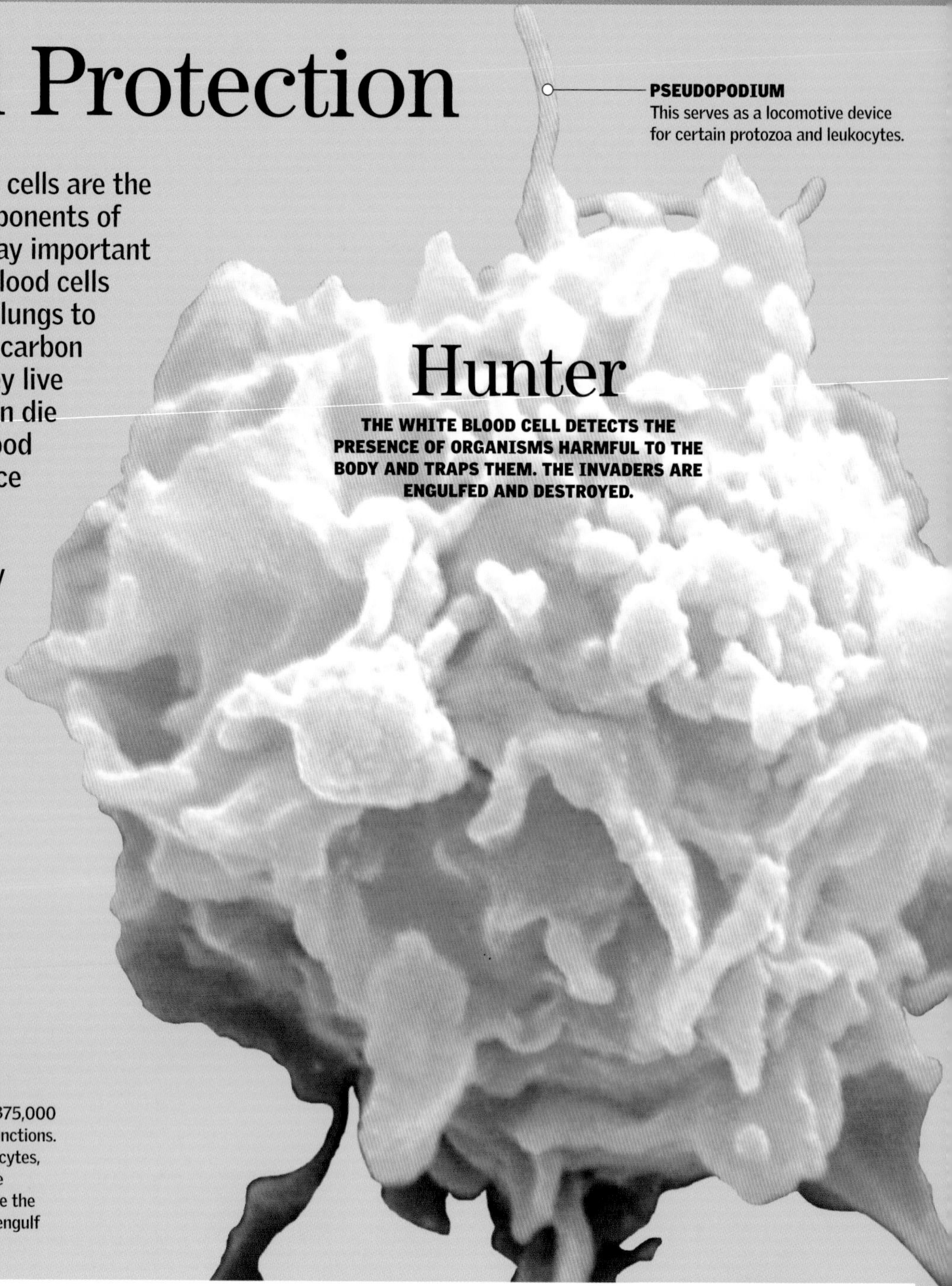

PSEUDOPODIUM
This serves as a locomotive device for certain protozoa and leukocytes.

## Hunter

**THE WHITE BLOOD CELL DETECTS THE PRESENCE OF ORGANISMS HARMFUL TO THE BODY AND TRAPS THEM. THE INVADERS ARE ENGULFED AND DESTROYED.**

## White Blood Cells

These cells occur mainly in the blood and circulate through it to fight infections or foreign bodies, but they can occasionally attack the normal tissues of their own body. They are part of the human body's immune defense. For each white blood cell in the blood, there are 700 red blood cells. White blood cells, however, are larger. Unlike the red ones, they have a nucleus. By changing shape, they can go through capillary walls to reach tissues and hunt their prey.

### Anatomy of a White Blood Cell

In a single drop of blood, there can be about 375,000 white blood cells with different shapes and functions. They are divided into two groups: the granulocytes, which have granules in the cytoplasm, and the agranulocytes, which do not and which include the lymphocytes and the monocytes. Monocytes engulf the invader, ingest it, and then digest it.

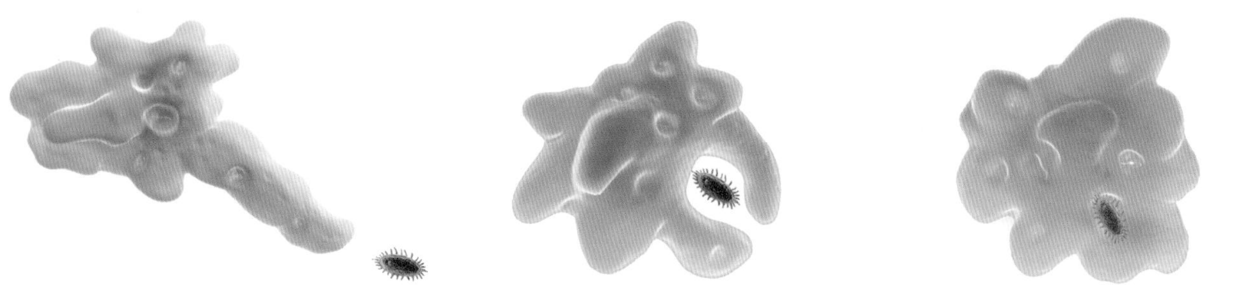

**1** White blood cells can come out of blood vessels and move between the tissues. When they detect an intruder, they approach to hunt it down.

**2** The cell stretches, forming a pseudopodium, or false leg, which pushes against the medium, and it then propels the rest of the cell to advance toward the bacterium.

**3** It traps the bacterium and destroys it. During the fight against the infection, millions of white blood cells may die and appear as pus.

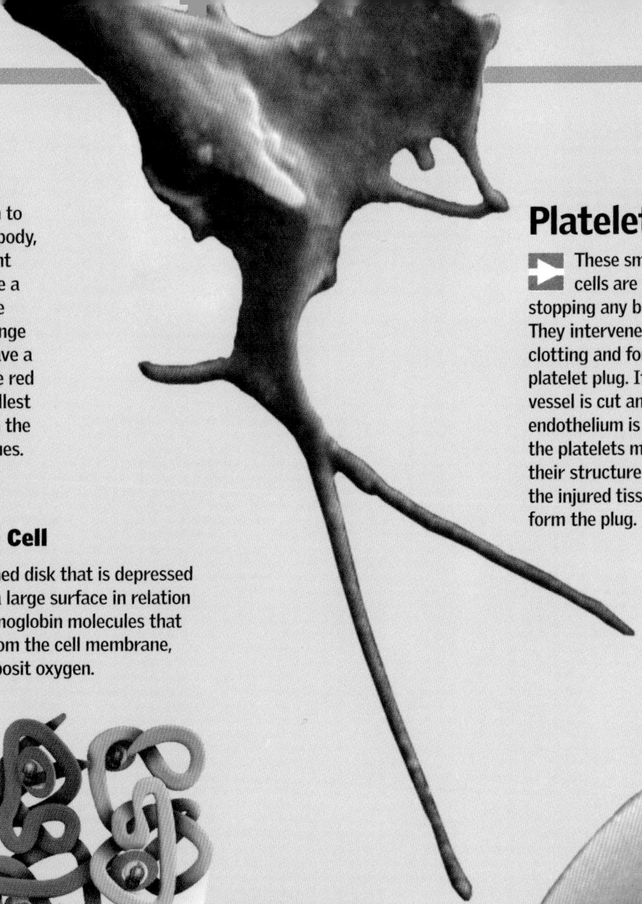

## Red Blood Cells

The main carriers of oxygen to the cells and tissues of the body, red blood cells make up 99 percent of the cells in the blood. They have a biconcave shape so that they have a larger surface for oxygen exchange in the tissues. In addition, they have a flexible membrane that allows the red blood cells to go through the smallest blood vessels, obtain oxygen from the lungs, and discharge it in the tissues. The cells do not have a nucleus.

### Anatomy of a Red Blood Cell

The cell has the shape of a flattened disk that is depressed in the center. This shape gives it a large surface in relation to its volume. In this way, the hemoglobin molecules that transport oxygen are never far from the cell membrane, which helps them pick up and deposit oxygen.

**HEMOGLOBIN**
Formed by a heme group (with iron, which will give blood its red color) and globin, a globular protein.

**OXYHEMOGLOBIN**
Formed when the hemoglobin takes up oxygen and gives blood its color.

## Platelets

These small cells are key to stopping any bleeding. They intervene in blood clotting and form a platelet plug. If a blood vessel is cut and the endothelium is affected, the platelets modify their structure and join the injured tissue to form the plug.

**1** Platelets accumulate and form a plug in the wound.

**2** The red blood cells close in. Together with a protein network they form the blood clot. The white blood cells fight the infection.

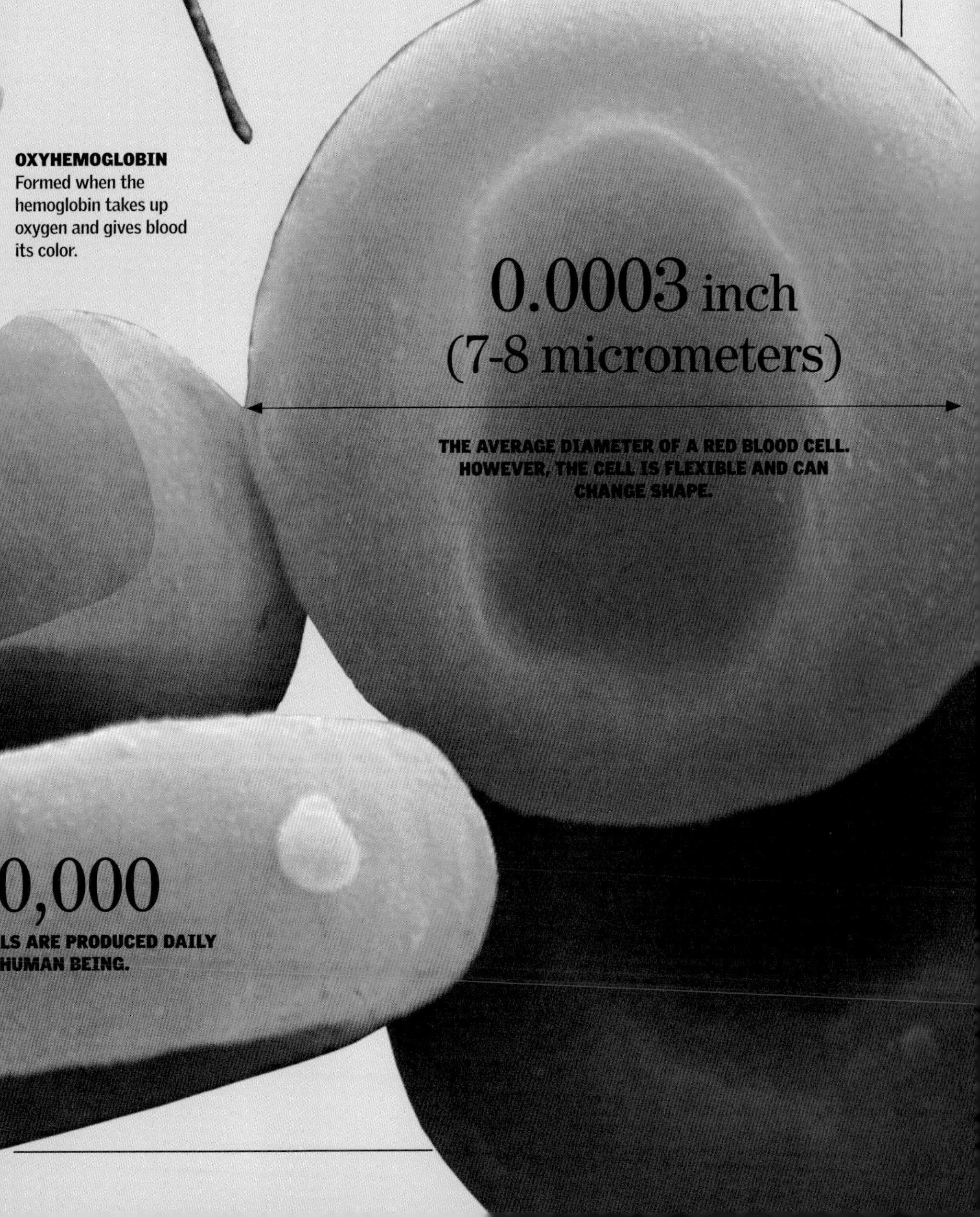

0.0003 inch (7-8 micrometers)

**THE AVERAGE DIAMETER OF A RED BLOOD CELL. HOWEVER, THE CELL IS FLEXIBLE AND CAN CHANGE SHAPE.**

200,000

**RED BLOOD CELLS ARE PRODUCED DAILY BY A HUMAN BEING.**

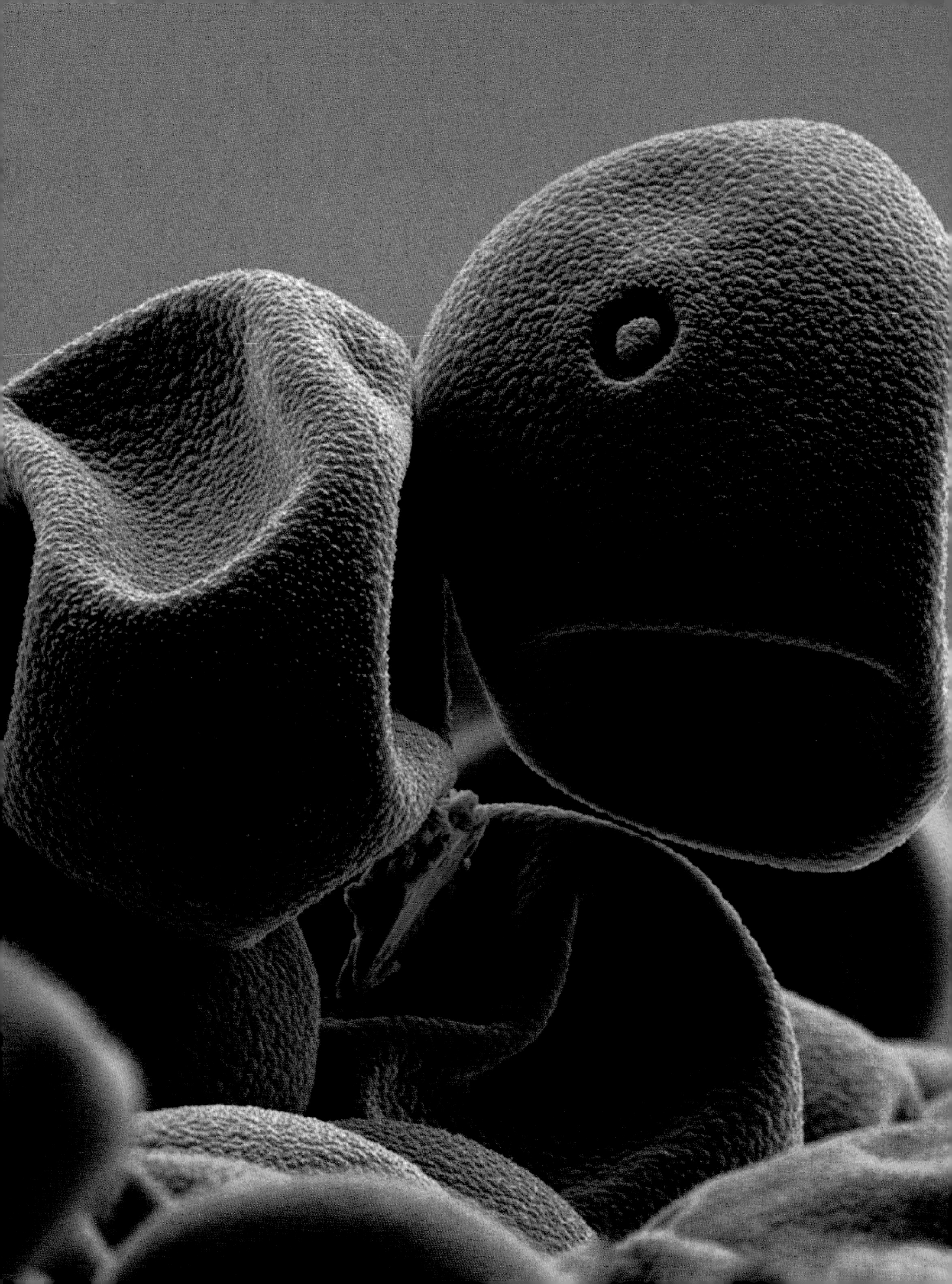

# The Most Common Diseases

Allergies are the body's response to a foreign substance, called an allergen. The most common are pollen, mites, animal dander, and nut proteins. In this chapter, we tell you which are the most common diseases that humankind suffers from today—some of them worse than others—their symptoms, and how they can be avoided. The information, written in an accessible and understandable way, is accompanied by pictures and full-color images that reveal, for example, how metastasis happens and how the AIDS virus attacks. Turn the page and you will discover astounding aspects of human disease.



**POLLEN** (opposite)
An enlarged photo of the pollen of Timothy grass (*Phleum pratense*), one of the most allergenic and best-known grasses because it causes hay fever.

# Cancer

The word "cancer" describes a group of more than 200 diseases caused by uncontrolled cell division. The genes of normal cells change so that regular cell death (apoptosis) does not take place, and the tissues grow much larger than normal. Some factors, such as tobacco use and excessive exposure to different types of radiation, can notably increase the chances of developing cancer. In other cases, the genes that alter the normal functions of cells can be inherited.

## Common Symptoms

Although they are not always indicators of cancer, unusual bleeding, unexplained changes in weight, indigestion, and difficulty swallowing can be signs of tumors.

## How it Behaves

In general, cancer consists of the abnormal growth of cells. When the cells of a tissue undergo disorderly and accelerated cell division, they can invade other, healthy tissues in the body and often destroy them. Instead of undergoing a controlled and programmed cell death (apoptosis), cancer cells continue proliferating. They can form a lump or bulge in an organ, called a tumor. Tumors are termed malignant if they are formed by cancer cells; otherwise they are called benign.

## Phases of Cancer

Before the definitive formation of cancer, there are two prior noncancerous stages: hyperplasia and dysplasia. The cell volume increases as the cells undergo uncontrolled cell division. The proliferation can be detected through studies done under a microscope (biopsies).

**1 HYPERPLASIA**
Although the cell structure remains normal, the tissue increases in size. Hyperplasia is reversible.

**2 DYSPLASIA**
The tissue loses its normal appearance. Like hyperplasia, this stage can be detected with microscopic tests.

**3 CANCER**
The cells grow uncontrollably and settle in one place. If they migrate and spread to other parts of the body, it is called metastasis.

## Breast Cancer

One in nine women develops this disease, which causes the most deaths among women. The risk of breast cancer increases with age. The most common symptom is the appearance of a small lump in the breast, which can be removed early with surgery. Other symptoms of cancer are the appearance of blood in the nipple and dimples in the breast skin. A mammogram is usually used to detect cancer. If the results of this study are positive, then treatment can begin early.

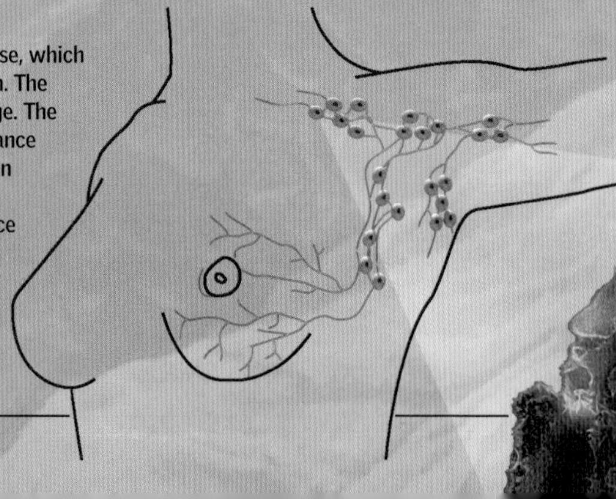

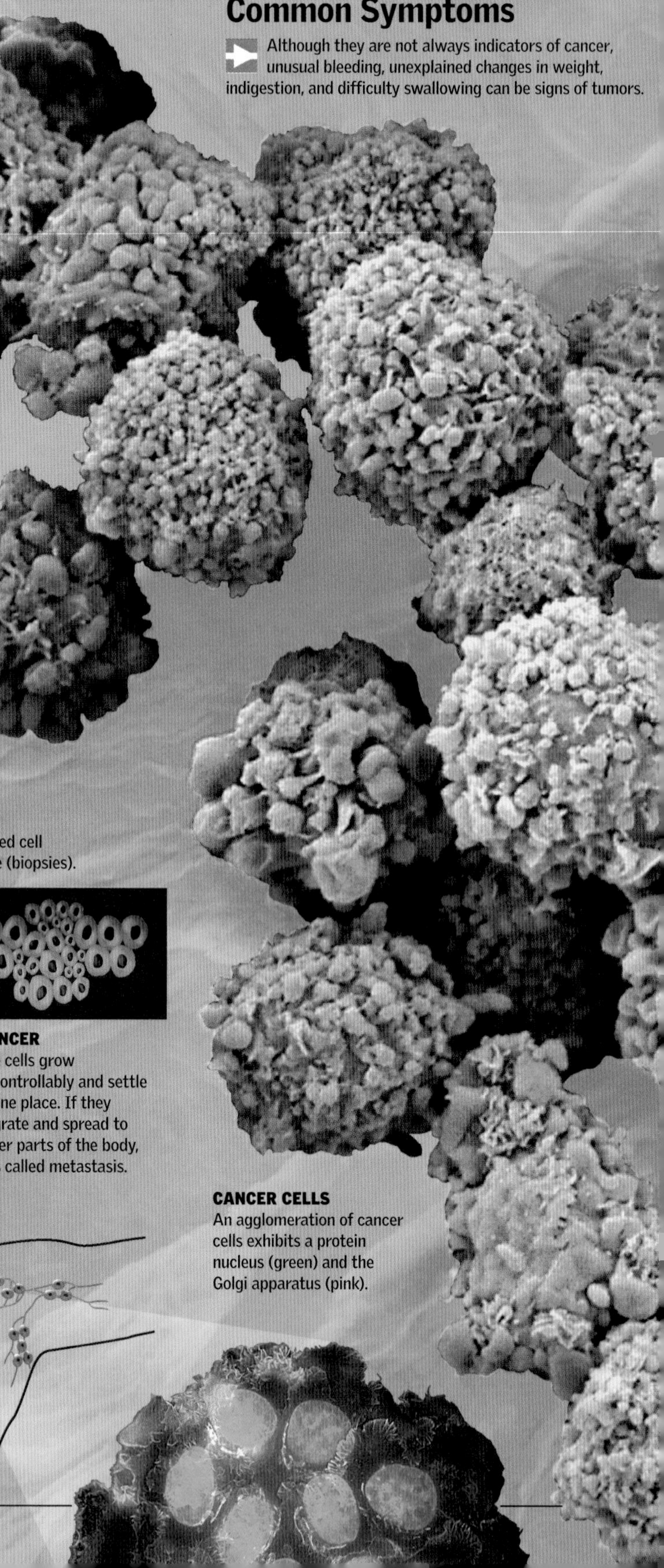

**CANCER CELLS**
An agglomeration of cancer cells exhibits a protein nucleus (green) and the Golgi apparatus (pink).

## Metastasis

Metastasis occurs when cancer cells pass from their original proliferation site to another that they were not in direct contact with (e.g., from the lungs to the brain). To achieve this migration, the cells build their own circulatory and feeding systems. This allows them to penetrate the blood vessels (intravasation) and survive after extravasation. Only one in every 1,000 cells can survive the complex intermediate processes, but if metastasis does occur, it is almost irreversible and causes irreparable damage.

### Metastasis: Step by Step

**1 ANGIOGENESIS**
The cancer cells divide and diversify. They form their own blood vessels to receive nutrients and oxygen.

**2 INTRAVASATION**
After passing through the basal membrane, the metastatic cells invade the blood vessels of the body and enter the bloodstream.

**3 MIGRATION**
The cells travel through the bloodstream and move to a new organ, different from the one with the original tumor.

**4 INTERACTION**
Cancer cells interact with the lymphocytes in the bloodstream. Their adhesion to platelets leads to the formation of tumorous embolisms.

**5 INVASION**
Before migrating and producing the secondary tumor in a new organ, the cells adhere to the basal membrane of the blood vessels.

**6 EXTRAVASATION**
The cells break the membrane, and the final migration takes place. They deposit themselves in metastatic form and begin angiogenesis to arrange for a capillary system that can provide them with nutrients. From there, they begin their growth.

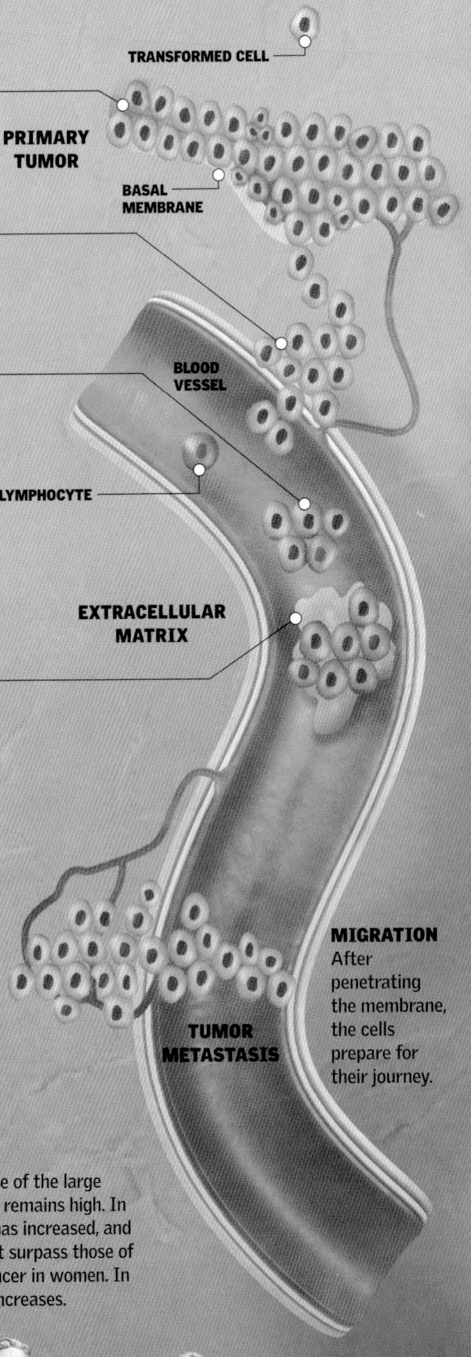


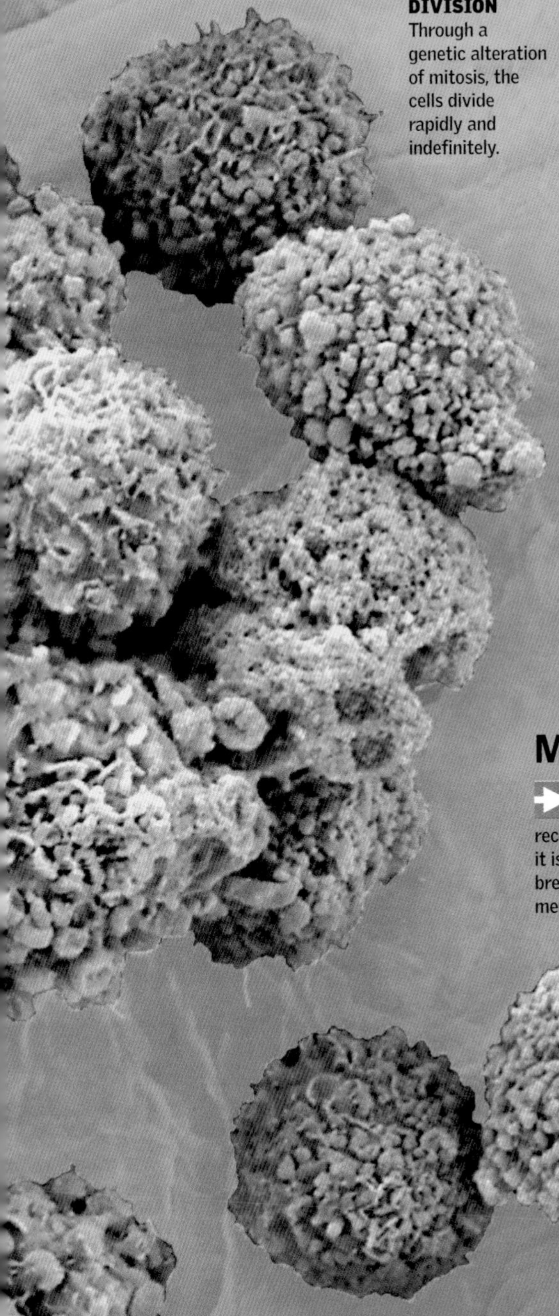

**CHAOTIC DIVISION**
Through a genetic alteration of mitosis, the cells divide rapidly and indefinitely.

**TUMORS**
are produced when the cancer cells group and form agglomerates. Tumors can be benign (noncancerous) or malignant.

## Most Frequent Cancers

The most common cancer is lung cancer. Because of the large smoking population, the incidence of this cancer remains high. In recent years, the frequency of lung cancer in women has increased, and it is possible that cases of lung cancer in women might surpass those of breast cancer, currently the most common type of cancer in women. In men, prostate cancer becomes more common as age increases.

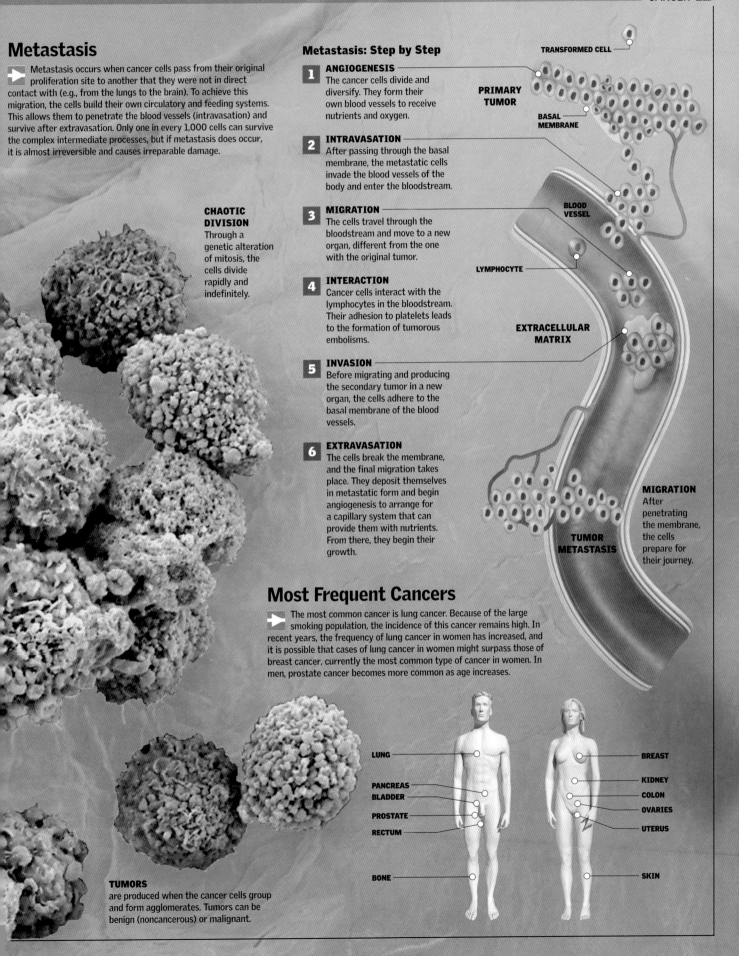

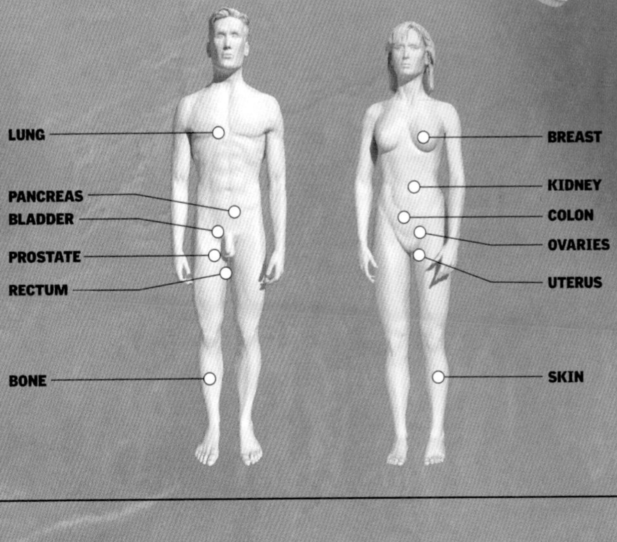

# Neurological Problems

Diseases that directly affect the brain cause structural, biochemical, or electrical changes in the brain or the spinal cord. When some of these diseases (Alzheimer's, Parkinson's, multiple sclerosis) affect the body, different symptoms appear, such as memory and reasoning disorders, tremors, rigidity of movements, paralysis, or loss of sensation. The challenge for science is to discover a way to reverse them. So far, the symptoms can only be reduced.

## Language

The language region of the brain also deteriorates. People who suffer from Alzheimer's tend to have trouble carrying out and expressing complex reasoning. Language disorders include lack of initiative in speaking and slowness to respond to the listener.

## Memory

Becomes progressively damaged. In the beginning, close relatives might not be recognized. Later, memory loss is complete.

## Alzheimer's Disease

Alzheimer's disease, which has no cure, affects mostly persons over 60 years of age. Age and the aging process are determining factors. The cortex of the brain suffers atrophy, which is permanent because nerve cells cannot regenerate. In a brain affected by Alzheimer's, the abnormal deposit of amyloid protein forms neuritic (senile) plaques in the brain tissue. Tangles of degeneration (neurofibrillary tangles) form, which progressively damage the brain's functioning.

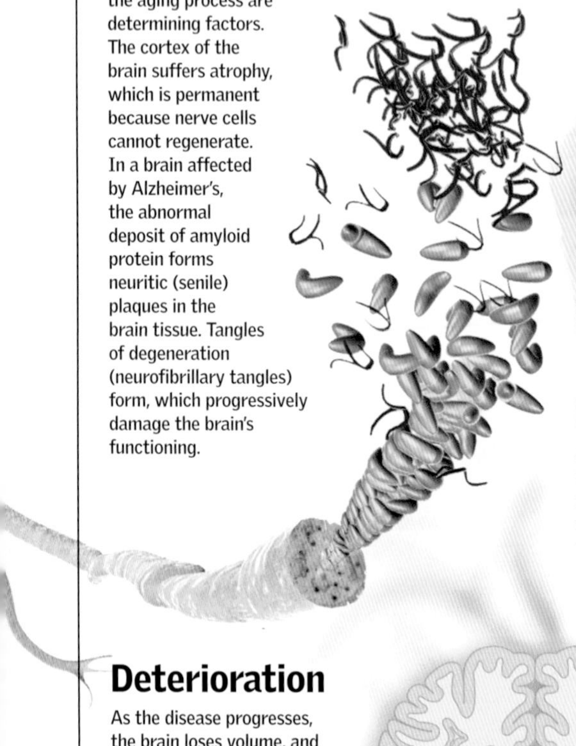

**MICROTUBULES**
These help transmit nerve impulses throughout the body. Alzheimer's disease causes disintegration of the microtubules.

## Neurons

Alzheimer's disease causes the appearance of senile plaques and tangles of degeneration that damage the neurons.

## Deterioration

As the disease progresses, the brain loses volume, and the sections of the cortex that carry out different processes are progressively damaged. The areas of the cortex shrink.

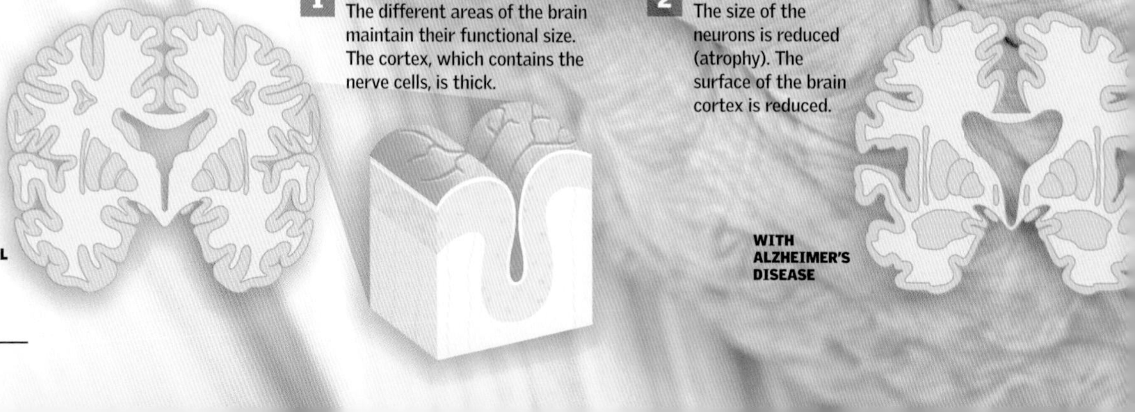

NORMAL BRAIN

WITH ALZHEIMER'S DISEASE

**1 HEALTHY CORTEX**
The different areas of the brain maintain their functional size. The cortex, which contains the nerve cells, is thick.

**2 DAMAGED CORTEX**
The size of the neurons is reduced (atrophy). The surface of the brain cortex is reduced.

**Motor cortex**

## Symptoms of Alzheimer's Disease

The first manifestations of the disease are linked to the loss of ability for verbal expression. There is also a gradual loss of memory as the disease progresses. In later phases, people with Alzheimer's can become incapable of taking care of themselves because of damage to the motor cortex.

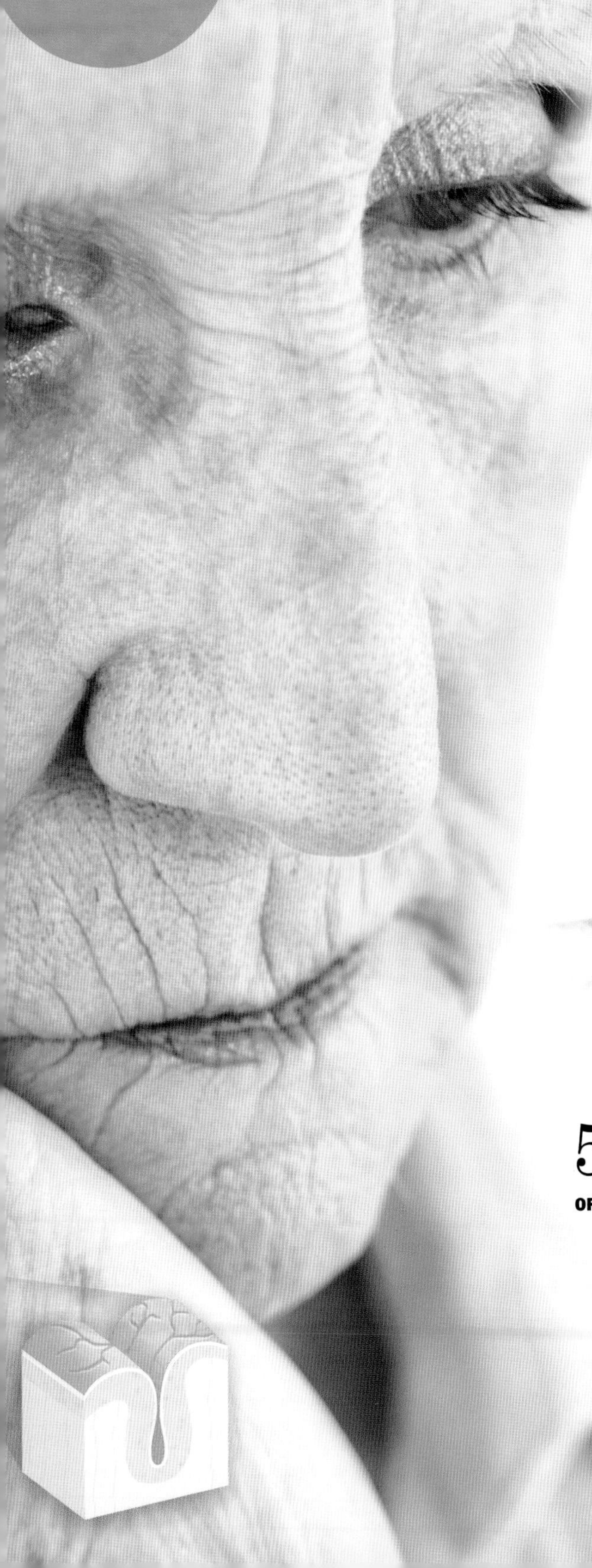

## Parkinson's Disease

Parkinson's disease is a degenerative disease that attacks one in 200 people, mostly over 60 years of age. This neurological disorder, which affects more men than women, progressively deteriorates the central nervous system. The cause of the disease is unknown. Its appearance is related to the reduction of dopamine in certain brain structures. Among the main noticeable effects are tremors, muscle rigidity, and a slowing of body movements. Parkinson's also causes complications in speech, walking, and carrying out daily chores. Progressively tremors in the arms and legs occur, followed by facial inexpressiveness and repetition of movements.

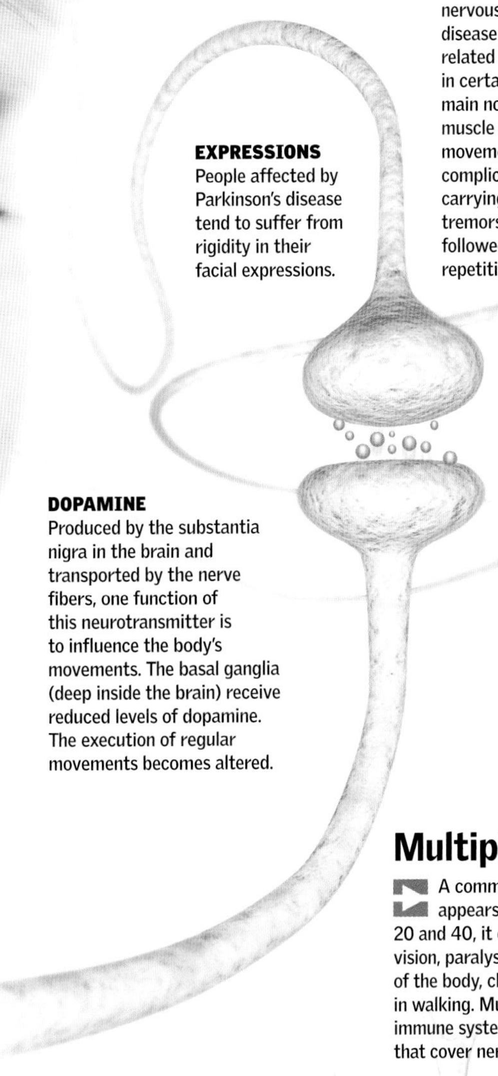

**EXPRESSIONS**
People affected by Parkinson's disease tend to suffer from rigidity in their facial expressions.

**ELECTRICAL CONDUCTION**
This occurs inside each neuron, preceding the interneural synapse. In Parkinson's disease, the connections and their ability to function are reduced dramatically.

**DOPAMINE**
Produced by the substantia nigra in the brain and transported by the nerve fibers, one function of this neurotransmitter is to influence the body's movements. The basal ganglia (deep inside the brain) receive reduced levels of dopamine. The execution of regular movements becomes altered.

**SYMPTOMS**
Muscle rigidity and slowing movement. Body posture is characterized by a forward bending of the head and trunk.

## Multiple Sclerosis

A common neurological disorder that appears sometime between the ages of 20 and 40, it can cause distorted or double vision, paralysis of the lower limbs or one-half of the body, clumsy movements, and difficulty in walking. Multiple sclerosis occurs when the immune system damages the layers of myelin that cover nerve fibers.

**50 percent**

**OF PERSONS OVER 80 SUFFER FROM NEUROLOGICAL DISEASES.**

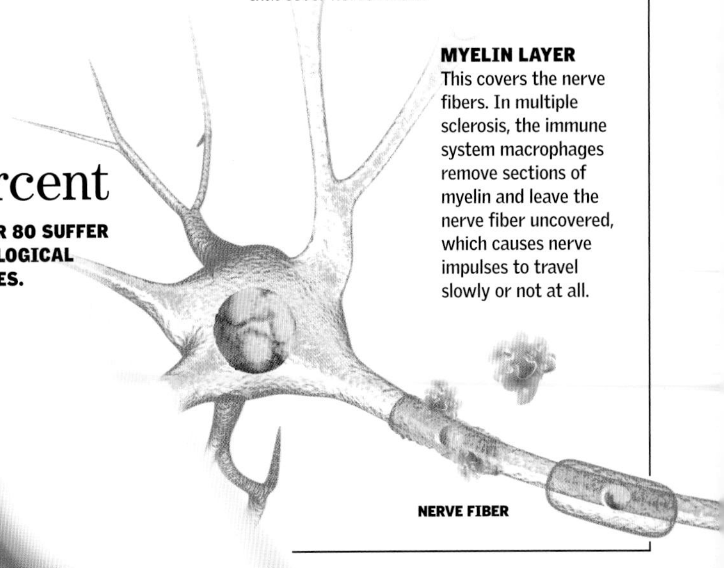


**MYELIN LAYER**
This covers the nerve fibers. In multiple sclerosis, the immune system macrophages remove sections of myelin and leave the nerve fiber uncovered, which causes nerve impulses to travel slowly or not at all.

# Bone Degeneration

Because joints are made to function in very specific ways, any abnormal movement tends to cause injury. Some injuries can result from falling or being struck, while others can be caused by degeneration of the joint. The general term for inflammation of the joints is arthritis. In the bones, the loss of bone mass is called osteoporosis and is usually related to aging.

## Osteoarthritis

Osteoarthritis, the most common form of arthritis, is the process of progressive erosion of the joint cartilage. Unlike rheumatoid arthritis, which can affect other organs, osteoarthritis affects only the joints, either in a few specific joints or throughout the body. The joint degeneration of osteoarthritis could worsen due to congenital defects, infections, or obesity. Because cartilage normally erodes with age, osteoarthritis affects people close to 60 years of age.

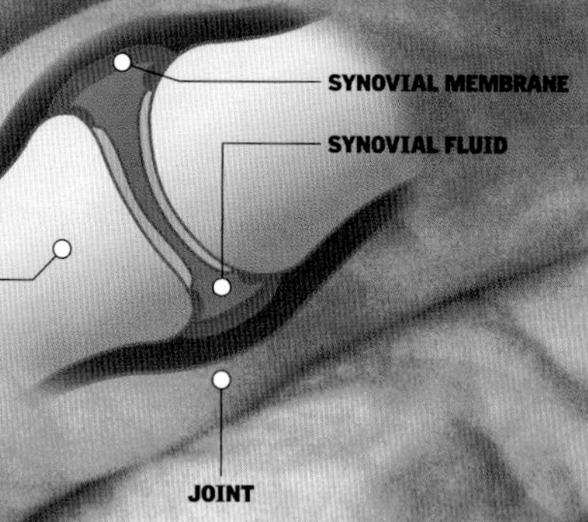


**STRUCTURE**
The joint is normally made up of cartilage that is lubricated by the synovial fluid to allow ease of movement.

### Phases of the Disease

**1 DETERIORATION**
Osteoarthritis causes progressive damage of the cartilage. When the cartilage cells die, cracks appear on the surface of the bone. From this moment on, the synovial fluid begins to leak. Later the fluid enters the cartilage and causes it to degenerate.

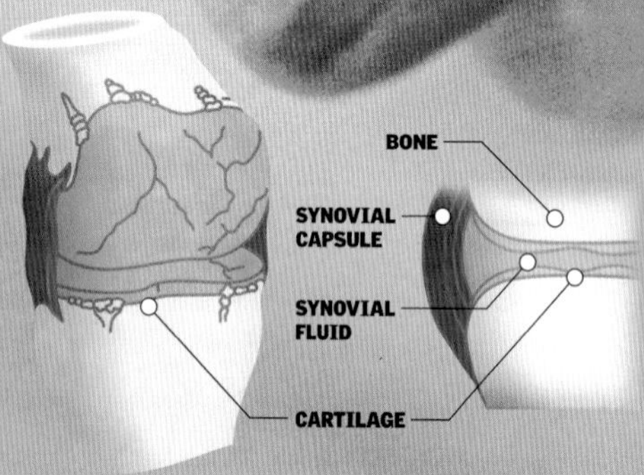


**30%**
**OR MORE OF THE MINERAL DENSITY OF THE BONE IS LOST THROUGH THE DEGENERATION OF OSTEOPOROSIS.**

**2 BONE FRACTURE**
The cartilage is worn away down to the bone and breaks its surface. From this erosion, a hole appears. New blood vessels begin to grow. To fill the gap, a plug develops that is made up of fibrocartilage.

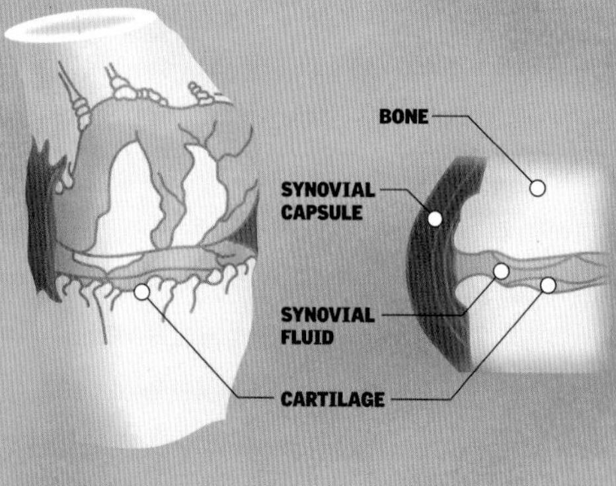


**3 EXPOSED BONE**
The plug disappears and leaves the bone surface exposed. If the surface fractures become deeper, the synovial fluid can enter the bone marrow and form a cyst surrounded by weakened bone. Osteophytes (bone spurs) can appear.

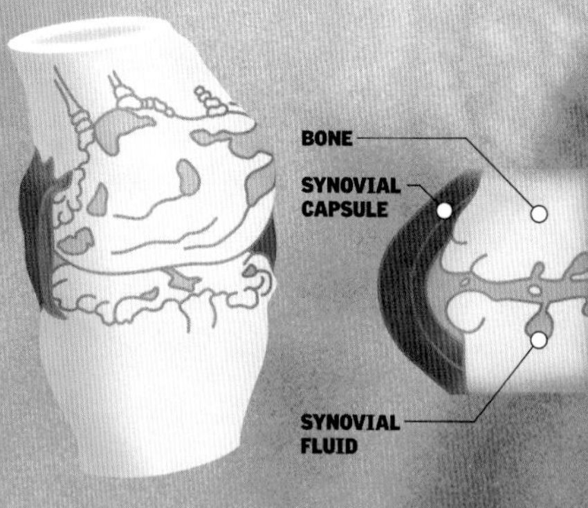

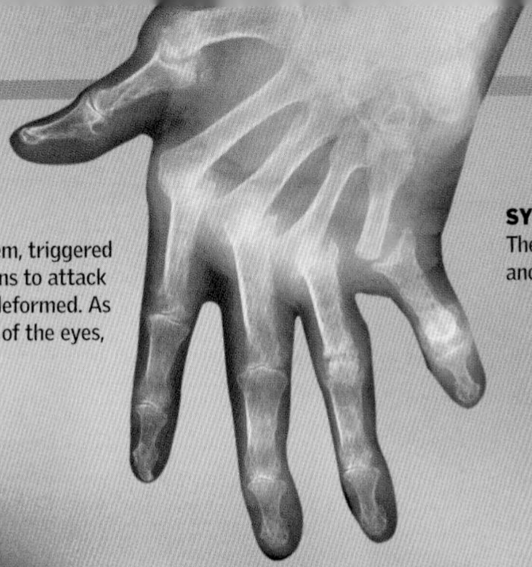

## Rheumatoid Arthritis

In this autoimmune disease, the immune system, triggered by some antigen in a predisposed person, begins to attack the body's tissues. The joints become inflamed and deformed. As rheumatoid arthritis develops over time, the tissues of the eyes, skin, heart, nerves, and lungs may be affected.

**SYMPTOMS**
The typical symptoms are fatigue, anorexia, and muscle and joint pain.

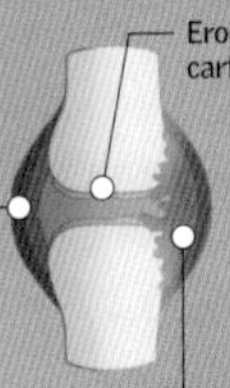


## Symptoms of Osteoarthritis

The most common signs of the degeneration of the joint cartilage are the deformation and swelling of the joints. Some cases might include numbness and limited movement of the joint.

## Osteoporosis

Between the fifth and sixth decade of life, the bones tend to become more porous and to decrease in thickness. Both men and women lose bone mass, even if they are healthy. The levels of estrogen decrease rapidly in women after menopause, leading to osteoporosis in many cases. In men, the reduction in testosterone is gradual, and the likelihood of suffering from osteoporosis is lower.

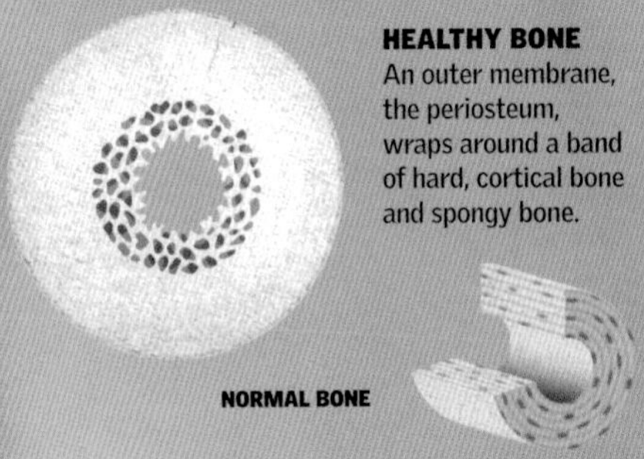


**HEALTHY BONE**
An outer membrane, the periosteum, wraps around a band of hard, cortical bone and spongy bone.

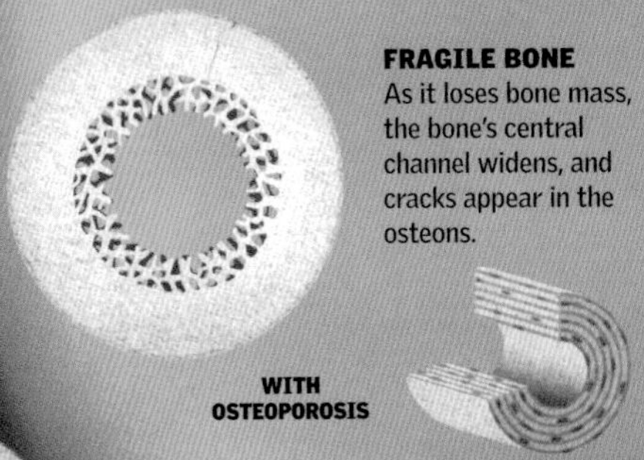


**FRAGILE BONE**
As it loses bone mass, the bone's central channel widens, and cracks appear in the osteons.

**REDUCED MASS**
Osteoporosis generates a decrease of total bone mass. As a consequence, pores appear that could weaken the bone.

**PORES**
These appear on the bone surface as a consequence of tissue degeneration and the progressive erosion of the bone.

**SURFACE**
It is more susceptible to fracture as the bone loses rigidity because of injury to the bone cells.

# Gout

**IS CAUSED BY HIGH LEVELS OF URIC ACID IN THE BLOOD. THE ACID IS DEPOSITED IN THE JOINTS, CAUSING INFLAMMATION. PRIMARY GOUT IS DUE TO A CONGENITAL METABOLIC ERROR, AND SECONDARY GOUT IS CAUSED BY ANY OTHER METABOLIC DISORDER.**

# Circulatory Conditions

Among the most frequent diseases that affect the circulatory system are those that result from blockages of the arteries and veins. The buildup of fat in the arteries can lead to arteriosclerosis, which blocks the supply of blood to the tissues. In many cases, as in a myocardial infarction, there are no warning signs. This could lead to the death of the tissue that loses blood supply. Certain drugs can be used to dilate blocked blood vessels.

## Vena Cava

**THE SUPERIOR VENA CAVA TAKES THE BLOOD FROM THE HEAD AND ARMS TO THE RIGHT ATRIUM. THE INFERIOR VENA CAVA TAKES DEOXYGENATED BLOOD RETURNING FROM THE LOWER TRUNK AND LIMBS TO THE RIGHT ATRIUM.**

## Aorta

**THE LARGEST BLOOD VESSEL IN THE BODY, WITH AN INTERNAL DIAMETER OF 1 INCH (2.5 CM). IT TAKES BLOOD WITH FRESH OXYGEN TO ALL PARTS OF THE BODY.**

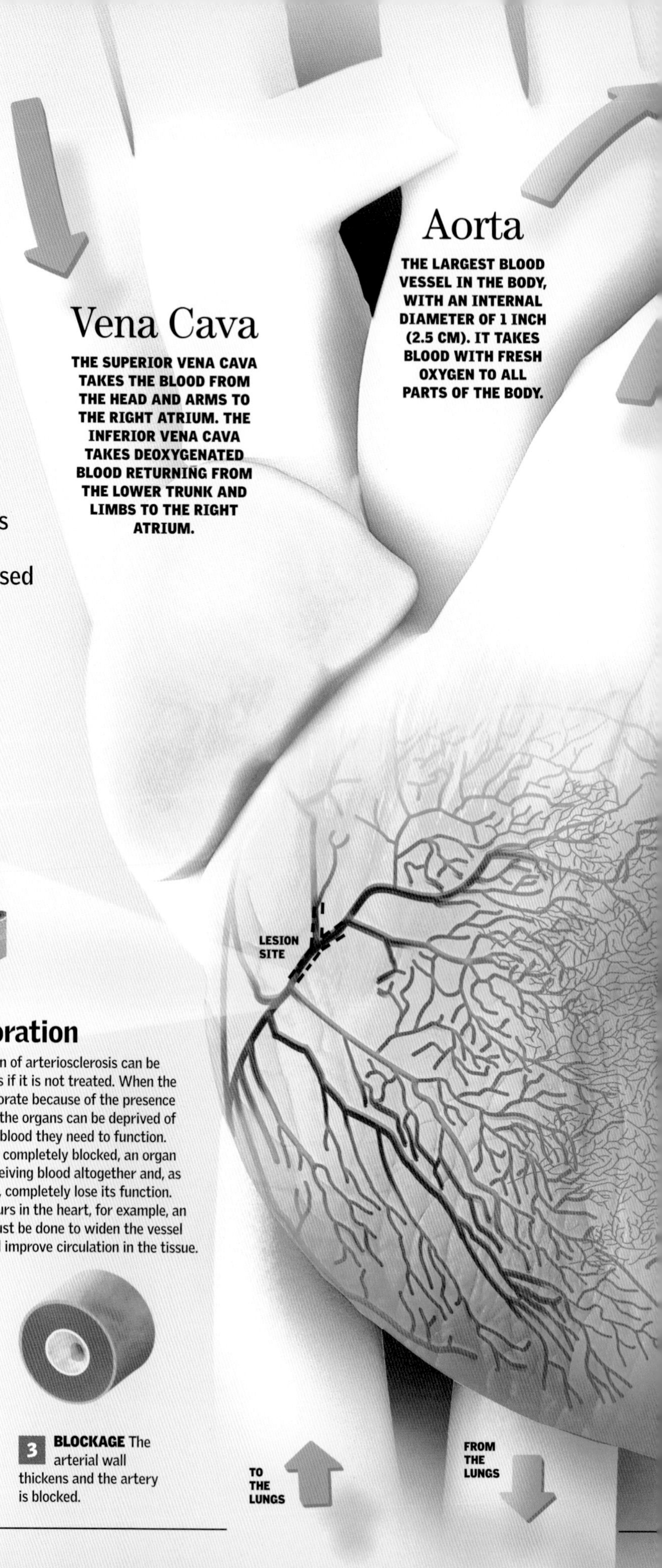


## Arteriosclerosis

Arteriosclerosis of the cardiac blood vessels, or heart disease, is caused by a narrowing of the arteries as cholesterol, cells, and other substances accumulate in the lining of these vessels. Arterial obstruction is gradual; it begins when excess fats and cholesterol build up in the blood. These substances infiltrate the lining of the arteries to create microscopic damage sites. Atheromata form, which in turn develop into fatty masses called plaque. The appearance of these plaques thickens the arterial walls and prevents the normal flow of blood, thus reducing the blood flow.

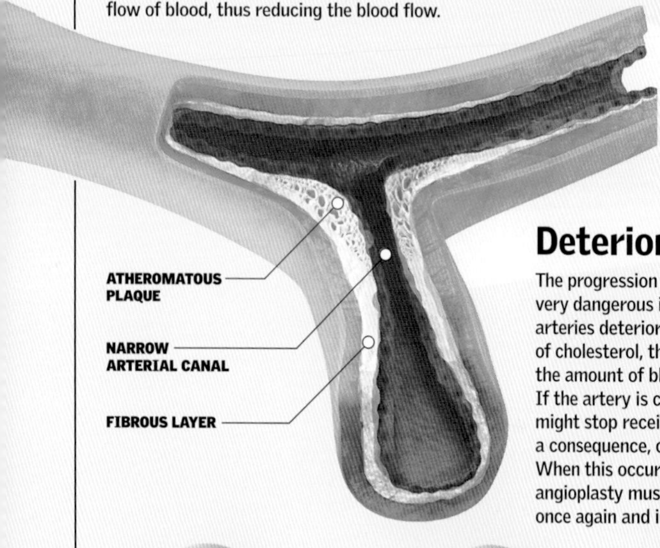


## Deterioration

The progression of arteriosclerosis can be very dangerous if it is not treated. When the arteries deteriorate because of the presence of cholesterol, the organs can be deprived of the amount of blood they need to function. If the artery is completely blocked, an organ might stop receiving blood altogether and, as a consequence, completely lose its function. When this occurs in the heart, for example, an angioplasty must be done to widen the vessel once again and improve circulation in the tissue.

1 **FREE** Without the formation of fatty plaque, the blood flows normally.

2 **WITH ATHEROMATOUS PLAQUE** Inside this plaque, cholesterol and other substances accumulate.

3 **BLOCKAGE** The arterial wall thickens and the artery is blocked.

# Pulmonary Artery

**BRANCHES OUT FROM THE RIGHT VENTRICLE. EACH BRANCH TAKES DEOXYGENATED BLOOD TO THE LUNGS. THE PULMONARY ARTERY IS THE ONLY ARTERY THAT TRANSPORTS DEOXYGENATED BLOOD.**

## Pulmonary Hypertension

When the blood pressure in the pulmonary artery increases, the walls thicken. The blood flow from the heart is reduced.

## Pectoral Angina

Chest pains could be a warning sign that the cardiac muscle is not receiving enough blood to keep up with the demands of the work it is doing. In pectoral angina, very strong chest pains occur because of the arteries that are obstructed by arteriosclerosis.

### TREATMENT

Nitroglycerin, a drug that dilates blood vessels, can be used to relieve the effects of pectoral angina.

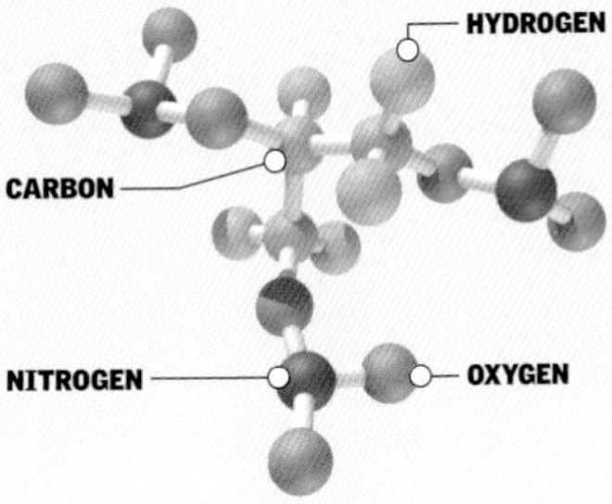


### WIDENING VESSELS

To reestablish adequate blood flow, drugs such as nitroglycerin can be used. The narrow blood vessels dilate, and the heart does not have to work so hard.

**1 BEFORE** The narrowed blood vessels do not provide adequate blood flow to the heart.

**2 AFTER** With the drugs applied, the walls of the blood vessels relax and widen.

## Heart Attack

An infarction usually happens suddenly, almost without warning. The pain in the chest area can be like angina but generally is more severe and does not go away with rest. A person who suffers an attack experiences excessive sweat, weakness, and, in some cases, loss of consciousness. The attack could be a direct consequence of the lack of blood volume. If the artery begins to fill with fat after the partial obstruction of a blood vessel by a plaque, a lesion in its wall may form, resulting in the formation of a thrombus that could block the blood vessel. This could deprive a portion of the myocardium of oxygen, which would then produce a heart attack.

### HOW IT HAPPENS

A blockage in a coronary artery prevents blood from reaching the muscle. If the blockage is complete, the blood-deprived muscle dies.

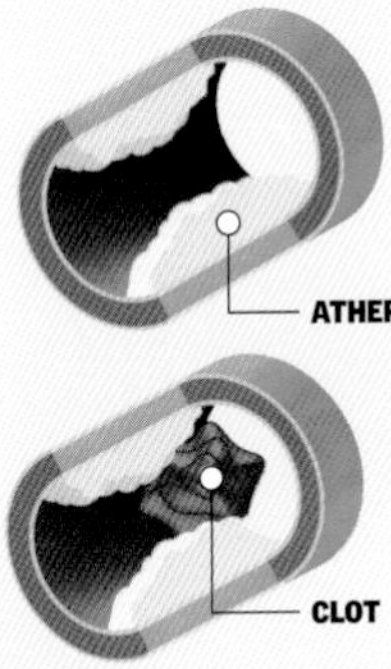


**1 ATHEROMA** The inner wall of the artery accumulates fat, producing an atheroma.

**2 INFARCTION** A clot forms. The myocardium stops receiving blood, and this region dies.

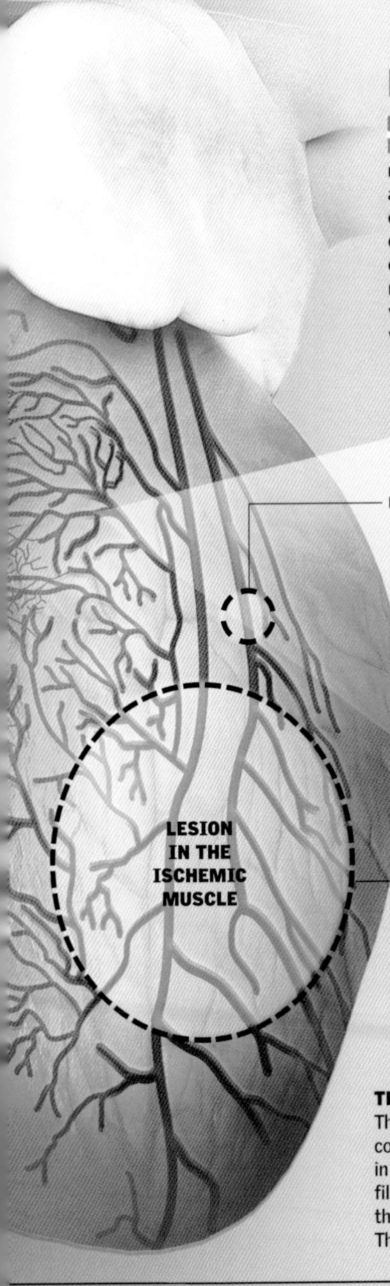


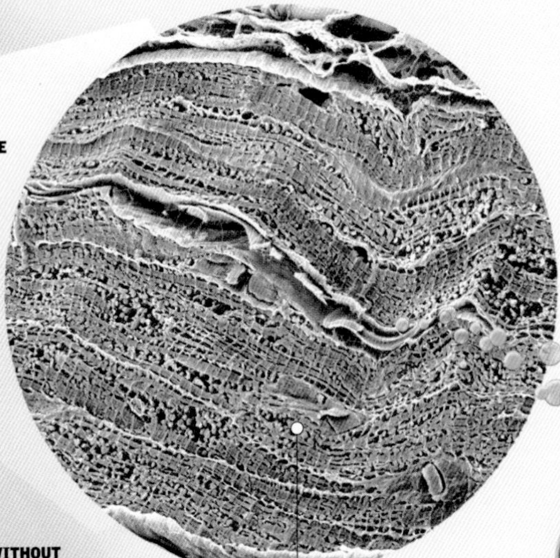


### DETECTION

When a heart attack occurs, the muscle fibers release enzymes into the bloodstream.

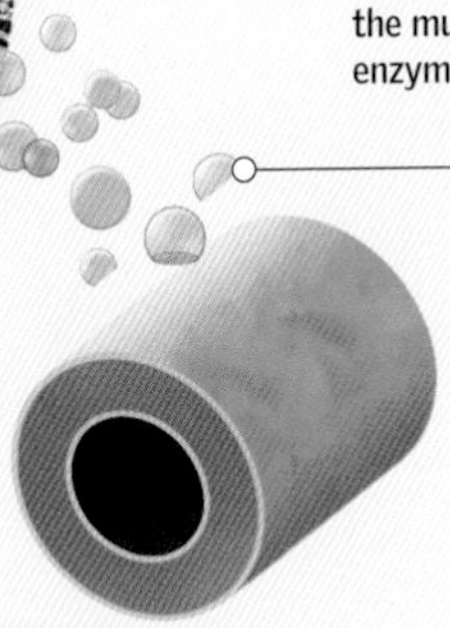

**ENZYMES** These make it possible to estimate the severity of the attack. If enzyme levels are high, it was severe.

### THROMBUS IN THE ARTERY

This forms when blood platelets come into contact with collagen in the lining of the artery. Fibrous filaments appear that interact with the platelets, and the clot grows. The artery becomes blocked.

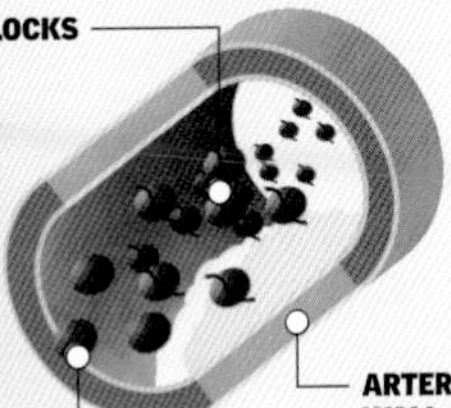


## Thrombosis

Unlike a natural clot that forms to prevent blood loss from an injured blood vessel, in arteriosclerosis, the blood vessels are already damaged. This causes a predisposition to form a thrombus when an atheroma ruptures. In thrombosis, unlike arteriosclerosis, clots form that in many cases can migrate through the bloodstream and lodge somewhere away from the original site.

# Respiratory Infections

In many cases, respiratory-tract obstructions can cause severe complications. Although bronchitis is more often related to a viral or bacterial infection, the chronic form is associated with the consumption of tobacco, because the smoking habit has severe consequences for the respiratory system. In cases of pneumonia or complications associated with the respiratory tract, bacteria or other airborne microorganisms are usually responsible for the infection.

## Acute Bronchitis

An inflammation of the bronchi that develops suddenly, it can result from an infection of the respiratory tract or exposure to toxins, irritants, or atmospheric pollutants. Acute bronchitis is usually caused by a virus. The common symptoms are cough, which increases the need to salivate, and in some instances a high fever. In acute bronchitis, the tissues and membranes of the bronchi become inflamed, and the air passages narrow. The amount of mucus increases, causing congestion.

**HOW IT HAPPENS**

The disease usually affects the large- and medium-sized bronchi. In children or older persons, the infection can expand and inflame the bronchioles and lung tissue.

**1 HEALTHY**

The air passage is wide enough for an adequate flow of air. The mucus does not obstruct the passage.

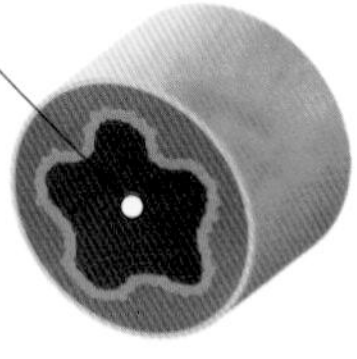


**2 BRONCHITIS**

The lining and tissues of the bronchi are inflamed. The air passage narrows, and mucus builds up.

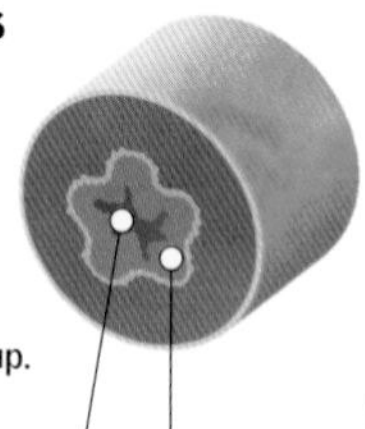


## Chronic Bronchitis

The most common cause of chronic bronchitis is the irritation of the bronchi by chemical substances. The effect of tobacco, which contains nicotine, is another primary factor in the development of chronic bronchitis. The typical symptoms are cough with phlegm, hoarseness, and difficulty breathing. One effect of smoking is an excessive production of mucus, followed by enlargement of the mucous glands and dysfunction of the cilia. Thus, the respiratory tract can be affected and can even function as a medium for the growth of some bacteria. In some cases, chronic bronchitis can be brought on by recurrent episodes of acute bronchitis.

## Bronchi

**THE LUNG HAS TWO MAIN BRONCHI THAT BRANCH OUT FROM THE TRACHEA. THESE TWO BRONCHI BRANCH OUT FURTHER INTO AN INTRICATE NETWORK OF BRONCHIAL BRANCHES THAT PROVIDE SPACE FOR THE PASSAGE OF AIR IN THE LUNGS.**

AORTIC ARCH

PULMONARY VEIN

## Pneumonia

This disease causes the inflammation of the smallest bronchioles and the alveolar tissue. In 1976, a bacteria was detected, *Legionella pneumophila* (pictured), that causes a severe, rapidly spreading form of pneumonia.

**INFECTED BRONCHI**

From the inhalation of irritant chemicals, the glands that secrete mucus become enlarged. This increases the production of mucus that cannot be eliminated from the respiratory tract. Serious breathing difficulties follow.

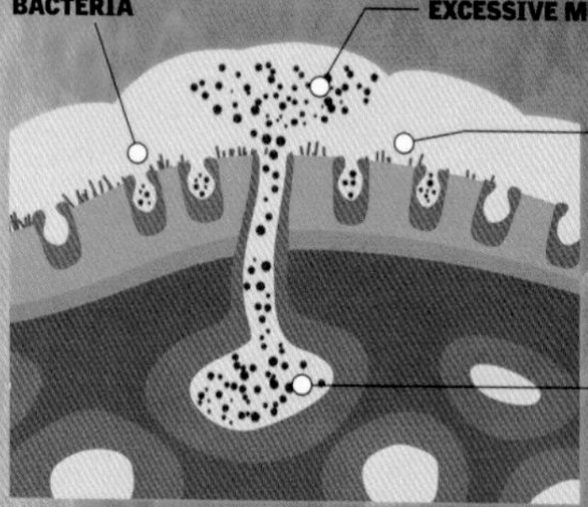


## Cilia

**ARE SMALL HAIRS LOCATED IN THE BRONCHI. THE MUCUS IN THE RESPIRATORY TRACT IS EXPELLED BY THE CILIA.**

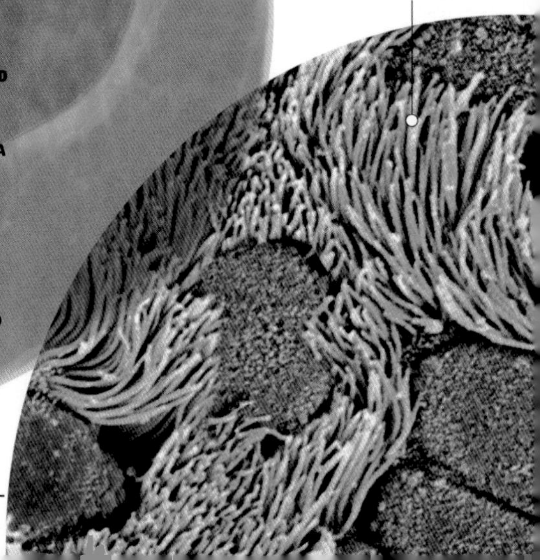

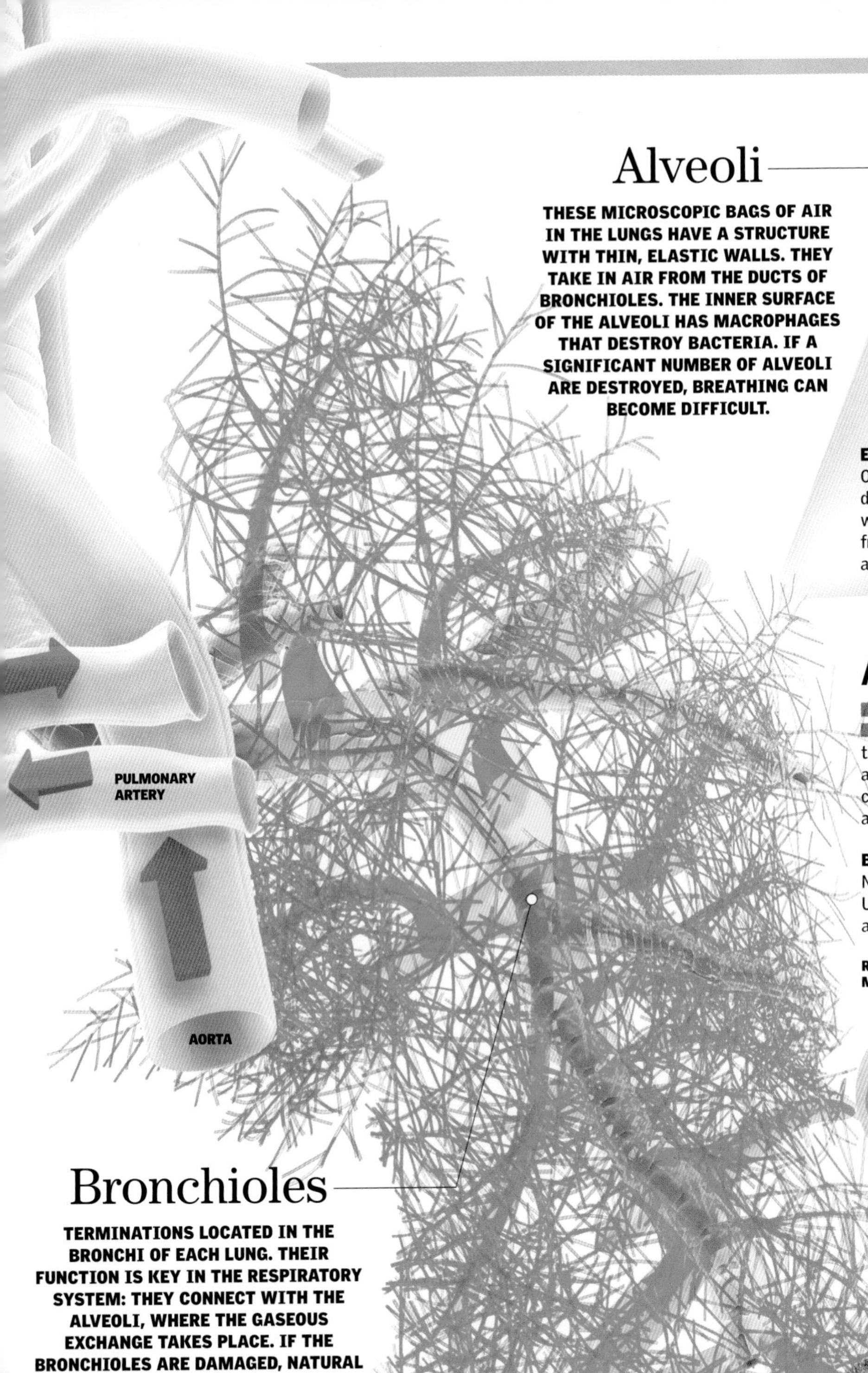


## Alveoli

**THESE MICROSCOPIC BAGS OF AIR IN THE LUNGS HAVE A STRUCTURE WITH THIN, ELASTIC WALLS. THEY TAKE IN AIR FROM THE DUCTS OF BRONCHIOLES. THE INNER SURFACE OF THE ALVEOLI HAS MACROPHAGES THAT DESTROY BACTERIA. IF A SIGNIFICANT NUMBER OF ALVEOLI ARE DESTROYED, BREATHING CAN BECOME DIFFICULT.**

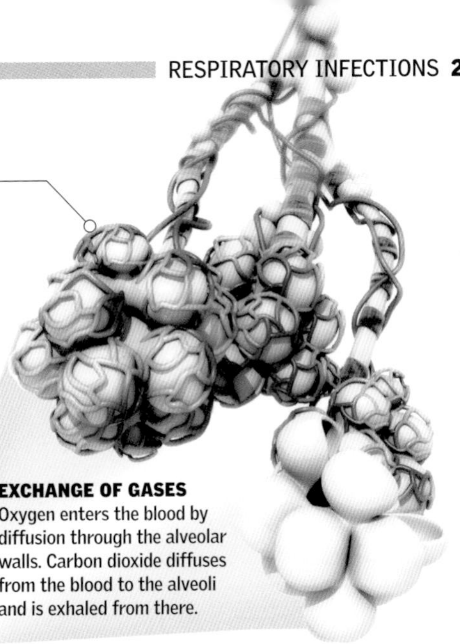

**EXCHANGE OF GASES**
Oxygen enters the blood by diffusion through the alveolar walls. Carbon dioxide diffuses from the blood to the alveoli and is exhaled from there.

## Asthma

Asthma attacks are characterized by recurrent episodes of an out-of-breath feeling caused by the constriction of the airways. The smallest bronchi and bronchioles become inflamed and filled with mucus, causing difficulty in breathing. The most recurrent type of asthma is the allergic kind, which develops in childhood.

**BRONCHIOLE CONTRACTION**
Normally the bronchiole is relaxed, and the air flows freely. Upon contraction of the muscle, there is a shortage of air and the space for its circulation is limited.

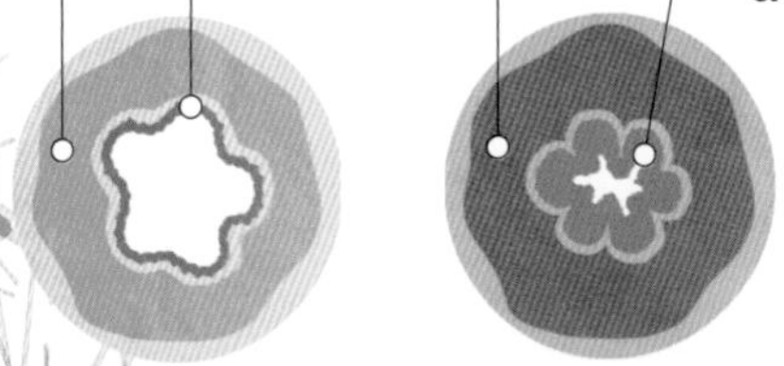


## Bronchioles

**TERMINATIONS LOCATED IN THE BRONCHI OF EACH LUNG. THEIR FUNCTION IS KEY IN THE RESPIRATORY SYSTEM: THEY CONNECT WITH THE ALVEOLI, WHERE THE GASEOUS EXCHANGE TAKES PLACE. IF THE BRONCHIOLES ARE DAMAGED, NATURAL BREATHING IS IMPEDED.**

## Lung Cancer

The growth of cancer cells (pictured) is caused by tobacco in 90 percent of cases. Of the 4,000 chemical substances in tobacco, 40 components have a carcinogenic potential that causes the appearance of atypical cells. When these cells grow, cancer cells multiply.

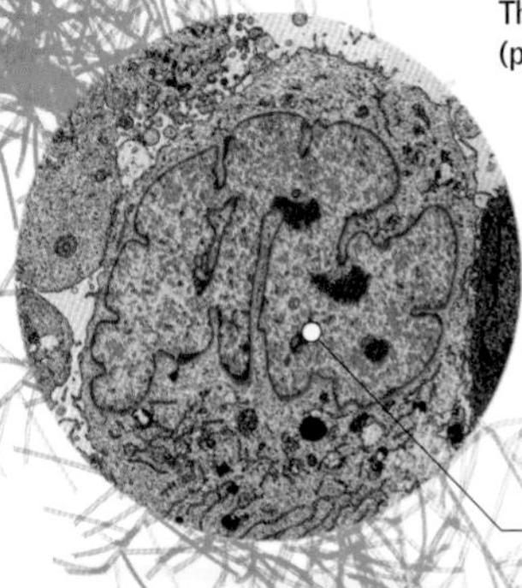

**NUCLEUS**
This is elongated and projects extensions outside the cell wall. It contains chromatin, DNA, and proteins.

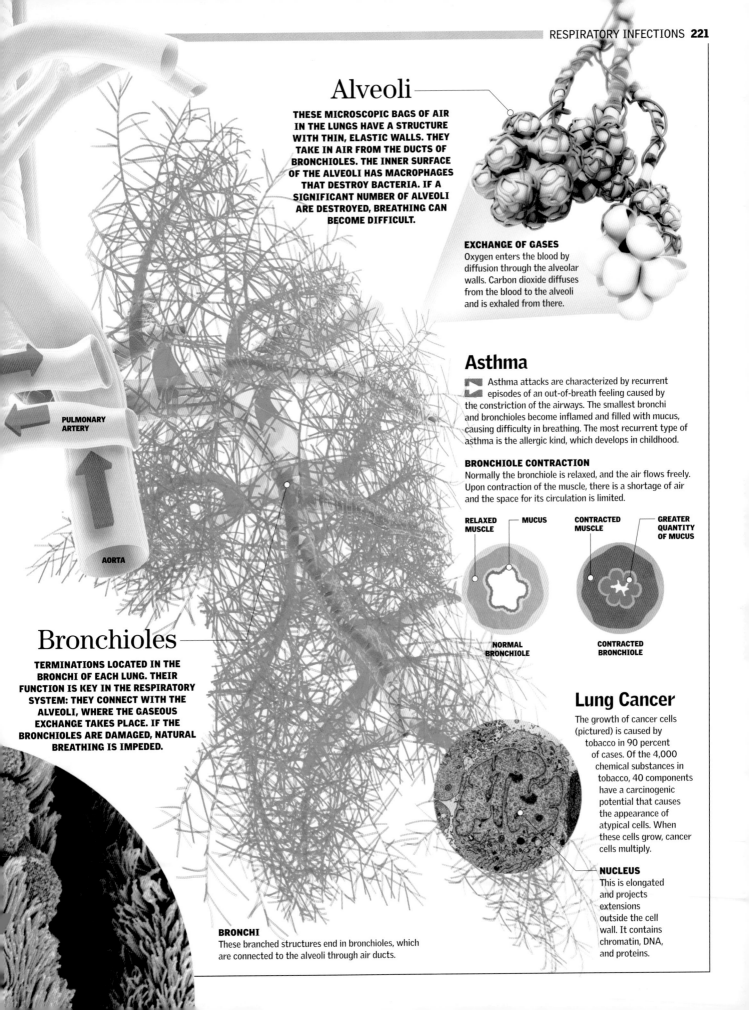

**BRONCHI**
These branched structures end in bronchioles, which are connected to the alveoli through air ducts.

# Excesses in the Digestive System

Diseases that affect the organs of the digestive system, such as the stomach, pancreas, and liver, find their origin in alcoholic drinks, poor nutrition, or bacteria that break down the layers of tissue and harm the organs. Diseases, such as cirrhosis, hepatitis B, gallstones, and ulcers, can lead to irreparable damage in different parts of the body.

## Cirrhosis

This liver disease causes fibrosis and dysfunction of the liver. The main causes are chronic alcoholism and infection with the hepatitis C virus. Cirrhosis can cause a buildup of fluid in the abdomen (ascites), clotting disorders, increased blood pressure in the hepatic veins of the digestive tract, with dilation and risk of rupture, and confusion or changes in the level of consciousness (hepatic encephalopathy). Some symptoms are edema in the lower limbs, bloody vomit, jaundice (yellowish skin), generalized weakness, weight loss, and kidney disorders.

**A** **A FATTY LIVER**
This can develop as a result of excessive alcohol consumption. The liver contains fat cells that infiltrate, become larger, and enlarge the liver.

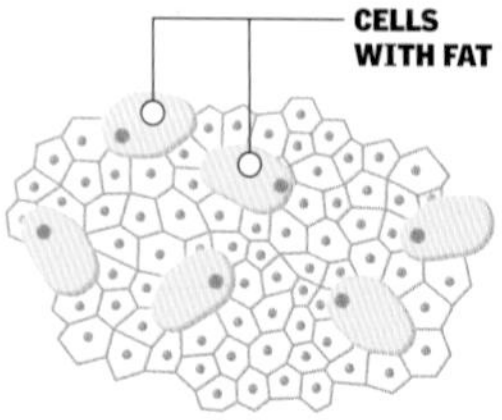


DAMAGED CELLS

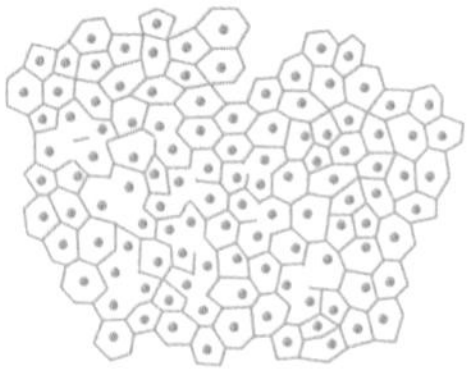

**B** **ALCOHOLIC HEPATITIS**
Alcohol consumption induces enzymes to produce acetaldehyde, which generates inflammation. This damages the hepatic cells, impairing normal liver function.

**C** **CIRRHOSIS**
Bands of damaged tissue separate the cells. This stage of destruction is irreversible and can also stem from other causes, such as viral hepatitis.

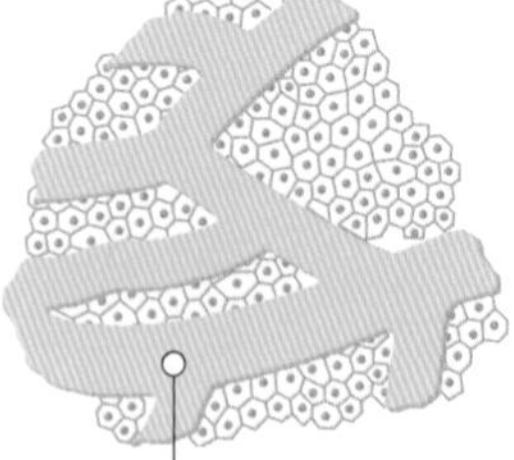


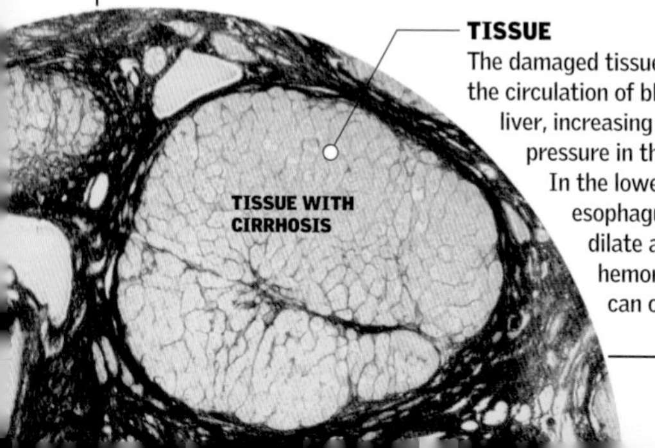


**TISSUE**
The damaged tissue affects the circulation of blood in the liver, increasing the blood pressure in the portal vein. In the lower part of the esophagus, the veins dilate and a digestive hemorrhage can occur.

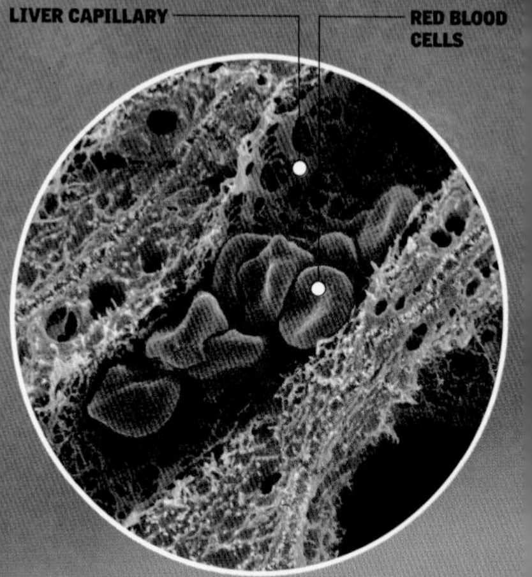


## Cleaning

Substances carried in the blood are modified during their passage through the liver, which cleans and purifies the blood supply, breaks down certain chemical substances, and synthesizes others.

## Gastritis

An inflammation of the mucous membrane of the stomach, it may have various causes, including alcohol consumption, anti-inflammatory medication, and smoking tobacco. It is also associated with *Helicobacter pylori* bacteria.

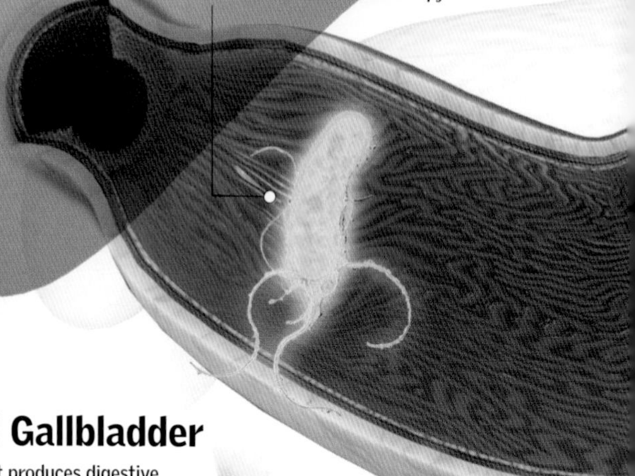

## Pancreas and Gallbladder

The pancreas is a gland that produces digestive enzymes and hormones. The gallbladder is a small sac full of bile (a substance produced by the liver), which it stores and releases into the duodenum (the upper portion of the small intestine) to help digest food.

**GALLBLADDER**
This organ stores digestive juices produced by the liver. Sometimes they solidify and form gallstones.

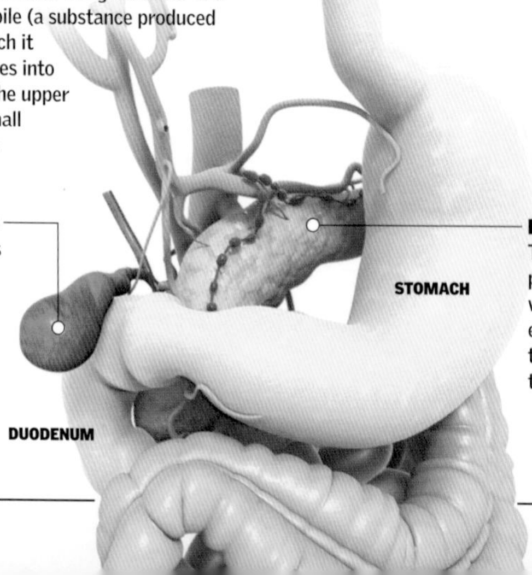


**PANCREAS**
This organ secretes pancreatic juices, which contain the enzymes necessary to digest foods, into the duodenum.

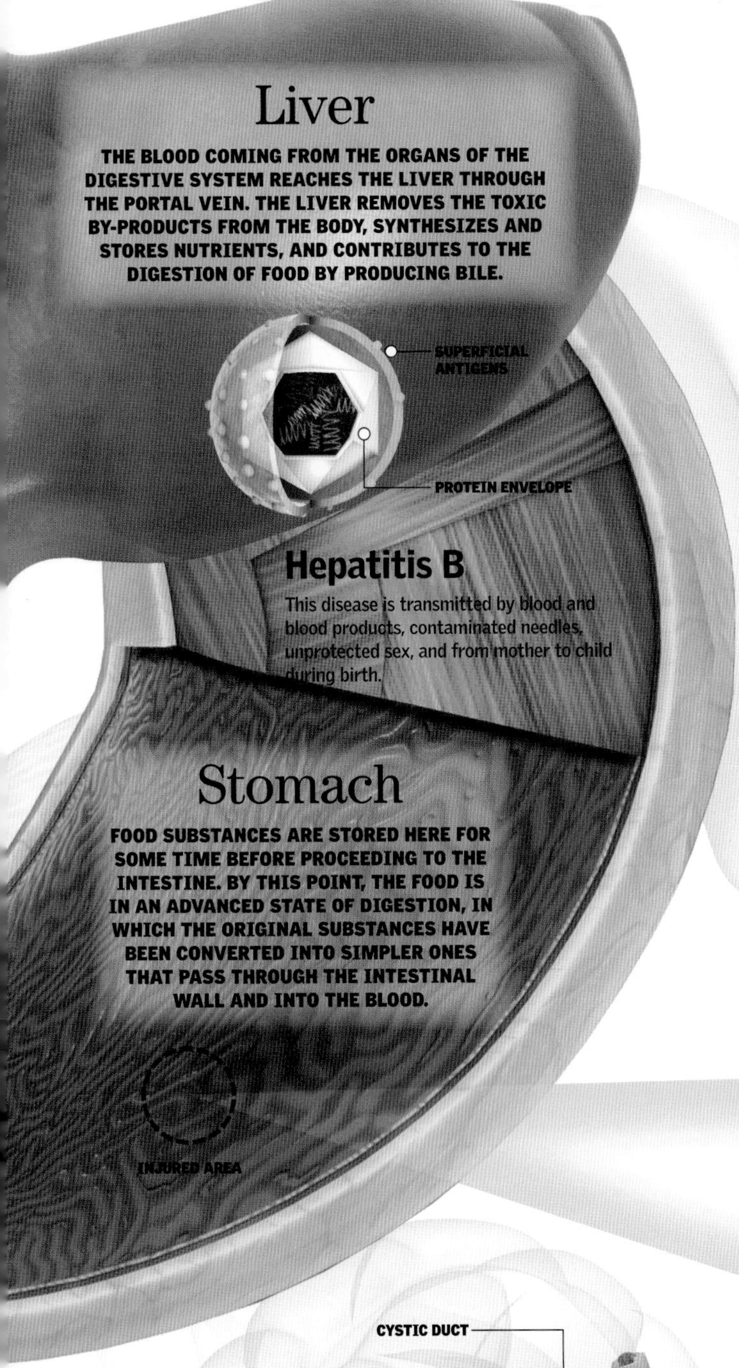


# Liver

**THE BLOOD COMING FROM THE ORGANS OF THE DIGESTIVE SYSTEM REACHES THE LIVER THROUGH THE PORTAL VEIN. THE LIVER REMOVES THE TOXIC BY-PRODUCTS FROM THE BODY, SYNTHESIZES AND STORES NUTRIENTS, AND CONTRIBUTES TO THE DIGESTION OF FOOD BY PRODUCING BILE.**

## Hepatitis B

This disease is transmitted by blood and blood products, contaminated needles, unprotected sex, and from mother to child during birth.

# Stomach

**FOOD SUBSTANCES ARE STORED HERE FOR SOME TIME BEFORE PROCEEDING TO THE INTESTINE. BY THIS POINT, THE FOOD IS IN AN ADVANCED STATE OF DIGESTION, IN WHICH THE ORIGINAL SUBSTANCES HAVE BEEN CONVERTED INTO SIMPLER ONES THAT PASS THROUGH THE INTESTINAL WALL AND INTO THE BLOOD.**

## Peptic Ulcer

A sore in the mucous membrane of the stomach or duodenum. Peptic ulcers are common, and one of the main causes is infection by the bacterium *Helicobacter pylori*. However, some are caused by the prolonged use of nonsteroidal anti-inflammatory agents, such as aspirin and ibuprofen. In some instances, stomach or pancreatic tumors can cause ulcers. The relationship between ulcers and certain types of foods or stress has not been clearly demonstrated. The main symptom is abdominal pain that is more common at night, when the stomach is empty, or two to three hours after eating.

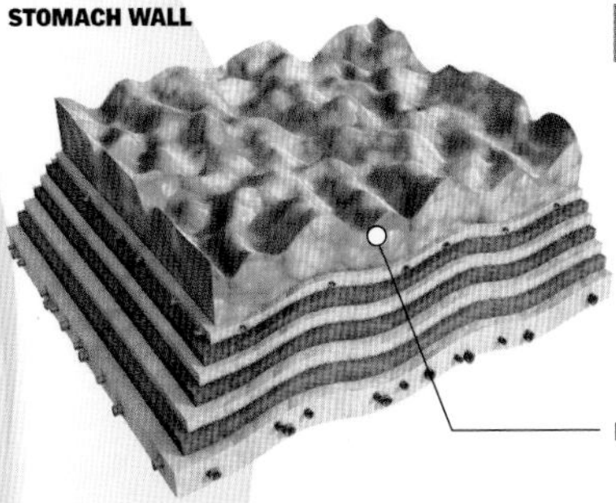


**1 EARLY STAGES**
When the barrier of protective mucosa is altered and the stomach juices come into contact with the cells of the mucosa, erosion occurs.

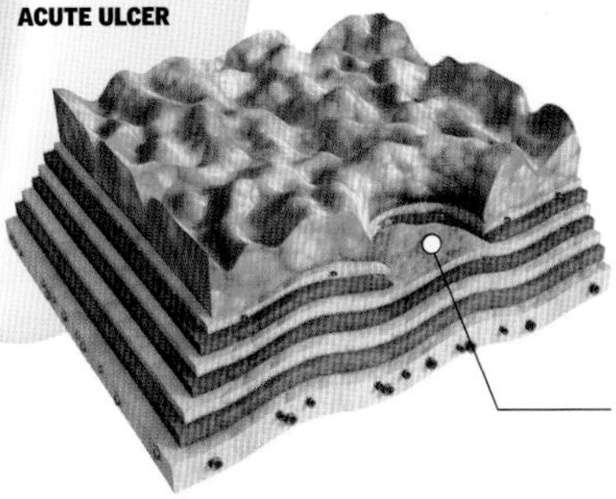


**2 DEEPENING**
The sore completely penetrates the mucosa, reaching the muscle layer of the mucosa and submucosa. An ulcer is formed.

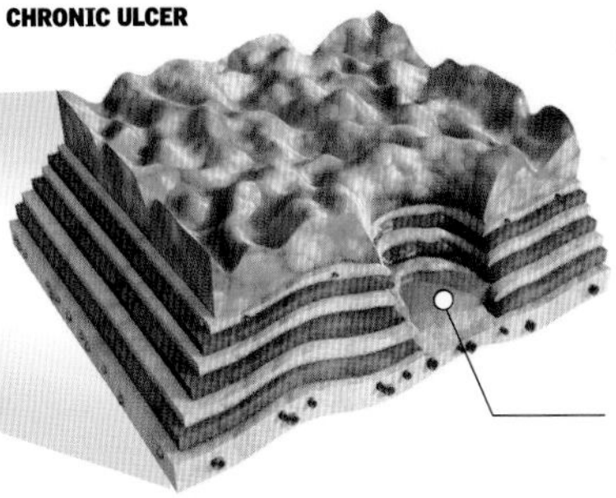


**3 COMPLICATIONS**
As the stomach wall is more deeply eroded, a large artery could be damaged enough to cause a hemorrhage. It could also lead to peritonitis.

## Gallstones

These growths form inside the gallbladder, an organ that stores the bile secreted by the liver. Bile is a solution of water, salts, lecithin, cholesterol, and other substances. If the concentration of these components changes, stones may form. They can be as small as a grain of sand or can grow to about 1 inch (3 cm) in diameter depending on how long they have been forming.

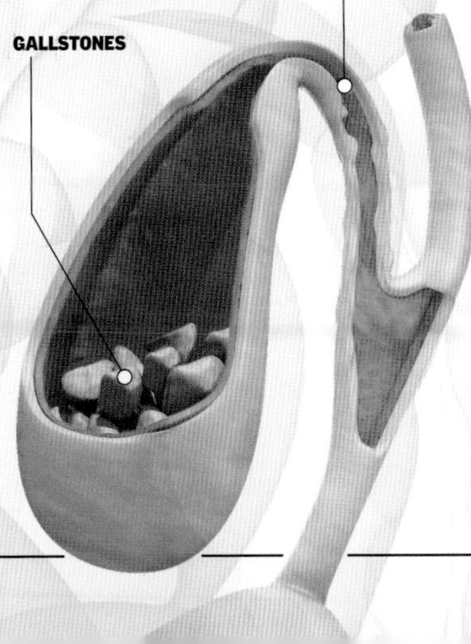


**1 OBSTRUCTION**
The bile is blocked from leaving the gallbladder by a gallstone. This causes pain and inflammation of the gallbladder.

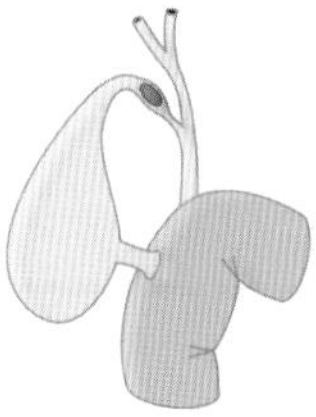

**2 INFLAMMATION**
The inflammation progresses by means of various mechanisms. The contents of the gallbladder can become infected and form pus.

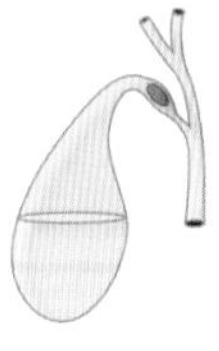

**3 RUPTURE**
If the process continues and the inflammation is very significant, the gallbladder could rupture.

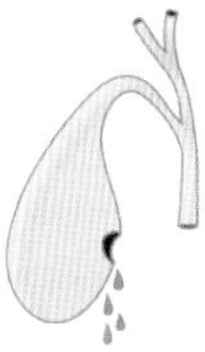

**4 CONTRACTION**
If the process is repeated, the gallbladder could shrink and lose its shape.

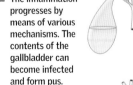

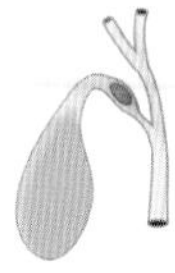

# Intestines and Colon

Intestinal infections and inflammations are among the most common disorders of the digestive system. In developing nations, an increase in infant mortality has been due to some of these diseases. Many are bacterial and can be treated with the ingestion of fluids or antibiotics, but others can be caused by a problem in the digestive system.

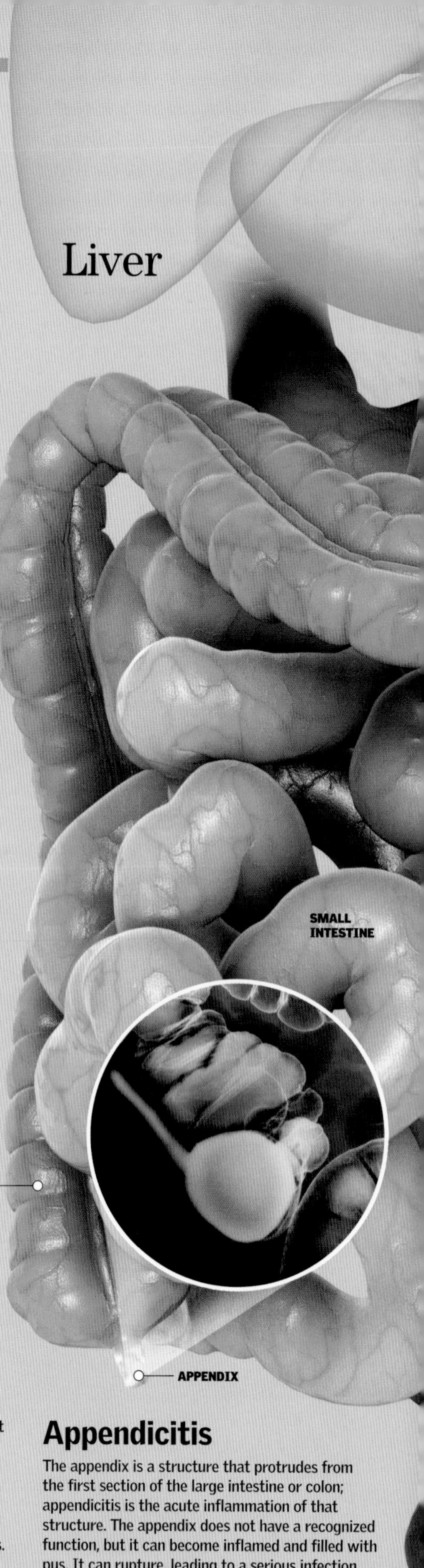


## Intestinal Infections

The most common intestinal infection is viral gastroenteritis, but it can also be caused by bacteria or protozoa. Almost all infections are transmitted by ingesting contaminated water or food. The most common symptoms are vomiting, diarrhea, and abdominal pain. Viral gastroenteritis is a self-limiting process that resolves itself in several days simply by replacing fluids to prevent dehydration, but other infections must be treated with antibiotics.

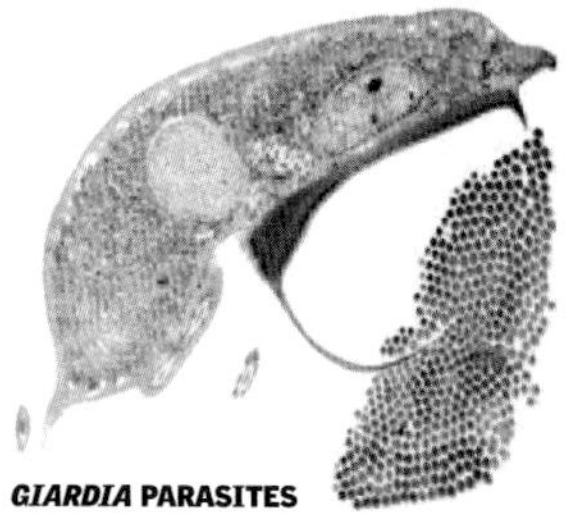

***GIARDIA* PARASITES**

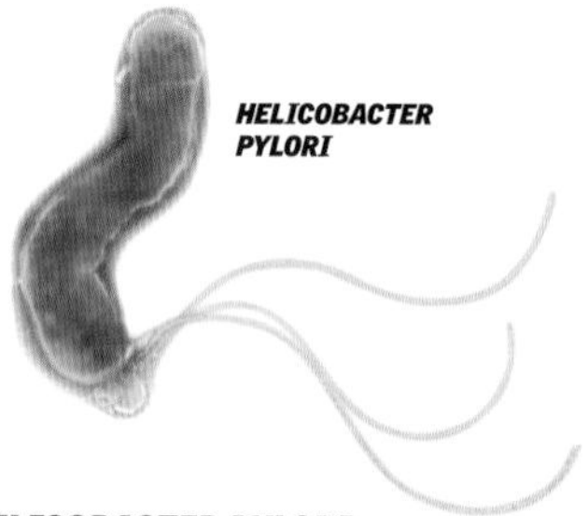


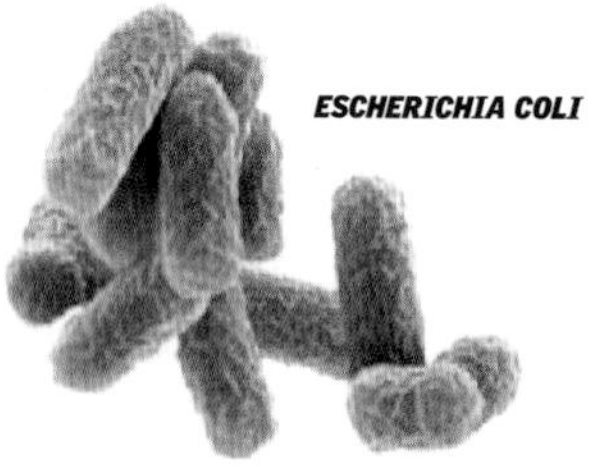


***HELICOBACTER PYLORI***

This bacterium causes gastritis and is usually found in the mucous tissue of the stomach. It can also cause ulcers in the duodenum and may be involved in causing stomach cancer.

***ESCHERICHIA COLI***

These bacteria are part of the normal intestinal flora. Some strains produce a toxin that can cause diarrhea and even be deadly for a susceptible victim, such as a baby or an elderly person.

## Hemorrhoids

These dilatations of veins occur in the venous plexus in the mucosa of the rectum and anus. If the affected veins are in the superior plexus, they are called internal hemorrhoids. Those of the inferior venous plexus are located below the anorectal line and are covered by the outer skin. The drainage system in the area lacks any valves.

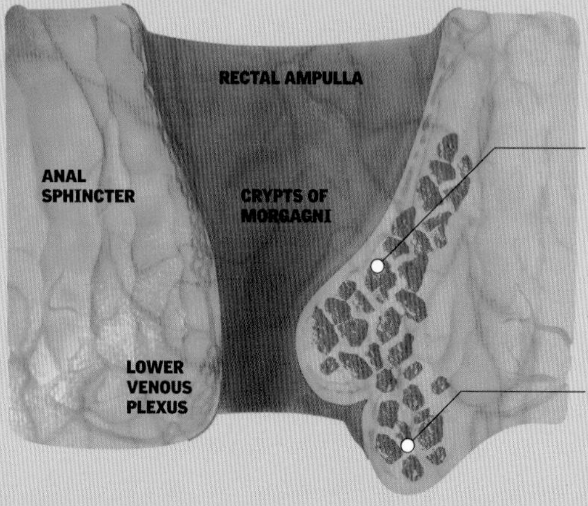


### Types of Hemorrhoids

There are two types of hemorrhoids: internal and external.

1 **INTERNAL**
Classified according to grades. Grade I hemorrhoids are located in the submucous tissue and bleed bright red blood. Grade II hemorrhoids protrude during defecation but recede once the pushing stops. Grade III come out while defecating, and Grade IV are irreducible and are always prolapsed.

2 **EXTERNAL**
Come from the inferior hemorrhoidal plexus. They can swell and cause pain and also become ulcerated and bleed. Thrombosis can be resolved.

## Appendicitis

The appendix is a structure that protrudes from the first section of the large intestine or colon; appendicitis is the acute inflammation of that structure. The appendix does not have a recognized function, but it can become inflamed and filled with pus. It can rupture, leading to a serious infection in the abdominal cavity (peritonitis). If this occurs, the person must get immediate medical attention.

## Intestinal Inflammation

Intestinal inflammations include ulcerative colitis and Crohn's disease. They can be caused by an attack of the immune system on the body's own tissues or by genetic predisposition. Symptoms include fever, blood loss, abdominal pain, and diarrhea. These conditions can be diagnosed with X-rays, a colonoscopy, or a biopsy of the intestinal tissue. The treatment might include anti-inflammatory drugs.

**GASTRIC VILLI**
This image shows the walls of the duodenum where the gastric villi can be seen.

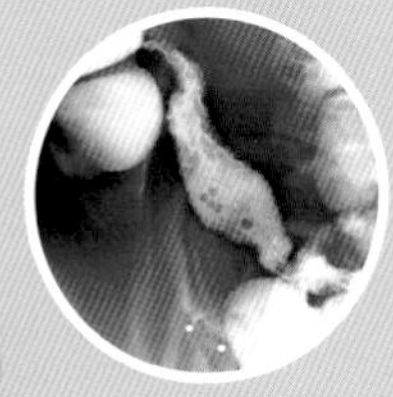

### Colitis

Ulcerative colitis is an inflammatory disease of the colon and rectum. It is characterized by the inflammation and ulceration of the colon's inner wall. Typical symptoms include diarrhea (sometimes bloody) and frequent abdominal pain.

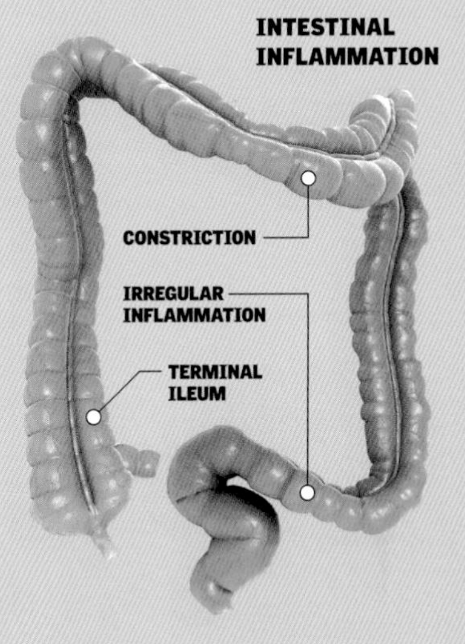


### Crohn's Disease

Crohn's disease is a chronic autoimmune condition in which the individual's immune system attacks its own intestine, causing inflammation.

### Ulcer

A peptic ulcer is a sore, or chronic erosive lesion, of the lining of the stomach or the duodenum (the first section of the small intestine). Peptic ulcers are common and can originate from a bacterial infection or in some cases from the prolonged use of anti-inflammatory drugs.

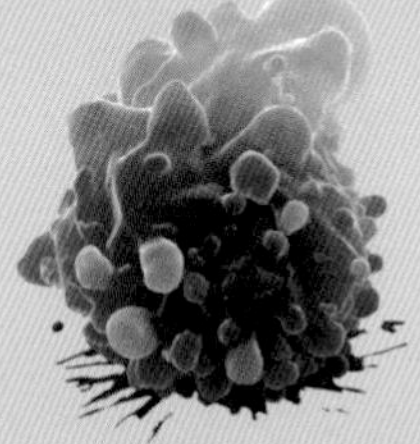

### Colon Cancer

This type of cancer is one of the most common in industrialized nations. Risk factors include family medical history, intestinal polyps, and advanced age. The symptoms are blood in the stool, a change in intestinal habits, and abdominal pain. People over 50 years of age should be evaluated by their doctor to check for the presence of blood in the stool (as seen in the photo), and if this test is positive a colonoscopy should be performed.

### Diverticulitis

The inflammation or infection of a pouch, called a diverticulum, formed in the wall of the large intestine (colon). It is believed to be caused by the slow movement of food through the intestines, which builds up a constant pressure. This increases and pushes on the inside walls of the colon, forming pouches. Ingested food or stool becomes trapped in a pouch, leading to inflammation and infection.

**1 HARD, DRY STOOL**
Bulky, soft stool passes easily through the colon. But if the stool is hard and dry, the force of the contractions increases, putting more pressure on the walls of the colon.

**2 DIVERTICULA**
Increasing pressure against the inner intestinal lining forms pouches in weak spots of the muscle wall. These pouches can then become inflamed, causing pain and distension.

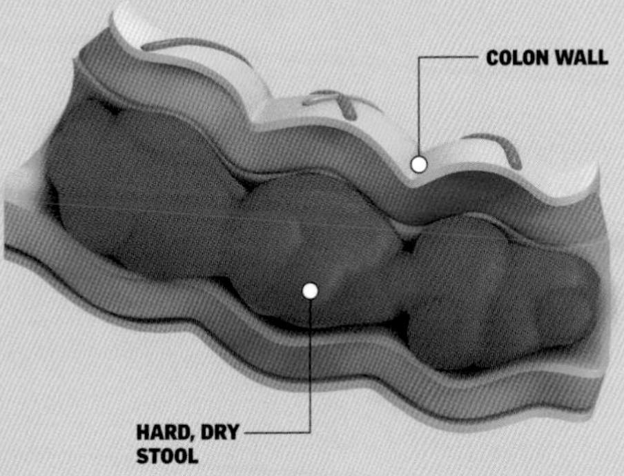


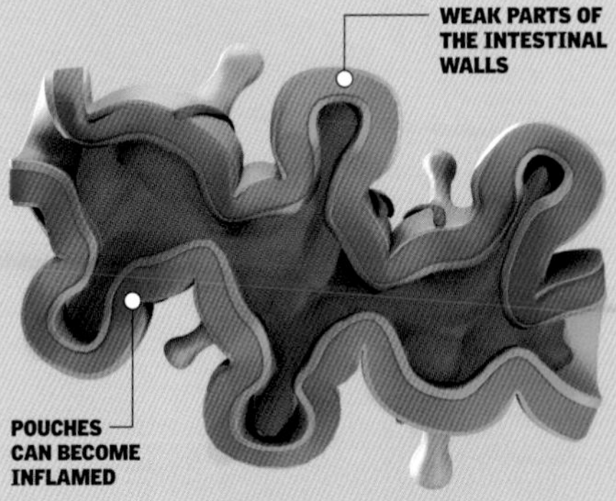


### Obstruction

Cause: the obstruction of the appendix's inner opening by fecal matter or ingested foreign bodies (bones, etc.). The appendix continues secreting intestinal fluids, which causes pressure to build up inside it, until it ulcerates and finally becomes infected with bacteria.

# Allergies: A Modern Evil

Sneezing and watery eyes, rashes and skin irritation, swelling, and itching. These are just some of the most common symptoms of allergies, a condition that affects millions of people throughout the world, especially in developed countries. What is the cause of allergies? The immune system does not function properly: it overreacts, attacking foreign substances that normally would not cause any harm. These invaders, called allergens, might include pollen, mold, and dust mites, among many other possibilities.

## An Attack on an Innocent

In developed countries, the percentage of the population affected by allergies has increased. One reason for this epidemic of modern times is the obsession with cleanliness. This means that the body, from infancy, is not exposed to enough dirt to train the immune system, which then reacts inappropriately to any foreign substance, no matter how harmless. Upon the first exposure to an allergen, the immune system becomes sensitized. In subsequent exposures, an allergic reaction occurs, which can range from a skin rash to various breathing problems. The reaction varies from person to person.

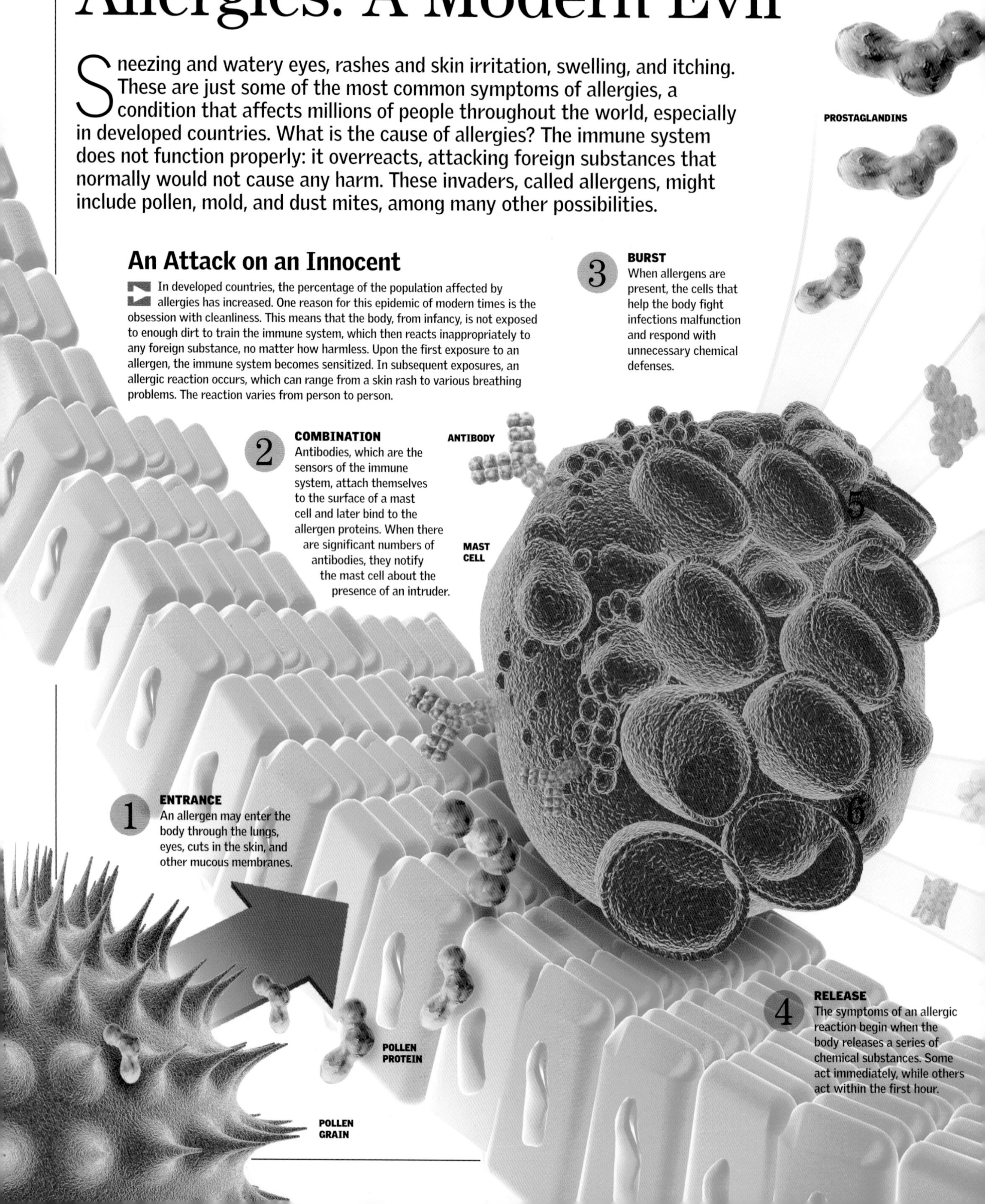


**1 ENTRANCE**
An allergen may enter the body through the lungs, eyes, cuts in the skin, and other mucous membranes.

**2 COMBINATION**
Antibodies, which are the sensors of the immune system, attach themselves to the surface of a mast cell and later bind to the allergen proteins. When there are significant numbers of antibodies, they notify the mast cell about the presence of an intruder.

**3 BURST**
When allergens are present, the cells that help the body fight infections malfunction and respond with unnecessary chemical defenses.

**4 RELEASE**
The symptoms of an allergic reaction begin when the body releases a series of chemical substances. Some act immediately, while others act within the first hour.

## No Help from Fall

Rhinitis and asthma, like the other respiratory allergies, increase with the arrival of fall. They are incapacitating, and they exact an enormous cost in terms of lost work and missed school days. The cold, in turn, irritates the respiratory tract, making it more susceptible to infections, especially viral ones. Changes in the respiratory mucous membranes and the immune system activate or reactivate the allergies. A cold, for example, can trigger a bronchial asthma attack. Moreover, the lack of ambient ventilation because of the cold weather and the concentration of indoor allergens, such as mites and fungi, increase and contribute to triggering this disease.

# Test

**THE MOST EFFECTIVE WAY TO IDENTIFY THE ALLERGENS RESPONSIBLE IS THROUGH A SERIES OF PRICKS ON THE PATIENT'S ARM TO INOCULATE THEM WITH DROPS OF ALLERGEN SOLUTIONS. THIS TEST CAN IDENTIFY THE CAUSE OR CAUSES OF THE ILLNESS AND ITS TREATMENT.**

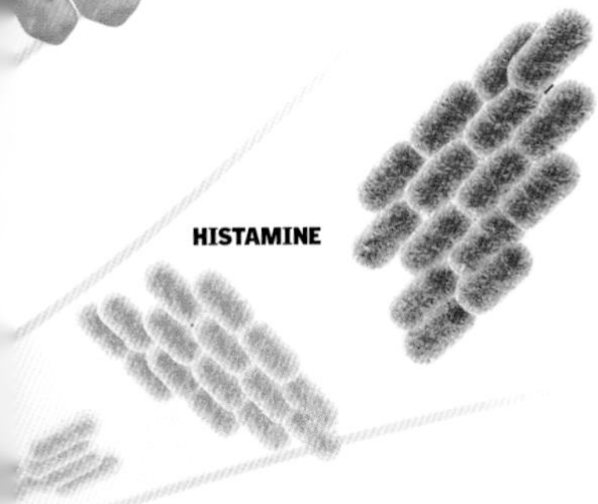

5 **FIRST RESPONSE**
Prostaglandins, leukotrienes, and histamine act on the nerve endings to produce itching. They also affect blood pressure and muscle contractions, and they act on the glands to produce mucus, vasodilatation, and, later, congestion.

# Asthma

**THIS ILLNESS HAS GROWN BY 50 PERCENT IN THE LAST 10 YEARS. CURRENTLY, IT IS ESTIMATED THAT BETWEEN 100 AND 150 MILLION PERSONS SUFFER FROM THIS DISEASE, AND ALTHOUGH IT IS MORE FREQUENT IN YOUNG CHILDREN, BETWEEN 3 PERCENT AND 7 PERCENT OF THE ADULT POPULATION COULD BE AFFECTED.**

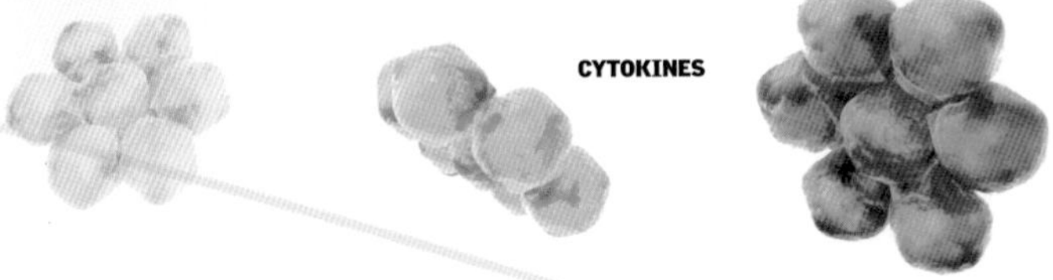


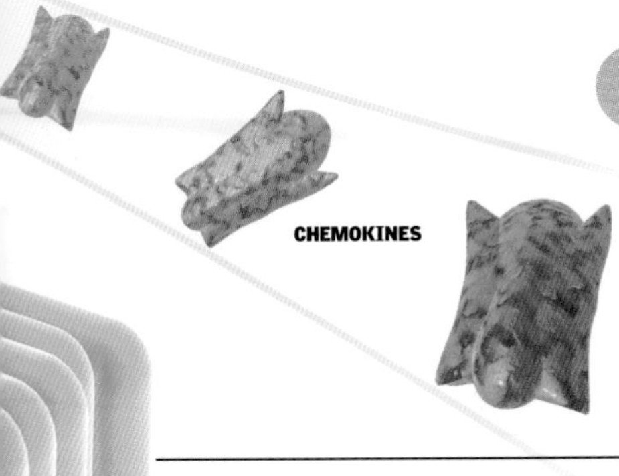


6 **SUBSEQUENT RESPONSES**
Cytokines and chemokines, which slowly damage the tissue and recruit other cells, are strongly related to the symptoms of acute and chronic asthma.

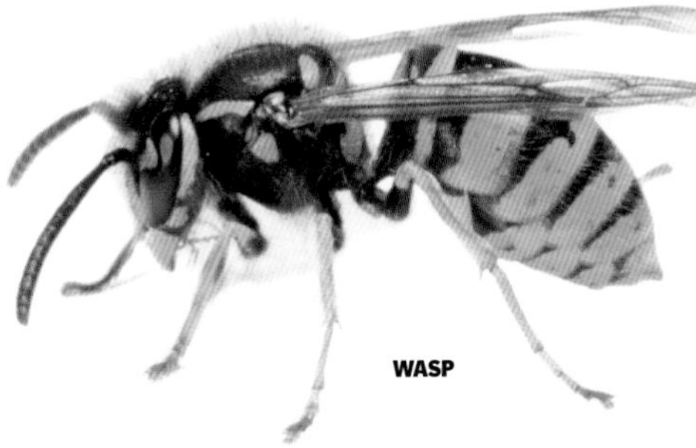


## Best-Known Allergens

Among all the substances that can produce an allergic reaction, these are the most important:

**POLLEN**: Minuscule grains released by plants during their reproductive process. They cause hay fever and breathing problems.

**DUST MITES**: Small insects that live inside the home. They cause allergies and asthma.

**WASP STINGS**: Some people have an excessive, even deadly, allergic reaction to the sting of a wasp or other insects.

**PEANUTS**: The allergy to this food is rapidly growing. In a few cases, it can be fatal.

**RAGWEED**: A type of weed that is one of the main causes of allergies in the United States. It produces intense rhinoconjunctivitis and, more rarely, asthma. Its pollen is very potent and is the cause of the allergic reaction.

## Allergies by Level of Development

Allergies, like obesity, are epidemics of modern times. The more industrialized a country, the greater the affected population. In contrast, in developing regions, such as Africa and Latin America, the number of people affected is much lower. In remote regions, allergies are almost nonexistent.

# AIDS

Acquired Immune Deficiency Syndrome (AIDS) is still considered one of the most important epidemics of the 21st century. Some 40 million people are infected with HIV (human immunodeficiency virus), the virus that causes AIDS; most of them are in Africa. Scientific research is aimed at finding a remedy to stop the development of the virus, but until now they have only produced therapies that slow viral activity.

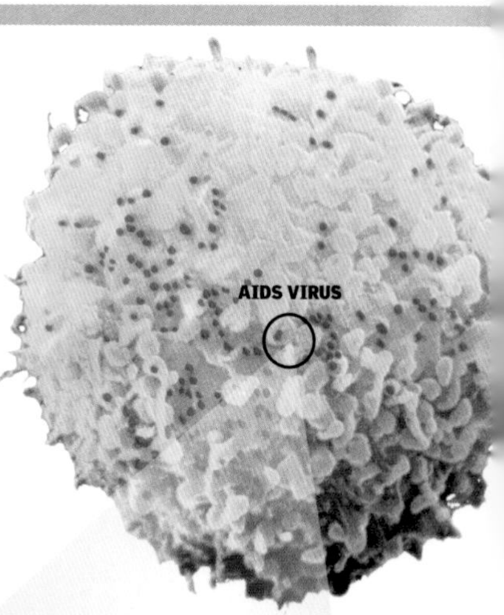


**CD4-POSITIVE T LYMPHOCYTE**
Immune system cell that defends the body against infections.

## The AIDS Virus

Human immunodeficiency virus (HIV) is the cause of AIDS. This virus destroys a type of white blood cell, the CD4 T lymphocyte, through the interaction of the viral DNA with the lymphocyte's DNA. These lymphocytes are essential to the immune system's fight against infections. For this reason, persons infected with HIV can suffer severe diseases, and even minor conditions, such as a cold, might be difficult to cure. However, not all those infected with HIV suffer from AIDS, which is the final stage of the disease. A person with HIV is seropositive. When the level of CD4-positive T lymphocytes goes below 200 cells per 1 $mm^3$ of blood, the disease progresses to the stage of AIDS.

## History and Evolution

The "age of AIDS" began on June 5, 1981. The U.S. Centers for Disease Control found patients with pneumonia that simultaneously suffered from Kaposi sarcoma, a malignant tumor of the skin. It was noted that all the patients had a notable depletion of CD4-positive T lymphocytes. Unprotected sex and the use of needles with infected blood were the typical causes at that time. Today mother-to-child transmission and transfusions of blood and blood products play an important role.

**ENLARGED VIRUS**

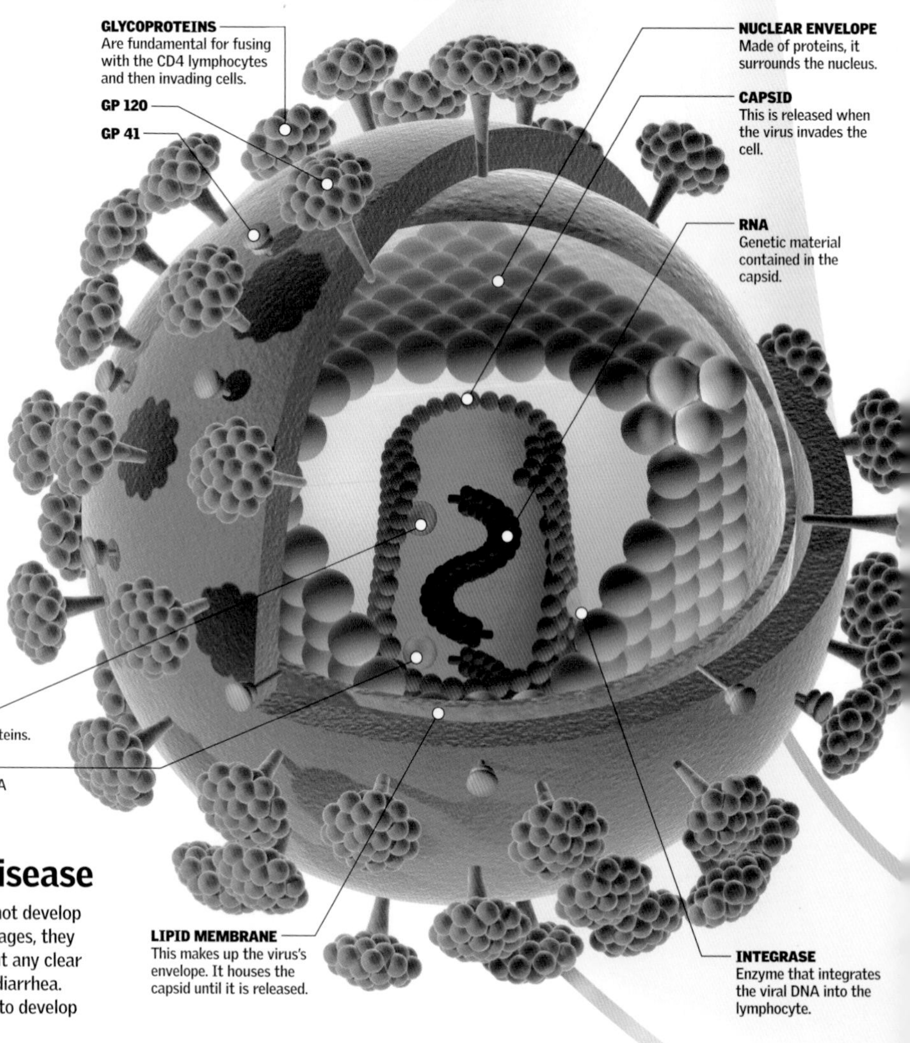

GLYCOPROTEINS
Are fundamental for fusing with the CD4 lymphocytes and then invading cells.
GP 120
GP 41
NUCLEAR ENVELOPE
Made of proteins, it surrounds the nucleus.
CAPSID
This is released when the virus invades the cell.
RNA
Genetic material contained in the capsid.
PROTEASE
Enzyme that synthesizes viral proteins.
REVERSE TRANSCRIPTASE
Enzyme that synthesizes viral DNA from the RNA it uses as a mold.
LIPID MEMBRANE
This makes up the virus's envelope. It houses the capsid until it is released.
INTEGRASE
Enzyme that integrates the viral DNA into the lymphocyte.

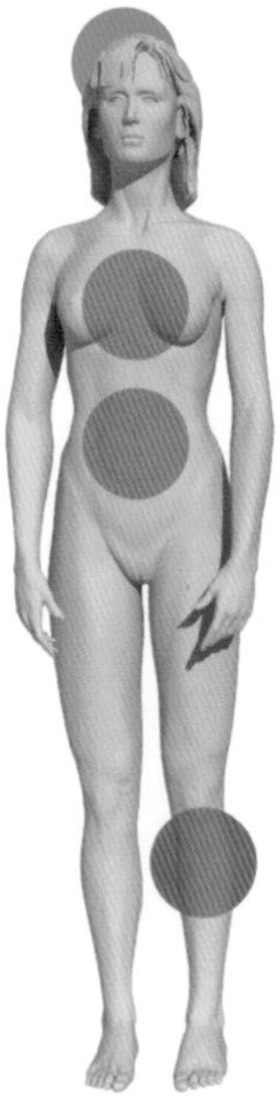

## Symptoms of the Disease

Many people infected with the virus do not develop symptoms for several years. In earlier stages, they might lose weight and have fever without any clear cause and in later stages have frequent diarrhea. Those severely infected are predisposed to develop various infections and cancers.

**BRAIN** If damaged, it can cause vision problems, weakness, and paralysis.

**LUNGS** The most common disease that can be contracted is pneumonia.

**SKIN** The appearance of Kaposi's sarcoma, brown and blue spots on the skin, is generally associated with AIDS.

**DIGESTIVE SYSTEM** Persistent diarrhea due to an infection of the gastrointestinal tract by parasites such as *Giardia lamblia* can result.

## How the AIDS Virus Works

The virus uses its layer of proteins to attach to the cell that will harbor it. A specific protein (gp120) fuses with a receptor on the CD4-positive T lymphocyte. After the immune system loses many cells, the body is left susceptible to many diseases. Ten years might pass from the time of infection until the development of full-blown AIDS.

**1 VIRAL STRUCTURE**
Before attachment, the virus's envelope contains a capsid that carries the genetic material. With this material, which contains RNA, the virus will begin to act on the lymphocyte's DNA. The envelope that covers the capsid is made of protein.

7 **OUTWARD PUSH**
The new virus model begins to come out of the infected cell. It takes part of the cell membrane with it.

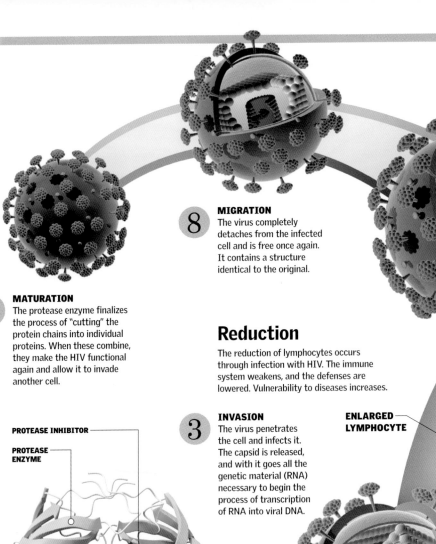

8 **MIGRATION**
The virus completely detaches from the infected cell and is free once again. It contains a structure identical to the original.

9 **MATURATION**
The protease enzyme finalizes the process of "cutting" the protein chains into individual proteins. When these combine, they make the HIV functional again and allow it to invade another cell.

## Reduction

The reduction of lymphocytes occurs through infection with HIV. The immune system weakens, and the defenses are lowered. Vulnerability to diseases increases.

**VIRAL PROTEIN**
This is synthesized by cellular mechanisms.

**PROTEASE**

6 **SYNTHESIS**
Viral protein chains begin to be synthesized. The protease cuts these chains and converts them into individual proteins.

3 **INVASION**
The virus penetrates the cell and infects it. The capsid is released, and with it goes all the genetic material (RNA) necessary to begin the process of transcription of RNA into viral DNA.

**ENLARGED LYMPHOCYTE**

5 **INTEGRATION**
Integrase inserts the viral DNA into the DNA strand of the lymphocyte. The normal activity of the white blood cell changes because of the new DNA.

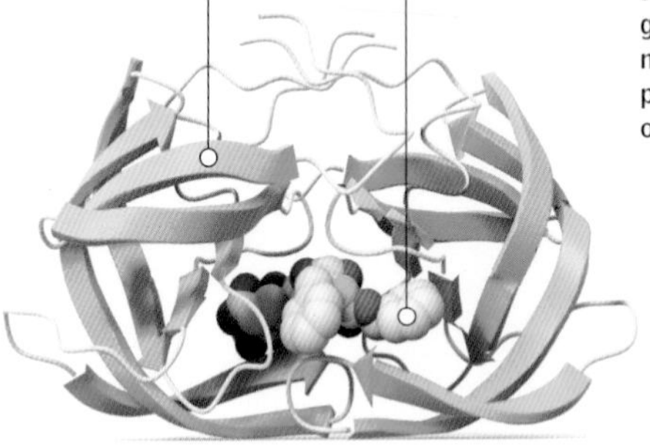


**HIV DNA**
Formed by reverse transcriptase from the RNA content in the capsid.

**LYMPHOCYTE NUCLEUS**

**LYMPHOCYTE DNA**

## Protease Inhibitor

The drug used to prevent the action of the protease (spheres) attaches to the protease enzyme of the HIV (yellow). The protease inhibitor's power lies in stopping or slowing the formation of specific proteins that are necessary for the synthesis and function of viral DNA. In many cases, protease inhibitor drugs are combined with other medicines, such as antiretroviral drugs.

**CAPSID**
This contains the elements necessary to synthesize viral DNA.

**HIV RNA**

**REVERSE TRANSCRIPTASE**

**PROTEASE**

**INTEGRASE**

2 **ATTACHMENT**
Through certain receptors on the cell surface, the virus's proteins can link with proteins on the CD4-positive T lymphocyte. The glycoprotein gp120 that covers the virus enables it to fuse with the lymphocyte.

4 **TRANSCRIPTION**
The RNA serves as a mold to synthesize viral DNA. Reverse transcriptase produces the DNA, preparing it to be inserted into the structure of the CD4 T lymphocyte.

# Advanced Technology

Technology, in the service of medicine, has permitted the understanding and prevention of many serious diseases thanks to the study of early diagnostic techniques, such as magnetic resonance imaging and positron emission tomography, which provide images of the interior of the body. Future decades, however, promise to bring even more exciting developments. In this chapter, we will tell you about exciting developments like nanomedicine, whose main objective is to cure diseases from inside the body. For this purpose, devices smaller in diameter than a human hair have been developed. Among other dreams in the minds of scientists is that of preventing the degeneration of nerve cells. Enjoy the fascinating information offered in this chapter!



**VIRTUAL REALITY** (opposite)
This image shows a microscopic submarine, small enough to travel through an artery.

# Early Diagnosis

There are various methods of examining the body to search for possible diseases. One of the most novel procedures is positron emission tomography (PET), which is able to detect the formation of a malignant tumor before it becomes visible through other methods. It is also useful for evaluating a person's response to a specific treatment and for measuring heart and brain function.

## X-Rays

The simple emission of X-rays consists of sending out short electromagnetic waves. After passing through the body, they reach a photographic film and create shadow images. The denser structures, like bone, absorb more X-rays and appear white, whereas the softer tissues appear gray. In other cases, a fluid must be used to fill hollow structures and generate useful images. To examine the digestive tract, for example, a barium sulfate mixture must be ingested.

**CONTRAST**
Is introduced through an enema made of barium. It allows the structures of the digestive tract to be distinguished in detail.

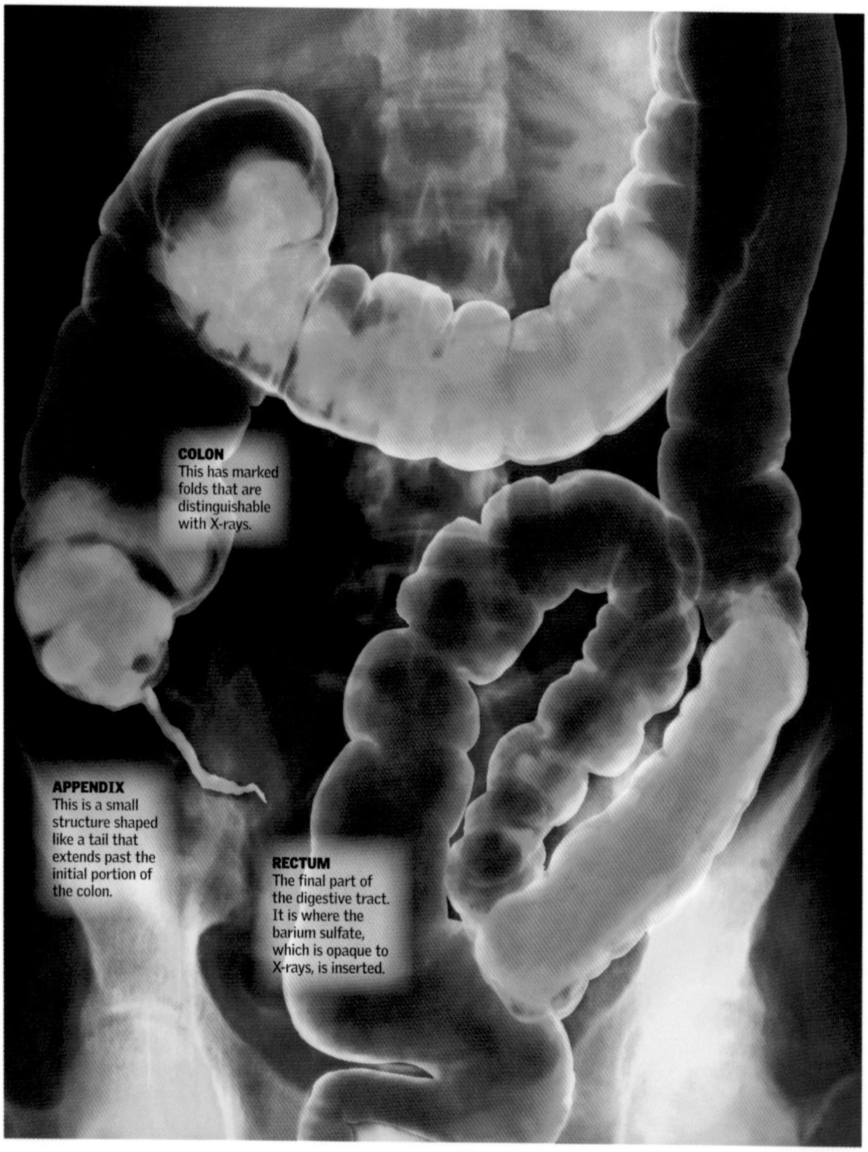

COLON
This has marked folds that are distinguishable with X-rays.
APPENDIX
This is a small structure shaped like a tail that extends past the initial portion of the colon.
RECTUM
The final part of the digestive tract. It is where the barium sulfate, which is opaque to X-rays, is inserted.

## Scanning Methods

The different techniques for exploring the body aim to detect possible anomalies in the organs and tissues. The latest developments, such as magnetic resonance imaging and positron emission tomography, have surpassed classic X-ray methods. It is now possible to obtain detailed images of tissues and of the metabolic activity of tumor cells.

## 3-D Magnetic Resonance Imaging

This technique permits greater detail and is used mostly to monitor fetuses.

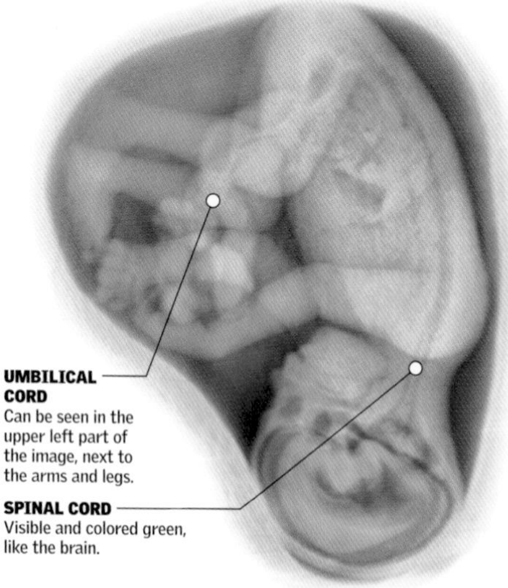

UMBILICAL CORD
Can be seen in the upper left part of the image, next to the arms and legs.
SPINAL CORD
Visible and colored green, like the brain.

## Ultrasound

A device called a transducer emits extremely high frequency sound waves. The transducer is passed back and forth over the part of the body being examined. The sound waves return to the transducer as an echo and are analyzed by a computer.

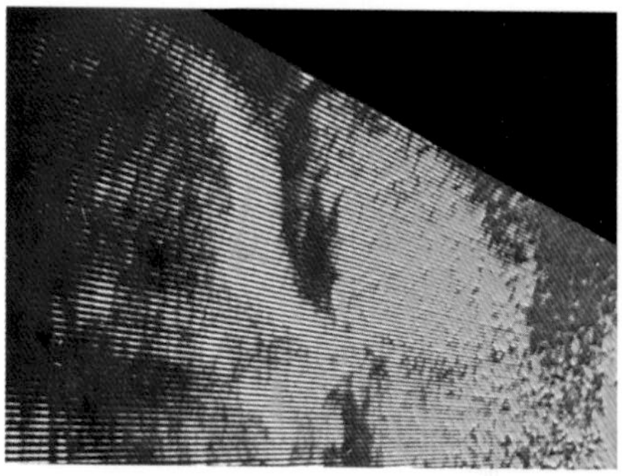

ULTRASOUND SCAN

## Encapsulated Camera

A miniature camera enters the body through a capsule and takes detailed pictures of the digestive tract. It travels using the natural movements of the intestinal walls.

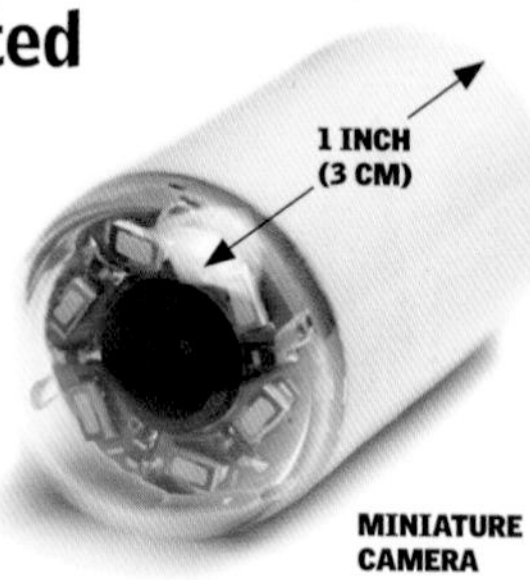

## Positron Emission Tomography

This technology enables doctors to obtain detailed information about metabolic issues, such as the cell activity of a tumor. When combined with computerized tomography, it provides high-quality images and advanced knowledge regarding diseases such as cancer. This way, it may be possible to detect an illness before it spreads.

**WHEN IT IS SUITABLE TO USE**
It should be used for patients with coronary or brain diseases and to detect cancer.

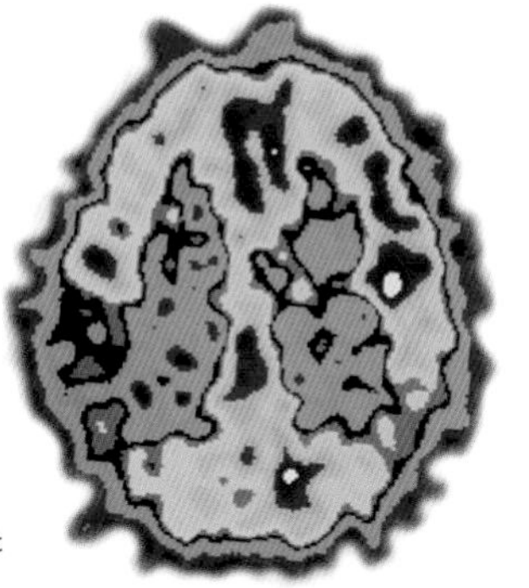

**METABOLIC ACTIVITY**
This scan shows the activity in a brain with Alzheimer's disease. There are few zones with high activity (red); most are low (blue-green).

### How it Works

**1 INJECTION**
The patient receives a dose of radioactive glucose, or FDG, which is taken up by affected organs.

**2 POSITRONS**
The active tumors take up large amounts of glucose. When the FDG decays, it emits positrons.

**3 GAMMA RAYS**
These are emitted when the positrons collide with electrons and are annihilated.

**4 IMAGES**
A computer receives the rays and converts them into images that provide details about possible tumors.

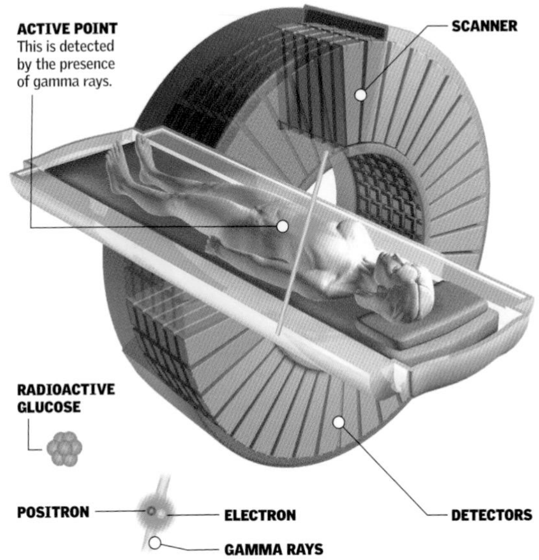


## Computerized Tomography

Computerized tomography (CT) provides information about regions denser than those typically penetrated by X-rays. The tomography covers each millimeter of the body's contour, providing many images of cross sections of the body. By combining these images, a three-dimensional grayscale picture of a particular organ can be obtained.

**WHEN IT IS SUITABLE TO USE**
It should be used when images of internal organs of the body are needed.

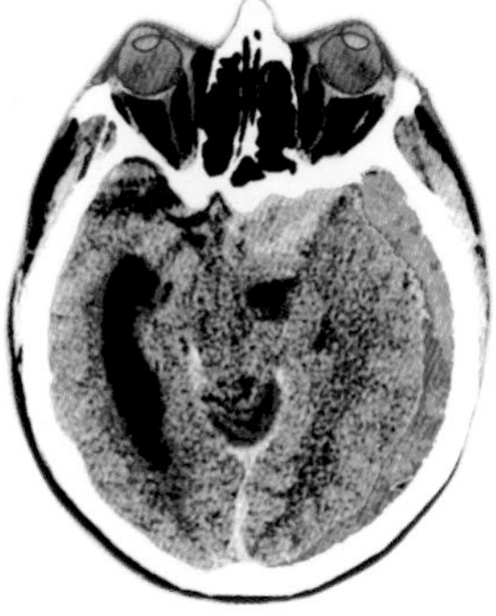

**INTERNAL HEMORRHAGE**
In this CT scan, a hematoma (in orange) can be seen that was formed from a blood clot after an injury to the membranes surrounding the brain.

### How it Works

**1 SCAN**
The patient enters the tomography machine through an opening that divides the body contour into sections.

**2 X-RAY TUBE**
This tube rotates simultaneously with the detector to completely X-ray the patient.

**3 RECEPTION**
The detectors sense the intensity of the rays as they pass over each point of the body.

**4 IMAGE**
The information is processed by a computer that integrates the data into images.

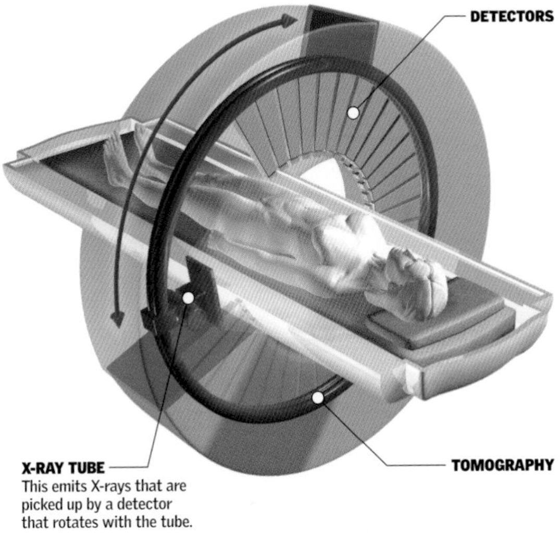

DETECTORS
X-RAY TUBE
This emits X-rays that are picked up by a detector that rotates with the tube.
TOMOGRAPHY

## Magnetic Resonance Imaging

A technique that uses a cylindrical chamber capable of producing a magnetic field 40,000 times stronger than the Earth's. Unlike X-rays, magnetic resonance allows imaging of soft tissues (like fat) and from every angle. It provides the most detailed images and is used most frequently for examining the brain.

**BRAIN**
The fibers of the nerve cells that transmit electrical signals are shown in color.

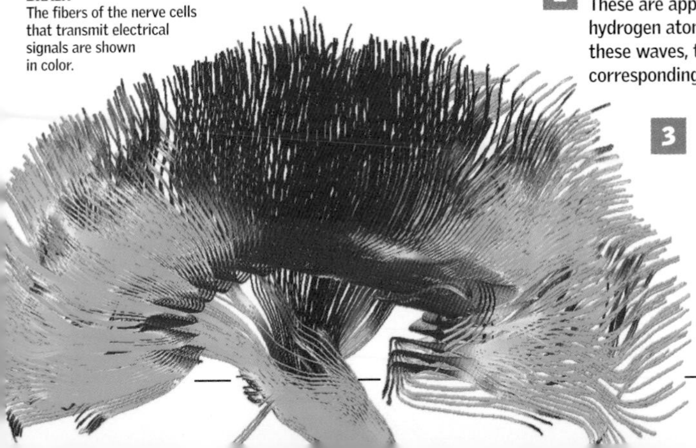

### How it Works

**1 MAGNETIC FIELD**
This acts on the hydrogen atoms of the body when the patient enters the magnetized chamber.

**2 RADIO WAVES**
These are applied to the hydrogen atoms. Upon receiving these waves, they emit a corresponding radio wave.

**3 PROCESSING**
A computer receives and processes the signals emitted by the atoms and then builds an image from them.

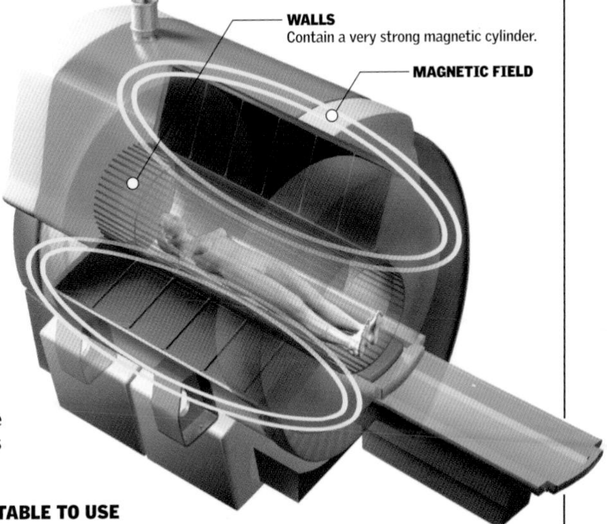


**WHEN IT IS SUITABLE TO USE**
It should be used when the anatomy of the softest tissues, which X-rays cannot reveal, need to be examined.

# Laser Surgery

Surgeries performed with laser beam techniques are much simpler than traditional procedures. Lasers are frequently used in eye surgery. They can close blood vessels in the retina. Lasers can also burn papillomas (benign epithelial tumors) and excise precancerous lesions from the mouth without scarring. Currently lasers are used to break down kidney stones and to open clogged arteries.

## Laser Angioplasty

When fatty deposits (atheromas) accumulate in the arteries, plaque forms, and the internal channel for blood flow narrows. Laser angioplasty can be used to eliminate this plaque. In this operation, a catheter with a small balloon is used. The balloon is introduced into the artery and is inflated to momentarily cut off the circulation. The plaque is removed easily by a laser emitter located at the tip of the catheter. The laser angioplasty operation is quick, and the patient's recovery period is usually short. Laser angioplasty is recommended when only one artery is blocked.

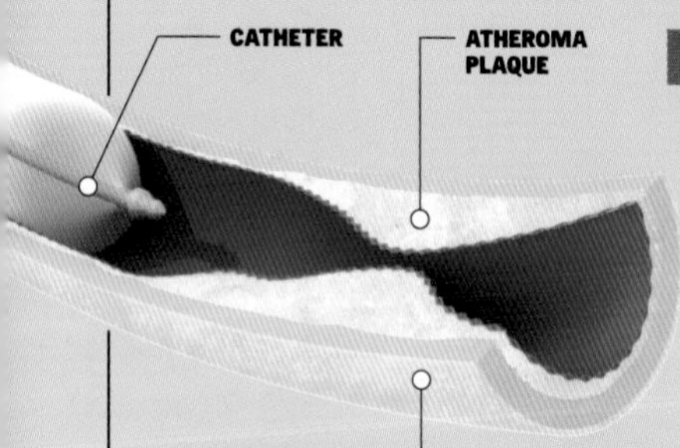


### 1 INCISION

An incision is made in an arm or leg or in an artery. The catheter is introduced. Guided by X-rays or ultrasound, it is directed toward the location of the blocked artery.

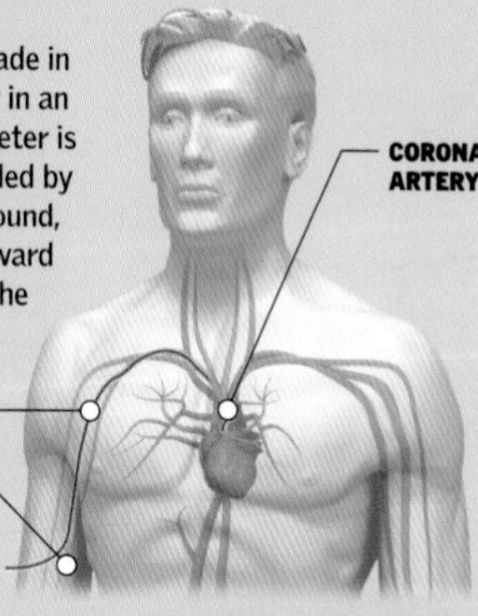


**BEFORE THE OPERATION**
An incision is made in the patient's arm or leg.

### 2 LOCATION

The catheter is directed toward the location of the blockage. The balloon is inflated, and it presses against the arterial walls. Blood circulation is temporarily cut off to permit the removal of plaque.

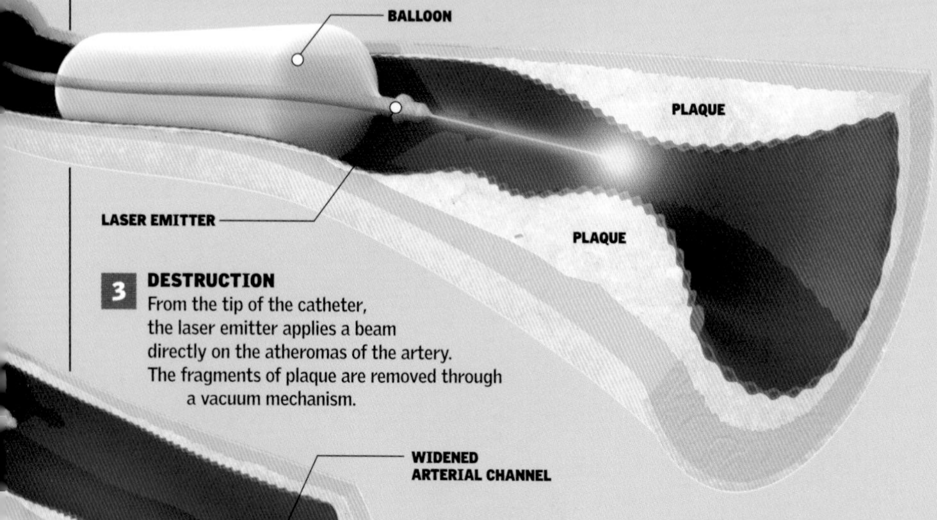


### 3 DESTRUCTION

From the tip of the catheter, the laser emitter applies a beam directly on the atheromas of the artery. The fragments of plaque are removed through a vacuum mechanism.

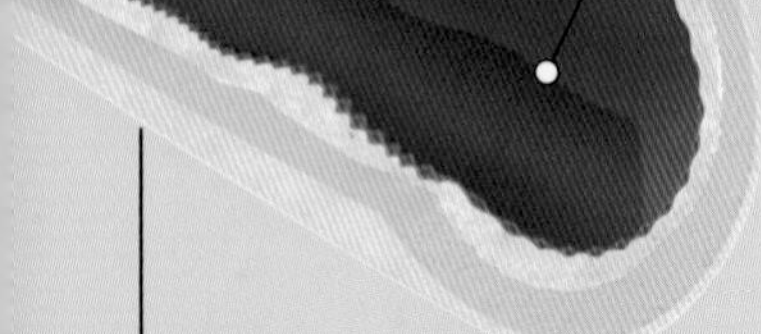


### 4 CONTROL

Once the process of destroying the plaque is finished, the blood pressure is checked on both sides of the arterial wall to ensure that it is equal. The catheter with the balloon is then removed. The recovery period is short; the patient only needs a brief postoperative rest period.

## Pupil Contraction

The pupil plays an important role in regulating the light that enters the eye. In a normally functioning eye, light enters through the pupil, passes through the cornea and the lens, and finally reaches the retina. When the ambient light is intense, the pupil contracts. This causes the eye to receive less light and prevents glare. The contraction of the pupil is a reflex action.

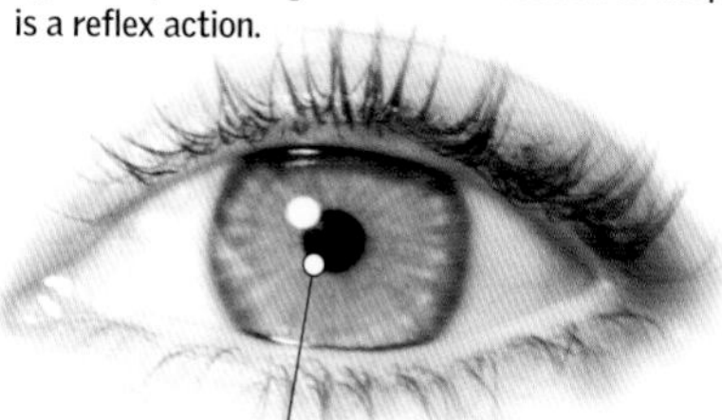


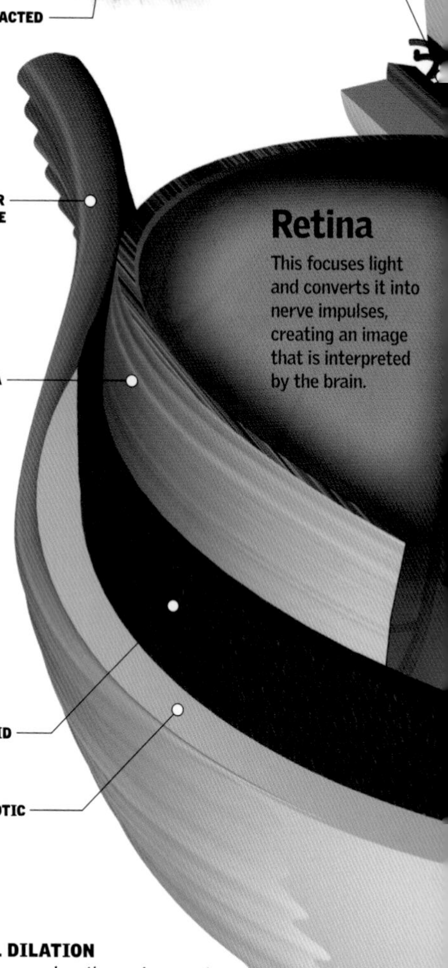


### Retina

This focuses light and converts it into nerve impulses, creating an image that is interpreted by the brain.

**PUPIL DILATION**
This occurs when the environment is dark or poorly lit. The reflex dilation movement allows the eye to receive more light through the pupil.

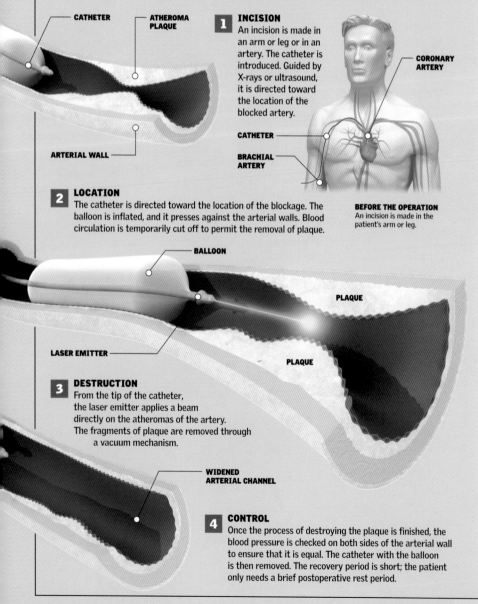

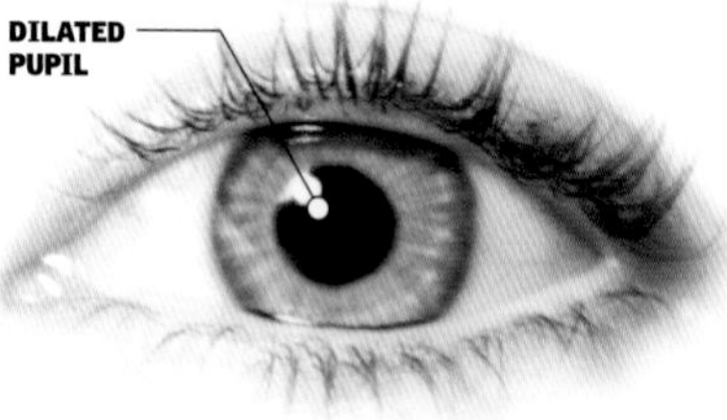

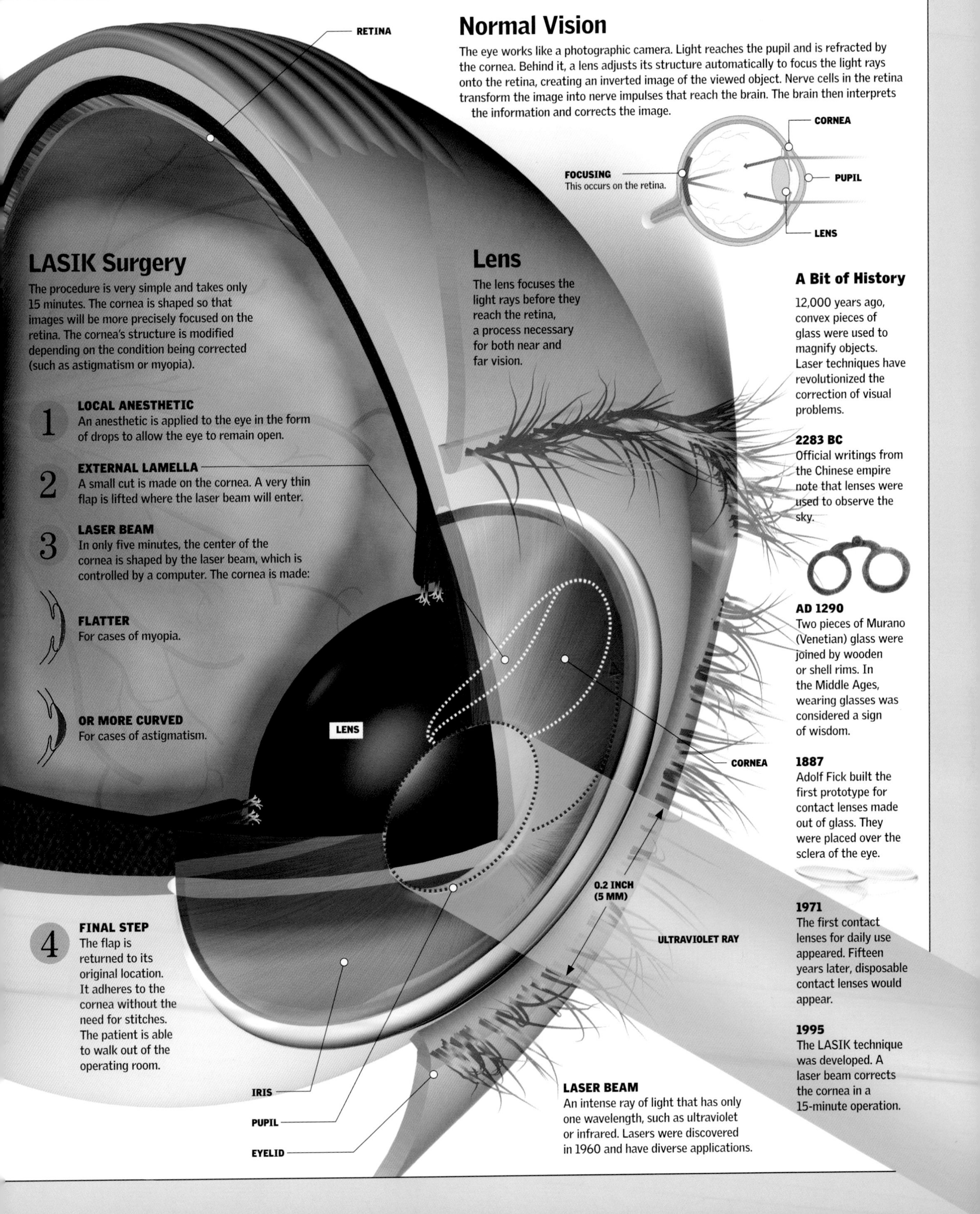


## Normal Vision

The eye works like a photographic camera. Light reaches the pupil and is refracted by the cornea. Behind it, a lens adjusts its structure automatically to focus the light rays onto the retina, creating an inverted image of the viewed object. Nerve cells in the retina transform the image into nerve impulses that reach the brain. The brain then interprets the information and corrects the image.

## LASIK Surgery

The procedure is very simple and takes only 15 minutes. The cornea is shaped so that images will be more precisely focused on the retina. The cornea's structure is modified depending on the condition being corrected (such as astigmatism or myopia).

1 **LOCAL ANESTHETIC**
An anesthetic is applied to the eye in the form of drops to allow the eye to remain open.

2 **EXTERNAL LAMELLA**
A small cut is made on the cornea. A very thin flap is lifted where the laser beam will enter.

3 **LASER BEAM**
In only five minutes, the center of the cornea is shaped by the laser beam, which is controlled by a computer. The cornea is made:

**FLATTER**
For cases of myopia.

**OR MORE CURVED**
For cases of astigmatism.

4 **FINAL STEP**
The flap is returned to its original location. It adheres to the cornea without the need for stitches. The patient is able to walk out of the operating room.

## Lens

The lens focuses the light rays before they reach the retina, a process necessary for both near and far vision.

**LASER BEAM**
An intense ray of light that has only one wavelength, such as ultraviolet or infrared. Lasers were discovered in 1960 and have diverse applications.

### A Bit of History

12,000 years ago, convex pieces of glass were used to magnify objects. Laser techniques have revolutionized the correction of visual problems.

**2283 BC**
Official writings from the Chinese empire note that lenses were used to observe the sky.

**AD 1290**
Two pieces of Murano (Venetian) glass were joined by wooden or shell rims. In the Middle Ages, wearing glasses was considered a sign of wisdom.

**1887**
Adolf Fick built the first prototype for contact lenses made out of glass. They were placed over the sclera of the eye.

**1971**
The first contact lenses for daily use appeared. Fifteen years later, disposable contact lenses would appear.

**1995**
The LASIK technique was developed. A laser beam corrects the cornea in a 15-minute operation.

# Transplants

When the possibilities for treating certain diseases run out, the only remaining alternative is to replace the sick organ with another one through a transplant. The organs can come from a live person (as long as it does not cause harm to that person, as in the case of kidney donation) or from a donor corpse. Today the most novel transplant is the face transplant, which involves working with many nerves and is highly complex.

## The Mouth and Nose of Another Person

The operation for replacing the damaged face (generally due to burns) is still in its developing stage. The first recorded case of a successful transplant was that of Isabelle Dinoire, a French woman who lost her nose, her chin, and her lips when she was savagely attacked by her dog in 2005. The surgery was partial, and it restored those parts she had lost with skin donated by a sick woman suffering a case of cerebral coma. The complex operation included the ligation of blood vessels and nerves between the donating tissue and the beneficiary.

### The Nerves

These can only be joined through microsurgery. The operation is very complicated because the face is full of nerve endings.

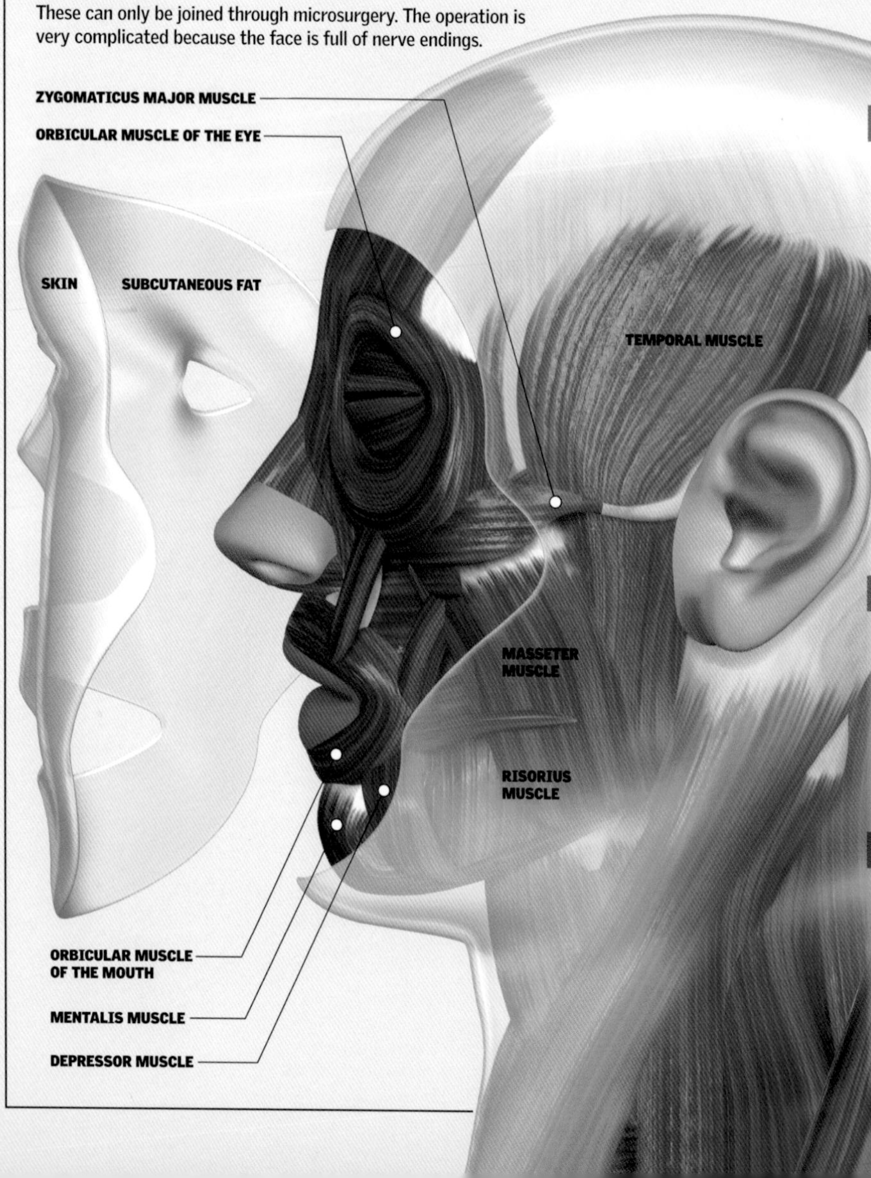


## Organ Transplants

Of the two types of transplant operations (organs and tissues), organ transplants are by far the more difficult. They require complex surgeries to achieve the splicing of vessels and ducts. Tissue transplants are simpler: cells are injected, to be implanted later.

### Types of Transplants

**ALLOGRAFT**: Consists of the donation of organs from one individual to another genetically different individual of the same species.

**AUTOGRAFT**: A transplant in which the donor and the recipient are the same person. The typical case is a skin graft from a healthy site to an injured one.

**ISOGRAFT**: A transplant in which the donor and the recipient are genetically identical.

**XENOGRAFT**: A transplant in which the donor and the recipient are of different species (e.g., from a monkey to a human). This type generates the strongest rejection response by the body of the recipient.

**1 REMOVAL**
The skin of the patient's face is removed. A wide range of injuries can be treated with this surgery. The transplant can be partial or total. In France, a woman attacked by a dog lost her nose, lips, and chin and underwent a partial face transplant to recover these parts.

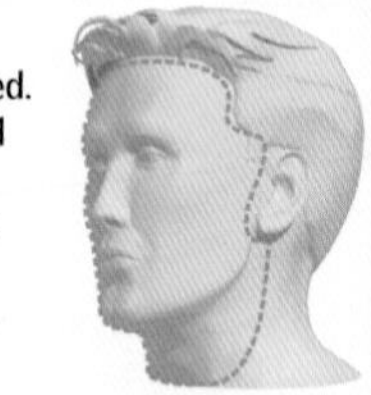

**2 PREPARATION**
Since the face is a complex framework of blood vessels, capillaries, arteries, and veins, care must be taken during the insertion of the new face. The original muscles and nerves are left on the patient. Blood vessels are cut before the surgery. Later they will be joined to the donated skin.

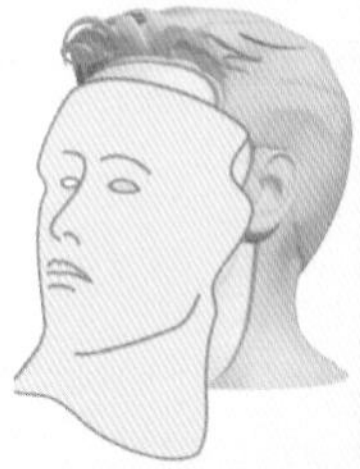

**3 ALIGNMENT**
The surgeons position the donated skin, aligning it exactly over the face of the patient. Through microsurgery, the blood vessels and nerves are connected to the new tissue. As the blood begins to circulate, the face takes on a progressively pinker color, characteristic of tissue with normal blood supply.

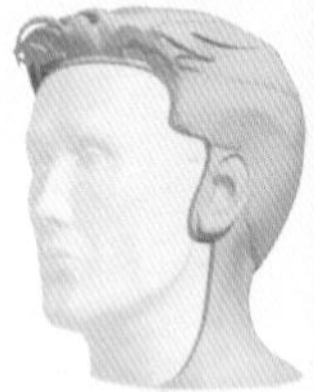

**4 RESTORATION**
The skin is sutured, as shown in the image. The areas should normalize within 14 days. After the surgery, the patient usually requires psychological treatment to better cope with the idea that he or she now possesses a "hybrid" face, with his or her own bone structure, but the skin and fatty tissues of someone else.

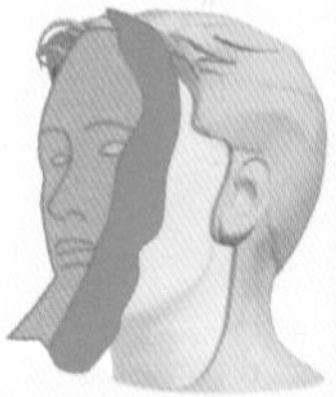

# Heart Transplant

Heart transplant is, in general, the preferred treatment for heart failure when it is deemed that the possibility of survival and the quality of life cannot improve with any other traditional therapeutic alternative. The problem lies in establishing when other medical options should be discarded on the basis of this criterion. According to the American Heart Association, the clearest indications that such a transplant must be carried out are: cardiogenic shock, severe symptoms of ischemia that limit daily activity, and ventricular arrhythmias.

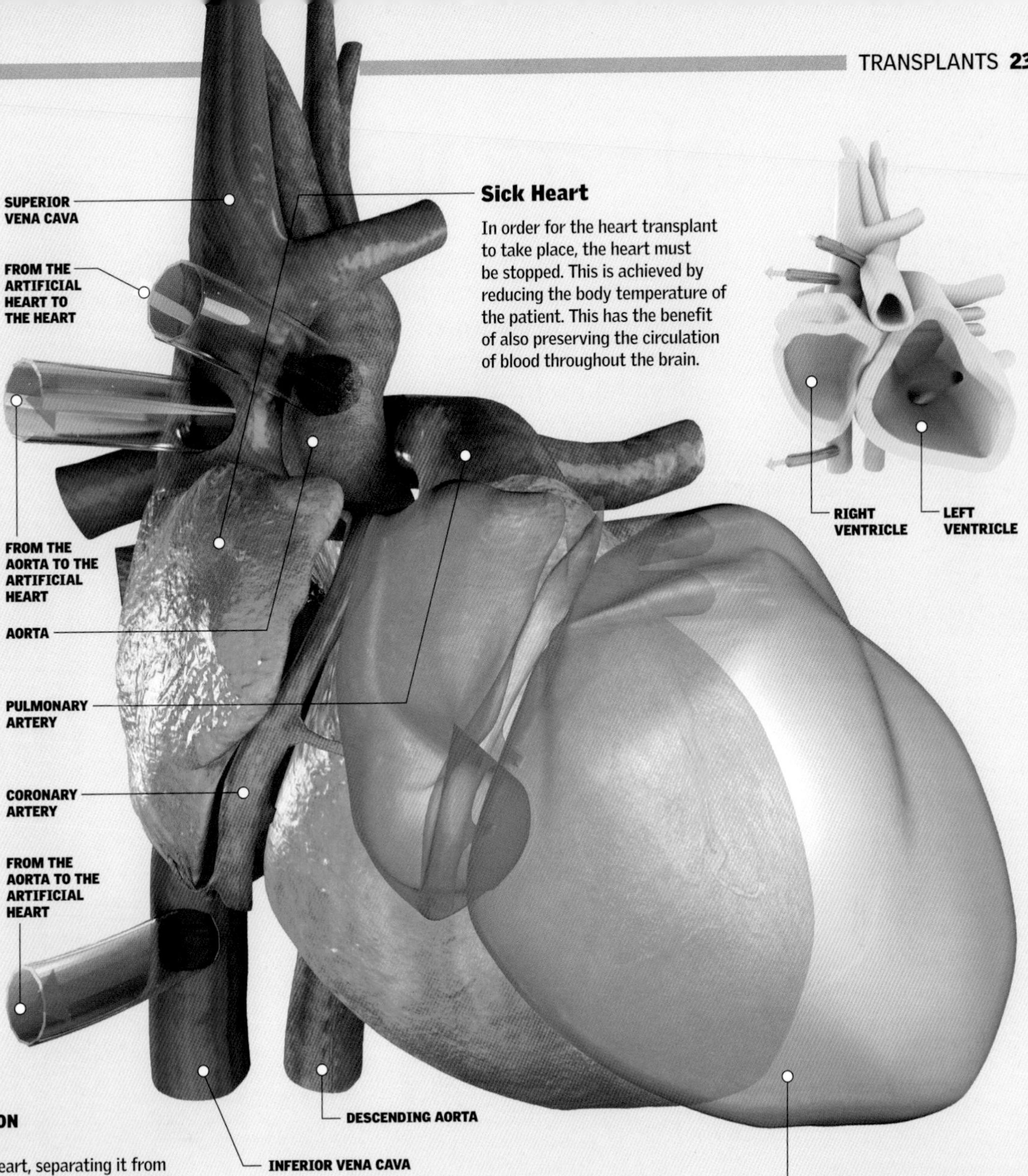


## Sick Heart

In order for the heart transplant to take place, the heart must be stopped. This is achieved by reducing the body temperature of the patient. This has the benefit of also preserving the circulation of blood throughout the brain.

**1 THE INCISION**
Once the patient is under the effects of the anesthesia, the surgeon carries out an incision in the middle of the patient's chest and proceeds to open up the sternum. He then opens the pericardium until the sick heart is left in plain view.

**2 PUMP**
Once the pulmonary and cardiac functions of the patient have been substituted by an external artificial pump, called a heart-lung machine, the aorta is clamped. This is the doctor's cue for the heart exchange to happen.

**3 EXTRACTION AND INSERTION OF DONOR HEART**
The surgeon removes the sick heart, separating it from the aorta and the pulmonary arteries. He then inserts the donated heart in its place. He sutures the left atrium in first, then sutures the septum, continuing all the way to the rim of the right atrium wall.

**4 RESTORATION OF BLOOD FLOW**
The pulmonary artery and the aorta are sutured to the donor heart. The aorta must be unclamped at this time. The surgeon checks for possible bleeding, and if the thermal and hemodynamic condition of the patient so permit, he proceeds to disconnect the patient from the artificial heart-lung machine.

**5 INTENSIVE CARE**
With the help of drainage tubes, the surgeon proceeds to close the breast incision. Finally, the patient, under constant surveillance, is transported to the intensive care unit. Once the postoperative period is over, the patient is released and begins a supervised ambulatory program in which he or she resumes physical movement such as walking.

**EXIT**
The graft is considered to be successful when the new heart contracts forcefully and evenly.

## Donated Heart

The donated heart must be the adequate size, taking into account the beneficiary's needs. In general, when a donor has an average weight and height, his or her heart most probably will work well on the majority of heart-transplant beneficiaries.

# Liver Transplant

People who suffer advanced, irreversible, life-threatening hepatic conditions now have the possibility of an attempted liver transplant. The most typical liver transplant cases are those of people who suffer chronic hepatitis or primary biliary cirrhosis, an autoimmune disease. Patients must not be infected in any way and cannot be suffering from any cardiac or pulmonary disease at the time.

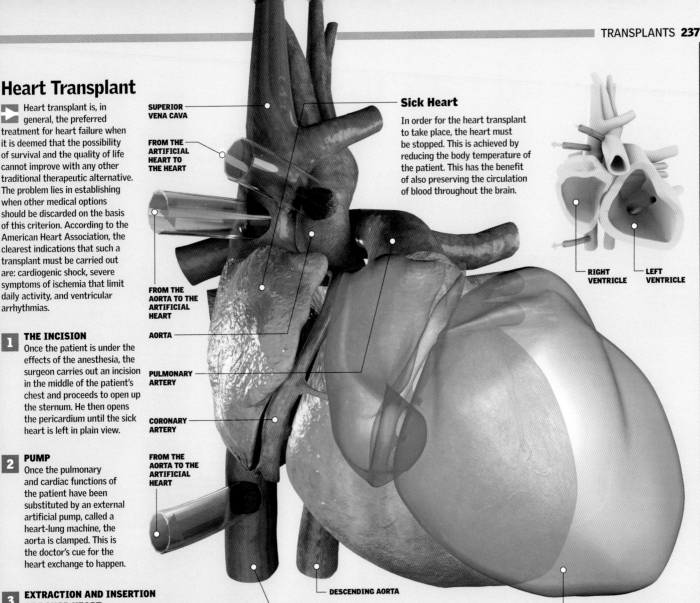

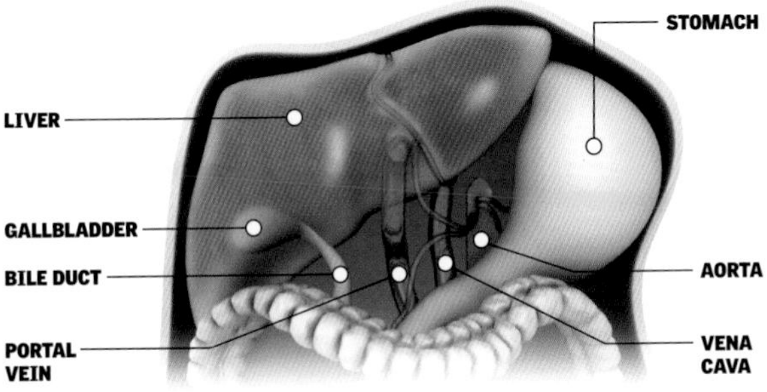


**1 DONATED LIVER**
The organ, along with all its blood vessels and its bile duct, is removed immediately after the death of the donor.

**2 THE NEW LIVER**
Is fused with the vena cava and the rest of the blood vessels. The opposite ends of the bile duct are sutured. A probe is inserted inside the reconstructed bile duct to drain the blood and the bile.

# Artificial Organs

The search for alternative solutions to save human lives has reached its maximum development thus far with the construction of artificial organs. The AbioCor artificial heart is currently being improved, and it is expected that soon it will have a useful life of five years. Similarly bionics has made it possible for blind people to perceive images through impulses transmitted to the brain by video footage from a camera that acts as a retina.

## The Development of Bionics

Advancements in bionics have begun to fulfill the wish that has been searched for in recent years: artificial organs literally identical to the natural ones—that is, organs that will not come with a limited useful life like other electronic devices. The world has already witnessed 16 successful bionic eye implants, and bionic arms are currently under development. Jesse Sullivan, the first bionic man, is able to control his artificial arms with his brain: the nerves of the lost arms were embedded in his chest, and when the patient thinks about closing his fist, a portion of the muscles in his chest contract, and the electrodes that detect the muscle activity "tell" the bionic arm to close the fist.

**A BIONIC EYE**
A microchip is placed at the back of the human eye. It is connected to a miniature video camera, which captures images that the chip later processes. The information is then sent as impulses to the brain, which interprets them.

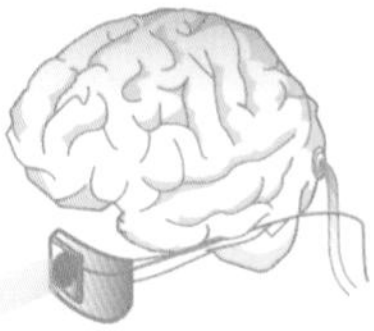

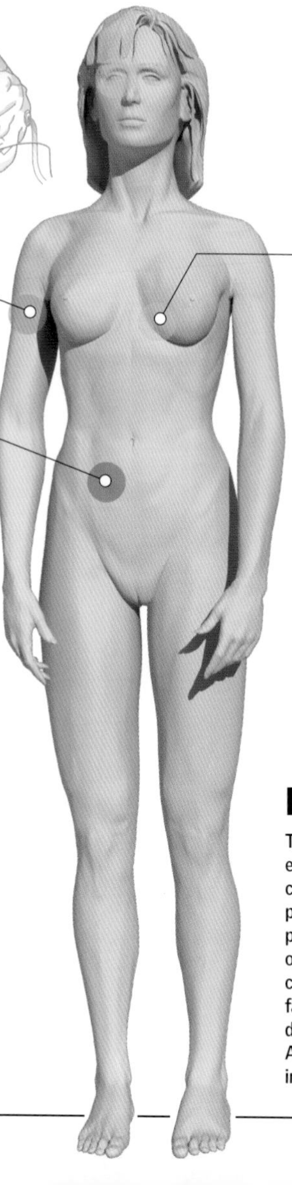

**ARMS**
Today surgeries for prosthetics are common. The possibility of implanting joints that could be controlled by the brain was achieved with the case of Jesse Sullivan in 2001.

**ARTIFICIAL KIDNEY**
Research to improve dialysis is still active. The patient is connected to a machine that removes impurities and toxic elements from the blood in the event of renal failure.

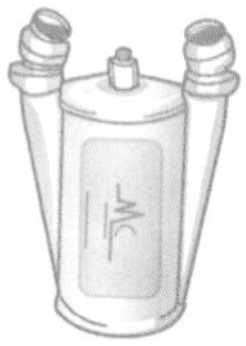

**ARTIFICIAL LUNG**
It consists of an intravenous device that permits breathing. It is inserted in a vein in the leg and is later positioned inside the vena cava, the largest vein for blood return to the heart. Fibrous membranes introduce oxygen into the body and remove carbon dioxide from it. Although not intended for prolonged use, it helps provide information that can guide future studies.

## Machines of Life

There are currently machines that can replace damaged bodily functions. Scientific developments and advances in bionics have created devices that can functionally replace organs with great effectiveness. The successful development of these machines has allowed organ activity to be restored in patients who would otherwise have lost it forever. The clear disadvantage of these devices, however, is that the patient must be permanently attached to the machine in order to avoid any risk. To overcome this limitation, organ transplants are being sought more and more frequently. The latest medical advancements led to the creation of artificial organs, such as the artificial lung and heart, which can perform essential functions of a patient's body without requiring him/her to be connected to a bulky machine.

## Artificial Heart

AbioCor was a milestone in the development of the artificial heart. Unlike its predecessor, the Jarvik-7, AbioCor is the first mechanical heart that can be totally self-contained in the patient's body. It functions almost exactly the same as a natural heart. It has two ventricles and two valves that regulate blood circulation. The AbioCor heart is powered without the need for cables or tubes that pass through the patient's body.

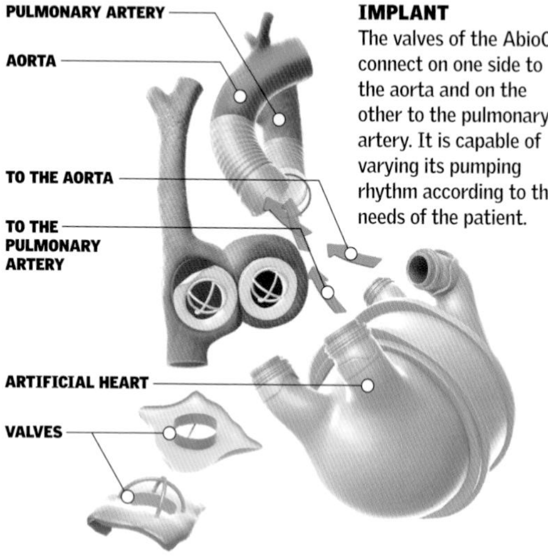


**IMPLANT**
The valves of the AbioCor connect on one side to the aorta and on the other to the pulmonary artery. It is capable of varying its pumping rhythm according to the needs of the patient.

### History of the Artificial Heart

**1 JARVIK-7**
Robert Jarvik designed the first artificial heart, which was placed inside a patient in 1982. The Jarvik-7 functioned with an external air compressor, which provided power.

JARVIK-7

**2 ABIOCOR**
Unlike its predecessor, the AbioCor does not require an external power source. It was the first artificial heart to be fully implanted into a patient. It is still being developed, and scientists are attempting to extend its useful life to five years. It has already been authorized for use in the United States.

## Heart 2006

The AbioCor heart was designed especially to support a patient's circulatory system and to prolong the lives of people who would otherwise die from cardiovascular failure. The heart, developed by Abiomed, is completely implantable in the body.

**PUMPING SYSTEM**
The heart developed by Abiomed is based on a hydraulic pump located at the center. Powered by a battery, the artificial heart reproduces the natural heart's performance almost identically. The deoxygenated blood goes to the lungs, and the oxygenated blood goes to the body.

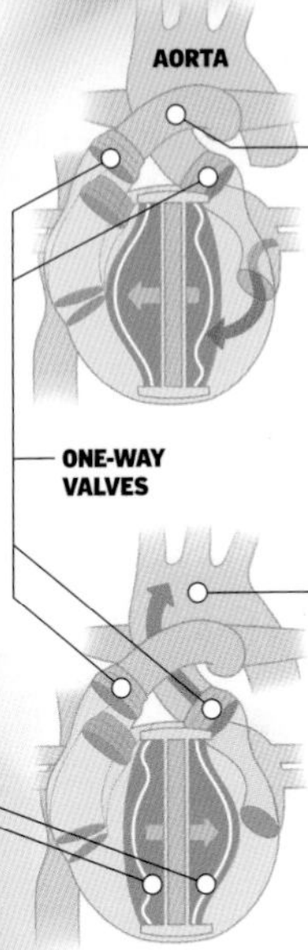


**1 TO THE LUNGS**
The blood lacking in oxygen flows to the lungs. It is pushed by a hydraulic pump and two membranes.

**2 TO THE BODY**
The oxygen-rich blood flows to the body. A cardiac rhythm is established to pump the blood according to the needs of the patient.

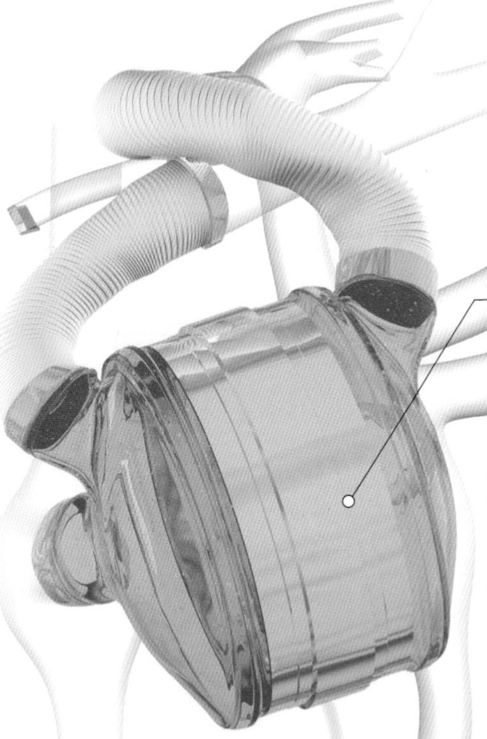

## 5 Abiocor Heart

Is made up of two ventricles with valves. Each ventricle pushes 2 gallons (8 l) of blood a minute and emits 100,000 beats in a day. The right ventricle pushes the blood toward the lungs, and the left one pushes it toward the rest of the vital organs and the body. The operation of the mechanical heart replicates that of a natural heart. It is made of titanium and plastic.

### 2 WITHOUT INCISIONS

The transcutaneous energy transfer (TET) system allows the battery to transfer energy to an internal battery through the skin. This way, potential infections caused by maintaining an opening in the abdomen are avoided.

### 4 INTERNAL BATTERY

This recharges directly from the external battery. It allows the patient a certain degree of autonomy, since it can run for an hour and a half without needing to connect to the external battery at the waist.

### 1 EXTERNAL BATTERIES

These prevent the use of tubes, and the patient does not have to be immobilized. This source of power eliminates the need to connect to external machines to recharge the batteries. The device is worn at the waist and is portable.

### 3 CONTROL SYSTEM

This regulates the rhythm with which the artificial heart pumps the blood. Depending on the needs of the patient, it can be increased or decreased. The internal control system is an electronic device capable of detecting any type of anomaly and making it known so that the patient can act on it.

# Nanomedicine

The prefix "nano" indicates the scale on which the latest scientific developments are taking place: one billionth of a meter. From nanotechnology, advances have appeared in what is called nanomedicine. The main objective of this variant of nanotechnology is to obtain cures for diseases from inside the body and at a cellular or molecular level. Devices smaller than the diameter of a human hair have even been developed.

## Nano-scaffolds for Regenerating Organs

The latest developments regarding the possibility of creating organs starting from a patient's own cells have demonstrated that by 2014 it may be possible to obtain a natural kidney simply through cellular regeneration rather than through a transplant. Beginning with biodegradable nanomolds, different organs could be created. The latest developments were able to produce a regenerated bladder in 1999. After being created, it was implanted successfully in seven patients. The procedure was done by doctor Anthony Atala of Wake Forest University. A section of a kidney that secretes a substance similar to urine has already been produced. Millions of nephrons still need to be regenerated, however, to achieve a fully functional kidney.

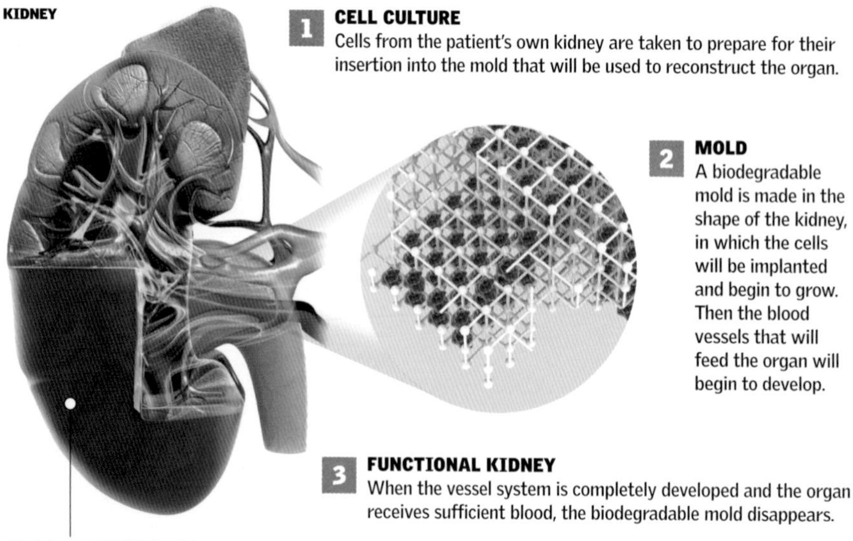


**1 CELL CULTURE**
Cells from the patient's own kidney are taken to prepare for their insertion into the mold that will be used to reconstruct the organ.

**2 MOLD**
A biodegradable mold is made in the shape of the kidney, in which the cells will be implanted and begin to grow. Then the blood vessels that will feed the organ will begin to develop.

**3 FUNCTIONAL KIDNEY**
When the vessel system is completely developed and the organ receives sufficient blood, the biodegradable mold disappears.

## Reconnecting Neurons

A group of scientists have developed a technique that allows nerve cells to regenerate. Chains of amino acids one thousandth the size of a red blood cell are used. Injected into the brain, these nanoparticles form a network over which the axons can stretch out and the connections may be able to be restored.

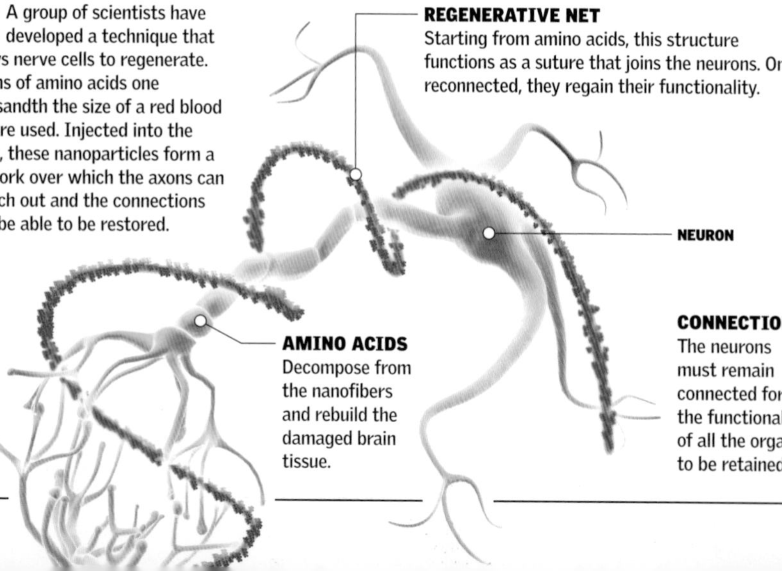


**REGENERATIVE NET**
Starting from amino acids, this structure functions as a suture that joins the neurons. Once reconnected, they regain their functionality.

**AMINO ACIDS**
Decompose from the nanofibers and rebuild the damaged brain tissue.

**CONNECTION**
The neurons must remain connected for the functionality of all the organs to be retained.

## Nanotechnology

By working at the scale of a nanometer ($10^{-9}$ meters), nanotechnology can currently be used in numerous areas of electronics, optics, and biomedicine. This state-of-the-art development builds devices so small that they can only be measured on the molecular scale. Today the most important and safest advances are the nanodevices used to detect cancer in its early stages. The nanoparticles can be between 100 and 10,000 times smaller than a human cell. Their size is similar to that of the larger biological molecules, such as enzymes. Nanoparticles smaller than 50 nanometers can easily enter any cell, while those smaller than 20 nanometers can move outside the blood vessels and circulate throughout the body.

### Microscopic Motor

Smaller in diameter than a hair and 100 times thinner than a sheet of paper, micromotors are the basis for tiny machines that could travel through the body and destroy tumors or bacteria in their paths.

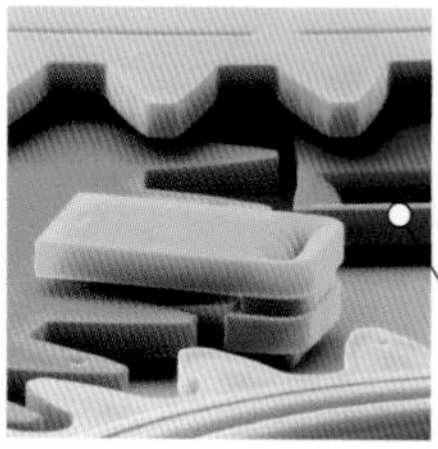

### Nanotubes

Nanotubes are structures whose diameter is on the order of a nanometer and whose length reaches up to a millimeter. They are the most resistant fibers known, between 10 and 100 times stronger than steel.

**ELEMENTAL FORM**
Like graphite and diamond, nanotubes are a basic form of carbon. They are used in heavy industry.

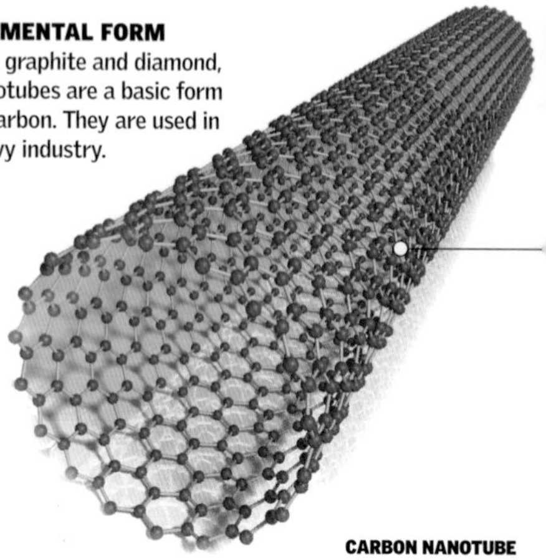


### Nanotechnologic Molecule

Each sphere of the molecule represents an atom: carbon in yellow, hydrogen in green, and sulfur in orange. It is based on fullerenes.

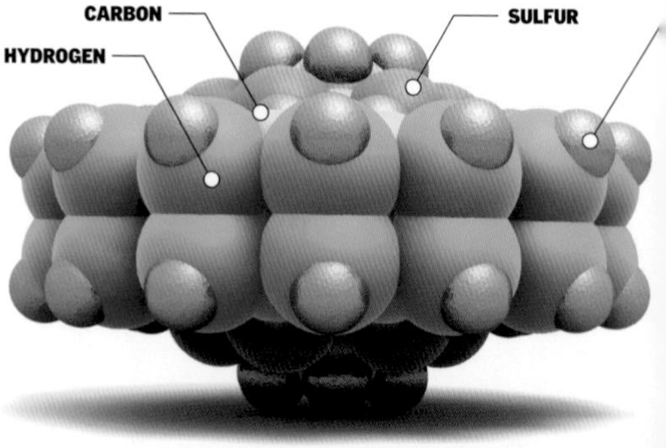

**MULTIPLES OF A METER IN DESCENDING ORDER**

METER
DECIMETER
CENTIMETER
MILLIMETER
MICROMETER
NANOMETER
ANGSTROM
PICOMETER
FEMTOMETER
ATTOMETER
ZEPTOMETER
YOCTOMETER

30,000 nanometers

## Scales

Nanotechnologies can reach unimaginably small dimensions. The developments achieved to this day have been at the level of a micrometer, which corresponds to a fraction of a cell, and of a nanometer, which corresponds to a particle (about the size of five molecules of water) scale.

**MILLIMETER**
Equivalent to a thousandth of a meter. Abbreviated mm. $10^{-3}$m.

**MICROMETER**
Equivalent to a millionth of a meter. Abbreviated µm. $10^{-6}$m.

**NANOMETER**
Equivalent to a billionth of a meter. Abbreviated nm. $10^{-9}$m.

**ANGSTROM**
Equivalent to one ten billionth of a meter. Abbreviated Å. $10^{-10}$m.

**A RELATIONSHIP OF SCALES**
The relation between the diameter of a stem cell and that of a nanoparticle is similar in proportion to the relation between the diameter of a tennis ball and that of a small asteroid.

20,000 nanometers

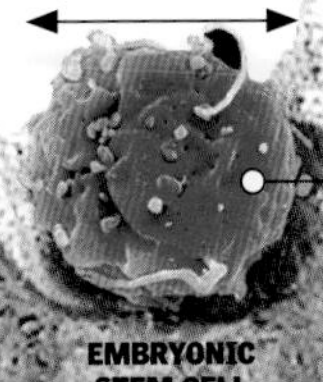


# Nanoparticles

The use of nanoparticles to combat diseases such as cancer has been carried out successfully in rats by scientists Robert Langer and Omid Farokhzad. The nanoparticles are one thousandth the size of the period at the end of this sentence. They are made up of carbon polymers that directly attack the cancer cells and destroy them without harming surrounding healthy cells. They act like guided missiles. This approach would make it possible to surpass the complications of chemotherapy. It is estimated that its full development will be complete in 2014.

**1 NANOSHIELDS**
The small carbon "bombs" detect the cancer cells and go directly toward them. They adhere to the tumor and prepare for their second phase: unloading.

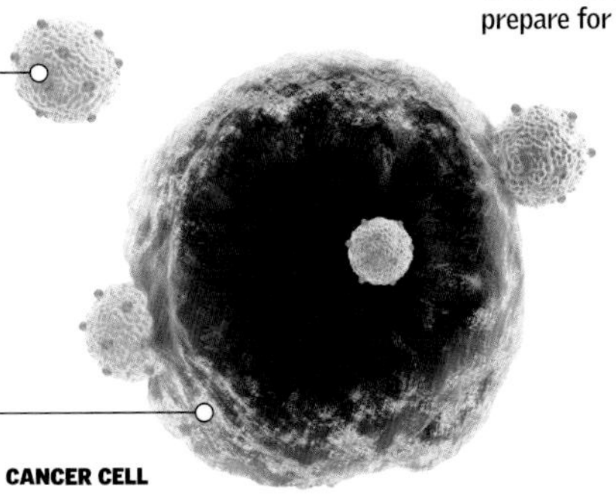


**2 UNLOADING**
Once the nanoparticles have entered the tumor, they release their carbon load, which contains instructions to destroy the cell.

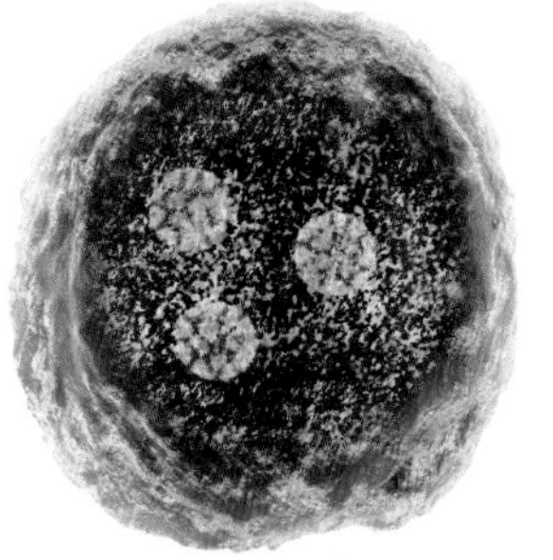

**3 EXPLOSION**
The attacked tumor cells are destroyed, and they die. Unlike chemotherapy, the surrounding healthy cells are not harmed.

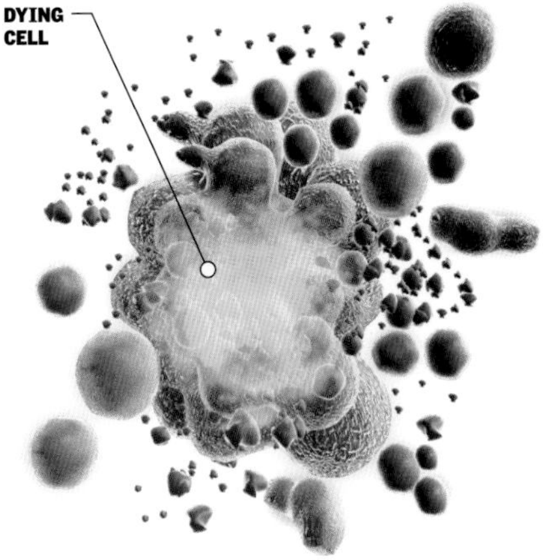


### Nanoparticles and Cells

To understand the scale at which nanoscopy works, we can compare the particles involved: a nanoparticle is to a cell what a grain of sand is to a football stadium.

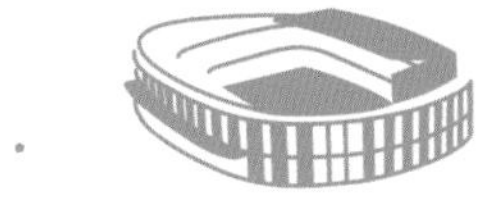

A GRAIN OF SAND IN A FOOTBALL STADIUM

# Nanoscopic Beams

Small microscopic and flexible beams that are built with semiconductors using lithographic techniques. These beams are covered with molecules capable of adhering to specific DNA. If a cancer cell secretes its molecular products, the antibodies placed on the flexible beams will bind to the secreted proteins. This generates a change in the physical properties of these beams, and researchers can read and interpret this information in real time.

**1 ATTACK**
The cancer cell secretes proteins to infect the organism.

**2 DEFENSE**
The antibodies attract the proteins. The nanobeam varies and provides information about the presence of cancer.

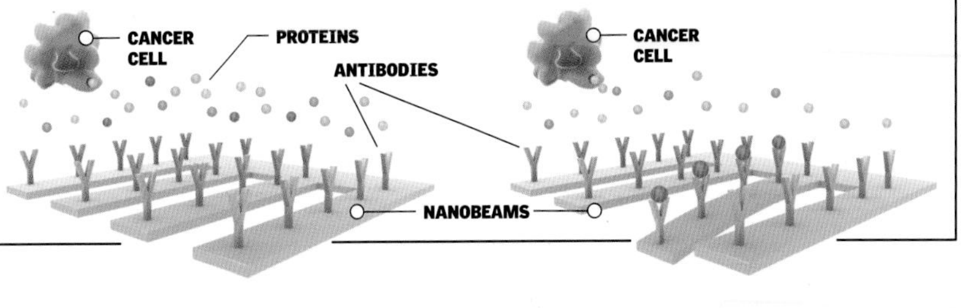

# En Route to Eternity

The dream of an eternal body seems to dominate scientific study today. The possibility of building a nerve system from a network of cables, proposed through developments in neuroscience, and the building of metallic muscle systems are two examples of steps that are being taken in that direction. According to some specialists, the future promises the creation of a bionic body, without ties to flesh. In this scenario, every health problem could be solved though metallic implants. There is even a study that explores the possibility of repairing DNA after cell death to assure the eternal youth of cells.

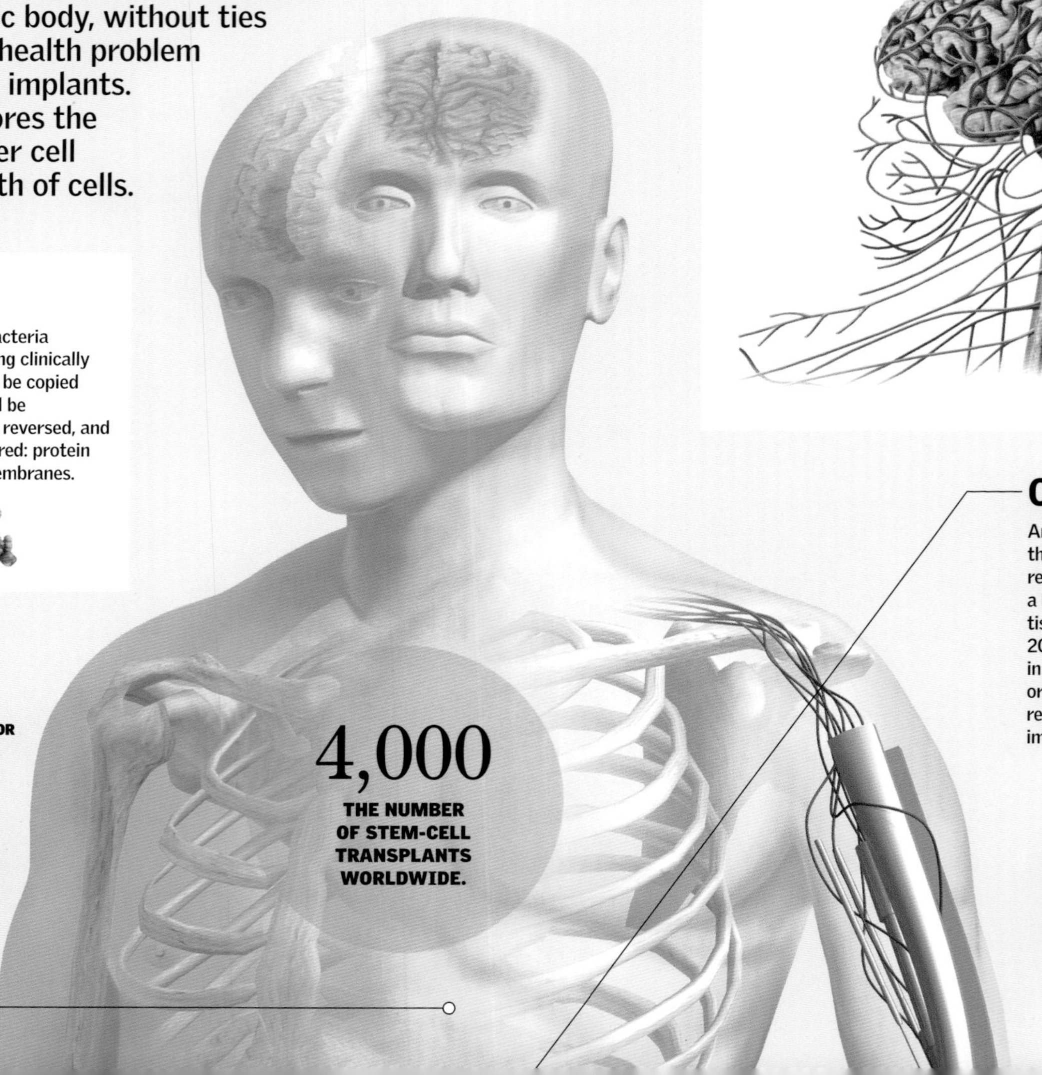

4,000
THE NUMBER OF STEM-CELL TRANSPLANTS WORLDWIDE.

## DNA Repair

Biologist Miroslav Radman discovered that the bacteria Deinococcus radiodurans can be revived after being clinically dead, through the repair of its DNA. If DNA could be copied rapidly and the genome of dead human cells could be reconstructed, then the death of cells could be reversed, and all their organic functions could be restored: protein synthesis, lipids, and membranes.

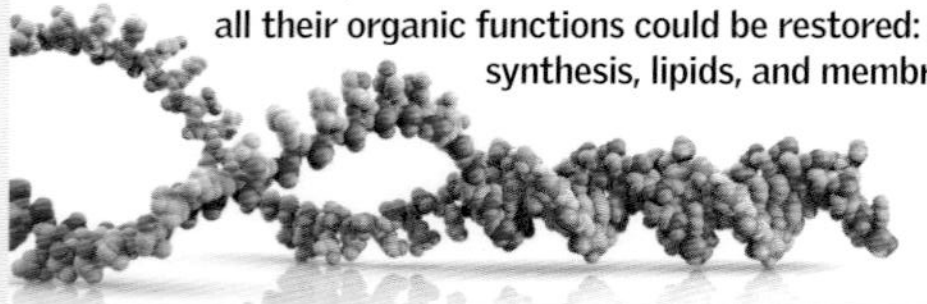

## Artificial Organs

Today efforts continue to design artificial organs that could replace organs that have been damaged or affected by severe diseases. By 2008, the Abiomed company plans to have the AbioCor heart developed to perfection. Although initial trials have been unsuccessful, Abiomed plans to design a heart that lasts for at least five years. Even so, it is very expensive: at least $100,000.

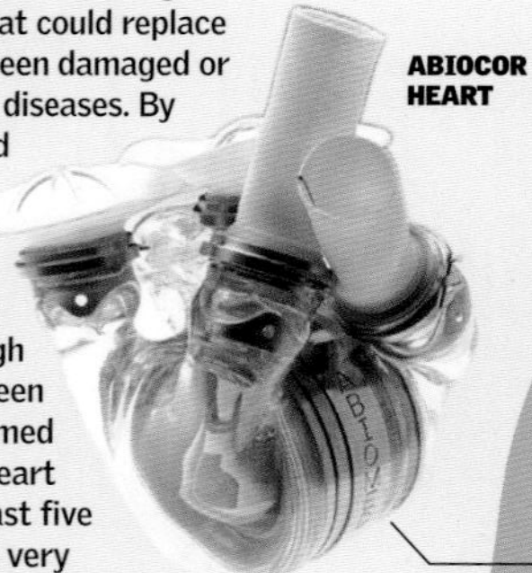
ABIOCOR HEART

## Self-Healing Cells

The dream of having a body in which there is no degeneration of nerve cells is on its way to becoming a reality: neuroscientist John Donoghue of Brown University is trying to re-create the nervous system through optical fibers. These fibers would be used to transmit brain impulses. In the future, the body would be a perfect network of fibers that would be degeneration-proof. Any problem linked to the nerve system could be eliminated because the cables would substitute for the nerves.

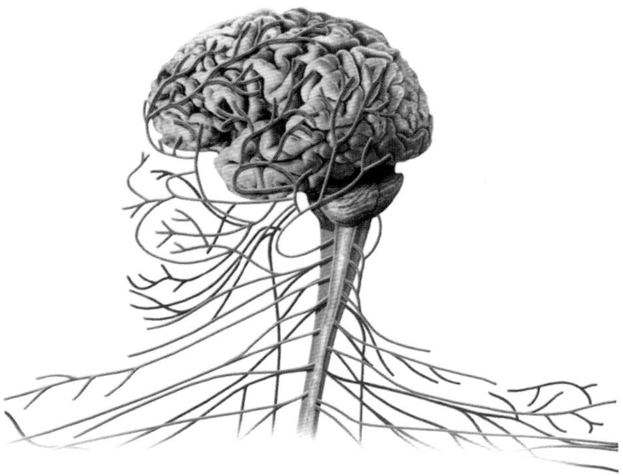

## Organ Regeneration

Anthony Atala of Wake Forest University is the leading pioneer of organ-regeneration research. In 1999, he was able to re-create a bladder from cells extracted from other tissue. Atala and his team estimate that by 2014 great advancements will have occurred in the regeneration of the most complex organ: the kidney. Once a kidney can be regenerated, transplants and artificial organ implants will become a thing of the past.

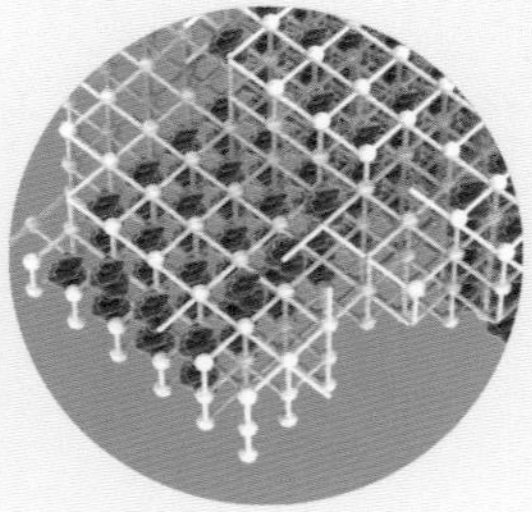

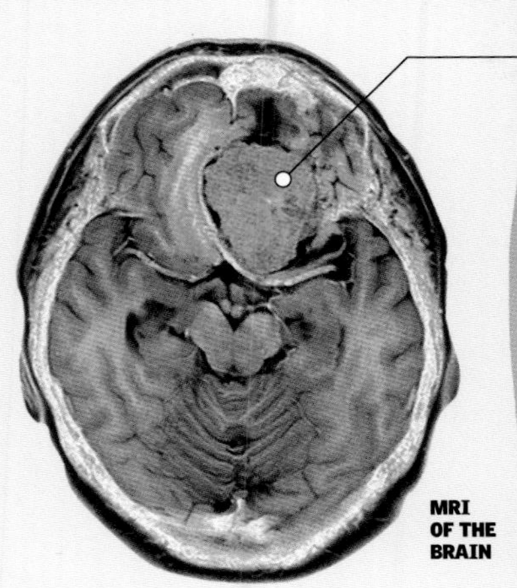

MRI OF THE BRAIN

**TUMOR**
A meningioma appears in the magnetic resonance image. It can be removed with surgery.

## Scan of the Body

The images obtained from magnetic resonance imaging (MRI) permit exploration of the body in 360 degrees. The most extensive use of MRI, however, is in the complete scanning of the brain to obtain a variety of images, which permits doctors to observe both the surface and the inside of the brain. Magnetic resonance imaging is one of the most commonly employed medical techniques for obtaining accurate images of the different organs of the body. Its resolution capability permits even the scanning of tissues.

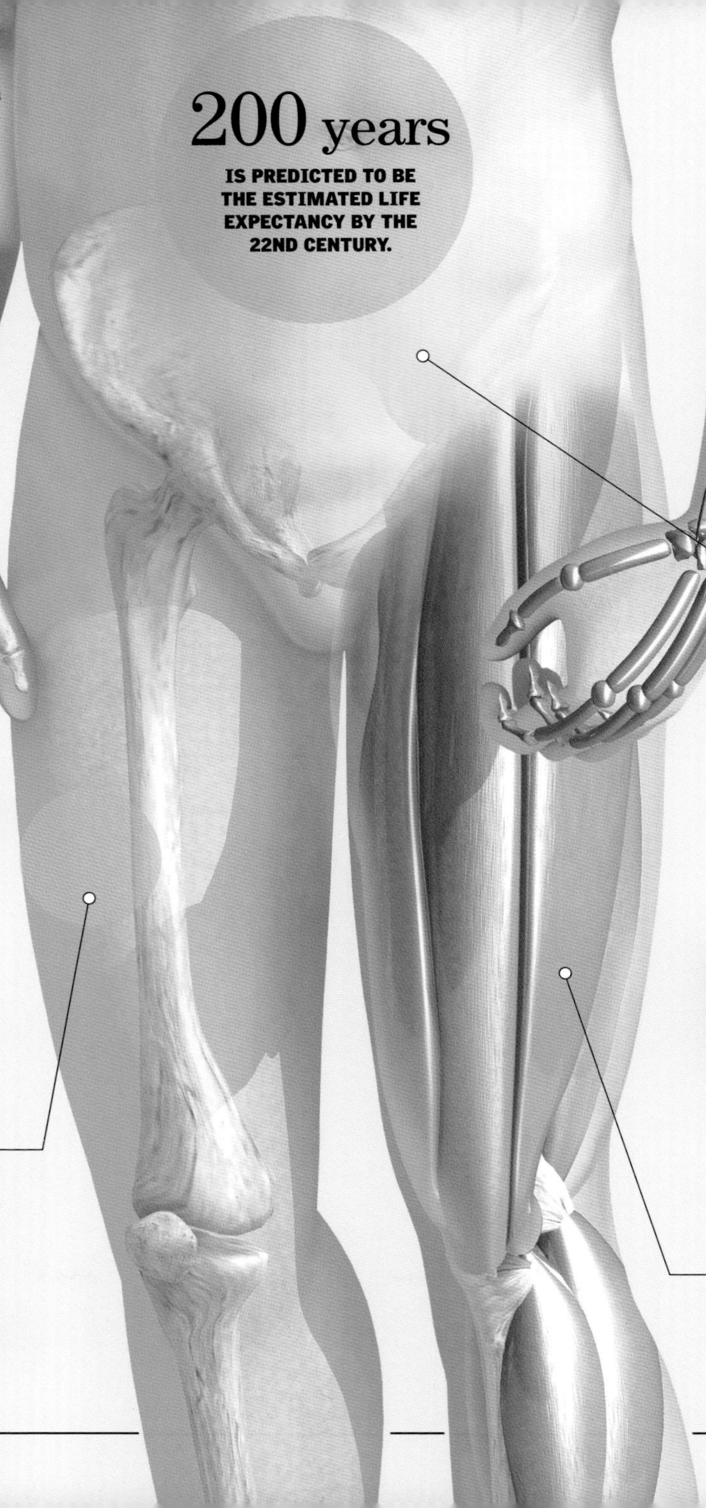

**200 years**

**IS PREDICTED TO BE THE ESTIMATED LIFE EXPECTANCY BY THE 22ND CENTURY.**

## Graft

The latest skin grafts are called autografts. They can save lives after severe injuries suffered by serious burn victims. From a small sample of healthy tissue, damaged tissue can be regenerated in three weeks through a cell culture.

**SIZE**
The required sample is as small as a postage stamp.

**CULTURE**
The creation of new cells takes place in a plastic container with a gel that provides nutrients for the epithelial cells.

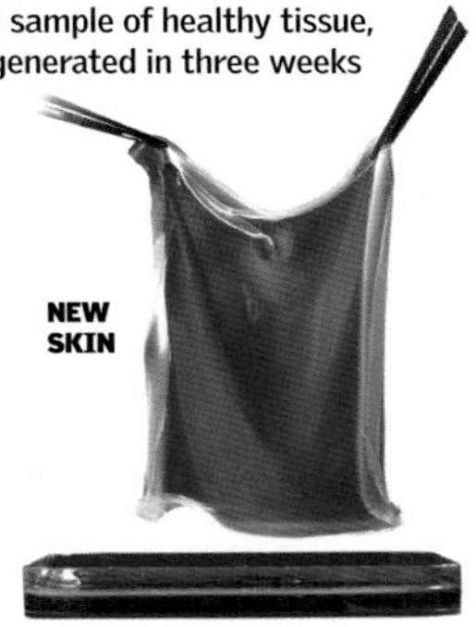


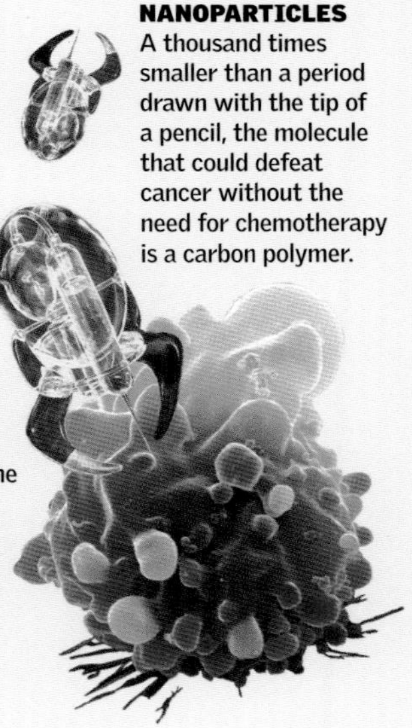

**NANOPARTICLES**
A thousand times smaller than a period drawn with the tip of a pencil, the molecule that could defeat cancer without the need for chemotherapy is a carbon polymer.

**INVASION**
Once it has detected a cancer cell, the nanoparticle penetrates it and unloads its carbon. The sick cell is destroyed.

## Cancer

If the studies by Robert Langer of the Massachusetts Institute of Technology (MIT) and Omid Farokhzad of Harvard University prove to be beneficial, chemotherapy will become a thing of the past. By using particles the size of amino acids (nanoparticles), cancer cells could be eliminated without harming the healthy cells located near them. Like guided missiles, the molecules go directly to the infected cells. Carbon polymers have been successfully tested in rats to eliminate the cancerous cells by penetrating them and injecting their content. Tests are still in development, and scientists estimate that by 2014 it will be possible to apply these new drugs.

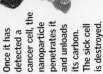

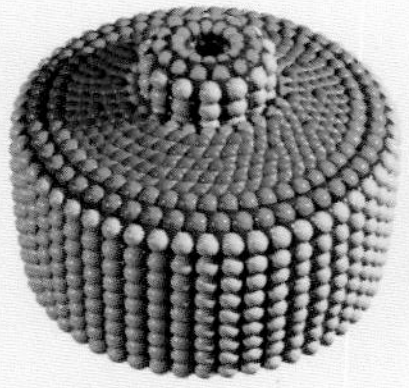

NANOPARTICLE

**60,000**

**NANOMETRIC MOLECULES WOULD BE NECESSARY TO COVER THE DIAMETER OF A HUMAN HAIR. BUILDING DEVICES AT THIS LEVEL COULD SPEED UP ALL KINDS OF TREATMENT.**

## Bionic Limbs

In 2005, the Rehabilitation Institute of Chicago performed the implantation of an artificial arm controlled by the brain. The University of Texas has been investigating an elastic metal that could replace natural muscles. The artificial muscular system is 100 times stronger and more resistant than human tissue. If these developments work out, they would provide a way of successfully replacing damaged joints.

# GLOSSARY

**Acid**
Type of chemical compound that, in solution, increases the concentration of hydrogen ions and combines with bases to form salts. DNA, vinegar, and lemon juice are weak acids.

**Adaptation**
A particular characteristic of an organism's structure, physiology, or behavior that enables it to live in its environment.

**Adrenaline**
Hormone secreted primarily by the adrenal medulla of the adrenal glands. It constricts blood vessels and is used as a medicine.

**Agonist**
Chemical product that, in addition to combining with a receptor (such as an antagonist), stimulates it, producing an observable effect. The term is also applied to a muscle that carries out a specific movement.

**Allele**
Variation of a gene in the population that codifies a specific trait. A diploid cell contains an allele from each parent for each characteristic. For example, the gene for eye color can have brown and blue alleles.

**Allergen**
Substance or material capable of provoking an allergic reaction.

**Alzheimer's Disease**
A specific type of breakdown of the nervous system that causes cognitive disorders. It is related to advanced age.

**Amino Acid**
Organic chemical whose molecular composition includes an amino group (derived from ammonia) and a carboxyl group (a radical that characterizes organic acids). It is one of the 20 chemical compounds that living beings use to form proteins.

**Angiogenesis**
Growth (normal or abnormal, depending on the circumstances) of new blood vessels in an organ or tissue.

**Antagonist**
Substance that inhibits or interferes with the action of other substances (hormones or enzymes). The term is also applied to muscles that, in the same anatomical region, act in opposite directions.

**Anthropologist**
Scientist who studies human beings from the viewpoint of their social and biological relationships.

**Antigen**
Substance that causes an immune response, such as the production of antibodies, when introduced into the body.

**Aorta**
The largest artery in the body, it starts in the left ventricle of the heart. It is called the thoracic aorta until it reaches the diaphragm, and below that it is called the abdominal aorta, where it later bifurcates into the iliac arteries.

**Aortic Arch**
Curve in the aortic artery near its origin at the heart. The arch has the shape of a shepherd's crook.

**Apparatus**
Complex of organs that fulfills one function. In the physiology of the human body it is also used as a synonym for system, for example, the digestive apparatus, reproductive apparatus, or respiratory apparatus.

**Archaeologist**
Scientist who studies human history based on the objects humans have left behind, such as buildings, ceramics, and weapons.

**Arterial Hypertension**
Elevated blood pressure, above 140 millimeters of mercury (systolic) and above 90 mm (diastolic).

**Artery**
Each one of the blood vessels through which the blood travels from the heart to supply the whole body.

**Arthritis**
Inflammation of the joint that could be the result of several causes.

**Arthroscopy**
Surgical procedure used by orthopedic surgeons to inspect, diagnose, and treat problems in the joints. It consists of making a small incision and inserting an arthroscope, an instrument the size of a pencil that contains a small lens and a lighting system to magnify and illuminate the interior. The light is transmitted via fiber optics to the end of the arthroscope, and the interior of the joint can be observed via a miniature television camera.

**Artificial Fertilization**
Technique for fertilizing ovules. It is usually done in vitro, after which the fertilized ovule is implanted.

**Articulation**
Joint between two bones of the body.

**Artificial Selection**
As opposed to natural selection, human intervention in the process of speciation; breeding animals or plants to improve their traits is an example.

**Atherosclerosis**
The accumulation of lipids (especially cholesterol) in the internal walls of the arteries; one of the main causes of diseases of the circulatory system.

**ATP**
Adenosine triphosphate. A molecule produced primarily by mitochondria that functions as the primary energy source for the cells.

**Atrium**
The name for each of the two chambers of the heart that receive blood from the veins.

**Autonomous Nervous System**
Part of the nervous system that regulates involuntary processes (heart rhythm, pupil dilation, stomach contractions, etc.). It includes the sympathetic and parasympathetic systems.

**Bacteriophage**
Virus that only infects bacteria; used as a vector in genetic engineering.

**Bacterium**
Microscopic organism that divides in two to reproduce. There are bacteria that are innocuous, pathogenic, and even beneficial to the human body.

**Basal Metabolism**
Activity level of the body functions during rest or while fasting.

**Bioballistics**
Recombinant genetic technique that consists of shooting small metal projectiles covered with DNA into a cell to penetrate the nucleus and recombine the genes in the desired manner.

**Biologist**
Scientist who studies living beings.

**Bioprospecting**
The taking of tissue samples from living beings to find genes that can be patented to obtain economic benefits.

**Blastocyst**
Cell mass, resulting from the division of the morula, that gives rise to the embryo.

**Bones**
Rigid structures, rich in calcium, that make up the skeleton.

**Calcification**
Fixation of calcium, an essential trace element for the formation of bones.

**Cancer**
Disease caused by the appearance and uncontrolled growth of a mass of abnormal tissue (malignant tumor).

**Carpal**
The structure of the wrist, composed of eight connected bones arranged in two rows. On the side toward the arm it joins with the cubital and radial bones, and on the side toward the hand it joins with the metacarpal bones.

**Cartilage**
Flexible skeletal tissue consisting of isolated groups of cells within a collagenous matrix.

**Celiac Artery**
Artery that brings blood from the heart to the stomach and the other organs of the abdomen.

**Cell**
Smallest independent unit that forms part of a living being.

**Cell Membrane**
Flexible envelope of all living cells that contains the cytoplasm. The membrane regulates the exchange of water and gases with the exterior.

**Cell Nucleus**
The central part of a cell, it contains the chromosomes and regulates the cell's activity. In some cells it is well differentiated. Other cells, such as some bacteria and red blood cells, have no nucleus.

**Cellular Membrane**
Flexible covering of all living cells that contains the cytoplasm. It is semipermeable and regulates the interchange of water and gases with the outside.

**Central Nervous System**
Structure made up of the brain and the spinal cord.

**Cerebral Cortex**
Made up of gray matter present on the surface of the brain. It is the largest part of the central nervous system. Many of the most advanced functions take place in this cortex.

**Chimera**
In Greek mythology, a monster with the head of a lion, the body of a goat, and the tail of a serpent. In genetics, the hypothetical creation of one being from the parts of others.

**Cholesterol**
Unsaturated lipid found in the body's tissues and in blood plasma. It is also found in elevated concentrations in the liver, spinal cord, pancreas, and brain. Cholesterol is ingested through some foods and is synthesized by the liver, then passed to the blood as HDL cholesterol, considered protective, or as LDL cholesterol, which in excess leads to the development of atherosclerosis.

**Chromatin**
Complex substance in the cell nucleus composed of nucleic acid and proteins.

**Chromosomal Crossover**
A step during meiosis that corresponds to the protein essential for the body's proper functioning.

**Chromosome**
Sequence of DNA coiled inside the nucleus of a cell. One cell usually has more than one chromosome, and together they make up the genetic inheritance of an individual.

**Cilia**
Small cellular appendages shaped like hairs and used for locomotion in a liquid medium.

**Cilium**
Tiny hairlike protuberance on a cell with a locomotive function in a liquid medium.

**Clone**
A living being that is identical to another. It also refers to parts, such as organs or fragments of DNA that are identical.

**Cloning**
Action of producing a clone.

**Coagulation**
Organic process in which the blood turns from a liquid to a solid state and whose normal purpose is to stop bleeding.

**Coevolution**
When more than one species evolve together, and the changes in one cause the others to undergo modifications in mutual adaptation.

**Coccyx**
Bone formed by the fusion of the last vertebrae. At its base it articulates with the sacral bone. In human beings and other vertebrates that do not have a tail, it is an actual bone.

**Conception**
The union of a sperm with an egg.

**Coronal**
A name given to the frontal bone, located at the anterior and superior part of the cranium. At birth the frontal bone or coronal is divided into two halves, which fuse over time. In medicine this can also refer to a suture that joins the frontal bone with the two parietal bones.

**Coronary Arteries**
A pair of arteries, originating in the aortic artery, that branch out and supply blood to the heart.

**Cortex**
The gray material present in most areas of the brain. It is the largest part of the central nervous system. The majority of the most advanced functions occur in the cortex.

**Corticoids**
Hormonal steroids produced by the adrenal gland cortex. Corticoids can be produced artificially. They have a therapeutic application as anti-inflammatory drugs.

**Cystoscope**
Apparatus used to explore the inner surface of the bladder.

**Cytoplasm**
Watery or gelatinous substance that contains organelles and makes up most of the interior of the cell, except for the nucleus.

**Cytosine**
One of the four bases that make up the DNA molecule.

**Dermatophytosis**
Infection in the skin caused by some species of fungi.

**Descendant**
Family member belonging to later generations, such as a child, grandchild, or great-grandchild.

**Designer Baby**
Human baby selected as an embryo based on a set of genetic traits chosen before its birth.

**Diabetes**
Chronic disease characterized by elevated levels of blood glucose due to metabolic disorders.

**Diaphragm**
Respiratory muscle between the thorax and the abdomen.

**Digestion**
The set of processes through which the digestive system converts food into substances that can be assimilated by the organism.

**Diploid**
A cell with two complete sets of chromosomes. It is denoted by the symbol 2n.

**Dislocation**
The displacement of any bone from its normal position in a joint.

**DNA**
Deoxyribonucleic acid. A molecule in the shape of a double helix containing encoded genetic information of an individual.

**DNA Footprint**
The identification of a person by DNA; used in forensics.

**DNA Sequencing**
Obtaining the structure of bases that make up DNA. The long DNA chain is often divided into smaller fractions for study.

**Dominance**
Functional attribute of the genes by which they manifest their effect, regardless of the effect of the allele that accompanies them.

**Dominant Gene**
Gene that, when present in a pair of alleles, is always manifested.

**Double Helix**
Shape of two spirals in geometric space. The DNA chain has this shape.

**Dyslexia**
Disorder, sometimes genetically based, that causes difficulties with reading, writing, and speech.

**Ejaculation**
The action of expelling semen.

**Embryo**
Product of fertilization of the egg by a sperm. It can develop into an adult organism.

**Emulgent Arteries**
Arteries that bring blood from the heart to the kidneys, also called renal arteries.

**Endocardium**
Membrane that lines the walls of the heart. It consists of two layers: an exterior, consisting of connective tissue, and an interior, of endothelial tissue.

**Endometrium**
Mucous membrane covering the inner walls of the uterus.

**Endoplasmic Reticulum**
Group of narrow channels that transport various types of substances and molecules from one point to another inside a cell.

**Endothelial**
Organic tissue that lines wall-like structures within the body, such as those of the pleura or of blood vessels.

**Enzyme**
Protein that helps regulate the chemical processes in a cell, usually triggering or accelerating a reaction.

**Erythrocytes**
Red blood cells; they carry oxygen.

**Erythropoiesis**
The creation of red blood cells, stimulated by the action of a protein called erythropoietin.

***Escherichia coli***
Abundant bacteria, often used in genetic experiments.

**Estrogens**
Female hormones produced by the ovaries and by the adrenal glands. They stimulate the growth of cells in the endometrium, the ovaries, and the breasts.

**Eugenics**
Science that seeks to improve humankind by selecting and controlling human genes. Its objectives are highly controversial.

**Evolution**
Gradual change in a species or organism; not necessarily an improvement. It was theorized by Darwin in his famous book *On the Origin of Species*.

**Extinction**
The disappearance of all specimens of one or more species.

**Fertilization**
Fusion of a male gamete with a female gamete, forming a zygote, which can develop into a new individual.

**Fetus**
The human body in gestation, after the third month and until birth.

**Filler DNA**
Long, repeated sequences of DNA that do not provide genetic information. Also called junk DNA.

**Flagellum**
Filament-like structure that is found on some bacteria and is used for locomotion.

**Follicle**
Sac-shaped gland located in the skin or in the mucous membranes which usually surrounds the base of a hair.

**Forensics**
Scientific discipline of the study of evidence of a crime.

**Fossil**
All traces of past life, even those that have not been petrified.

**FSH**
Follicle stimulating hormone. Female hormone involved in the ovulation process.

**Fungus**
Live unicellular or multicellular organism belonging to the Fungi kingdom.

**Gamete**
Reproductive cell, also called sex cell, such as sperm and eggs.

**Gene**
Unit of information of a chromosome; sequence of nucleotides in a DNA molecule that carries out a specific function.

**Generation**
A "level" in the history of a family or species. There is one generation between parents and children.

**Gene Therapy**
Treatment of certain diseases of genetic origin by replacing the patient's defective gene(s) with the correct gene(s) to cure the disease.

**Genetic Disease**
Disease caused partially or wholly by a genetic disorder.

**Genetic Engineering**
The study of the application of genetics in relation to technological uses.

**Geneticist**
Scientist who studies genetics.

**Genetic Mutation**
Error in the copying of a cell's DNA. A few mutations can be beneficial and intensify the cell's original qualities. Mutations are believed to have generated the evolution of species. Most give rise to closed evolutionary lines.

**Genetics**
The study of DNA and genes.

**Genetic Trait**
Physical trait transmitted to an organism's descendants, such as hair color and height. The interchange of genes between diploid chromosomes causes the recombination of genes.

**Genome**
The entire complex of chromosomes and their genes; the totality of the genetic material in a cell or individual.

**Genotype**
Genetic constitution of a single cell or an organism with reference to a single characteristic or set of characteristics; the sum of all the genes present in an individual.

**Gland**
Organ that has the function of producing secretions that can be expelled through the skin or mucous membranes (salivary glands or sweat glands, for example) or into the bloodstream (the thyroid, for example).

**Graft**
Implantation into an organism of a portion of live tissue that comes from another organism or another part of the same one. Also, the portion of tissue to be implanted.

**Haploid**
From the Greek term haplous, "one"; a cell with only one set of chromosomes, unlike diploid cells. Gametes are haploid cells.

**Helix**
Geometric spiral shape equivalent to a curve along a cylindrical surface; the shape in which the DNA molecule is curled.

**Hemoglobin**
Protein (globin) associated with a porphyrin that contains iron (heme group) and is found inside the red blood cells; it transports oxygen.

**Hemophilia**
A group of hereditary diseases caused by the lack of a clotting factor (the most important being Factors VIII and IX). Its most common symptom is spontaneous hemorrhaging.

**Hemostatic**
Substance or agent that halts hemorrhaging.

**Heredity**
In genetics, all types of genetic material passed on by the parents to a descendant.

**Hippocampus**
Part of the brain that governs the memory.

**Holocrine**
Gland with an exclusively secretory function or whose secretion consists of disintegrated cells of the gland itself, such as the sebaceous glands.

**Homeostasis**
Complex of self-regulatory phenomena that keep the composition and the properties of the body's internal environment constant. It is said that homeostasis is reached when the body's internal environment contains the optimum concentrations of gases, nutrients, ions, and water; when its temperature is optimum; and when the volume of fluids is optimum for the life of the cells.

**Hormone**
Glandular secretion with the function of stimulating, inhibiting, or regulating the action of other glands, systems, or organs of the body.

**Immune System**
Set of processes centered on the blood and the lymphatic system that is activated to defend the human body against diseases.

**Innominate Bones**
A pair of bones, one in each hip, which join the sacrum and the coccyx to form the pelvis. They consist of the fusion of the iliac, the ischium, and the pubic bones.

**Insulin**
Hormone secreted by the pancreas that is responsible for the metabolism of glucose in the body.

**Ion**
Atom of an element or a molecule that is electrically charged because it has gained or lost electrons from its normal configuration.

**Joint**
The area where a bone or a skeletal organ comes together with another.

**Karyotype**
Ordering of the chromosomes of a cell according to shape, number, or size.

**Keratin**
Protein found in skin, hair, and nails.

**Laser**
From the acronym for Light Amplification by Stimulated Emission of Radiation, it is a luminous artificial emission of variable frequency. Its energy can be controlled because of the coherence of its beams.

**Leukocyte**
White blood cell. A component cell of the blood, its main function is to defend the body from infectious agents.

**Ligase**
Protein used by geneticists to join sections of DNA.

**Lipids**
Organic chemical compounds formed mostly by hydrogen and carbon. Cholesterol and edible oils are the best known.

**Lobes**
Rounded protuberances of organs, such as the liver, the lungs, or the brain.

**Lymph**
Liquid that moves through the lymphatic system.

**Lymphatic System**
Ensemble of lymphatic vessels and ganglia that is independent of blood flow. It acts as a regulator of osmotic equilibrium in the body and as an activator of the immune system.

**Lymphocyte**
Belongs to the group of white blood cells. It is present in the blood and in the lymphatic system.

**Lymphoma**
Neoplastic disease that originates in the lymphatic system.

**Lysosome**
Part of the cell that breaks down and reuses worn-out proteins. It is also a potent antibacterial.

**Meiosis**
Type of cell division in which two successive divisions of the nucleus of a diploid cell create four haploid nuclei. As a result of this mechanism, gametes or spores are produced.

**Meristem**
Tissue with cells that produce other cells by cellular division.

**Metabolism**
Complex of chemical reactions that take place continuously within cells to synthesize complex substances from simpler substances or to degrade a substance into simpler substances. An example is the digestive process. The activity level of body functions at rest and while fasting is called basal metabolism.

**Metacarpal**
Middle part of the skeletal structure of the hand, between the wrist (carpal bones) and the phalanges. It consists of five bones, which are the largest bones of the hand.

**Metastasis**
Spreading of a cancerous tissue, making it capable of attacking organs other than the one from which it originated.

**Metatarsal**
Part of the skeletal structure of the foot, between the tarsus (posterior part of the foot) and the phalanges (toes). It consists of five bones and is usually called the sole of the foot.

**Micturition**
Act of urinating, or expelling urine.

**Mitochondria**
Cellular organelle that combines food and oxygen to produce energy for the cell.

**Mitochondrial DNA**
Small amount of DNA contained in the mitochondria of the cell.

**Mitosis**
Division of a cell in which two identical cells are formed from the parent cell.

**Molecule**
Minimum quantity into which a substance can be divided without losing its chemical properties.

**Monozygotic Twins**
Twins who develop from a single zygote that splits in two, forming two genetically identical individuals.

**Morula**
Early stage in the development of a multicellular organism, made up of 16 to 64 cells. It gives rise to the blastocyst.

**Mucous Membrane**
Covering of body cavities that communicate with the exterior (such as the nose). A mucous membrane contains numerous single-celled glands that secrete mucus.

**Mummy**
Human corpse preserved by artificial methods, which can be preserved for long periods of time. The genetic study of mummies provides much evidence about life in the past.

**Muscles**
Organs composed of fibers capable of contracting.

**Mycosis**
Infection caused by fungi.

**Myocardium**
Muscular part of the heart, between the pericardium and the endocardium.

**Nanotechnology**
Industrial technology that permits the fabrication of microscopic devices.

**Natural Selection**
Process in which only the organisms that are best adapted thrive and evolve. This selection is carried out without human intervention.

**Neurotransmitter**
Chemical substances responsible for the transmission of the nerve impulse through the neuron synapses.

**Nitrogenous Base**
Type of chemical compound. Four distinct types of bases in DNA make up the genetic code, according to their combinations.

**Nucleic Acid**
Molecule that carries genetic information about the cell. There are two types: DNA and RNA.

**Nucleus**
The part of the cell that contains the DNA with its genetic information.

**Organ**
Any part of the body that accomplishes a function.

**Organelle**
Any organ of a cell, including mitochondria, ribosomes, and lysosomes. They carry out specific functions.

**Osmosis**
Movement of a liquid through a selectively permeable membrane.

**Ovulation**
Release of the mature egg from the ovary through the fallopian tube.

**Ovule**
Female gamete, or sex cell.

**Oxyhemoglobin**
Hemoglobin of arterial blood that is loaded with oxygen.

**Oxytocin**
Female hormone produced by the hypothalamus, it is transported to the hypophysis and is later released into the bloodstream. In women, it is responsible for, among other functions, the milk ejection reflex and uterine contractions.

**Pancreas**
Organ that produces insulin, located below the stomach.

**Papillae**
Conical protuberances, usually sensory, formed on the skin or mucous membranes (especially the tongue) by the branching of nerves and blood vessels.

**Parkinson's Disease**
Neurological disorder caused by a deficit of the neurotransmitter called dopamine.

**PCR (Polymerase Chain Reaction)**
Technique for multiplying fragments of DNA using polymerase.

**Pectoral Angina**
Oppressive pain located in the retrosternal region, caused by an insufficient flow of oxygenated blood to the cardiac muscle.

**Pericardium**
Pair of membranes that surround the heart.

**Phagocytes**
Cells found in blood and tissue. They capture bacteria or any other kind of noxious particles and "phagocytize," or "eat," them, absorbing them into their cytoplasm and later digesting them.

**Phalanges**
Bones of the fingers and toes. They extend to the metacarpal bones in the hand and the metatarsals in the foot. Starting from the metacarpals and the metatarsals, they are sequentially numbered: first, second, and third phalanges (of each finger or toe). The word "phalanges" commonly designates the first phalanges, or each of the jointed parts of the fingers or toes.

**Phenotype**
In biology, the visible manifestation of a genotype in a certain environment.

**Phylogenetics**
The study of evolutionary relationships between the various species, reconstructing the history of their speciation.

**Physiology**
Study of the functions of the organism.

**Platelet**
Cellular component of blood that takes part in the clotting process.

**Polymer**
Macromolecule consisting of repeated structural units, called monomers.

**Popliteus**
Section of the leg opposed to, or behind, the knee.

**Preimplantation Genetic**
Diagnosis Method of in vitro selection of embryos based on preferred genetic conditions. They are then implanted in the uterus for normal development.

**Progesterone**
Female hormone involved in the menstrual cycle and gestation.

**Protein**
Substance that makes up parts of the cells. It is formed by one or more chains of amino acids and is fundamental to the constitution and functioning of the essentials of life, such as enzymes, hormones, and antibodies.

**Protozoa**
Microscopic, unicellular, heterotrophic organisms that live in an aqueous medium and reproduce through bipartition.

**Radioactivity**
Energy given off by certain chemical elements; it can cause genetic alterations or even diseases such as cancer.

**Ranine Artery**
Artery that branches out toward the front of the tongue.

**Recessive Gene**
Gene that, even though present, might or might not be manifested in a pair of alleles depending on the presence of a dominant gene.

**Recombinant DNA**
Sequence that contains a combination originating from one or more organisms.

**Reflex**
Automatic and involuntary reaction of the nervous system that is produced in response to a stimulus.

**Replica**
Exact or nearly exact copy of an original. A virus creates replicas of itself after invading a cell.

**Repressor**
Protein that binds to a DNA chain in order to stop the functioning of a gene.

**Reproduction**
Sexual or nonsexual creation of other organisms of the same species. The fertilization of gametes is sexual, whereas parthenogenesis is not.

**Respiration**
The act and effect of inhaling air, to take in the substances that the body requires, such as oxygen, and after processing them exhaling unneeded substances, such as carbon dioxide.

**Restriction Enzyme**
Protein in certain bacteria that can cut the DNA molecule.

**Ribosome**
Part of a cell that reads the instructions of the genes and synthesizes the corresponding proteins.

**Ribs**
Long and curved bones. They originate at the back of the body at the spinal column and curve forward. They are called "true" if they end at the sternum and "false" if they remain floating without completely enclosing the rib cage.

**RNA**
Ribonucleic acid, similar to DNA but used to transport a copy of DNA code to the ribosome, where proteins are manufactured.

**RNA Polymerase**
Enzyme that serves as a catalyst for synthesizing an RNA molecule based on DNA code.

**Saturated Fats**
Fats of animal origin that are involved in nutrition.

**Schwann Cells**
Cells that produce myelin, a fatty insulating substance that prevents electrical signals from losing strength as they move away from the body of the neuron.

**Selective Breeding**
The production of plants or animals that display the results of artificial selection of their genetic traits. Agronomists, veterinarians, and geneticists use selective breeding to improve certain species and breeds or varieties to achieve, for example, greater productivity and crop yields.

**Semen**
The spermatozoa and fluids produced in the male genital organs. It is often called sperm.

**Sensation**
Physiological process of receiving and recognizing stimuli produced by vision, hearing, smell, taste, touch, or the body's spatial orientation.

**Sex Cells**
Special cells, also called gametes, with a reproductive function. Some examples are ovules, spermatozoa, and pollen.

**Sleep**
State of repose characterized by inactivity or suspension of the senses and voluntary motion. The cerebral activity called dreaming takes place during sleep.

**Somatotropin**
Human growth hormone, secreted by the pituitary gland.

**Speciation**
Evolutionary process in which a new species is formed, for various reasons, from another species.

**Species**
The lowest unit of classification in evolution. It was originally defined according to the phenotype of each individual. The field of genetics has raised new questions about what constitutes a species.

**Spermatozoon (Sperm)**
Male gamete or sex cell.

**Spinal**
Relating to the spine.

**Spinal Bulbar**
Part of the cerebral trunk that goes from the annular protuberance to the cranium's occipital foramen.

**Spine**
The neuroskeletal axis that runs along the medial dorsal of the body and consists of a series of short bones called vertebrae, which are arranged in a column and jointed with each other.

**Spirillum**
Flagellated bacterium with a helical or spiral form.

**Spore**
Reproductive cell of a fungus.

**Stem Cell**
Cell with the ability to develop into a specific type of cell or bodily tissue. Pluripotent stem cells can develop into any type of cell of the body.

**Sternum**
Bone of the anterior thorax, which joins the front of the ribs.

**Striated Muscle**
Muscle used for voluntary motion. Its muscle fibers show striations, or grooves.

**Subclavian Arteries**
Pair of arteries, one of which branches off from the brachiocephalic trunk (on the right side of the body) and the other from the aortic arc (on the left). They run toward the shoulder on each side and, after passing below the clavicle, become the axillary artery.

**Sugar**
Generic name of the organic chemical compounds known as carbohydrates.

**Synthesis**
Chemical process in which two or more molecules join to produce a larger one.

**System**
Complex of organs that participates in any of the principal functions of the body. A synonym of "apparatus."

**Systemic**
Describes a disorder that affects several organs or the body as a whole.

**Tarsal**
The skeletal structure of the leg between the foot and the metatarsal. It consists of seven bones that constitute the posterior part of the foot.

**Telomerase**
Protein for repairing the telomere of a chromosome. Found only in certain cells.

**Telomere**
DNA sequence at the end of a chromosome, it is shortened every time the cell divides. The number of times the cell can divide depends on the length of the telomere.

**Testosterone**
Androgenic hormone related to the primary as well as secondary male sexual traits. It is produced by the testicles and to a lesser extent by the adrenal glands and ovaries in women.

**Thrombus**
Solid mass of blood formed inside a vein or artery. If it travels through the circulatory system, it is called an embolus, which can cause an obstruction.

**Thymine**
One of the four bases that make up DNA, combining in different sequences to form genes.

**Tissue**
Group of identical cells that carry out a common function.

**Transcription**
Copying of the genetic code of the DNA into another molecule, such as RNA.

**Transgenic**
Describes plants or animals of one species that have undergone genetic modifications using one or more genes from another species.

**Transplant**
Insertion of live tissue into a living organism from another organism (living or not).

**Tumor**
Any alteration of the tissues that produces an increase in volume.

**Uterus**
Hollow viscera of the female reproductive system. It is located inside a woman's pelvis. In the uterus, or womb, either menstrual fluid is produced or a fetus develops until it is born.

**Vagus Nerve**
Also called the pneumogastric nerve, it is the 10th of the 12 cranial pairs. It originates in the brainstem and innervates the pharynx, esophagus, larynx, trachea, bronchia, heart, stomach, and liver.

**Vector**
In genetic engineering, the agent that introduces a new sequence of DNA into an organism. Viruses and bacteria are often used as vectors.

**Veins**
Blood vessels that bring blood from the entire body toward the heart.

**Ventricles**
Cavities of the heart that receive blood from their respective atrium (right or left) and pump it through the arteries.

**Vibrio**
Genus of elongated bacteria in the shape of a comma with a single cilium, such as the one that produces cholera.

**Virus**
Organism composed of DNA or RNA enclosed in a capsid, or protein structure. A virus can invade cells and use them to create more viruses.

**Viscera**
Organs located in the principal cavities of the body (such as the stomach or the liver within the abdominal cavity).

**Vitamins**
Organic substances present in food. The body ingests them to ensure the balance of various vital functions. There are different kinds of vitamins, designated with the letters A, B, C, etc.

**X Chromosome**
One of the chromosomes that determines a person's sex.

**Y Chromosome**
Chromosome that determines the male sex; passed on only from fathers to sons.

**Zona Pellucida**
Envelope that protects the egg; the sperm must negotiate it during fertilization.

**Zygote**
The diploid cell formed by the union of the sperm and the egg after fertilization; also called the ovum.

# INDEX

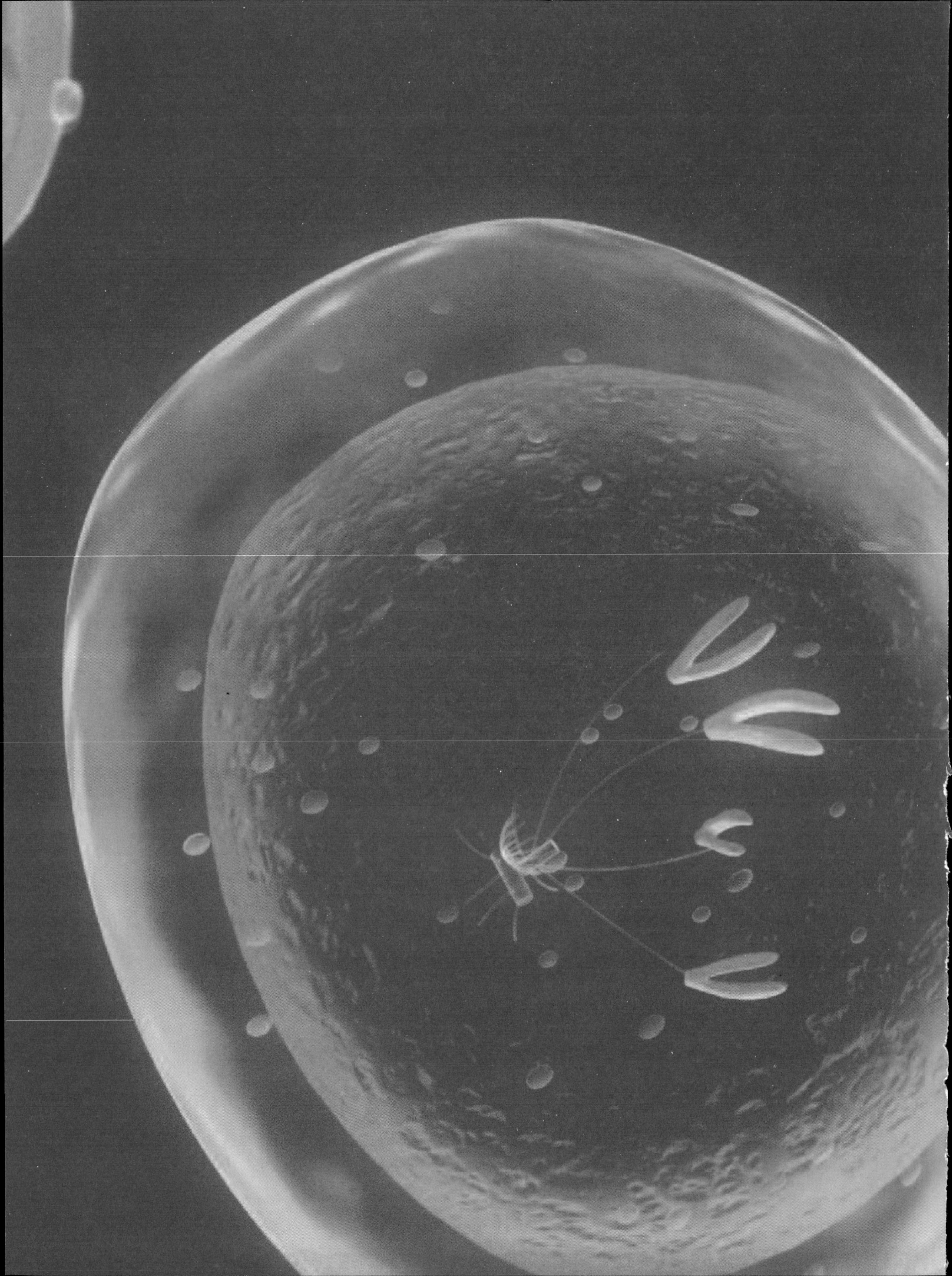